CLOUD AND PRECIPITATION MICROPHYSICS – PRINCIPLES AND PARAMETERIZATIONS

Numerous studies have demonstrated that cloud and precipitation parameterizations are essential components for accurate numerical weather prediction and research models on all scales, including the cloud scale, mesoscale, synoptic scale, and global climate scale.

This book focuses primarily on bin and bulk parameterizations for the prediction of cloud and precipitation at various scales. It provides a background to the fundamental principles of parameterization physics, including processes involved in the production of clouds, ice particles, rain, snow crystals, snow aggregates, frozen drops, graupels and hail. It presents complete derivations of the various processes, allowing readers to build parameterization packages, with varying levels of complexity based on information in this book. Architectures for a range of dynamical models are also given, in which parameterizations form a significant tool for investigating large non-linear numerical systems. Model codes are available online at www.cambridge.org/straka.

Written for researchers and advanced students of cloud and precipitation microphysics, this book is also a valuable reference for all atmospheric scientists involved in models of numerical weather prediction.

JERRY M. STRAKA received a Ph.D. in Meteorology from the University of Wisconsin, Madison in 1989. He then worked for a short time at the University of Wisconsin's Space Science and Engineering Center (SSEC) in Madison before joining the University of Oklahoma in 1990 where he is an Associate Professor of Meteorology. Dr Straka's research interests include microphysical modeling, severe thunderstorm dynamics, numerical prediction, radar meteorology, and computational fluid dynamics. He was co-director of the Verifications of the Origins of Rotation in Tornadoes Experiment (VORTEX I) and is a Member of the American Meteorological Society.

CLOUD AND PRECIPITATION MICROPHYSICS

Principles and Parameterizations

JERRY M. STRAKA
University of Oklahoma, USA

CAMBRIDGE
UNIVERSITY PRESS

CAMBRIDGE UNIVERSITY PRESS
Cambridge, New York, Melbourne, Madrid, Cape Town,
Singapore, São Paulo, Delhi, Tokyo, Mexico City

Cambridge University Press
The Edinburgh Building, Cambridge CB2 8RU, UK

Published in the United States of America by Cambridge University Press, New York

www.cambridge.org
Information on this title: www.cambridge.org/9780521297592

First published 2009
First paperback edition 2011

A catalogue record for this publication is available from the British Library

Library of Congress Cataloguing in Publication data

Straka, Jerry M.
Cloud and precipitation microphysics : principles and parameterizations / Jerry M. Straka.
p. cm.
ISBN 978-0-521-88338-2 (hardback)
1. Precipitation (Meteorology)–Measurement. 2. Cloud forecasting.
3. Precipitation forecasting. I. Title.
QC925.S77 2009
551.57–dc22

2009009612

ISBN 978-0-521-88338-2 Hardback
ISBN 978-0-521-29759-2 Paperback

This book is specially dedicated to Katharine, Karen, and Michael

"All men dream, but not equally. Those who dream by night in the dusty recesses of their minds wake in the day to find that it was vanity; but dreamers of the day are dangerous men, for they may act their dream with open eyes, to make it possible. This I did."

Seven Pillars of Wisdom (A Triumph) by T. E. Lawrence

Contents

Preface

Through the experience of the author and his interaction with others that teach cloud and precipitation physics at the University of Oklahoma over the course of at least the past 17 years, it became apparent that there were no current reference books or textbooks on the specific topic of the principles of parameterization of cloud and precipitation microphysical processes. This is despite the knowledge that the research community in numerical simulation models of clouds regularly uses microphysical parameterizations. Moreover, the operational community would find that numerical weather prediction models are not possible without microphysical parameterizations. Therefore, it is hoped that this book will be one that begins to fill this niche and provides a reference for the research and operational communities, as well as a textbook for upper-level graduate students.

Researchers and students should have a prerequisite of a basic graduate-level course in cloud and precipitation physics before using this book, though every effort has been made to make the book as self-contained as possible. The book provides a single source for a combination of the principles and parameterizations, where possible, of cloud and precipitation microphysics. It is not intended to be a comprehensive text on microphysical principles in the spirit of Pruppacher and Klett's book *Microphysics of Clouds and Precipitation*. Not every existing parameterization available is included in the book, as this would be an overwhelmingly daunting task, though every effort has been made to include the more common and modern parameterizations. There are some elegant, modern parameterizations that are not covered, though the reader will find references to them. Some simpler early parameterizations such as those used in one-moment parameterizations (mixing ratio of vapor or hydrometeor) are omitted for practical reasons, and because these are quickly becoming outdated. Some operational numerical weather-prediction modelers cling to these simpler microphysics parameterizations as their mainstay owing to their low memory overhead, and computational cost. Furthermore, an appendix of symbols was deemed to be essentially impossible to make user-friendly, as characters and symbols are recycled time and time again throughout the literature, and thus, they are recycled in this book. Admittedly, this is unfortunate for the reader. Hopefully variables are defined in enough detail where used so that what they represent can be easily understood. Enough material is presented for readers to make educated choices about the types of parameterizations they might find necessary for their work or interest. Every attempt has been made to include state-of-the-art science on the topic by drawing heavily from the peer-reviewed literature. Each chapter covers specific microphysical processes, and includes many theoretical principles on which the parameterization designs are based, where such principles exist. It should be interesting to the reader just how ad hoc some parameterizations actually are in reality and how poorly or well some of them perform.

Gratitude is extended to the publishers who have granted permission for the reproduction of figures throughout the text. Some of my own research is included in the book, and for the support of this work as well as time spent on this book, I acknowledge the National Science Foundation in the USA. First and foremost, however, this book would not have at all been possible without the contribution of various derivations and the often tedious and repeated editing provided by my wife and colleague, Dr. Katharine M. Kanak. Next I would like to thank Dr. Robert Ballentine for trusting in me as an undergraduate and graduate student and teaching me the finer points of numerical modeling. I also would like to thank my Ph.D. Advisor, Professor Pao K. Wang for stimulating my initial interest in cloud and precipitation physics, and in particular research on hail initiation and growth. In addition I extend a special thanks to Drs. Matthew Gilmore, Erik Rasmussen, Alan Shapiro, and Ted Mansell for many stimulating conversations about microphysics parameterizations, along with many others, too numerous to list, with whom I had various degrees of complex discussions on the principles and parameterizations presented in this book.

Special thanks are owed to Cambridge University Press Syndicate, and especially Dr. Susan Francis, Commissioning Editor, Earth and Planetary Science, for her guidance, assistance, and opinions in the production and the publication of this book. Diya Gupta, the assistant editor, was invaluable for guidance and help. Eleanor Collins, Production Editor, and Zoë Lewin, Copy Editor, were a pleasure to work with and helped tremendously with getting the book in its present form.

J. M. Straka
Norman, Oklahoma
May 2008

1
Introduction

1.1 Cloud and precipitation physics and parameterization perspective

Cloud and precipitation physics is a very broad field encompassing cloud dynamics, cloud microphysics, cloud optics, cloud electrification, cloud chemistry, and the interaction of cloud and precipitation particles with electromagnetic radiation (i.e. radar). The focus of this book is on a very specific aspect of cloud and precipitation physics: the development of various parameterizations of cloud and precipitation microphysical processes; and when possible the exploration of the basic theories necessary for their development. In numerical models, based on theory and observations, microphysical parameterizations are a means to represent sub-grid-scale microphysical processes using grid-scale information. Some of the parameterizations are quite complex, whilst others are quite simple. In the realm of the design of parameterizations of cloud and precipitation microphysics, complex schemes do not always provide more accurate results than simple schemes. The parameterizations of cloud and precipitation microphysical processes are essential components to numerical weather prediction and research models on all scales, including the cloud scale, mesoscale, synoptic scale, global, and climate scale. In particular, the accuracy of quantitative precipitation forecasts, as well as the representation of atmospheric and terrestrial radiation physical processes, depend significantly on the type of cloud and precipitation microphysics parameterizations used. More recently cloud models also have been used to simulate lightning, which depends on an accurate account of microphysical processes, hydrometeor amounts and locations.

The scales involved with cloud and precipitation microphysical processes range from the size of Aitken aerosol particles $O(10^{-2}$ mm) to giant aerosol particles $O(10^{0}$ mm) to ultra-giant aerosols and cloud particles $O(10^{1}$ mm) to drizzle and snow crystal particles $O(10^{2}$ mm) to rain, snow aggregate, and graupel particles $O(10^3$ to 10^4 mm) to hail particles $O(10^4$ to 10^5 mm). Thus to study theories and parameterizations of cloud and precipitation particle growth, nearly seven orders of magnitude in size must be covered. To put this in perspective, this is similar to studying the development of small wind swirls $O(10^{-1}$ m) to dust devils $O(10^0$ to 10^1 m) to cumulus clouds $O(10^2$ to 10^3 m) to convective clouds such as thunderstorms $O(10^3$ to 10^4 m) to mesoscale phenomena such as large thunderstorm complexes and hurricanes $O(10^5$ m) to synoptic scale phenomena such as Rossby waves $O(10^6$ m) all relative to one another. With these vast scale differences it is no wonder that theories and parameterizations of microphysical processes can be so difficult to develop and be accurate enough for research and operational model usefulness.

This chapter begins with a brief description of the types of cloud and precipitation parameterization methodologies available. The complexity of a parameterization is governed by theoretical equations that can be derived and observations that are used as needed. Descriptions of warm and cold rain processes for physically consistent and complete microphysical parameterizations are presented next. Then, hydrometeors and their characteristics such as phase, size, concentration, content, and structure are discussed. This information is essential as different parameterizations treat hydrometeors differently based on these characteristics depending on the complexity required in a model.

1.2 Types of microphysical parameterization models

1.2.1 Lagrangian trajectory parameterization models

Lagrangian trajectory parameterization models are the type of models that can incorporate the most detailed microphysical information based on observations, physical experiments, and theoretical considerations of any parameterization model for hydrometeor growth described in this book. Particles grow following three-dimensional trajectories in a prescribed or radar-deduced flow field, which can be provided by multiple Doppler radar analyses. Then with approximations for temperature, vapor, and liquid-water content and/or ice-water content, growth of individual precipitating hydrometeors is predicted using equations of varying complexity as described throughout the book. Some Lagrangian trajectory models can be quite comprehensive, whereas others are very simple. In addition to predicting the growth of individual hydrometeors in Lagrangian trajectory models, the growth of hydrometeor packets can be predicted, though this is done less commonly.

1.2.2 Bin parameterization models

Bin parameterization models are often considered the type of parameterization most able to represent, for example, rain distribution evolutions in rain clouds. They have bins (i.e. small divisions) representing the spectrum of drops from very small cloud droplet sizes (4 μm) to larger raindrops (4 to 8 mm) for parameterizations of rain formation. Each bin is usually exponentially larger than the previous size/mass bin owing to the wide spectrum of liquid-water drops that are possible, which ranges over three orders of magnitude. For liquid-water drop sizes, bins often will increase by 2, $2^{1/2}$, $2^{1/3}$, or $2^{1/4}$ times the previous size bin over 36, 72, or 144 bins (or any number required for a converged solution), starting with particles of about 2 to 8 μm in diameter and increasing to a size that contains the spectra of rain, ice, snow, graupel, and hail. Bin parameterization size-spectra can also be made for other hydrometeor species including ice crystals, snow crystals, snow aggregates, graupel, frozen drops, and hail with similar bin spacing. Some models also have bins for aerosols and track solute concentrations. A shortcoming of bin models is the excessively large computation resources needed to make use of them [both Computer Processing Unit (CPU) and memory], except for two-dimensional models (both axisymmetric and slab-symmetric) as well as smaller-domain three-dimensional models. At a minimum, the number concentration must be predicted with these schemes, though mixing ratio and reflectivity can be predicted or calculated. Considering number concentration with mixing ratio prediction improves the results against using just number concentration for analytical test problems as will be demonstrated in a later chapter.

1.2.3 Bulk parameterization models

Bulk microphysical model parameterizations are some of the most popular schemes available owing to much reduced computational cost compared to most bin models for use in three-dimensional models. These microphysical parameterizations are based on number distribution functions such as mono-dispersed, negative exponential, gamma, and log-normal distributions, to name a few, for each hydrometeor species' size distribution. These distributions are normalizable and integratable over complete size distributions of diameter from zero to infinity, or partial distributions (most common with the gamma distribution) from diameters of 0 to D_1 meters or D_2 to ∞ meters or even D_1 to D_2 meters. Typically, mixing ratio and number concentration are predicted with these parameterizations. Whilst reflectivity can be predicted, it can be used to obtain an estimate of the gamma distribution-shape parameter of the size distribution as a function of time. Some of the simpler dynamical models also predict two and/or three moments, including the slope or characteristic diameter of the distributions as well as, or in place of, number concentration. Most one- through three-dimensional dynamical models developed during the period from the 1970s through the mid 1980s generally predicted only hydrometeor mixing ratio for bulk microphysical parameterizations. As computer power increased in the mid 1980s, the number of species predicted and the number of moments predicted slowly increased to the point where most models utilized two moments and eventually some used three moments (Milbrandt and Yau 2005a, b).

1.2.4 Hybrid bin parameterization models

Hybrid bin parameterizations have many of the qualities of both bulk and microphysical model parameterizations, however growth and loss parameterizations are done differently than direct integration of spectrum interactions such as, for example, collection of one hydrometeor species by another species. Instead, the mixing-ratio and number-concentration distribution functions are converted to bins and computations are done with a bin model; then results are converted back to bulk microphysical model parameterization mixing ratios and number concentrations as described by some distribution function. Typically, lookup tables are made to reduce the computation overhead. With these models an attempt is made to capture the "supposed" accuracy of bin models in a bulk microphysical model parameterization without the memory storage of the full bin model. One shortcoming with hybrid bin parameterizations compared to bin parameterizations is that the bin parameterization solution is not carried from timestep to timestep, in particular the bin parameterization size spectra.

1.3 Warm-rain parameterizations

Warm-rain processes include the development of precipitating rain without the presence of ice water. However, clouds can have both warm-rain processes and cold-rain processes occurring simultaneously, both in the same and in different locations. Following closely the ideas put forth by Cotton and Anthes (1989), the basic physics that need to be included in a warm-rain parameterization are the following in some fashion or other. These concepts are to some extent based on bin parameterizations of warm-rain processes, but are quickly becoming more commonplace in bulk microphysical model parameterizations. The processes are shown in Fig. 1.1 (Braham and Squires 1974):

- The nucleation of droplets on aerosol particles
- Condensation and evaporation of cloud droplets as well as drizzle and raindrops
- The development of a mature raindrop spectrum by collection of other liquid species (including cloud droplets, drizzle, and raindrops themselves)
- The inclusion of breakup of raindrops
- The occurrence of self-collection in the droplet and drop spectra
- The differential sedimentation of the various liquid drop species within the species, for example, rain from different sources.

Cotton and Anthes (1989) used some of these concepts, to which some processes have been added here. They also argue from bin model results, that a possible and perhaps attractive approach to parameterizing these physics is to separate the liquid-water spectrum into two separate species: cloud droplets and raindrops. Others, such as Saleeby and Cotton (2004, 2005), include both small-cloud droplet ($D < 100\ \mu m$) and large-cloud droplet ($D > 100\ \mu m$) species or liquid-water habits. Similarly, Straka *et al.* (2009a) include a drizzle species category between the cloud droplet and rain categories. Straka *et al.* (2009b) also include a large raindrop species or category to account for melting graupel and small hail that become drops that do not immediately break up.

1.4 Cold-rain and ice-phase parameterizations

The microphysical parameterization of cold-rain processes and the ice phase of water is significantly more difficult than that for warm-rain parameterizations, and warm-rain processes. However, for both warm-rain and ice parameterizations, conversion processes are tremendously burdensome in a theoretical and parameterization design perspective. Cotton and Anthes (1989) point out that many of the ice microphysical processes are not parameterizable in terms of results from bin models, theoretical consideration, or empirical fits to observations without considerable uncertainty. Nevertheless, models continue to grow apace in complexity with more degrees of freedom to accommodate data as these become available. Again following Cotton and Anthes (1989), to as reasonable an extent as possible, the following processes should be included in some fashion or other. In addition, the processes are shown in Fig. 1.1.

- Homogeneous freezing of cloud drops into ice crystals.
- Primary, heterogeneous ice nucleation mechanism such as contact freezing, deposition, sorption, and immersion freezing nucleation.

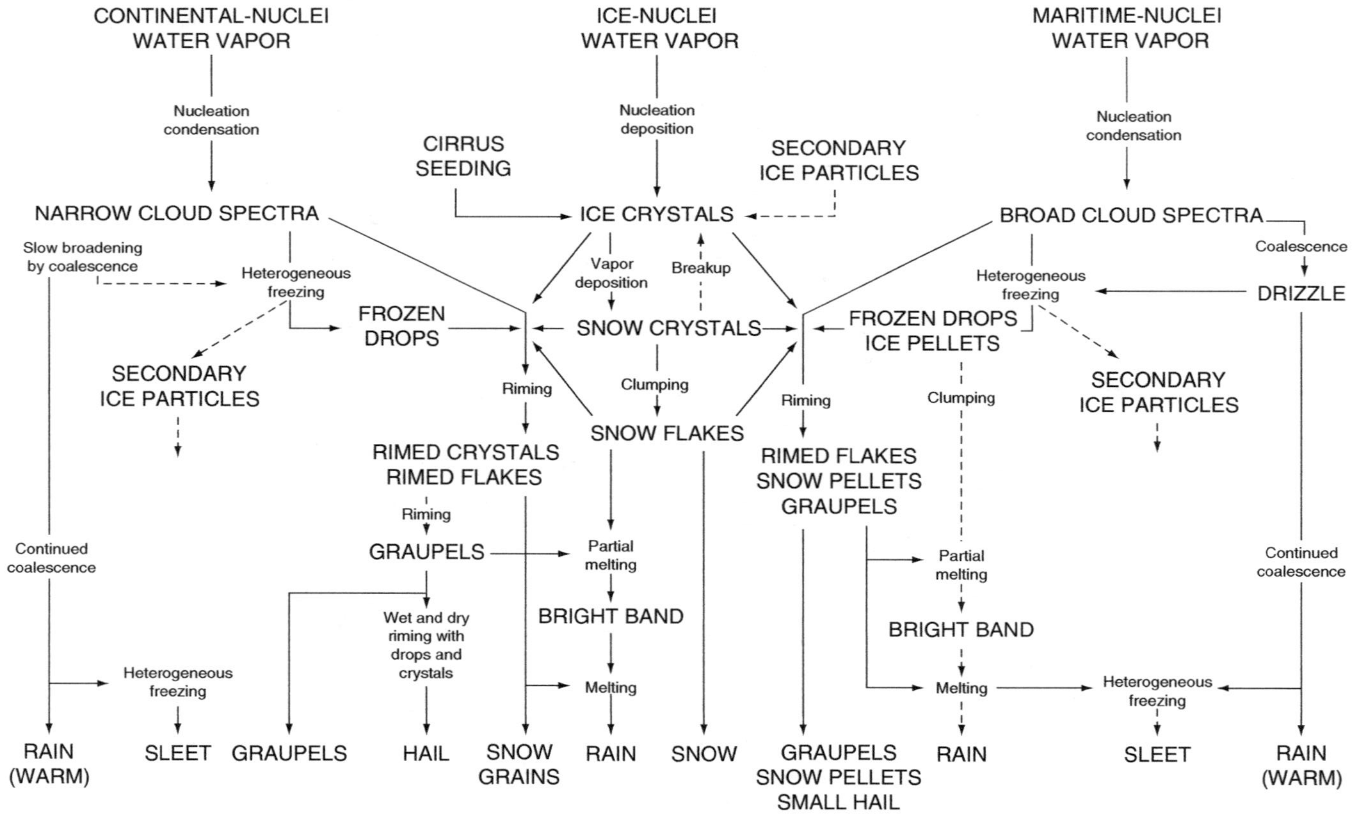

Fig. 1.1. The major types of precipitation elements, and the physical processes through which they originate and grow, are shown in this flow diagram. Computer models of many of these processes have now been developed compared to 30 years ago, although inadequate basic knowledge about process kinetics still has tended to restrict their complete and appropriate application. (From Braham and Squires 1974; courtesy of the American Meteorological Society.)

- Secondary ice nucleation mechanisms such as rime-splintering ice production and mechanical fracturing of ice.
- Vapor deposition and sublimation of ice particles.
- Riming and density changes of ice particles.
- Aggregation of ice crystals to form snow aggregates.
- Graupel initiation by freezing of drizzle and subsequent heavy riming.
- Graupel initiation by heavy riming of ice crystals.
- Freezing of raindrops, with smaller particles becoming graupel particle embryos owing to riming, and larger particles possibly becoming hail embryos.
- Graupel and frozen drops becoming hail embryos by collecting rain or heavy riming.
- Wet and dry growth of hail.
- Temperature prediction of ice-water particles.
- Density changes in graupel and hail.
- Shedding from hail during wet growth and melting.
- Soaking of hail and graupel particles during wet growth and melting.
- Melting of ice-water particles.
- Mixed-phase liquid- and ice-water particles.
- Differential sedimentation of the various sub-ice species and within a given species.

Figure 1.1 shows the many possible physical processes hydrometeors can undergo when they fall to the ground as different types of precipitation. Continental and maritime nuclei represent different sizes of nuclei and different number concentrations. Specifically, continental nuclei exist at smaller sizes and larger numbers than maritime nuclei, in general.

1.5 Hydrometeor characteristics overview

1.5.1 Hail

List (1986) describes a weak association between hail size and shape. That is, hail 5 to 10 mm in diameter generally is spherical or conical, although disk shapes can be observed; hail with $10 < D < 20$ mm is ellipsoidal or conical; hail with $10 < D < 50$ mm is ellipsoidal, with lobes and other protuberances along the short axis; and hail with $40 < D < 100$ mm is spherical with small and large lobes and other protuberances. However, List (1986) found no relation when comparing protuberance size and number with hail size, the only exception being that larger hail tends to be more irregular in shape.

Another observation is that most hailstones are oblate (Barge and Isaac 1973). For example, 83% have axis ratios (axis ratio is a/b, where a is the minor axis and b is the major axis) between 0.6 and 1.0, 15% have axis ratios between 0.4 and 0.6, and less than 2% have axis ratios less than 0.4. Furthermore, the majority of hailstones observed at the ground have axis ratios of 0.8 (Knight 1986; Matson and Huggins 1980). Wet hail typically has an axis ratio of about 0.8, and spongy hail has an axis ratio of 0.6 to 0.8 (Knight 1986).

There is evidence that hailstones fall with their maximum dimensions in both the horizontal (Knight and Knight 1970; List *et al.* 1973; Matson and Huggins 1980) and the vertical (Knight and Knight 1970; Kry and List 1974; List 1986). List (1986) suggested that ellipsoidal hailstones 10 to 50 mm in diameter typically fall most stably when oriented in the vertical. Hailstones also can exhibit gyrating motions (List *et al.* 1973; Kry and List 1974; List 1986) and tumbling motions (List *et al.* 1973; Knight and Knight 1970; Matson and Huggins 1980). The structure of hail can vary from porous to solid to spongy. The outer shell can be dry or wet, which is in part related to the rate at which the hail spins and environmental conditions. Hail density typically varies from about 400 to 900 kg m^{-3} for a hail diameter smaller than 10 mm and from 700 to 900 kg m^{-3} for hail that has larger diameter.

Hail distributions can be represented with some form of negative-exponential (Marshall–Palmer) or gamma distribution (Ulbrich and Atlas 1982; Ziegler *et al.* 1983). Ziegler *et al.* (1983) show particularly good matches of hail number to size for the gamma size distribution, but not for the Marshall–Palmer size distribution for two different datasets (collections A and B) of observations from a hail collection in Oklahoma (Figs. 1.2, 1.3). In contrast, Cheng and English (1983) show very good matches to negative-exponential distributions (Marshall–Palmer) for two datasets (July 27, 1980 and July 28, 1980) from Canada (Fig. 1.4). It should be noted that Cheng *et al.* (1985) also found an association between observed hail distributions and inverse exponential fits. Both of these studies were conducted in continental regimes, so generalities concerning the accuracy of size-distribution functions should be made with caution. Finally, total hail number concentrations range from 10^{-2} to 10^{1} m^{-3} or greater for hail diameters of 5 to 25 mm and from 10^{-6} to 10^{-2} m^{-3} for hail diameters larger than 25 to 80 mm (Auer 1972; Pruppacher and Klett 1981).

Hail usually falls very quickly compared to other hydrometeors with updraft speed dictating to some degree the maximum hail size. In addition, in order for hail particles to make it to the ground as hailstones, they must be larger than about 1 cm at the melting level (Rasmussen and Heymsfield 1987b). Larger hailstones of 3 to 4 cm require so much heat to melt that they may change only moderately in diameter (<10 to 20%) from their size at the melting level on the way to the ground.

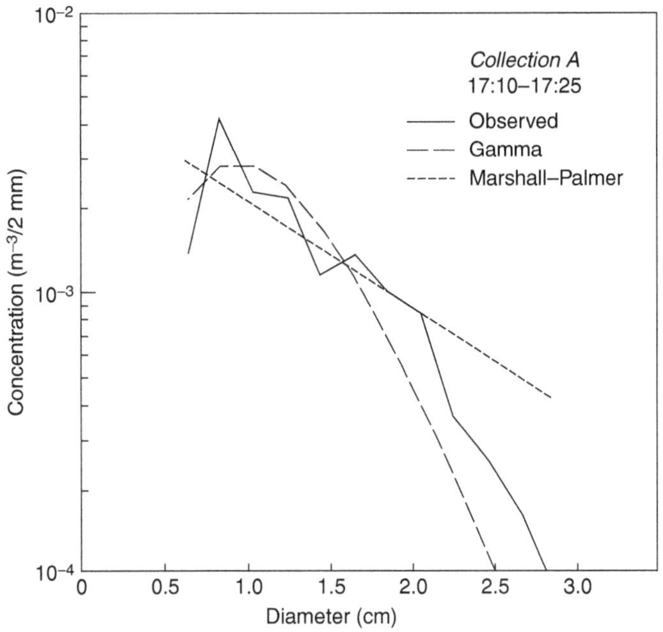

Fig. 1.2. The curve of the observed (solid) hail concentration spectra, with superimposed gamma (long dash) and Marshall–Palmer (short dash) curves fits to collection A. (From Ziegler *et al.* 1983; courtesy of the American Meteorological Society.)

1.5.2 Graupel

Graupel diameters range from $0.5 < D < 5$ mm, and very low-density graupel are sometimes of larger size. The density of graupel can range from 100 to 900 kg m^{-3}, and size distributions generally can be represented by negative-exponential and gamma distributions. Moreover, number concentrations are on the order of 1 to 10 m^{-3} or higher (Auer 1972; Pruppacher and Klett 1981). The shapes of graupel can be spherical, conical, or can be highly irregular, with axis ratios both larger and smaller than unity (Bringi *et al.* 1984; Aydin and Seliga 1984). In a modeling study by Bringi *et al.* (1984), graupel less than 1 mm are assumed to be spherical, graupel with $1 < D < 4$ mm are conical with axis ratio, $a/b = 0.5$, and graupel with $4 < D < 9$ mm are conical with $a/b = 0.75$. Both smaller and larger particles might be spherical or irregular in shape (e.g. lump graupel: highly irregular-shaped rimed crystals and aggregates) based on *in situ* observations.

Low-density graupel sometimes is conical in shape, which might be explained by low-density riming of planar, ice-crystal edges (Fig. 1.5; Knight and Knight 1973). Figure 1.5 shows a planar ice crystal (denoted by a

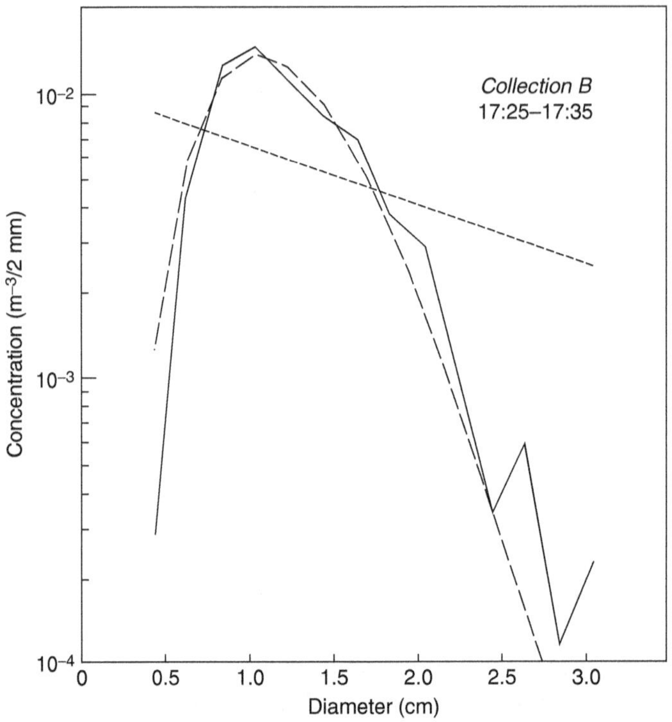

Fig. 1.3. The curve of the observed (solid) hail concentration spectra, with superimposed gamma (long dash) and Marshall–Palmer (short dash) curves fits to collection B. (From Ziegler *et al.* 1983; courtesy of the American Meteorological Society.)

horizontal line) that is riming on its edges. The schematic at the bottom of Fig. 1.5 shows the resulting conical shapes of the embryonic graupel. In general though, graupel tend to be relatively smooth in comparison with some hailstones. The fall orientation of graupel is not known with any certainty; some hypothesize that the larger of these hydrometeors probably tumble, though conical graupel may have a preferential fall orientation (List and Schemenaur 1971; Pruppacher and Klett 1981). Some graupel may fall with their largest axis in the horizontal, whereas others may fall with their largest axis in the vertical.

Graupel and small frozen drops generally melt completely as they fall to the ground (except in mountainous regions, particularly in summer), though they do not shed any water as they melt and become water drops (Mason 1956; Drake and Mason 1966; and Rasmussen and Pruppacher 1982). A review of the characteristics of graupel particles in Northeastern Colorado cumulus congestis clouds is given by Heymsfield (1978).

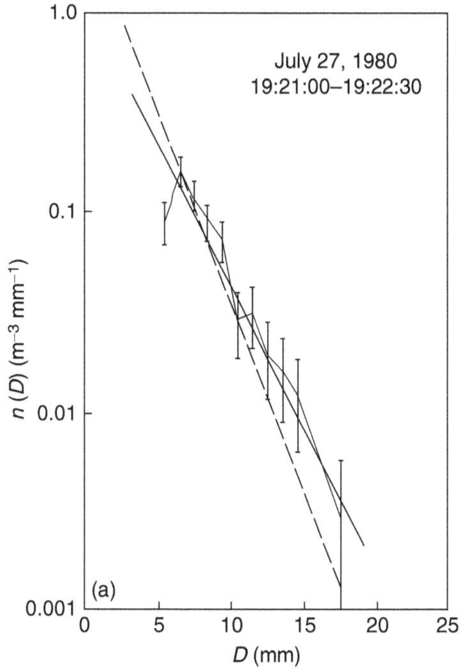

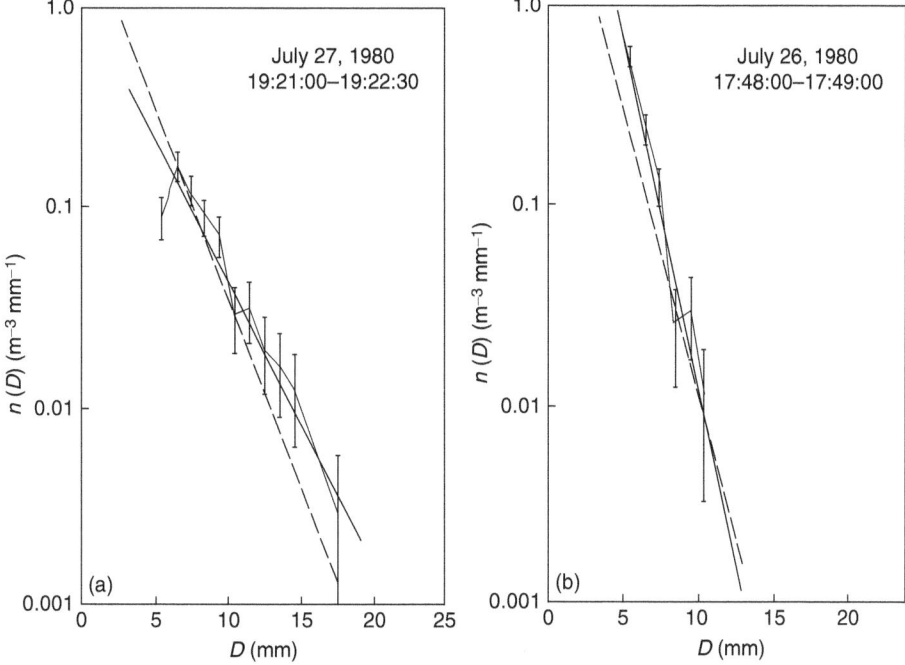

Fig. 1.4. Examples of measured hailstone size spectra and approximated exponential distributions obtained by least-square regression (solid line) and the method suggested by Federer and Waldvogel (dashed line) for (a) 27 July, (b) 26 July (from Cheng and English 1983; courtesy of the American Meteorological Society).

1.5.3 Ice crystals and snow aggregates

The density of ice crystals and aggregates varies from 50 to 900 $\mathrm{kg\,m^{-3}}$ depending on habit, size, and riming, with higher-density values expected for solid ice structures such as plates and wetted particles. Aggregates are usually two to five millimeters in diameter, whilst ice crystals are typically 50 to 2000 microns in diameter. Fallspeeds asymptote to about 1 $\mathrm{m\,s^{-1}}$ for aggregates and fallspeed seems to only have a weak dependence on size once snow aggregates become larger than a couple of millimeters. The size distributions of snow crystals and snow aggregates are well represented by the negative-exponential distribution (Gunn and Marshall 1958; Fig. 1.6), except for sizes smaller than about 1 mm in diameter, with total number concentrations on the order of 1 to 10^4 $\mathrm{m^{-3}}$ for aggregates, and 10 to 10^9 $\mathrm{m^{-3}}$ at the extreme for individual crystals at colder temperatures. However, it should be noted that snow- and ice-crystal concentrations as large as 10^7 $\mathrm{m^{-3}}$ have been

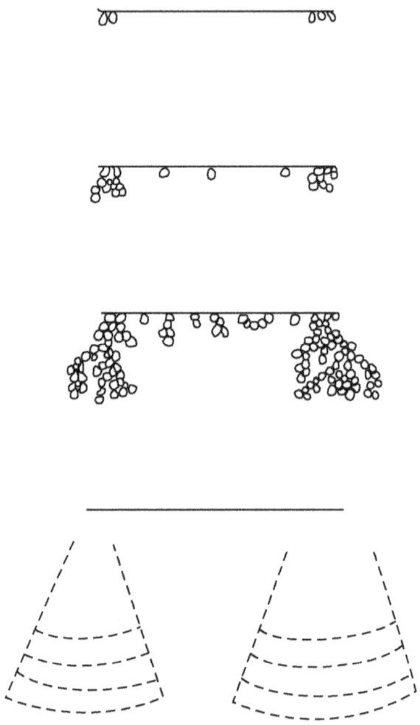

Fig. 1.5. Diagram showing the suggested mode of origin of conical graupel. (From Knight and Knight 1973; courtesy of the American Meteorological Society.)

found on a regular basis at temperatures of –4 to –25 °C (Hobbs 1974), with no temperature dependence. In recent years, gamma distributions have been used for parameterization of snow particles (e.g. Schoenberg-Ferrier 1994). Values of number concentrations of crystals are often as large as 10^4 m^{-3} at temperatures warmer than 0 to –10 °C (Pruppacher and Klett 1981; 1997). The diameters of large aggregates can be $D \sim 20$ to 50 mm, whereas the diameter of large crystals typically can be $D \sim 1$ to 5 mm. The shapes of aggregates are nearly spherical to extremely oblate, and the approximate shapes of crystals can vary from extreme prolates and oblates to spheroids (Pruppacher and Klett 1981). Typically, thin plates are found at temperatures of 0 to –4 °C (Fig. 1.7b), needles and hollow columns from –4 to –9 °C (Fig. 1.7a), sectors from –5 to –10 °C and –16 to –22 °C, while dendrites (Fig. 1.7c) are found at greater than water saturation at temperatures of –12 to –16 °C (Cotton 1972b); finally an assortment of columns, side planes, and other shapes are found from temperatures of –22 to –70 °C. General aspects of ice aggregation are discussed in Hosler and Hallgren (1961), and in Kajikawa and Heymsfield (1989) for cirrus clouds. Figure 1.8 (Fletcher 1962)

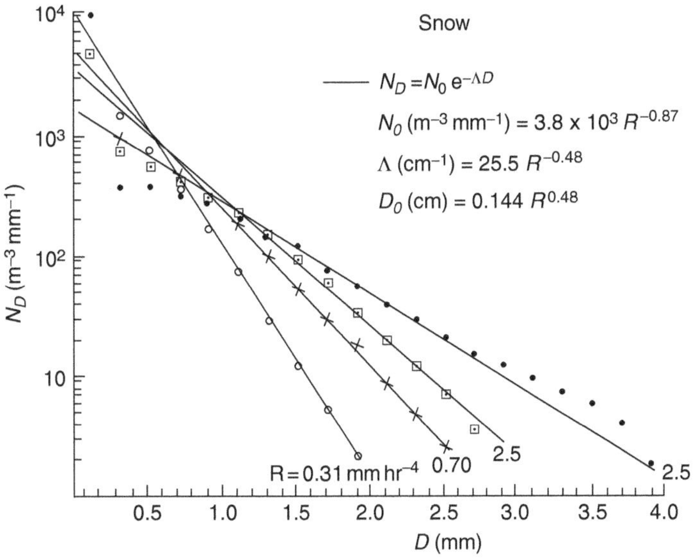

Fig. 1.6. The size distribution of snowflakes in terms of drops produced by melting the snowflakes. (From Gunn and Marshall 1958; courtesy of the American Meteorological Society.)

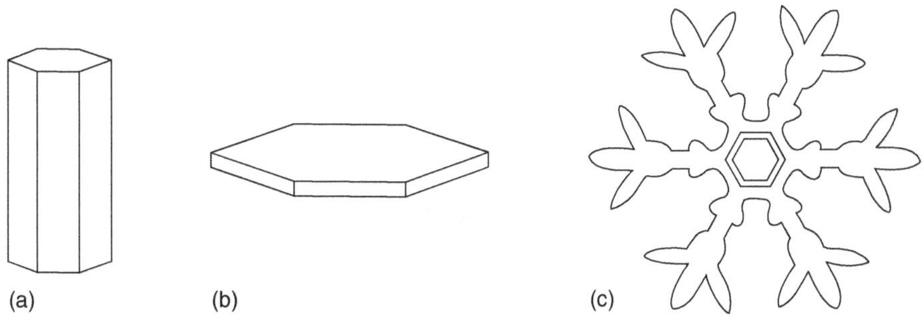

Fig. 1.7. Schematic representation of the main shapes of ice crystals: (a) columns, (b) plates, and (c) dendrites. (From Rogers and Yau 1989; courtesy of Elsevier.)

shows the conditions under which many forms of ice crystal exist. In recent years, Bailey and Hallet (2004) showed unique ice-crystal forms at temperatures colder than –20 °C. Most individual crystals fall with their largest dimension horizontally oriented unless there are pronounced electric fields, which can orient small crystals vertically. Aggregates can fall in a horizontal orientation or may tumble. Both above and below the melting layer, aggregates rarely break up (Otake 1970).

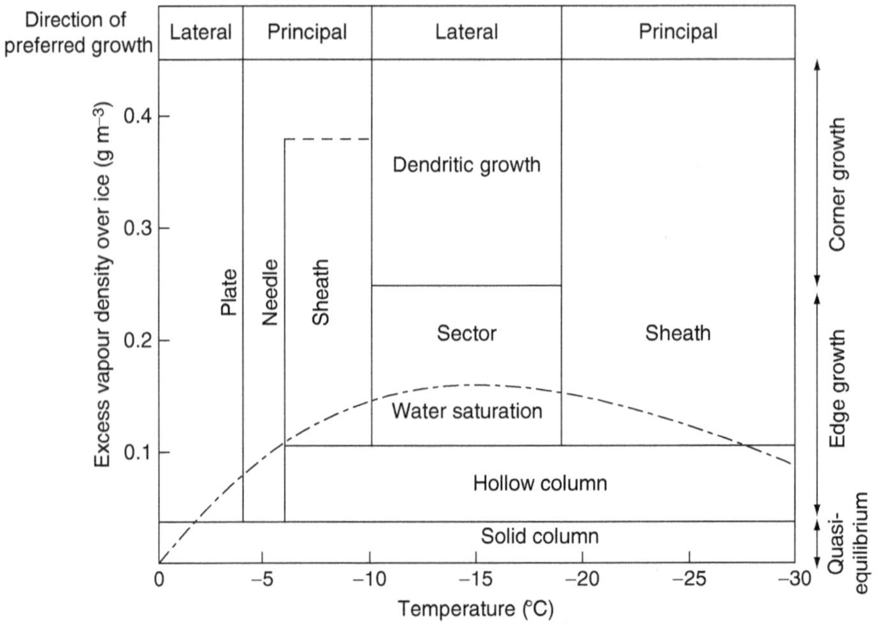

Fig. 1.8. Kobayashi's diagram of crystal habit as a function of temperature and excess vapor density over ice saturation. (From Rogers and Yau 1989; courtesy of Elsevier.)

1.5.4 Cloud droplets, drizzle, and raindrops

Precipitating liquid particles include cloud droplets, drizzle and raindrops. Cloud droplets and drizzle are highly spherical, when cloud droplets are less than 81 μm in diameter and when drizzle has $D < 400$ μm. Larger sized particles are raindrops. A characteristic that sets raindrops apart from other liquid precipitating particles is the dependence of the raindrop axis ratio on diameter. Some models now consider both small and large cloud droplet modes (Saleeby and Cotton 2005).

Axis ratios commonly are related to drop sizes through equivalent diameter D_e (this is the diameter achieved by assuming the particle is a sphere) (Pruppacher and Klett 1981); several relations exist, including Beard (1976) where $a/b = 1.03–0.062D_e$; D_e is in mm (Pruppacher and Beard 1970; Pruppacher and Pitter 1971). Studies by Jones (1959), Jameson and Beard (1982), and Goddard *et al.* (1982) show that, in heavier rain events, a large range in axis ratios might be expected with even prolates possible (though the latter are likely transient shape oscillations). An average of typical rain axis ratios are given by Andsager *et al.* (1999) as shown in Fig. 1.9 (note that axis ratio is given as α in the figure). Recent estimates show that fluctuations in

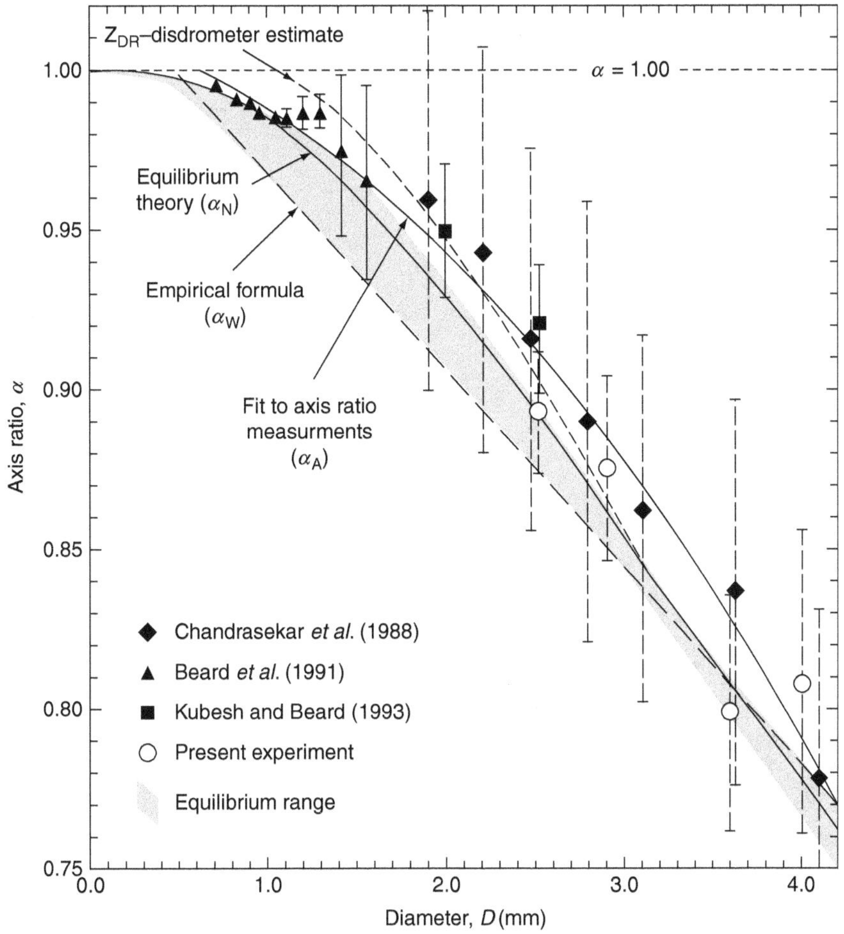

Fig. 1.9. Raindrop axis ratios as a function of diameter. Shown are mean axis ratios (symbols) and standard deviations (vertical lines) from aircraft observations by Chandrasekar *et al.* (diamonds), the laboratory measurements of Beard *et al.* (1991; triangles), Kubesh and Beard (1993; squares), and present experiments (circles). Curves are shown for the numerical equilibrium axis ratio (α_N) from Beard and Chuang (1987), the radar-disdrometer-derived axis ratios of Goddard and Cherry (1984), the empirical formula (α_W) from the wind tunnel data of Pruppacher and Beard (1970), and the present fit to axis ratio measurements (α_A). The shaded region covers the range from previous estimates of the equilibrium axis ratio. (From Andsager *et al.* 1999; courtesy of the American Meteorological Society.)

axis ratios are greater than the older empirical data suggest (Fig. 1.9). In particular, particles are likely to be less oblate, especially for larger diameters, however this may be due to drop shape oscillations.

Nearly two decades after the empirical data in Fig. 1.9 were obtained, raindrop axis ratio and its functional form, as well as the importance of drop

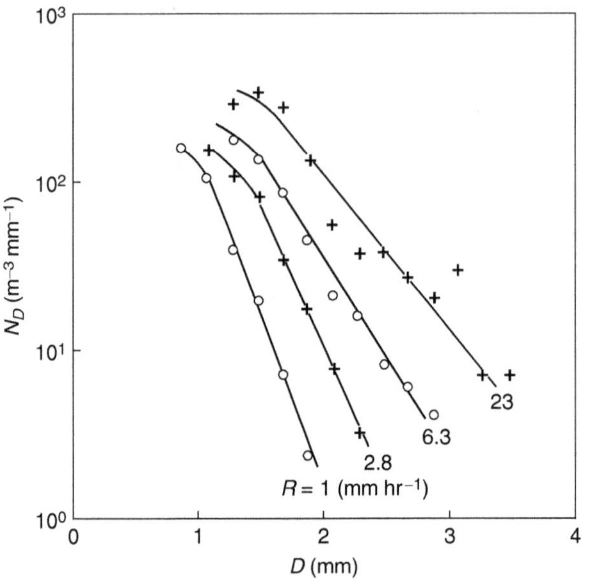

Fig. 1.10. Distribution of number versus diameter for raindrops recorded at Ottawa, summer 1946. Curve A is for rate of rainfall 1.0 mm hr^{-1}, curves B, C, D, for 2.8, 6.3, 23.0 mm hr^{-1}. $N_D\delta D$ is the number of drops per cubic meter, of diameter between D and $D + \delta D$. (From Marshall and Palmer 1948; courtesy of the American Meteorological Society.)

oscillations, came under renewed scrutiny in the 1990s (e.g. Feng and Beard 1991; Beard *et al.* 1991; Tokay and Beard 1996; Bringi *et al.* 1998; and Andsager *et al.* 1999). Nevertheless, for diameters greater than 1 mm, drops become increasingly oblate with size. Raindrops generally fall with their minor axis oriented in the vertical, though a rare few drops might be temporarily elongated vertically, possibly because of oscillations, collisions, or both.

Raindrop size distributions can be approximated by negative-exponential (Marshall and Palmer 1948; Fig. 1.10 and Fig. 1.11 show two datasets) or gamma distributions (Ulbrich 1983) for mean droplet spectra, but extreme local variations from these are observed (e.g. Rauber *et al.* 1991; Young 1993; Sauvageot and Lacaux 1995; and Joss and Zawadski 1997). Even though the negative-exponential distribution functions fit observed raindrop distributions well, there is indication that gamma distributions may provide a better fit, especially for raindrops of sizes less than 1 mm in diameter (Fig. 1.10 and Fig. 1.11). Schoenberg-Ferrier (1994) also noted this and suggested using gamma distributions for parameterization of raindrop distributions. However, it should be stated that observations of raindrop distributions are highly variable (Ulbrich 1983). In general, the largest raindrops have diameters of

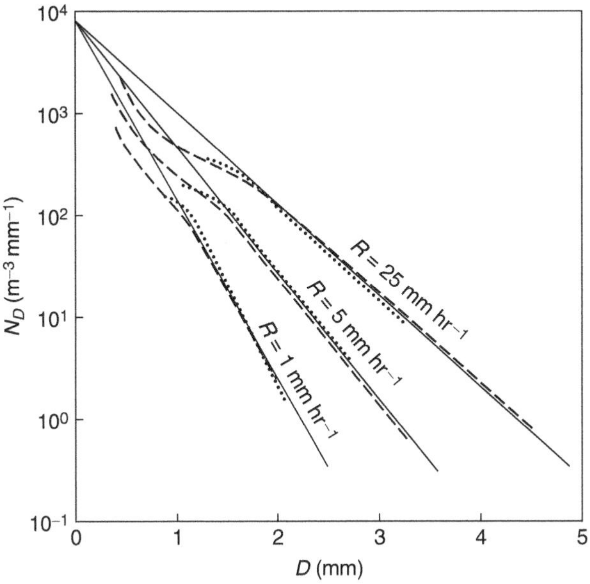

Fig. 1.11. Distribution function (solid straight lines) compared with results of Laws and Parsons (1943; broken lines) and Ottawa observations (dotted lines). (From Marshall and Palmer 1948; courtesy of the American Meteorological Society.)

3 to 5 mm. However, values as large as $D = 6$ to 8 mm for a very few drops have been documented in nature (e.g. Rauber *et al.* 1991). Total number concentrations of raindrops in general are present in concentrations on the orders of 10^3 to 10^5 m^{-3}, but very large drops are usually present in much lower concentrations. Finally, Gunn and Kinzer (1949) give the terminal velocity of fall for water droplets in stagnant air, whilst Kinzer and Gunn (1951) look at terminal velocity with more generality.

1.6 Summary

The book is organized starting with microphysical foundation material, followed by nucleation, saturation adjustments, vapor diffusion growth, collection growth, drop–breakup, conversion, hail growth, melting, parameterization limitations, and various dynamical model designs. It is impossible to include all of the work in the many thousands of papers in the literature, though an effort has been made to include in the book some of the most important developments through the past forty years of microphysical parameterization modeling. Not only are new parameterizations discussed, but also some older ones (Kinzer and Gunn, 1951). Where appropriate the chapters start with some basic theoretical considerations before discussion of the different parameterizations. It should be noted that details of aerosols are a vast and complex subject that is exceptionally well covered in Pruppacher and Klett (1981, 1997) and that they will not be covered in this book, except in developing equations for cloud condensation nuclei and ice nuclei as needed for heterogeneous nucleation. Ice microdynamics is found in an information-packed, easy flowing book by Pao K. Wang (2002), with all the latest on detailed ice-crystal modeling. Aerosol scavenging and riming are both covered exceedingly well in Pao K. Wang's book as well as modeling cirrus clouds.

2

Foundations of microphysical parameterizations

2.1 Introduction

In this chapter the foundations of bulk, bin, and hybrid bulk–bin microphysical parameterizations will be presented with more significant focus on the first. Many aspects of these microphysical parameterizations require functional relationships to describe attributes of populations of hydrometeors so that specific equations for source and sink terms for different hydrometeor species or habits can be integrated for mixing ratio and concentration. In addition equations for the prognostication of reflectivity, mass weighted riming rate, elapsed time of riming, and mass weighted rime density are presented or derived. In addition a diagnostic equation for the shape parameter in the gamma distribution is developed. Then number density functions and moment generators are presented for bin microphysical parameterizations.

2.2 Background

This part of chapter two derives heavily from Flatau *et al.* (1989) in presenting some of the concepts of microphysical parameterization fundamentals. First the probability density function will be defined and explained. It is essential for parameterization work that the probability density function be integratable. We start with $f(D)$ as the probability of number of particles of a certain size. It is preferred that the degrees of freedom of $f(D)$ be restrained to a small number of observable quantities. If not, it may be possible or likely that certain parameterizations cannot be created. Finally it is necessary that $f(D)$ needs to be readily normalizable.

The spectral number density function $n(D)$ is the concentration of particles per unit size interval from D to ΔD. The total number concentration is usually written as N_{T}, and results from integrating the spectral density function over a given size interval such as zero to infinity. But it can, in theory, be any interval of positive definite numbers. The slope intercept of a spectral number density function is given as N_0. The slope of the particle distribution is λ, and one over λ is called the characteristic diameter D_n of the particle distribution.

A mathematical function that is used repeatedly in microphysical parameterizations is the gamma function and is the solution to the following integral:

$$\Gamma(\xi) = \int_0^\infty t^{\xi-1} \exp(-t) \mathrm{d}t \quad [\mathrm{Re}\ \xi > 0]. \tag{2.1}$$

This is the complete gamma function. There are also incomplete gamma functions given by

$$\gamma(\xi, y) = \int_0^x t^{\xi-1} \exp(-t) \mathrm{d}t \quad [\mathrm{Re}\ \xi > 0], \tag{2.2}$$

and

$$\Gamma(\xi, y) = \int_x^\infty t^{\xi-1} \exp(-t) \mathrm{d}t \quad [\mathrm{Re}\ \xi > 0]. \tag{2.3}$$

In this work, x needs to be made non-dimensional, and since y is usually diameter dependent, the scaling diameter is the characteristic diameter that is used to normalize y. There are also several special cases of the gamma distribution, such as what will be called the complete gamma distribution here,

$$\frac{\Gamma(k)}{\mu\, \xi^k} = \int_0^\infty t^n \exp(-\xi t^\mu) \mathrm{d}t, \tag{2.4}$$

where k is given by

$$k = \frac{n+1}{\mu}, \tag{2.5}$$

where $n > -1$, $\mu > 0$, and $\xi > 0$.

A number of general relationships need to be considered. These include:

$$n(D) = N_\mathrm{T} f(D), \tag{2.6}$$

$$n(D)\mathrm{d}D = N_\mathrm{T} f(D)\mathrm{d}D, \tag{2.7}$$

and

$$f(D) = f(x)\frac{\mathrm{d}x}{\mathrm{d}D}, \tag{2.8}$$

$$f(D)\mathrm{d}D = f(x)\mathrm{d}x. \tag{2.9}$$

Now, the derivation of the characteristic diameter is presented. First define,

$$x = \frac{D}{D_n} = D\lambda, \tag{2.10}$$

where $\lambda = 1/D_n$ is the slope of the distribution.

Next, use (2.8) and (2.10) to write

$$f(D) = \frac{\mathrm{d}(D/D_n)}{\mathrm{d}D} f(x), \tag{2.11}$$

which becomes

$$f(D) = \frac{1}{D_n} f(x). \tag{2.12}$$

This last relation will be very important later for some derivations with different spectral density functions.

2.3 Power laws

For bulk microphysical parameterizations it is advantageous to use power laws for mass, density, terminal velocity, and other necessary variables if possible, as these are amenable to integration with the various forms of the gamma function and log-normal spectral density functions. With bin microphysical parameterizations, it is not so important that power laws be used because there are no spectral density functions used.

2.3.1 Mass–diameter (or length)

For many parameterizations (e.g. collection growth) a mass–diameter relationship is needed. A power law relating these quantities is used and is given by

$$m(D) = aD^b. \tag{2.13}$$

Some examples of power laws for mass–diameter relationships, in SI units, from Pruppacher and Klett (1997) are:

Hexagonal plate (Pla)	$m(D) = 156.74D^{3.31}$
Skeleton (C1h)	$m(D) = 6.08D^{2.68}$
Sector (P1b)	$m(D) = 2.898D^{2.83}$
Broad branch (P1c)	$m(D) = 1.432D^{2.79}$
Stellar (P1d)	$m(D) = 0.145D^{2.59}$

Dendrite (Ple)	$m(D) = 2.37 \times 10^{-2} D^{2.29}$
Needle (N1a)	$m(D) = 1.23 \times 10^{-3} L^{1.8}$
Long column (N1e)	$m(D) = 3.014 \times 10^{-3} L^{1.8}$
Solid column (Cle)	$m(D) = 4.038 L^{2.6}$
Hollow column (C1f)	$m(D) = 9.23 \times 10^{-3} L^{1.8}$

where $m(D)$ is mass of a particle of diameter D; a_L is the constant leading coefficient, which for a sphere is $\rho_x\, \pi/6$, where ρ_x is the density of the sphere; and b_L is the power, which is a constant, given as $b_L = 3$ for a sphere. The variables in parentheses are the Magono and Lee (1966) classification identifier. For columnar-like crystals, a power law for mass–length is used in the same fashion as that for a sphere,

$$m(L) = a_L L^{b_L}. \tag{2.14}$$

2.3.2 Diameter–thickness (or length)

For most observed ice crystals diameter (D)–thickness (H) or diameter (D)–length (L) relationships are specified, such as

$$H(D) = a_H D^{b_H} \text{ or } D(L) = a_L L^{b_L}. \tag{2.15}$$

Some examples, following Pruppacher and Klett (1997), include, in cm,

Hexagonal plate (Pla)	$H(D) = 1.41 \times 10^{-2} D^{0.474}$
Broad branches (Plb)	$H(D) = 1.05 \times 10^{-2} D^{0.423}$
Dendrites plc-r, plc-s (Pld)	$H(D) = 9.96 \times 10^{-3} D^{0.415}$
Solid thick plate (Clg)	$H(D) = 0.138 D^{0.778}$
Solid columns (Cle)	$D(L) = 0.578 L^{0.958}$
Hollow columns (Clf)	$D(L) = 0.422 L^{0.892}$
Needle (Nle)	$D(L) = 3.527 \times 10^{-2} L^{0.437}$

2.3.3 Density–diameter (or length)

For most particles, densities are constant (such as for raindrops) or the density can be written as a power law in terms of diameter. This is particularly true for ice crystals, with values given in a later chapter. The relationship is simply,

$$\rho(D) = e D^f. \tag{2.16}$$

Some examples, following Heymsfield (1972) and Pruppacher and Klett (1997) include, in CGS units,

Hexagonal plate	$\rho(D) = 0.900$
Dendrites	$\rho(D) = 0.588 D^{-0.377}$

Stellar with broad branches	$\rho(D) = 0.588D^{-0.377}$
Stellar with narrow branches	$\rho(D) = 0.46D^{-0.482}$
Column	$\rho(D) = 0.848D^{-0.014}$
Bullet	$\rho(D) = 0.78D^{-0.0038}$

2.3.4 *Terminal velocity–diameter (or length)*

Terminal velocities are also generally given as power-law relationships,

$$V_{Tx}(D_x) = c_x D_x^{d_x}, \tag{2.17}$$

where D_x is the diameter of some hydrometeor species of habit, x; V_{Tx} is the terminal velocity; c_x is the leading coefficient of the power law; and d_x is the power. Some examples of terminal-velocity–diameter relationships include the following, after Heymsfield and Kajikawa (1987), and Pruppacher and Klett (1997), in cm s^{-1} unless noted,

Hexagonal plate (Pla)	$V_T(D) = 155.86D^{0.86}$
Crystal with broad branches (P1b)	$V_T(D) = 190D^{0.81}$
Dendrites (Plc-r, Plc-s, Pld)	$V_T(D) = 58D^{0.55}$
Sphere (in m s^{-1}), such as graupel, and hail	$V_T(D) = \left(\frac{4}{3}\frac{\rho g}{c_d \rho_o}\right)^{0.5} D^{0.5}$
Rain (in m s^{-1})	$V_T(D) = 842D^{0.8}$
Snow aggregates (in m s^{-1})	$V_T(D) = 4.83607122D^{0.25}$

The terminal velocity used in the microphysical equations and sedimentation rates equation can be based on the mass weighted mean value for mixing ratio, Q; number weighted mean for N_T; and reflectivity weighted mean for reflectivity, Z. Following Milbrandt and Yau (2005a, b), if the reflectivity overshoots the precipitation front of Q or N_T because the flux of reflectivity is larger than either that for Q or N_T, then the Z is set to zero. The same is true if N_T overshoots Q; N_T is set to zero.

2.4 Spectral density functions

2.4.1 *Gamma distribution*

One of the most common distribution functions used in microphysical parameterizations is what will be called herein the complete gamma distribution, written as

$$f(x) = \frac{\mu}{s} x^{\nu\mu - 1} \exp(-\alpha x^{\mu}), \tag{2.18}$$

where α, ν, and μ are shape parameters and s is a scaling parameter. Ultimately it is desireable to obtain an equation for $n(D)$ based on this equation.

First, it will be important to consider the distribution for D over the interval from zero to infinity, but other intervals could be considered too. These will be discussed later.

The first concept that must be maintained is that the distribution can be normalized, or that a scaling factor exists such that

$$\int_0^\infty f(D)\mathrm{d}D = 1. \tag{2.19}$$

Recalling (2.10), (2.12), and (2.18), (2.19) can be written as

$$\int_0^\infty \frac{\mu}{s}\frac{D^{\nu\mu-1}}{D_n}\frac{1}{D_n}\exp\left(-\alpha\left[\frac{D}{D_n}\right]^{\mu}\right)\mathrm{d}D = 1, \tag{2.20}$$

which can be written in a readily integratable form as

$$\int_0^\infty \frac{\mu}{s}\left(\frac{D}{D_n}\right)^{\nu\mu-1}\exp\left(-\alpha\left[\frac{D}{D_n}\right]^{\mu}\right)\mathrm{d}\left(\frac{D}{D_n}\right) = 1. \tag{2.21}$$

The scaling factor, s, for this case can be found by integrating (2.21),

$$\frac{\mu}{s}\frac{\Gamma\left(\frac{\mu\nu-1+1}{\mu}\right)}{\mu\alpha^{\left(\frac{\mu\nu-1+1}{\mu}\right)}} = 1, \tag{2.22}$$

and solving for s,

$$s = \frac{\Gamma(\nu)}{\alpha^{\nu}}. \tag{2.23}$$

Thus, using the scaling factor s, then the function $f(D)$ is written using (2.10), (2.12) and (2.18) as

$$f(D) = \frac{\alpha^{\nu}\mu}{\Gamma(\nu)}\left(\frac{D}{D_n}\right)^{\nu\mu-1}\frac{1}{D_n}\exp\left(-\alpha\left[\frac{D}{D_n}\right]^{\mu}\right). \tag{2.24}$$

Next, incorporating the definition of $n(D)$, (2.6), the following can be written for the spectral density function for the "complete gamma" distribution where $\nu \neq 1$, $\mu \neq 1$ and $\alpha \neq 1$,

$$n(D) = \frac{N_{\mathrm{T}}\alpha^{\nu}\mu}{\Gamma(\nu)}\left(\frac{D}{D_n}\right)^{\nu\mu-1}\frac{1}{D_n}\exp\left[-\alpha\left(\frac{D}{D_n}\right)^{\mu}\right] \tag{2.25}$$

Notice now that there are five free parameters in the above. These include D_n, N_T, μ, α, and ν. These values must be estimated from observations or derived from observations. Often, it is not known what they should be and educated guesses have to be made. Fortunately, N_T can be observed, diagnosed or predicted, D_n can be diagnosed or predicted, and ν can be diagnosed by predicting reflectivity Z, as is shown later.

Similarly, the "modified gamma" distribution, where $\nu \neq 1$, $\mu \neq 1$ and $\alpha = 1$ in (2.25), may be expressed as

$$n(D) = \frac{N_T \mu}{\Gamma(\nu)} \left(\frac{D}{D_n} \right)^{\nu\mu - 1} \frac{1}{D_n} \exp\left[- \left(\frac{D}{D_n} \right)^{\mu} \right]. \tag{2.26}$$

The "gamma" distribution function can be greatly simplified and made more useable by assuming $\nu \neq 1$, $\mu = 1$ and $\alpha = 1$ in (2.25), which results in the gamma function

$$n(D) = \frac{N_T}{\Gamma(\nu)} \left(\frac{D}{D_n} \right)^{\nu - 1} \frac{1}{D_n} \exp\left[- \left(\frac{D}{D_n} \right) \right]. \tag{2.27}$$

Many times this will be rewritten, especially in radar meteorology, in terms of the slope intercept of the distribution, or a quantity related to it, denoted as N_0,

$$n(D) = N_0 D^{\nu} \exp\left[- \left(\frac{D}{D_n} \right) \right]. \tag{2.28}$$

2.4.2 Exponential distribution

The complete gamma distribution with $\mu = \nu = \alpha = 1$ gives a special form called the negative-exponential, inverse-exponential, or Marshall and Palmer (1948) form of the gamma distribution,

$$n(D) = \frac{N_T}{D_n} \exp\left[- \left(\frac{D}{D_n} \right) \right] = N_0 \exp\left[- \left(\frac{D}{D_n} \right) \right], \tag{2.29}$$

where,

$$N_T = N_0 D_n. \tag{2.30}$$

The distribution given by Laws and Parsons (1943) was made infamous by Marshall and Palmer (1948) when they showed that rain distributions could be well defined in the mean by this form of the gamma distribution and later the same was shown to be true for some cases of snow aggregates.

Finally, Cheng and English (1983) found using observations that, in the mean, even hail was well described by exponential distributions using the following assumptions. First, the slope intercept in units of (m^{-3} mm^{-1}) is related to λ in units of (mm^{-1}), and the equations are written for the slope intercept as a function of λ,

$$N_0 = 115\lambda^{3.63}. \tag{2.31}$$

Therefore, the distribution (2.29) can be written using N_0 and using λ in place of $1/D_n$ as

$$n(D) = N_0 \exp(-D\lambda). \tag{2.32}$$

2.4.3 Half-normal distribution

Next there is a special case of gamma distribution called the half-normal distribution, where $\alpha = 1$, $\mu = 2$, $\nu = 1/2$,

$$n(D) = \frac{2N_\mathrm{T}}{D_n\Gamma(1/2)}\exp\left[-\left(\frac{D}{D_n}\right)^2\right] = \frac{2N_\mathrm{T}}{D_n\pi}\exp\left[-\left(\frac{D}{D_n}\right)^2\right]. \tag{2.33}$$

This function is integrated over the interval $D(0, \infty)$, thus, half-normal. Equation (2.33) is required in the derivations using the log-normal distributions.

2.4.4 Normal distribution

When the gamma probability distribution function is integrated on the interval $D \in (-\infty, \infty)$, one can show that

$$n(D) = \frac{N_\mathrm{T}}{D_n\sqrt{2\pi}}\exp\left[-\frac{1}{2}\left(\frac{D}{D_n}\right)^2\right], \tag{2.34}$$

which is a normal distribution. This function is not obviously useable owing to the limits of integration used to derive it. With some algebra, however, it can be remapped in the interval of $D(0, \infty)$.

2.4.5 Log-normal distribution

With the log-normal distribution, the natural log of the diameter is normally distributed. This distribution has been used by a few investigators including Clark (1976) and Nickerson *et al.* (1986). To derive the log-normal distribution start with $f_{\mathrm{nor}}(x)$ for a normal distribution,

$$f_{\mathrm{nor}}(x) = \frac{1}{\sqrt{2\pi}}\exp\left(-\frac{x^2}{2}\right). \tag{2.35}$$

Now let the following transformation hold that maps $x \in (-\infty, \infty)$ into $D \in (-\infty, \infty)$,

$$x = \frac{1}{\sigma} \ln\left(\frac{D}{D_n}\right), \quad x \in (-\infty, \infty) \tag{2.36}$$

where σ is a parameter, x is a function of diameter D, and D_n is a scaling diameter. Substitution of (2.36) into (2.35), results in

$$f_{\text{nor}}(x) = \frac{1}{\sqrt{2\pi}} \exp\left(-\frac{[\ln(D/D_n)]^2}{2\sigma^2}\right). \tag{2.37}$$

The following can be written according to continuous distribution theory, which is consistent with (2.8),

$$f_{\log}(D) = f_{\text{nor}}(x) \frac{\mathrm{d}x}{\mathrm{d}D}. \tag{2.38}$$

Taking the derivative of (2.36) with respect to D and multiplying the result by (2.37) results in the log-normal distribution function,

$$f_{\log}(D) = \frac{1}{\sqrt{2\pi}\sigma D} \exp\left(-\frac{[\ln(D/D_n)]^2}{2\sigma^2}\right). \tag{2.39}$$

Substitution of (2.39) into (2.6) gives the particle distribution spectrum,

$$n(D) = \frac{N_{\mathrm{T}}}{\sqrt{2\pi}\sigma D} \exp\left(-\frac{[\ln(D/D_n)]^2}{2\sigma^2}\right), \quad D \in (0, \infty). \tag{2.40}$$

2.5 Gamma distributions

In the rest of this section a detailed examination of the complete gamma distribution (2.25) with $\nu \neq 1$, $\mu \neq 1$ and $\alpha \neq 1$, the modified gamma distribution (2.26) with $\nu \neq 1$, $\mu \neq 1$ and $\alpha = 1$, and the gamma distribution (2.27) with $\nu \neq 1$, $\mu = 1$ and $\alpha = 1$ is given. The Marshall–Palmer distribution form can be obtained readily by setting $\nu = 1$ in (2.27) and thus is not shown. In the next section, the log-normal distribution will be examined.

2.5.1 Moments

The moments M can be found with the following relationship where I is the degree of the moment and subscript x is the hydrometeor species. The moments for the complete gamma distribution (2.25) where, $\nu_x \neq 1$, $\mu_x \neq 1$ and $\alpha_x \neq 1$, are given by

$$M(I_x) = \int_0^\infty D^{I_x} \frac{\alpha_x^{\nu_x} \mu_x}{\Gamma(\nu_x)} \left(\frac{D}{D_{nx}}\right)^{\nu_x \mu_x - 1} \exp\left(-\alpha_x \left[\frac{D_x}{D_{nx}}\right]^{\mu_x}\right) \mathrm{d}\left(\frac{D_x}{D_{nx}}\right)$$

$$= \frac{\alpha_x^{\nu_x} \Gamma\left(\frac{\nu_x \mu_x + I_x}{\mu_x}\right)}{\alpha_x^{\left(\frac{\nu_x \mu_x + I_x}{\mu_x}\right)} \Gamma(\nu_x)} D_n^{I_x}. \tag{2.41}$$

For the modified gamma distribution (2.26), where $\nu_x \neq 1$, $\mu_x \neq 1$ and $\alpha_x = 1$, the moments are

$$M(I_x) = \int_0^\infty D^{I_x} \frac{1}{\Gamma(\nu_x)} \left(\frac{D}{D_{nx}}\right)^{\nu_x \mu_x - 1} \exp\left(-\left[\frac{D_x}{D_{nx}}\right]\right) \mathrm{d}\left(\frac{D_x}{D_{nx}}\right) = \frac{\Gamma\left(\frac{\nu_x \mu_x + I_x}{\mu_x}\right)}{\Gamma(\nu_x)} D_{nx}^{I_x}. \tag{2.42}$$

For the gamma function (2.27), $\nu_x \neq 1$, $\mu_x = 1$ and $\alpha_x = 1$, the moments are

$$M(I_x) = \int_0^\infty D^{I_x} \frac{1}{\Gamma(\nu_x)} \left(\frac{D}{D_{nx}}\right)^{\nu_x - 1} \exp\left(-\left[\frac{D_x}{D_{nx}}\right]\right) \mathrm{d}\left(\frac{D_x}{D_{nx}}\right) = \frac{\Gamma(\nu_x + I_x)}{\Gamma(\nu_x)} D_{nx}^{I_x}. \tag{2.43}$$

For the negative exponential distribution

$$M(I_x) = \int_0^\infty D^{I_x} \exp\left(-\left[\frac{D_x}{D_{nx}}\right]\right) \mathrm{d}\left(\frac{D_x}{D_{nx}}\right) = \Gamma(1 + I_x) D_{nx}^{I_x}. \tag{2.44}$$

2.5.2 The zeroth moment is number concentration

The zeroth moment ($I = 0$) is simply the number concentration, but is derived here,

$$N_{\mathrm{T}x} = \int_0^\infty n(D_x) \mathrm{d}D_x. \tag{2.45}$$

Using the complete gamma distribution (2.25) to substitute into (2.45) gives

$$N_{\mathrm{T}x} = \int_0^\infty \frac{N_{\mathrm{T}x} \alpha^{\nu_x} \mu_x}{\Gamma(\nu_x)} \left(\frac{D_x}{D_{nx}}\right)^{\nu_x \mu_x - 1} \exp\left(-\alpha_x \left[\frac{D_x}{D_{nx}}\right]^{\mu_x}\right) \mathrm{d}\left(\frac{D_x}{D_{nx}}\right). \tag{2.46}$$

Upon integration using (2.4)–(2.5) the following is obtained,

$$N_{Tx} = \frac{N_{Tx}\alpha^{\nu_x}\mu_x\Gamma\left(\frac{\nu_x\mu_x}{\mu_x}\right)}{\Gamma(\nu_x)\mu_x\alpha^{\left(\frac{\nu_x\mu_x}{\mu_x}\right)}} = N_{Tx}. \tag{2.47}$$

For the modified gamma function, it can be written,

$$N_{Tx} = \frac{N_{Tx}\mu_x\Gamma\left(\frac{\nu_x\mu_x}{\mu_x}\right)}{\Gamma(\nu_x)\mu_x} = N_{Tx}; \tag{2.48}$$

and for the gamma distribution,

$$N_{Tx} = \frac{N_{Tx}\Gamma(\nu_x)}{\Gamma(\nu_x)} = N_{Tx}. \tag{2.49}$$

Finally for the negative-exponential distribution,

$$N_{Tx} = \frac{N_{Tx}\Gamma(1)}{\Gamma(1)} = N_{Tx}. \tag{2.50}$$

Often the zeroth moment is predicted in modern cloud models. In many cloud models (e.g. Liu and Orville 1969; Lin *et al.* 1983; Rutledge and Hobbs 1983, 1984; Straka and Mansell 2005), the zeroth moment is diagnosed from D_n and N_0 with Marshall–Palmer distributions assumed.

2.5.3 Number-concentration-weighted mean diameter

The number-concentration-weighted mean diameter is a simple measure of diameter. Assuming a spherical hydrometeor, it is found for the complete gamma distribution (2.25), where,

$$\bar{D}_x^{N_T} = \frac{\int_0^\infty D_x n(D_x)\mathrm{d}D_x}{\int_0^\infty n(D_x)\mathrm{d}D_x}. \tag{2.51}$$

Substitution of (2.25) into (2.51) results in

$$\bar{D}_x^{N_{Tx}} = \frac{D_{nx}\int_0^\infty \left(\frac{D_x}{D_{nx}}\right)\frac{N_{T_x}\alpha_x^{\nu_x}\mu_x}{\Gamma(\nu_x)\mu_x}\left(\frac{D_x}{D_{nx}}\right)^{\nu_x\mu_x-1}\exp\left(-\alpha_x\left[\frac{D}{D_n}\right]^{\mu_x}\right)\mathrm{d}\left(\frac{D_x}{D_{nx}}\right)}{\int_0^\infty \frac{N_{T_x}\alpha_x^{\nu_x}\mu_x}{\Gamma(\nu_x)\mu_x}\left(\frac{D_x}{D_{nx}}\right)^{\nu_x\mu_x-1}\exp\left(-\alpha_x\left[\frac{D}{D_n}\right]^{\mu_x}\right)\mathrm{d}\left(\frac{D_x}{D_{nx}}\right)}. \tag{2.52}$$

Simplifying using (2.4)–(2.5) gives

$$\bar{D}_x^{N_{Tx}} = D_{nx} \frac{\alpha_x^{\nu_x} \Gamma\left(\frac{\nu_x \mu_x + 1}{\mu_x}\right)}{\alpha_x^{\left(\frac{\nu_x \mu_x + 1}{\mu_x}\right)} \Gamma(\nu_x)}. \tag{2.53}$$

For the modified gamma distribution (2.26), this can be written as

$$\bar{D}_x^{N_{Tx}} = D_{nx} \frac{\Gamma\left(\frac{\nu_x \mu_x + 1}{\mu_x}\right)}{\Gamma(\nu_x)}; \tag{2.54}$$

for the gamma distribution (2.27),

$$\bar{D}_x^{N_{Tx}} = D_{nx} \frac{\Gamma(\nu_x + 1)}{\Gamma(\nu_x)}; \tag{2.55}$$

and lastly for the negative exponential distribution,

$$\bar{D}_x^{N_{Tx}} = D_{nx} \frac{\Gamma(2)}{\Gamma(1)} = D_{nx}. \tag{2.56}$$

Next, there is a need to find a relation to diagnose D_n. As will be shown later, D_n can be predicted from the first and second moment in terms of mass. The method to diagnose D_n is derived later.

2.5.4 Mass-weighted mean diameter

The mass-weighted mean diameter for the complete gamma distribution is found as follows,

$$\bar{D}_x^{m} = \frac{\int_0^\infty D_x m(D_x) n(D_x) \mathrm{d}D}{\int_0^\infty m(D_x) n(D_x) \mathrm{d}D}. \tag{2.57}$$

Substitution of (2.25) into (2.57) results in

$$\bar{D}_x^{m} = \frac{\int_0^\infty D_x a_x D_x^{b_x} \frac{N_{Tx} \alpha_x^{\nu_x} \mu_x}{\Gamma(\nu)} \left(\frac{D_x}{D_{nx}}\right)^{\nu_x \mu_x - 1} \exp\left(-\alpha_x \left[\frac{D_x}{D_{nx}}\right]^{\mu_x}\right) \mathrm{d}\left(\frac{D_x}{D_{nx}}\right)}{\int_0^\infty a_x D_x^{b_x} \frac{N_{Tx} \alpha_x^{\nu_x} \mu_x}{\Gamma(\nu)} \left(\frac{D_x}{D_{nx}}\right)^{\nu_x \mu_x - 1} \exp\left(-\alpha_x \left[\frac{D_x}{D_{nx}}\right]^{\mu_x}\right) \mathrm{d}\left(\frac{D_x}{D_{nx}}\right)}, \tag{2.58}$$

or after dividing by D_{nx},

$$\bar{D}_x^{m} = \frac{D_{nx}^{b_x+1} \int_0^\infty a_x \left(\frac{D_x}{D_{nx}}\right)^{bx+1} \frac{N_{Tx} \alpha_x^{\nu x} \mu_x}{\Gamma(\nu)} \left(\frac{D_x}{D_{nx}}\right)^{\nu_x \mu_x - 1} \exp\left(-\alpha_x \left[\frac{D_x}{D_{nx}}\right]^{\mu_x}\right) \mathrm{d}\left(\frac{D_x}{D_{nx}}\right)}{D_{nx}^{b_x} \int_0^\infty a_x \left(\frac{D_x}{D_{nx}}\right)^{bx} \frac{N_{Tx} \alpha_x^{\nu x} \mu_x}{\Gamma(\nu)} \left(\frac{D_x}{D_{nx}}\right)^{\nu_x \mu_x - 1} \exp\left(-\alpha_x \left[\frac{D_x}{D_{nx}}\right]^{\mu_x}\right) \mathrm{d}\left(\frac{D_x}{D_{nx}}\right)}. \tag{2.59}$$

Applying (2.4)–(2.5) results in

$$\bar{D}_x^m = D_{nx}^{b_x} \frac{\left[\frac{\Gamma\left(\frac{b_x+v_x\mu_x+1}{\mu_x}\right)}{\alpha_x^{\left(\frac{b_x+v_x\mu_x+1}{\mu_x}\right)}} \right]}{\left[\frac{\Gamma\left(\frac{b_x+v_x\mu_x}{\mu_x}\right)}{\alpha_x^{\left(\frac{b_x+v_x\mu_x}{\mu_x}\right)}} \right]}. \tag{2.60}$$

For the modified gamma distribution,

$$\bar{D}_x^m = D_{nx}^{b_x} \frac{\Gamma\left(\frac{b_x+v_x\mu_x+1}{\mu_x}\right)}{\Gamma\left(\frac{b_x+v_x\mu_x}{\mu_x}\right)}; \tag{2.61}$$

for the gamma distribution, the mass-weighted mean is

$$\bar{D}_x^m = D_{nx}^{b_x} \frac{\Gamma(b_x + v_x + 1)}{\Gamma(b_x + v_x)}; \tag{2.62}$$

and for the negative exponential distribution,

$$\bar{D}_x^m = D_{nx}^{b_x} \frac{\Gamma(b_x + 2)}{\Gamma(b_x + 1)}. \tag{2.63}$$

2.5.5 Mean-volume diameter

The mean-volume diameter can be shown to be equal to the following for any distribution,

$$D_{\mathrm{MV}x} = \left(\frac{6\rho Q_x}{\pi \rho_x N_{\mathrm{T}x}} \right)^{1/3}, \tag{2.64}$$

where ρ is the density of air, Q_x is the mixing ratio of the hydrometeor species, ρ_x is the density of the hydrometeor species, and $N_{\mathrm{T}x}$ is the number concentration of the hydrometeor species.

2.5.6 Effective diameter

The effective diameter is the ratio of the integral of $D^3 n(D)$ to the integral of $D^2 n(D)$, and is primarily used in radiation physics parameterizations and related calculations and can be expressed as

$$D_{\mathrm{eff}} = \frac{\int\limits_0^\infty D_x^3 \frac{N_{Tx}\alpha_x^{\nu_x}\mu_x}{\Gamma(\nu)}\left(\frac{D_x}{D_{nx}}\right)^{\nu_x\mu_x-1}\exp\left(-\alpha_x\left[\frac{D_x}{D_{nx}}\right]^{\mu_x}\right)\mathrm{d}\left(\frac{D_x}{D_{nx}}\right)}{\int\limits_0^\infty D_x^2 \frac{N_{Tx}\alpha_x^{\nu_x}\mu_x}{\Gamma(\nu)}\left(\frac{D_x}{D_{nx}}\right)^{\nu_x\mu_x-1}\exp\left(-\alpha_x\left[\frac{D_x}{D_{nx}}\right]^{\mu_x}\right)\mathrm{d}\left(\frac{D_x}{D_{nx}}\right)}. \tag{2.65}$$

After dividing and multiplying through by the appropriate powers of D_{nx},

$$D_{\mathrm{eff}} = \frac{D_{nx}^3 \int\limits_0^\infty \left(\frac{D_x}{D_{nx}}\right)^3 \frac{N_{Tx}\alpha_x^{\nu_x}\mu_x}{\Gamma(\nu)}\left(\frac{D_x}{D_{nx}}\right)^{\nu_x\mu_x-1}\exp\left(-\alpha_x\left[\frac{D_x}{D_{nx}}\right]^{\mu_x}\right)\mathrm{d}\left(\frac{D_x}{D_{nx}}\right)}{D_{nx}^2 \int\limits_0^\infty \left(\frac{D_x}{D_{nx}}\right)^2 \frac{N_{Tx}\alpha_x^{\nu_x}\mu_x}{\Gamma(\nu)}\left(\frac{D_x}{D_{nx}}\right)^{\nu_x\mu_x-1}\exp\left(-\alpha_x\left[\frac{D_x}{D_{nx}}\right]^{\mu_x}\right)\mathrm{d}\left(\frac{D_x}{D_{nx}}\right)}, \tag{2.66}$$

which can be integrated and written for the complete gamma distribution to obtain

$$D_{\mathrm{eff}} = D_{nx}\frac{\left[\dfrac{\Gamma\left(\frac{3+\nu_x\mu_x}{\mu_x}\right)}{\alpha_x^{\left(\frac{3+\nu_x\mu_x}{\mu_x}\right)}}\right]}{\left[\dfrac{\Gamma\left(\frac{2+\nu_x\mu_x}{\mu_x}\right)}{\alpha_x^{\left(\frac{2+\nu_x\mu_x}{\mu_x}\right)}}\right]}. \tag{2.67}$$

Now for the modified gamma distribution the effective diameter is

$$D_{\mathrm{eff}} = D_{nx}\frac{\Gamma\left(\frac{3+\nu_x\mu_x}{\mu_x}\right)}{\Gamma\left(\frac{2+\nu_x\mu_x}{\mu_x}\right)}; \tag{2.68}$$

and for the gamma distribution the effective diameter is

$$D_{\mathrm{eff}} = D_{nx}\frac{\Gamma(3+\nu_x)}{\Gamma(2+\nu_x)}. \tag{2.69}$$

Lastly for the negative-exponential distribution,

$$D_{\mathrm{eff}} = D_{nx}\frac{\Gamma(4)}{\Gamma(3)} = D_{nx}\frac{3!}{2!} = 3D_{nx}. \tag{2.70}$$

2.5.7 *Modal diameter*

The modal diameter is the diameter for which the distribution has a maximum. The modal diameter is found by differentiating $f(D)$, setting the result equal to zero, and solving for the diameter, which will be the modal diameter.

The derivative of the complete gamma distribution [(2.25) with $\nu_x \neq 1$, $\mu_x \neq 1$ and $\alpha_x \neq 1$] with respect to D_x, is used to begin the derivation of the modal diameter,

$$f(D_x) = \frac{\alpha_x^{\nu_x} \mu_x}{\Gamma(\nu_x)} \left(\frac{D_x}{D_{nx}} \right)^{\nu_x \mu_x - 1} \frac{1}{D_{nx}} \exp\left(-\alpha_x \left[\frac{D_x}{D_{nx}} \right]^{\mu_x} \right). \tag{2.71}$$

The derivative is taken with respect to D_x and set equal to zero,

$$\frac{\mathrm{d}}{\mathrm{d}D_x} \left\{ D_x^{\nu_x \mu_x - 1} \exp\left(-\alpha_x \left[\frac{D_x}{D_{nx}} \right]^{\mu_x} \right) \right\} = 0. \tag{2.72}$$

The derivative is expanded:

$$\begin{aligned} &D_x^{\nu_x \mu_x - 1} \exp\left(-\alpha_x \left[\frac{D_x}{D_{nx}} \right]^{\mu_x} \right) \left(-\alpha_x \mu_x D_{nx}^{-\mu_x} D_x^{\mu_x - 1} \right) \\ &\quad + (\nu_x \mu_x - 1) D_x^{\nu_x \mu_x - 2} \exp\left(-\alpha_x \left[\frac{D_x}{D_{nx}} \right]^{\mu_x} \right) = 0. \end{aligned} \tag{2.73}$$

To solve for D_x,

$$D_{\mathrm{mod}} = D_x = D_{nx} \left(\frac{\nu_x \mu_x - 1}{\alpha_x \mu_x} \right)^{-\mu_x}. \tag{2.74}$$

Next the modified gamma distribution [(2.26) with $\nu_x \neq 1$, $\mu_x \neq 1$, $\alpha_x = 1$], can be used to obtain

$$D_{\mathrm{mod}} = D_x = D_{nx} \left(\frac{\nu_x \mu_x - 1}{\mu_x} \right)^{-\mu_x}, \tag{2.75}$$

and for the gamma distribution [(2.27) with $\alpha_x = \mu_x = 1$], the mode is simply

$$D_{\mathrm{mod}} = D_x = D_{nx}(\nu_x - 1). \tag{2.76}$$

For the negative-exponential distribution,

$$D_{\mathrm{mod}} = 0. \tag{2.77}$$

2.5.8 Median diameter

The median diameter is that diameter for which the distribution has half of the mass at smaller sizes and half the mass at larger sizes. The solution has to be solved numerically and is a function of the mass, m. The mass can be found from,

$$m_{\mathrm{T}} = \int_0^\infty m(D_x) n(D_x) \mathrm{d}D_x \tag{2.78}$$

If we start with total mass for a spherical hydrometeor (2.13), the complete gamma distribution (2.25) and divide through by D_{nx}, (2.78) becomes

$$m_{\rm T} = a_x \frac{D_{nx}^{b_x} N_{{\rm T}x} \alpha_x^{\nu_x} \mu_x}{\Gamma(\nu_x)} \int_0^\infty \left(\frac{D_x}{D_{nx}}\right)^{\nu_x\mu_x - 1} \left(\frac{D_x}{D_{nx}}\right)^{b_x} \exp\left[-\alpha_x \left(\frac{D_x}{D_{nx}}\right)^{\mu_x}\right] {\rm d}\left(\frac{D_x}{D_{nx}}\right), \quad (2.79)$$

which upon integration gives,

$$m_{\rm T} = \frac{a_x D_{nx}^{b_x+1} N_{{\rm T}x} \alpha_x^{\nu_x} \mu_x}{\Gamma(\nu_x)} \left[\frac{\Gamma\left(\frac{b_x+\nu_x\mu_x}{\mu_x}\right)}{\mu_x \alpha_x^{\left(\frac{b_x+\nu_x\mu_x}{\mu_x}\right)}}\right]; \quad (2.80)$$

and for the modified gamma distribution,

$$m_{\rm T} = \frac{a_x D_{nx}^{b_x+1} N_{{\rm T}x}}{\Gamma(\nu_x)} \Gamma\left(\frac{b_x + \nu_x \mu_x}{\mu_x}\right). \quad (2.81)$$

For the gamma distribution, the form is

$$m_{\rm T} = a_x D_{nx}^{b_x+1} N_{{\rm T}x} \frac{\Gamma(b_x + \nu_x)}{\Gamma(\nu_x)}; \quad (2.82)$$

and for the negative-exponential distribution

$$m_{\rm T} = a_x D_{nx}^{b_x+1} N_{{\rm T},x} \Gamma(b_x + 1). \quad (2.83)$$

For simplicity, the gamma distribution will be used to find the median diameter, D_0. It is found from integrating the equation for m (2.79; where $\alpha_x = \mu_x = 1$) divided by 2 and with new limits of integration; this is

$$\frac{m_{\rm T}}{2} = \int_0^{D_0} \frac{a_x D_{nx}^{b_x} N_{{\rm T}x}}{2\Gamma(\nu_x)} \left(\frac{D_x}{D_{nx}}\right)^{b_x + \nu_x - 1} \exp\left[-\left(\frac{D_x}{D_{nx}}\right)\right] {\rm d}\left(\frac{D_x}{D_{nx}}\right). \quad (2.84)$$

Solving this integral numerically gives the median diameter as

$$D_0 = 3.672 D_{nx}(\nu_x - 1). \quad (2.85)$$

2.5.9 The second moment is related to total surface area

The second moment, which is related to total surface area $A_{\rm T}$ of hydrometeors is often used in models with cloud and precipitation electrification parameterizations (e.g. Mansell *et al.* 2002, 2005). It is found by using the definition of surface area for a spherical hydrometeor and assuming the complete gamma distribution,

$$A_{\mathrm{T}} = \int_0^\infty \pi D_x^2 \frac{N_{\mathrm{T}x}\alpha_x^{\nu_x}\mu_x}{\Gamma(\nu)}\left(\frac{D_x}{D_{nx}}\right)^{\nu_x\mu_x-1} \exp\left(-\alpha_x\left[\frac{D_x}{D_{nx}}\right]^{\mu_x}\right) \mathrm{d}\left(\frac{D_x}{D_{nx}}\right). \tag{2.86}$$

Dividing by D_{nx} gives

$$A_{\mathrm{T}} = \pi D_{nx}^2 \frac{N_{\mathrm{T}x}\alpha_x^{\nu_x}\mu_x}{\Gamma(\nu)} \int_0^\infty \left(\frac{D_x}{D_{nx}}\right)^2 \left(\frac{D_x}{D_{nx}}\right)^{\nu_x\mu_x-1} \exp\left(-\alpha_x\left[\frac{D_x}{D_{nx}}\right]^{\mu_x}\right) \mathrm{d}\left(\frac{D_x}{D_{nx}}\right). \tag{2.87}$$

Making use of (2.4)–(2.5) gives

$$A_{\mathrm{T}} = \pi\alpha_x^{\nu_x} D_{nx}^2 N_{\mathrm{T}x} \frac{\mu_x}{\mu_x} \frac{\Gamma\left(\dfrac{\nu_x\mu_x+2}{\mu_x}\right)}{\Gamma(\nu_x)\alpha_x^{\left(\frac{\nu_x\mu_x+2}{\mu_x}\right)}} = \pi D_{nx}^2 N_{\mathrm{T}x} \frac{\alpha_x^{\nu_x}\Gamma\left(\dfrac{\nu_x\mu_x+2}{\mu_x}\right)}{\alpha_x^{\left(\frac{\nu_x\mu_x+2}{\mu_x}\right)}\Gamma(\nu_x)}. \tag{2.88}$$

For the modified gamma distribution the total surface area is simply

$$A_{\mathrm{T}} = \pi D_{nx}^2 N_{\mathrm{T}x} \frac{\Gamma\left(\dfrac{\nu_x\mu_x+2}{\mu_x}\right)}{\Gamma(\nu_x)}, \tag{2.89}$$

whilst that for the gamma distribution is just

$$A_{\mathrm{T}} = \pi D_{nx}^2 N_{\mathrm{T}x} \frac{\Gamma(\nu_x+2)}{\Gamma(\nu_x)}, \tag{2.90}$$

and for the negative-exponential distribution

$$A_{\mathrm{T}} = \pi D_{nx}^2 N_{\mathrm{T}x}\Gamma(3) = \pi D_{nx}^2 N_{\mathrm{T}x} 2! = 2\pi D_{nx}^2 N_{\mathrm{T}x}. \tag{2.91}$$

2.5.10 Total downward projected area

The total downward projected area A_{p} is defined as follows for a sphere or circular disk and assuming the complete gamma distribution

$$A_{\mathrm{p}} = \int_0^\infty \frac{\pi}{4} D_x^2 \frac{N_{\mathrm{T}x}\alpha_x^{\nu_x}\mu_x}{\Gamma(\nu_x)}\left(\frac{D_x}{D_{nx}}\right)^{\nu_x\mu_x-1} \exp\left(-\alpha_x\left[\frac{D}{D_n}\right]^{\mu_x}\right) \mathrm{d}\left(\frac{D_x}{D_{nx}}\right). \tag{2.92}$$

Dividing by D_{nx} gives

$$A_{\mathrm{p}} = D_{nx}^2 \int_0^\infty \frac{\pi}{4}\left(\frac{D_x}{D_{nx}}\right)^2 \frac{N_{\mathrm{T}x}\alpha_x^{\nu_x}\mu_x}{\Gamma(\nu_x)\mu_x}\left(\frac{D_x}{D_{nx}}\right)^{\nu_x\mu_x-1} \exp\left(-\alpha_x\left[\frac{D_x}{D_{nx}}\right]^{\mu_x}\right) \mathrm{d}\left(\frac{D_x}{D_{nx}}\right), \tag{2.93}$$

Making use of (2.4)–(2.5) gives

$$A_{\mathrm{p}} = \frac{\pi}{4} N_{\mathrm{T}x} D_{nx}^2 \frac{\alpha_x^{\nu_x} \Gamma\left(\frac{\nu_x \mu_x + 2}{\mu_x}\right)}{\alpha^{\left(\frac{\nu_x \mu_x + 2}{\mu_x}\right)} \Gamma(\nu_x)}. \tag{2.94}$$

The modified gamma distribution projected area is

$$A_{\mathrm{p}} = \frac{\pi}{4} N_{\mathrm{T}x} D_{nx}^2 \frac{\Gamma\left(\frac{\nu_x \mu_x + 2}{\mu_x}\right)}{\Gamma(\nu_x)}, \tag{2.95}$$

and for the gamma distribution the equation is

$$A_{\mathrm{p}} = \frac{\pi}{4} N_{\mathrm{T}x} D_{nx}^2 \frac{\Gamma(\nu_x + 2)}{\Gamma(\nu_x)}, \tag{2.96}$$

and the negative-exponential distribution is

$$A_{\mathrm{p}} = \frac{\pi}{4} N_{\mathrm{T}x} D_{nx}^2 \Gamma(3) = \frac{\pi}{2} N_{\mathrm{T}x} D_{nx}^2. \tag{2.97}$$

2.5.11 The third moment, mixing ratio and characteristic diameter

The mixing ratio Q_x of a hydrometeor species is related to mass, which is related to volume, and is written in terms of the third moment. For a spherical hydrometeor and the complete gamma distribution,

$$Q_x = \frac{1}{\rho} \int_0^\infty a_x D_x^{b_x} \frac{N_{\mathrm{T}x} \alpha_x^{\nu_x} \mu_x}{\Gamma(\nu_x)} \left(\frac{D_x}{D_{nx}}\right)^{\nu_x \mu_x - 1} \exp\left(-\alpha_x \left[\frac{D_x}{D_{nx}}\right]^{\mu_x}\right) \mathrm{d}\left(\frac{D_x}{D_{nx}}\right). \tag{2.98}$$

Dividing by D_{nx} produces

$$Q_x = a_x \frac{N_{\mathrm{T}x} \alpha_x^{\nu_x} \mu_x}{\Gamma(\nu_x)} \frac{D_{nx}^{b_x}}{\rho} \int_0^\infty \left(\frac{D_x}{D_{nx}}\right)^{b_x} \left(\frac{D_x}{D_{nx}}\right)^{\nu_x \mu_x - 1} \exp\left(-\alpha_x \left[\frac{D_x}{D_{nx}}\right]^{\mu_x}\right) \mathrm{d}\left(\frac{D_x}{D_{nx}}\right). \tag{2.99}$$

Making use of (2.4)–(2.5) results in

$$Q_x = a_x \frac{N_{\mathrm{T}x} \alpha_x^{\nu_x} \mu_x}{\Gamma(\nu_x)\ \mu_x} \frac{D_{nx}^{b_x}}{\rho} \frac{\Gamma\left(\frac{\nu_x \mu_x + b_x}{\mu_x}\right)}{\alpha^{\left(\frac{\nu_x \mu_x + b_x}{\mu_x}\right)}} = a_x \frac{N_{\mathrm{T}x} \alpha_x^{\nu}}{\Gamma(\nu_x)} \frac{D_{nx}^{b_x}}{\rho} \frac{\Gamma\left(\frac{\nu_x \mu_x + b_x}{\mu_x}\right)}{\alpha^{\left(\frac{\nu_x \mu_x + b_x}{\mu_x}\right)}}. \tag{2.100}$$

The mixing ratio for the modified gamma distribution is

$$Q_x = a_x N_{\mathrm{T}x} \frac{D_{nx}^{b_x}}{\rho} \frac{\Gamma\left(\frac{\nu_x \mu_x + b_x}{\mu_x}\right)}{\Gamma(\nu_x)}, \tag{2.101}$$

for the gamma distribution, the mixing ratio is

$$Q_x = a_x N_{\mathrm{T}x} \frac{D_{nx}^{b_x}}{\rho} \frac{\Gamma(v_x + b_x)}{\Gamma(v_x)}, \tag{2.102}$$

and for the negative-exponential distribution,

$$Q_x = a_x N_{\mathrm{T}x} \frac{D_{nx}^{b_x}}{\rho} \Gamma(1 + b_x). \tag{2.103}$$

Notice that there is a characteristic diameter D_{nx} to the b_x power in the equations, which for a spherical hydrometeor is equal to three. Thus, in this way, the mixing ratio is related to the third moment.

Mixing ratios of hydrometeor species are some of the variables that are almost always predicted in cloud models. From the mixing ratio, the first moment variable D_{nx} can be diagnosed with some algebra, for the complete gamma distribution,

$$D_{nx} = \left(\frac{Q_x \Gamma(v_x)}{\alpha_x^v N_{\mathrm{T}x}} \frac{\rho}{a_x} \frac{\alpha^{\left(\frac{v_x \mu_x + b_x}{\mu_x}\right)}}{\Gamma\left(\frac{v_x \mu_x + b_x}{\mu_x}\right)} \right)^{1/b_x}. \tag{2.104}$$

The equation for D_{nx} for the modified gamma distribution is

$$D_{nx} = \left(\frac{Q_x \Gamma(v_x)}{N_{\mathrm{T}x}} \frac{\rho}{a_x} \frac{1}{\Gamma\left(\frac{v_x \mu_x + b_x}{\mu_x}\right)} \right)^{1/b_x}, \tag{2.105}$$

and for the gamma distribution,

$$D_{nx} = \left(\frac{Q_x \Gamma(v_x)}{N_{\mathrm{T}x}} \frac{\rho}{a_x} \frac{1}{\Gamma(v_x + b_x)} \right)^{1/b_x}. \tag{2.106}$$

Lastly, for the negative-exponential distribution, the characteristic diameter is

$$D_{nx} = \left(\frac{Q_x}{N_{\mathrm{T}x}} \frac{\rho}{a_x} \frac{1}{\Gamma(1 + b_x)} \right)^{1/b_x}. \tag{2.107}$$

The value $\lambda = 1/D_{nx}$ is the slope of the distribution for the negative-exponential distribution.

2.5.12 The sixth moment is related to the reflectivity

The sixth moment is related to radar reflectivity as radar reflectivity is related to D^6 for Rayleigh scatterers in which the diameter is, normally, less than 1/16 of the radar wavelength, though a length 1/10 of the radar wavelength is often used. This moment is derived quite simply like the fourth and fifth moments. Starting with the definition for radar reflectivity, and using the complete gamma distribution gives

$$Z_x = D_{nx}^6 \int_0^\infty \left(\frac{D_x}{D_{nx}}\right)^6 \frac{N_{Tx}\alpha_x^{\nu_x}}{\Gamma(\nu_x)} \left(\frac{D_x}{D_{nx}}\right)^{\nu_x\mu_x-1} \exp\left(-\alpha_x \left[\frac{D_x}{D_{nx}}\right]^{\mu_x}\right) \mathrm{d}\left(\frac{D_x}{D_{nx}}\right). \tag{2.108}$$

Similarly, for the modified gamma distribution,

$$Z_x = D_{nx}^6 \int_0^\infty \left(\frac{D_x}{D_{nx}}\right)^6 \frac{N_{Tx}}{\Gamma(\nu_x)} \left(\frac{D_x}{D_{nx}}\right)^{\nu_x\mu_x-1} \exp\left(-\left[\frac{D_x}{D_{nx}}\right]^{\mu_x}\right) \mathrm{d}\left(\frac{D_x}{D_{nx}}\right), \tag{2.109}$$

and for the gamma distribution,

$$Z_x = D_{nx}^6 \int_0^\infty \left(\frac{D_x}{D_{nx}}\right)^6 \frac{N_{Tx}}{\Gamma(\nu_x)} \left(\frac{D_x}{D_{nx}}\right)^{\nu_x-1} \exp\left(-\left[\frac{D_x}{D_{nx}}\right]\right) \mathrm{d}\left(\frac{D_x}{D_{nx}}\right), \tag{2.110}$$

and for the negative-exponential distribution,

$$Z_x = D_{nx}^6 \int_0^\infty N_{Tx} \left(\frac{D_x}{D_{nx}}\right)^6 \exp\left(-\left[\frac{D_x}{D_{nx}}\right]\right) \mathrm{d}\left(\frac{D_x}{D_{nx}}\right). \tag{2.111}$$

The result from integration of (2.108) is the following equation,

$$Z_x = N_{Tx}\alpha_x^{\nu_x} D_{nx}^6 \frac{\Gamma\left(\dfrac{6+\nu_x\mu_x}{\mu_x}\right)}{\Gamma(\nu_x)\alpha_x^{\left(\frac{6+\nu_x\mu_x}{\mu_x}\right)}}, \tag{2.112}$$

and for the modified gamma distribution,

$$Z_x = N_{Tx} D_{nx}^6 \frac{\Gamma\left(\frac{6+\nu_x\mu_x}{\mu_x}\right)}{\Gamma(\nu_x)}, \tag{2.113}$$

whilst for the gamma distribution,

$$Z_x = N_{Tx} D_{nx}^6 \frac{\Gamma(6+\nu_x)}{\Gamma(\nu_x)}. \tag{2.114}$$

Lastly for the negative-exponential distribution,

$$Z_x = N_{Tx} D_{nx}^6 \Gamma(7) = N_{Tx} D_{nx}^6 6! = 720 N_{Tx} D_{nx}^6. \tag{2.115}$$

The radar reflectivity has generally been a diagnostic variable in cloud models, but more recently some have predicted radar reflectivity to gain insight as to the evolution of the shape parameter v as discussed below.

2.5.13 Rainfall rate

The rainfall rate or other hydrometeor fall rate can be computed with knowledge of particle terminal velocity V_T (remembering that d_x is the terminal velocity exponent in the power law for terminal velocity), and the liquid-water mixing ratio or content as is given below for the complete gamma function,

$$R_x = \frac{a_x c_x \alpha_x^{v_x}}{\Gamma(v_x)} N_{Tx} \int_0^\infty \left(\frac{D_x}{D_{nx}}\right)^{b_x+d_x} \left(\frac{D_x}{D_{nx}}\right)^{v_x\mu_x-1} \exp\left(-\alpha_x \left[\frac{D_x}{D_{nx}}\right]^{\mu_x}\right) \mathrm{d}\left(\frac{D_x}{D_{nx}}\right). \quad (2.116)$$

For the complete gamma distribution, the rate is then

$$R_x = N_{Tx} \frac{a_x c_x D_{nx}^{b_x+d_x} \alpha_x^{v_x} \Gamma\left(\frac{b_x+d_x+v_x\mu_x}{\mu_x}\right)}{\rho \alpha_x^{\left(\frac{b_x+d_x+v_x\mu_x}{\mu_x}\right)} \Gamma(v_x)}. \quad (2.117)$$

Next for the modified gamma distribution, the rainfall rate is

$$R_x = N_{Tx} \frac{a_x c_x D_{nx}^{b_x+d_x} \Gamma\left(\frac{b_x+d_x+v_x\mu_x}{\mu_x}\right)}{\rho \Gamma(v_x)}. \quad (2.118)$$

Then for the gamma distribution the rainfall rate is

$$R_x = N_{Tx} D_{nx}^{b_x+d_x} \Gamma(b_x + d_x + v_x) \frac{a_x}{\Gamma(v_x)}, \quad (2.119)$$

and for the negative-exponential distribution,

$$R_x = a_x N_{Tx} D_{nx}^{b_x+d_x} \Gamma(b_x + d_x + 1). \quad (2.120)$$

2.5.14 Terminal velocities

The terminal velocity used in the collection equations and sedimentation can be based on the mass-weighted mean value for Q, number-weighted mean for N_T, and reflectivity-weighted mean for Z. These are all given as follows for the complete gamma distribution first, the modified gamma distribution second, and the gamma distribution last.

2.5.14.1 Mixing-ratio-weighted terminal velocity

The form for mass-weighted (or mixing-ratio-weighted) mean terminal velocity is given by

$$\bar{V}_{TQ_x} = \frac{\int_0^\infty c_x D_x^{d_x} m(D_x) n(D_x) \mathrm{d}D_x}{\int_0^\infty m(D_x) n(D) \mathrm{d}D_x}. \tag{2.121}$$

For the complete gamma distribution,

$$\bar{V}_{TQ_x} = c_x D_{nx}^{d_x} \frac{\left[\dfrac{\Gamma\left(\frac{b_x + \nu_x \mu_x + d_x}{\mu_x}\right)}{\alpha_x^{\left(\frac{b_x + \nu_x \mu_x + d_x}{\mu_x}\right)}}\right]}{\left[\dfrac{\Gamma\left(\frac{b_x + \nu_x \mu_x}{\mu_x}\right)}{\alpha_x^{\left(\frac{b_x + \nu_x \mu_x}{\mu_x}\right)}}\right]}. \tag{2.122}$$

For the modified gamma distribution,

$$\bar{V}_{TQ_x} = c_x D_{nx}^{b_x} \frac{\Gamma\left(\dfrac{[b_x + \nu_x \mu_x + d_x]}{\mu_x}\right)}{\Gamma\left(\dfrac{[b_x + \nu_x \mu_x]}{\mu_x}\right)}, \tag{2.123}$$

and for the gamma distribution,

$$\bar{V}_{TQ_x} = c_x D_{nx}^{d_x} \frac{\Gamma(b_x + \nu_x + d_x)}{\Gamma(b_x + \nu_x)}. \tag{2.124}$$

Lastly for the negative-exponential distribution,

$$\bar{V}_{TQ_x} = c_x D_{nx}^{d_x} \frac{\Gamma(b_x + d_x + 1)}{\Gamma(b_x + 1)}. \tag{2.125}$$

2.5.14.2 Number-concentration-weighted terminal velocity

The equation form for the number-weighted terminal velocity is

$$\bar{V}_{TN_x} = \frac{\int_0^\infty c_x D^{d_x} n(D_x) \mathrm{d}D_x}{\int_0^\infty n(D_x) \mathrm{d}D_x}. \tag{2.126}$$

For the complete gamma distribution the number-weighted terminal velocity is

$$\bar{V}_{\mathrm{TN}_x} = c_x D_{nx}^{d_x} \frac{\alpha_x^{v_x} \Gamma\left(\dfrac{v_x \mu_x + d_x}{\mu_x}\right)}{\alpha_x^{\left(\frac{v_x \mu_x + d_x}{\mu_x}\right)} \Gamma(v_x)}. \tag{2.127}$$

For the modified gamma distribution the number-weighted terminal velocity is

$$\bar{V}_{\mathrm{TN}_x} = c_x D_{nx}^{d_x} \frac{\Gamma\left(\dfrac{[v_x \mu_x + d_x]}{\mu_x}\right)}{\Gamma(v_x)}, \tag{2.128}$$

and for the gamma distribution the number-weighted terminal velocity is

$$\bar{V}_{\mathrm{TN}_x} = c_x D_{nx}^{d_x} \frac{\Gamma(v_x + d_x)}{\Gamma(v_x)}. \tag{2.129}$$

For the negative-exponential distribution,

$$\bar{V}_{\mathrm{TN}_x} = c_x D_{nx}^{d_x} \Gamma(1 + d_x). \tag{2.130}$$

2.5.14.3 Reflectivity-weighted terminal velocity

The form for the reflectivity-weighted terminal velocity is

$$\bar{V}_{\mathrm{TZ}_x} = \frac{\int_0^\infty c_x D_x^{d_x} D_x^{2b_x} n(D_x) \mathrm{d}D_x}{\int_0^\infty D_x^{2b_x} n(D_x) \mathrm{d}D_x}. \tag{2.131}$$

For the complete gamma distribution the reflectivity-weighted terminal velocity is

$$\bar{V}_{\mathrm{TZ}_x} = c_x D_{nx}^{d_x} \frac{\alpha^{\left(\frac{[2b_x + v_x \mu_x]}{\mu_x}\right)} \Gamma\left(\dfrac{[2b_x + v_x \mu_x + d_x]}{\mu_x}\right)}{\alpha^{\left(\frac{[2b_x + v_x \mu_x + d_x]}{\mu_x}\right)} \Gamma\left(\dfrac{[2b_x + v_x \mu_x]}{\mu_x}\right)}, \tag{2.132}$$

and for the modified gamma distribution the reflectivity-weighted terminal velocity is

$$\bar{V}_{\mathrm{TZ}_x} = c_x D_{nx}^{d_x} \frac{\Gamma\left(\dfrac{[2b_x + v_x \mu_x + d_x]}{\mu_x}\right)}{\Gamma\left(\dfrac{[2b_x + v_x \mu_x]}{\mu_x}\right)}. \tag{2.133}$$

For the gamma distribution the number-weighted terminal velocity is

$$\bar{V}_{TZ_x} = c_x D_{nx}^{d_x} \frac{\Gamma(2b_x + \nu_x + d_x)}{\Gamma(2b_x + \nu_x)}. \tag{2.134}$$

The expression for the negative-exponential distribution is

$$\bar{V}_{TZ_x} = c_x D_{nx}^{d_x} \frac{\Gamma(2b_x + d_x + 1)}{\Gamma(2b_x + 1)}. \tag{2.135}$$

2.6 Log-normal distribution

Historically, the log-normal distribution (2.40) has not been often used (e.g. Chaumerliac *et al.* 1991). Therefore, only a subset of the number of quantities presented for the gamma spectral density function will be presented for the log-normal distribution. For integration, consult the Appendix.

2.6.1 Number-concentration-weighted mean diameter

The number-weighted mean diameter can be calculated from

$$\bar{D}_x^{N_T} = \frac{\int_0^\infty D_x n(D_x) \mathrm{d}D_x}{\int_0^\infty n(D_x) \mathrm{d}D_x} = \frac{1}{N_{Tx}} \int_0^\infty D_x n(D_x) \mathrm{d}D_x. \tag{2.136}$$

Substitution of (2.40) into (2.116) results in

$$\bar{D}_x^{N_T} = \frac{1}{\sqrt{2\pi}\sigma_x} \int_0^\infty \exp\left[-\frac{\ln(D_x/D_{nx})}{\sqrt{2}\sigma_x} \right]^2 \mathrm{d}D_x. \tag{2.137}$$

Division of all D_x terms by D_{nx} gives

$$\bar{D}_x^{N_T} = \frac{D_{nx}}{\sqrt{2\pi}\sigma_x} \int_0^\infty \exp\left[-\frac{\ln(D_x/D_{nx})}{\sqrt{2}\sigma_x} \right]^2 \mathrm{d}\left(\frac{D_x}{D_{nx}}\right). \tag{2.138}$$

Now letting $u = D_x/D_{nx}$,

$$\bar{D}_x^{N_T} = \frac{D_{nx}}{\sqrt{2\pi}\sigma_x} \int_0^\infty \exp\left[-\frac{\ln u}{\sqrt{2}\sigma_x} \right]^2 \mathrm{d}u. \tag{2.139}$$

By letting $y = \ln(u)$, $u = \exp(y)$, $du/u = dy$, so

$$\bar{D}_x^{N_T} = \frac{D_{nx}}{\sqrt{2\pi}\sigma_x} \int_{-\infty}^{\infty} \exp(y) \exp\left[-\frac{y}{\sqrt{2}\sigma_x}\right]^2 dy, \tag{2.140}$$

where the limits of the integral change as u approaches zero from positive values, and $\ln(u)$ approaches negative infinity. Likewise, for the upper limit, as u approaches positive infinity, $\ln(u)$ approaches positive infinity.

Now apply the following integral definition to (2.140) (see the appendix):

$$\int_{-\infty}^{\infty} \exp(2b'x) \exp(-a'x^2) dx = \sqrt{\frac{\pi}{a'}} \exp\left(\frac{b'^2}{a'}\right) \tag{2.141}$$

where $y = x$, $a' = 1/(2\sigma_x^2)$, $b' = 1/2$, and therefore (2.140) becomes the mass-weighted mean diameter,

$$\bar{D}_x^{N_T} = D_{nx} \exp\left(\frac{\sigma_x^2}{2}\right). \tag{2.142}$$

2.6.2 Effective diameter

The effective diameter is defined as

$$D_{\text{eff}} = \frac{\int_0^\infty D_x^3 n(D_x) dD_x}{\int_0^\infty D_x^2 n(D_x) dD_x}. \tag{2.143}$$

Substitution of (2.40) into (2.143) results in

$$D_{\text{eff}} = \frac{\int_0^\infty D_x^2 \exp\left(-\frac{[\ln(D_x/D_{nx})]^2}{2\sigma_x^2}\right) dD_x}{\int_0^\infty D_x \exp\left(-\frac{[\ln(D_x/D_{nx})]^2}{2\sigma_x^2}\right) dD_x}. \tag{2.144}$$

Dividing by D_{nx} gives

$$D_{\text{eff}} = D_{nx} \frac{\int_0^\infty \left(\frac{D_x}{D_{nx}}\right)^2 \exp\left(-\frac{[\ln(D_x/D_{nx})]^2}{2\sigma_x^2}\right) d\left(\frac{D_x}{D_{nx}}\right)}{\int_0^\infty \frac{D_x}{D_{nx}} \exp\left(-\frac{[\ln(D_x/D_{nx})]^2}{2\sigma_x^2}\right) d\left(\frac{D_x}{D_{nx}}\right)}. \tag{2.145}$$

Now letting $u = D_x/D_{nx}$,

$$D_{\text{eff}} = D_{nx} \frac{\int\limits_0^\infty u^2 \exp\left(-\frac{(\ln u)^2}{2\sigma_x^2}\right) \mathrm{d}u}{\int\limits_0^\infty u \exp\left(-\frac{(\ln u)^2}{2\sigma_x^2}\right) \mathrm{d}u}. \tag{2.146}$$

By letting $y = \ln(u)$, $u = \exp(y)$, $\mathrm{d}u/u = \mathrm{d}y$, so

$$D_{\text{eff}} = D_{nx} \frac{\int\limits_{-\infty}^\infty \exp(3y) \exp\left(-\frac{y^2}{2\sigma_x^2}\right) \mathrm{d}y}{\int\limits_{-\infty}^\infty \exp(2y) \exp\left(-\frac{y^2}{2\sigma_x^2}\right) \mathrm{d}y}, \tag{2.147}$$

where the limits of the integral change as stated above. Next the integral (2.141) is applied to both the numerator and the denominator, where $y = x$, and for the numerator, $a' = 1/(2\sigma_x^2), b' = 3/2$, and for the denominator, $a' = 1/(2\sigma_x^2), b' = 1$. The result is

$$D_{\text{eff}} = D_{nx} \exp\left(\frac{5\sigma_x^2}{2}\right). \tag{2.148}$$

2.6.3 Modal diameter

The mode of the distribution is obtained by taking the derivative of $n(D_x)/N_{\mathrm{T}}$, which is (2.40) divided by N_{T}, setting the result equal to zero, and then solving for D_x. Thus, making use of (2.40) it can be written

$$\frac{\mathrm{d}}{\mathrm{d}D_x}\left(\frac{n(D_x)}{N_{\mathrm{T}x}}\right) = \frac{\mathrm{d}}{\mathrm{d}D_x}\left[\frac{1}{\sqrt{2\pi}\sigma_x D_x} \exp\left(-\frac{[\ln(D_x/D_{nx})]^2}{2\sigma_x^2}\right)\right] = 0. \tag{2.149}$$

The derivative is expanded and simplified such that,

$$\frac{\mathrm{d}}{\mathrm{d}D_x}\left(\frac{n(D_x)}{N_{\mathrm{T}x}}\right) = \frac{-1}{\sqrt{2\pi}\sigma_x} \frac{1}{D_x^2}\left(\frac{[\ln(D_x/D_{nx})]^2}{\sigma_x^2} + 1\right) = 0. \tag{2.150}$$

Rearranging gives,

$$\ln\left(\frac{D_x}{D_{nx}}\right) = -\sigma_x^2. \tag{2.151}$$

Solving for D_x gives the mode of the distribution,

$$D_{\text{mod}} = D_{nx} \exp(-\sigma_x^2). \tag{2.152}$$

2.6.4 The second moment is related to total surface area

The second moment is related to the total area,

$$A_{\mathrm{T}x} = \int_0^\infty \pi D_x^2 n(D_x) \mathrm{d}D_x. \tag{2.153}$$

Substitution of (2.40) into (2.150) results in

$$A_{\mathrm{T}x} = \frac{N_{\mathrm{T}x}\pi}{\sqrt{2\pi}\sigma_x} \int_0^\infty D_x \exp\left[-\frac{\ln(D_x/D_{nx})}{\sqrt{2}\sigma_x}\right]^2 \mathrm{d}D_x. \tag{2.154}$$

Dividing all D_x terms by D_{nx} gives

$$A_{\mathrm{T}x} = \frac{N_{\mathrm{T}x}\pi D_{nx}^2}{\sqrt{2\pi}\sigma_x} \int_0^\infty \left(\frac{D_x}{D_{nx}}\right) \exp\left[-\frac{\ln(D_x/D_{nx})}{\sqrt{2}\sigma_x}\right]^2 \mathrm{d}\left(\frac{D_x}{D_{nx}}\right). \tag{2.155}$$

Now letting $u = D_x/D_{nx}$,

$$A_{\mathrm{T}x} = \frac{N_{\mathrm{T}x}\pi D_{nx}^2}{\sqrt{2\pi}\sigma_x} \int_0^\infty u \exp\left[-\frac{\ln(u)}{\sqrt{2}\sigma_x}\right]^2 \mathrm{d}u. \tag{2.156}$$

By letting $y = \ln(u)$, $u = \exp(y)$, $\mathrm{d}u/u = \mathrm{d}y$, so

$$A_{\mathrm{T}x} = \frac{N_{\mathrm{T}x}\pi D_{nx}^2}{\sqrt{2\pi}\sigma_x} \int_{-\infty}^\infty \exp(2y) \exp\left[-\frac{y}{\sqrt{2}\sigma_x}\right]^2 \mathrm{d}y, \tag{2.157}$$

where the limits of the integral change as before. If (2.141) is applied with $y = x$, $a' = 1/(2\sigma_x^2)$, $b' = 1$, (2.157) becomes the total area,

$$A_{\mathrm{T}x} = N_{\mathrm{T}x}\pi D_{nx}^2 \exp(2\sigma_x^2). \tag{2.158}$$

2.6.5 Total downward projected area

The total downward projected area can be found from

$$A_{\mathrm{p}x} = \int_0^\infty \frac{\pi}{4} D_x^2 n(D_x) \mathrm{d}D_x. \tag{2.159}$$

Substitution of (2.40) into (2.159) results in

$$A_{px} = \frac{N_{Tx}}{\sqrt{2\pi}\sigma_x}\frac{\pi}{4}\int_0^\infty D_x \exp\left(-\frac{[\ln(D_x/D_{nx})]^2}{2\sigma_x^2}\right)dD_x. \tag{2.160}$$

Dividing all D_x terms by D_{nx} gives

$$A_{px} = \frac{N_{Tx}D_{nx}^2}{\sqrt{2\pi}\sigma_x}\frac{\pi}{4}\int_0^\infty \left(\frac{D_x}{D_{nx}}\right) \exp\left(-\frac{[\ln(D_x/D_{nx})]^2}{2\sigma_x^2}\right)d\left(\frac{D_x}{D_{nx}}\right). \tag{2.161}$$

Letting $u = D_x/D_{nx}$,

$$A_{px} = \frac{N_{Tx}D_{nx}^2}{\sqrt{2\pi}\sigma_x}\frac{\pi}{4}\int_0^\infty u \exp\left(-\frac{(\ln u)^2}{2\sigma^2}\right)du. \tag{2.162}$$

By letting $y = \ln(u)$, $u = \exp(y)$, $du/u = dy$, so

$$A_{px} = \frac{N_{Tx}D_{nx}^2}{\sqrt{2\pi}\sigma_x}\frac{\pi}{4}\int_{-\infty}^\infty \exp(2y) \exp\left(-\frac{y^2}{2\sigma_x^2}\right)dy, \tag{2.163}$$

where the limits of the integral change as above. Now the integral (2.138) is applied to (2.160) where, $y = x$, $a' = 1/(2\sigma_x^2)$, $b' = 1/2$, thus, the expression for the total downward projected area is

$$A_{px} = \frac{\pi}{4}N_{Tx}D_{nx}^2 \exp(2\sigma_x^2). \tag{2.164}$$

2.6.6 The third moment, mixing ratio and characteristic diameter

The third moment or mixing ratio can be found from

$$Q_x = \frac{1}{\rho}\int_0^\infty m(D_x)n(D_x)dD_x, \tag{2.165}$$

where $m(D)$ can be defined by (2.13), and after substitution of (2.13) and (2.40) into (2.165) the result is,

$$Q_x = \frac{a_x N_{Tx}}{\rho\sqrt{2\pi}\sigma_x}\int_0^\infty D_x^{b_x-1} \exp\left(-\frac{[\ln(D_x/D_{nx})]^2}{2\sigma_x^2}\right)dD_x. \tag{2.166}$$

Dividing all D_x terms by D_{nx},

$$Q_x = \frac{a_x N_{Tx} D_{nx}^{b_x}}{\rho\sqrt{2\pi}\sigma_x}\int_0^\infty \left(\frac{D_x}{D_{nx}}\right)^{b_x-1} \exp\left(-\frac{[\ln(D_x/D_{nx})]^2}{2\sigma_x^2}\right)d\left(\frac{D_x}{D_{nx}}\right). \tag{2.167}$$

Now letting $u = D_x/D_{nx}$,

$$Q_x = \frac{a_x N_{Tx} D_{nx}^{b_x}}{\rho\sqrt{2\pi}\sigma_x} \int_0^\infty u^{b_x-1} \exp\left(-\frac{(\ln u)^2}{2\sigma_x^2}\right) du. \tag{2.168}$$

Letting $y = \ln(u)$, $u = \exp(y)$, $du/u = dy$, so

$$Q_x = \frac{a_x N_{Tx} D_{nx}^{b_x}}{\rho\sqrt{2\pi}\sigma_x} \int_{-\infty}^\infty \exp(b_x y) \exp\left(-\frac{y^2}{2\sigma_x^2}\right) dy, \tag{2.169}$$

where the limits of the integral change as above. Now, by applying the integral (2.141) with $y = x$, $a' = 1/(2\sigma_x^2)$, $b' = b_x/2$, the third moment or mixing ratio is

$$Q_x = \frac{a_x N_{Tx} D_{nx}^{b_x}}{\rho} \exp\left(\frac{b_x^2 \sigma_x^2}{2}\right). \tag{2.170}$$

2.6.7 The sixth moment is related to the reflectivity

The sixth moment or reflectivity can be expressed as

$$Z_x = \int_0^\infty D_x^{2b_x} n(D_x) dD_x. \tag{2.171}$$

Substituting (2.40) into (2.171),

$$Z_x = \frac{N_{Tx}}{\sqrt{2\pi}\sigma_x} \int_0^\infty D_x^{2b_x-1} \exp\left(-\frac{[\ln(D_x/D_{nx})]^2}{2\sigma_x^2}\right) dD_x. \tag{2.172}$$

Dividing all D_x terms by D_{nx},

$$Z_x = \frac{N_{Tx} D_{nx}^{2b_x}}{\sqrt{2\pi}\sigma_x} \int_0^\infty \left(\frac{D_x}{D_{nx}}\right)^{2b_x-1} \exp\left(-\frac{[\ln(D_x/D_{nx})]^2}{2\sigma_x^2}\right) d\left(\frac{D_x}{D_{nx}}\right). \tag{2.173}$$

Now letting $u = D_x/D_{nx}$

$$Z_x = \frac{N_{Tx} D_{nx}^{2b_x}}{\sqrt{2\pi}\sigma_x} \int_0^\infty u^{2b_x-1} \exp\left(-\frac{(\ln u)^2}{2\sigma_x^2}\right) du. \tag{2.174}$$

By letting $y = \ln(u)$, $u = \exp(y)$, $\mathrm{d}u/u = \mathrm{d}y$, so

$$Z_x = \frac{N_{Tx} D_{nx}^{2b_x}}{\sqrt{2\pi}\sigma_x} \int_{-\infty}^{\infty} \exp(2b_x y) \exp\left(-\frac{y^2}{2\sigma_x^2}\right) \mathrm{d}y, \tag{2.175}$$

where the limits of the integral change as above. Now the integral (2.141) is applied where $y = x$, $a' = 1/(2\sigma_x^2)$, $b' = b_x$, therefore, the expression for the sixth moment or the reflectivity is

$$Z_x = N_{Tx} D_{nx}^{2b_x} \exp(2b_x^2 \sigma_x^2). \tag{2.176}$$

2.6.8 Terminal velocities

The weighted terminal velocities are given again for Q_x, N_{Tx}, and Z_x, except this time for the log-normal distribution.

2.6.8.1 Mixing-ratio-weighted terminal velocity

$$\bar{V}_{TQ_x} = \frac{\int_0^\infty c_x D_x^{d_x} m(D_x) n(D_x) \mathrm{d}D_x}{\int_0^\infty m(D_x) n(D_x) \mathrm{d}D_x}. \tag{2.177}$$

Substitution of (2.13) and (2.40) into (2.177) results in

$$\bar{V}_{TQ_x} = \frac{c_x \int_0^\infty D_x^{b_x - 1 + d_x} \exp\left(-\frac{[\ln(D_x/D_{nx})]^2}{2\sigma_x^2}\right) \mathrm{d}D_x}{\int_0^\infty D_x^{b_x - 1} \exp\left(-\frac{[\ln(D_x/D_{nx})]^2}{2\sigma_x^2}\right) \mathrm{d}D_x}. \tag{2.178}$$

Dividing D_x by D_{nx},

$$\bar{V}_{TQ_x} = \frac{c_x D_{nx}^{b_x + d_x} \int_0^\infty \left(\frac{D_x}{D_{nx}}\right)^{b_x - 1 + d_x} \exp\left(-\frac{[\ln(D_x/D_{nx})]^2}{2\sigma_x^2}\right) \mathrm{d}\left(\frac{D_x}{D_{nx}}\right)}{D_{nx}^{b_x} \int_0^\infty \left(\frac{D_x}{D_{nx}}\right)^{b_x - 1} \exp\left(-\frac{[\ln(D_x/D_{nx})]^2}{2\sigma_x^2}\right) \mathrm{d}\left(\frac{D_x}{D_{nx}}\right)}. \tag{2.179}$$

Letting $u = D_x/D_{nx}$,

$$\bar{V}_{TQ_x} = \frac{c_x D_{nx}^{b_x + d_x} \int_0^\infty u^{b_x - 1 + d_x} \exp\left(-\frac{(\ln u)^2}{2\sigma_x^2}\right) \mathrm{d}u}{D_{nx}^{b_x} \int_0^\infty u^{b_x - 1} \exp\left(-\frac{(\ln u)^2}{2\sigma_x^2}\right) \mathrm{d}u}. \tag{2.180}$$

By letting $y = \ln(u)$, $u = \exp(y)$, $\mathrm{d}u/u = \mathrm{d}y$, so

$$\bar{V}_{\mathrm{T}Q_x} = \frac{c_x D_{nx}^{b_x+d_x} \int\limits_{-\infty}^{\infty} \exp[(b_x+d_x)y] \exp\left(-\frac{y^2}{2\sigma_x^2}\right) \mathrm{d}y}{D_{nx}^{b_x} \int\limits_{-\infty}^{\infty} \exp(b_x y) \exp\left(-\frac{y^2}{2\sigma_x^2}\right) \mathrm{d}y}, \tag{2.181}$$

where the limits of the integral change as above. The integral (2.141) is used so that in the numerator, $a' = 1/(2\sigma_x^2)$, $b' = (b_x + d_x)/2$, and in the denominator, $a' = 1/(2\sigma_x^2)$, $b' = b_x/2$, so that the mass-weighted mean terminal velocity in terms of Q_x is

$$\bar{V}_{\mathrm{T}Q_x} = \frac{c_x D_{nx}^{b_x+d_x} \exp\left[\frac{\sigma_x^2 (b_x+d_x)^2}{2}\right]}{D_{nx}^{b_x} \exp\left(\frac{\sigma_x^2 b_x^2}{2}\right)}. \tag{2.182}$$

2.6.8.2 Number-concentration-weighted terminal velocity

The number-concentration-weighted terminal velocity is given by

$$\bar{V}_{\mathrm{T}N_x} = \frac{\int\limits_0^{\infty} c_x D_x^{d_x} n(D_x) \mathrm{d}D_x}{\int\limits_0^{\infty} n(D_x) \mathrm{d}D_x}. \tag{2.183}$$

Substitution of (2.40) into (2.183) results in

$$\bar{V}_{\mathrm{T}N_x} = \frac{\int\limits_0^{\infty} c_x D_x^{d_x-1} \exp\left(-\frac{[\ln(D_x/D_{nx})]^2}{2\sigma_x^2}\right) \mathrm{d}D_x}{\int\limits_0^{\infty} D_x^{-1} \exp\left(-\frac{[\ln(D_x/D_{nx})]^2}{2\sigma_x^2}\right) \mathrm{d}D_x}. \tag{2.184}$$

Dividing D_x by D_{nx},

$$\bar{V}_{\mathrm{T}N_x} = \frac{c_x D_{nx}^{d_x} \int\limits_0^{\infty} \left(\frac{D_x}{D_{nx}}\right)^{d_x-1} \exp\left(-\frac{[\ln(D_x/D_{nx})]^2}{2\sigma_x^2}\right) \mathrm{d}\left(\frac{D_x}{D_{nx}}\right)}{\int\limits_0^{\infty} \left(\frac{D_x}{D_{nx}}\right)^{-1} \exp\left(-\frac{[\ln(D_x/D_{nx})]^2}{2\sigma_x^2}\right) \mathrm{d}\left(\frac{D_x}{D_{nx}}\right)}. \tag{2.185}$$

Letting $u = D_x/D_{nx}$,

$$\bar{V}_{\mathrm{T}N_x} = \frac{c_x D_{nx}^{d_x} \int\limits_0^{\infty} u^{d_x-1} \exp\left(-\frac{(\ln u)^2}{2\sigma_x^2}\right) \mathrm{d}u}{\int\limits_0^{\infty} u^{-1} \exp\left(-\frac{(\ln u)^2}{2\sigma_x^2}\right) \mathrm{d}u}. \tag{2.186}$$

By letting $y = \ln(u)$, $u = \exp(y)$, $\mathrm{d}u/u = \mathrm{d}y$, so

$$\bar{V}_{\mathrm{TN}_x} = \frac{c_x D_{nx}^{d_x} \int_{-\infty}^{\infty} \exp(d_x y) \exp\left(-\frac{y^2}{2\sigma_x^2}\right) \mathrm{d}y}{\int_{-\infty}^{\infty} \exp(0) \exp\left(-\frac{y^2}{2\sigma_x^2}\right) \mathrm{d}y}, \tag{2.187}$$

where the limits of the integral change as above. By applying the integral (2.141) so that in the numerator, $a' = 1/(2\sigma_x^2)$, $b' = d_x/2$, and in the denominator, $a' = 1/(2\sigma_x^2)$, $b' = 0$, the mass-weighted mean terminal velocity in terms of $N_{\mathrm{T}x}$ is

$$\bar{V}_{\mathrm{TN}_x} = c_x D_{nx}^{d_x} \exp\left(\frac{\sigma_x^2 d_x^2}{2}\right). \tag{2.188}$$

2.6.8.3 *Reflectivity-weighted terminal velocity*

The reflectivity (Z_x)-weighted terminal velocity is given by,

$$\bar{V}_{\mathrm{TZ}_x} = \frac{\int_0^{\infty} c_x D_x^{d_x} D_x^{2b_x} n(D_x) \mathrm{d}D_x}{\int_0^{\infty} c_x D_x^{2b_x} n(D_x) \mathrm{d}D_x}. \tag{2.189}$$

Substituting (2.40) into (2.189),

$$\bar{V}_{\mathrm{TZ}_x} = \frac{\int_0^{\infty} c_x D_x^{2b_x + d_x - 1} \exp\left(-\frac{[\ln(D_x/D_{nx})]^2}{2\sigma_x^2}\right) \mathrm{d}D_x}{\int_0^{\infty} D_x^{2b_x - 1} \exp\left(-\frac{[\ln(D_x/D_{nx})]^2}{2\sigma_x^2}\right) \mathrm{d}D_x}. \tag{2.190}$$

Dividing D_x by D_{nx},

$$\bar{V}_{\mathrm{TZ}_x} = \frac{c_x D_{nx}^{2b_x + d_x} \int_0^{\infty} \left(\frac{D_x}{D_{nx}}\right)^{2b_x + d_x - 1} \exp\left(-\frac{[\ln(D_x/D_{nx})]^2}{2\sigma_x^2}\right) \mathrm{d}\left(\frac{D_x}{D_{nx}}\right)}{D_{nx}^{2b_x} \int_0^{\infty} \left(\frac{D_x}{D_{nx}}\right)^{2b_x - 1} \exp\left(-\frac{[\ln(D_x/D_{nx})]^2}{2\sigma_x^2}\right) \mathrm{d}\left(\frac{D_x}{D_{nx}}\right)}. \tag{2.191}$$

Letting $u = D_x/D_{nx}$,

$$\bar{V}_{\mathrm{TZ}_x} = \frac{c_x D_{nx}^{2b_x + d_x} \int_0^{\infty} u^{2b_x + d_x - 1} \exp\left(-\frac{(\ln u)^2}{2\sigma_x^2}\right) \mathrm{d}u}{D_{nx}^{2b_x} \int_0^{\infty} u^{2b_x - 1} \exp\left(-\frac{(\ln u)^2}{2\sigma_x^2}\right) \mathrm{d}u}. \tag{2.192}$$

By letting $y = \ln(u)$, $u = \exp(y)$, $du/u = dy$, so

$$\bar{V}_{TZ_x} = \frac{c_x D_{nx}^{2b_x + d_x} \int_0^\infty \exp[(2b_x + d_x)y] \exp\left(-\frac{y^2}{2\sigma_x^2}\right) dy}{D_{nx}^{2b_x} \int_0^\infty \exp(2b_x y) \exp\left(-\frac{y^2}{2\sigma_x^2}\right) dy}, \tag{2.193}$$

where the limits of the integral change as above. By applying the integral (2.141) so that in the numerator, $a' = 1/(2\sigma_x^2)$, $b' = (2b_x + d_x)/2$, and in the denominator, $a = 1/(2\sigma_x)^2$, $b' = b_x$, the mass-weighted mean terminal velocity in terms of Z_x is

$$\bar{V}_{TZ_x} = \frac{c_x D_{nx}^{d_x} \exp\left[\frac{\sigma_x^2 (2b_x + d_x)^2}{2}\right]}{\exp(8\sigma_x^2 b_x^2)}. \tag{2.194}$$

2.7 Microphysical prognostic equations

2.7.1 Mixing ratio

A prognostic equation for the mixing ratio may be written as

$$\begin{aligned} \frac{\partial Q_x}{\partial t} = &- \frac{1}{\rho}\frac{\partial \rho u_i Q_x}{\partial x_j} + \frac{Q_x}{\rho}\frac{\partial \rho u_i}{\partial x_i} + \frac{\partial}{\partial x_i}\left(\rho K_h \frac{\partial Q_x}{\partial x_i}\right) \\ &+ \delta_{i3}\frac{1}{\rho}\frac{\partial\left(\rho \bar{V}_{TQ_x} Q_x\right)}{\partial x_i} + SQ_x, \end{aligned} \tag{2.195}$$

where K_h is the eddy mixing value and SQ_x are mixing ratio source terms.

2.7.2 Number concentration

The prognostic equation for number concentration may be written as

$$\begin{aligned} \frac{\partial N_{Tx}}{\partial t} = &- \frac{\partial u_i N_{Tx}}{\partial x_i} + N_{Tx}\frac{\partial u_i}{\partial x_i} + \frac{\partial}{\partial x_i}\left(K_h \frac{\partial N_{Tx}}{\partial x_i}\right) \\ &+ \delta_{i3}\frac{\partial\left(\overline{V}_T{}^{N_{Tx}} N_{Tx}\right)}{\partial x_i} + SN_{Tx}, \end{aligned} \tag{2.196}$$

where SN_{Tx} are number concentration source terms.

2.7.3 Characteristic diameter

The goal here is to show the steps that are needed to obtain an expression for a prognostic equation for the characteristic diameter, dD_{nx}/dt, which has not been well presented in prior works. For brevity, the equation is developed for the case of the gamma distribution (2.27) with $\alpha = \mu = 0$, following Passarelli (1978).

Having differentiated the definitions of water content χ_x and reflectivity Z_x with respect to time, t, the following are obtained, respectively,

$$\frac{d\chi_x}{dt} = \int_0^\infty D_x^{\delta_x} n(D_x) dD_x, \tag{2.197}$$

$$\frac{dZ_x}{dt} = 2\int_0^\infty D_x^{\delta_x} m(D_x) n(D_x) dD_x. \tag{2.198}$$

Also $dm(D_x)/dt \approx D^{d_x}$ has been used.

Dividing (2.198) by (2.197) and (2.197) by (2.198) and rearranging gives

$$2\frac{d\chi_x}{dt}\int_0^\infty D_x^{\delta_x} m(D_x) n(D_x) dD_x = \frac{dZ_x}{dt}\int_0^\infty D_x^{\delta_x} n(D_x) dD_x. \tag{2.199}$$

Substituting (2.27), the gamma distribution, canceling $N_{Tx}/\Gamma(\nu_x)$ from both sides, and substituting $m(D_x) = a_x D_x^{b_x}$ gives

$$\begin{aligned} 2\frac{d\chi_x}{dt}\int_0^\infty a_x D_x^{\delta_x + b_x}\left(\frac{D_x}{D_{nx}}\right)^{\nu_x - 1}\exp\left(-\left[\frac{D_x}{D_{nx}}\right]\right)d\left(\frac{D_x}{D_{nx}}\right) \\ = \frac{dZ_x}{dt}\int_0^\infty D_x^{\delta_x}\left(\frac{D_x}{D_{nx}}\right)\exp\left(-\left[\frac{D_x}{D_{nx}}\right]\right)d\left(\frac{D_x}{D_{nx}}\right), \end{aligned} \tag{2.200}$$

which, after multiplying the left side by $(D_x/D_{nx})^{\delta_x + b_x}$ and the right side by $(D_x/D_{nx})^{\delta_x}$ can be rewritten as

$$\begin{aligned} 2a_x D_{nx}^{\delta_x + b_x}\frac{d\chi_x}{dt}\int_0^\infty \left(\frac{D_x}{D_{nx}}\right)^{\delta_x + b_x + \nu_x - 1}\exp\left(-\left[\frac{D_x}{D_{nx}}\right]\right)d\left(\frac{D_x}{D_{nx}}\right) \\ = D_{nx}^{\delta_x}\frac{dZ_x}{dt}\int_0^\infty \left(\frac{D_x}{D_{nx}}\right)^{\delta_x + \nu_x - 1}\exp\left(-\left[\frac{D_x}{D_{nx}}\right]\right)d\left(\frac{D_x}{D_{nx}}\right). \end{aligned} \tag{2.201}$$

Then (2.201) can be integrated over the interval $(0, \infty)$, which results in

$$2a_x\lambda^{\delta_x + b_x}\Gamma(\delta_x + b_x + \nu_x)\frac{d\chi_x}{dt} = \lambda^{\delta_x}\Gamma(\delta_x + \nu_x)\frac{dZ_x}{dt}, \tag{2.202}$$

which can be rewritten in terms of dZ_x/dt as

$$\frac{dZ_x}{dt} = \frac{d\chi_x}{dt} 2a_x D_{nx}^{b_x} \frac{\Gamma(\delta_x + b_x + \nu_x)}{\Gamma(\delta_x + \nu_x)}. \tag{2.203}$$

The term Z_x can also be related to χ_x by dividing by

$$\chi_x \int_0^\infty m(D_x)^2 n(D_x) dD_x = Z_x \int_0^\infty m(D_x) n(D_x) dD_x. \tag{2.204}$$

Then (2.27) and $m(D_x) = a_x D_x^{b_x}$ can be substituted into (2.204). After canceling $N_{Tx}/\Gamma(\nu_x)$ from both sides, the result is

$$\begin{aligned} &\chi_x \int_0^\infty \left(a_x D_x^{b_x}\right)^2 \left(\frac{D_x}{D_{nx}}\right)^{\nu_x - 1} \exp\left(-\left[\frac{D_x}{D_{nx}}\right]\right) d\left(\frac{D_x}{D_{nx}}\right) \\ &= Z_x \int_0^\infty \left(a_x D_x^{b_x}\right) \left(\frac{D_x}{D_{nx}}\right)^{\nu_x - 1} \exp\left(-\left[\frac{D_x}{D_{nx}}\right]\right) d\left(\frac{D_x}{D_{nx}}\right). \end{aligned} \tag{2.205}$$

Multiplying the right-hand side by $(D_x/D_{nx})^{b_x}$ and the left-hand side by $(D_x/D_{nx})^{2b_x}$ gives

$$\begin{aligned} &\chi_x a_x^2 D_{nx}^{2b_x} \int_0^\infty \left(\frac{D_x}{D_{nx}}\right)^{2b_x + \nu_x - 1} \exp\left(-\left[\frac{D_x}{D_{nx}}\right]\right) d\left(\frac{D_x}{D_{nx}}\right) \\ &= Z_x D_{nx}^{b_x} \int_0^\infty \left(a_x D_x^{b_x}\right) \left(\frac{D_x}{D_{nx}}\right)^{b_x + \nu_x - 1} \exp\left(-\left[\frac{D_x}{D_{nx}}\right]\right) d\left(\frac{D_x}{D_{nx}}\right). \end{aligned} \tag{2.206}$$

Integrating (2.206) over the interval $(0, \infty)$, gives

$$\chi_x a_x^2 D_{nx}^{2b_x} \Gamma(2b_x + \nu_x) = Z_x a_x D_{nx}^{b_x} \Gamma(b_x + \nu_x). \tag{2.207}$$

Then, solving for Z_x gives

$$Z_x = 2a_x D_{nx}^{b_x} \frac{\Gamma(2b_x + \nu_x)}{\Gamma(b_x + \nu_x)} \chi_x. \tag{2.208}$$

Differentiating (2.208) by dt gives

$$\frac{dZ_x}{dt} = a_x \frac{\Gamma(2b_x + \nu_x)}{\Gamma(b_x + \nu_x)} \frac{d}{dt} \left(D_{nx}^{b_x} \chi_x\right), \tag{2.209}$$

and applying the product rule gives

$$\frac{dZ_x}{dt} = a_x \frac{\Gamma(2b_x + \nu_x)}{\Gamma(b_x + \nu_x)} \left[D_{nx}^{b_x} \frac{d\chi_x}{dt} + \chi_x b_x D_{nx}^{b_x - 1} \frac{d\lambda_x}{dt} \right]. \tag{2.210}$$

Rearranging and solving for dD_{nx}/dt gives

$$\frac{dD_{nx}}{dt} = \frac{D_{nx}^{b_x}}{\chi_x b_x} \left[\frac{\Gamma(2b_x + \nu_x + \delta_x)}{\Gamma(\delta_x + \nu_x)} \frac{\Gamma(b_x + \nu_x)}{\Gamma(2b_x + \nu_x)} - 1 \right] \frac{d\chi_x}{dt}. \tag{2.211}$$

The term $d\chi_x/dt$ is, for example, the rate of change of liquid-water content owing to vapor diffusion growth. By dividing the two χ_x terms (equivalent to multiplying by unity) in (2.211) by air density, ρ, a substitution for the mixing ratio can be made

$$\frac{dD_{nx}}{dt} = \frac{D_{nx}^{b_x}}{Q_x b_x} \left[\frac{\Gamma(2b_x + \nu_x + \delta_x)}{\Gamma(\delta_x + \nu_x)} \frac{\Gamma(b_x + \nu_x)}{\Gamma(2b_x + \nu_x)} - 1 \right] \frac{dQ_x}{dt}. \tag{2.212}$$

The term in brackets in (2.212) is defined as H such that

$$\frac{dD_{nx}^{b_x}}{dt} = \frac{D_{nx} H}{b_x Q_x} \frac{dQ_x}{dt}. \tag{2.213}$$

2.7.4 Reflectivity

Following Milbrandt and Yau (2005b), an equation for the prediction of Z_x is developed and then a diagnostic equation for $G(\nu_x)$ is derived from the prognostic equation for Z_x. The value of ν_x has been solved by iteration from $G(\nu_x)$. The derivation of dZ_x/dt starts with the definition of D_{nx}, as

$$D_{nx} = \left(\frac{\rho Q_x \Gamma(\nu_x)}{N_{Tx} a_x \Gamma(b_x + \nu_x)} \right)^{1/b_x}. \tag{2.214}$$

Now the definition for reflectivity for a spherical hydrometeor with $a_x = \pi/6\rho_x$, and $b_x = 3$, is simply given as

$$Z_x = \frac{N_{Tx} D_{nx}^6}{\Gamma(\nu_x)} \Gamma(6 + \nu_x). \tag{2.215}$$

Using (2.214) in (2.215) results in

$$Z_x = N_{Tx} \frac{\Gamma(6 + \nu_x)}{\Gamma(\nu_x)} \left(\frac{\rho Q_x \Gamma(\nu_x)}{N_{Tx}(\pi/6)\rho_x \Gamma(3 + \nu_x)} \right)^2. \tag{2.216}$$

Simplifying,

$$Z_x = \frac{[\Gamma(\nu_x)]^2 \Gamma(6 + \nu_x)}{[\Gamma(3 + \nu_x)]^2 \Gamma(\nu_x)} \frac{(\rho Q_x)^2}{N_{Tx}(\pi/6)^2 \rho_x^2}. \tag{2.217}$$

The derivative of (2.217) with respect to t is given by the following relationship,

$$Z_x = \frac{\Gamma(\nu_x)(5+\nu_x)(4+\nu_x)(3+\nu_x)\Gamma(3+\nu_x)}{(2+\nu_x)(1+\nu_x)(\nu_x)\Gamma(\nu_x)\Gamma(3+\nu_x)} \frac{(\rho Q_x)^2}{N_{Tx}(\pi/6)^2 \rho_x^2}, \tag{2.218}$$

or

$$Z_x = \frac{(5+\nu_x)(4+\nu_x)(3+\nu_x)}{(2+\nu_x)(1+\nu_x)(\nu_x)} \frac{(\rho Q_x)^2}{N_{Tx}(\pi/6)^2 \rho_x^2}. \tag{2.219}$$

Now defining $G(\nu_x)$,

$$G(\nu_x) = \frac{(5+\nu_x)(4+\nu_x)(3+\nu_x)}{(2+\nu_x)(1+\nu_x)(\nu_x)}, \tag{2.220}$$

thus,

$$Z_x = M_x(6) = G(\nu_x)\frac{(\rho Q_x)^2}{N_{Tx}(\pi/6)^2 \rho_x^2}. \tag{2.221}$$

The quotient rule,

$$\frac{\mathrm{d}}{\mathrm{d}x}\left(\frac{u}{v}\right) = \frac{1}{v}\frac{\mathrm{d}u}{\mathrm{d}x} - \frac{u}{v^2}\frac{\mathrm{d}v}{\mathrm{d}x}, \tag{2.222}$$

is used to derive $\mathrm{d}Z_x/\mathrm{d}t$ in terms of $G(\nu_x)$,

$$\frac{\mathrm{d}Z_x}{\mathrm{d}t} = \frac{G(\nu_x)\rho^2}{(\pi/6)^2\rho_x^2}\left[2\frac{Q_x}{N_{Tx}}\frac{\mathrm{d}Q_x}{\mathrm{d}t} - \left(\frac{Q_x}{N_{Tx}}\right)^2\frac{\mathrm{d}N_{Tx}}{\mathrm{d}t}\right]. \tag{2.223}$$

2.7.5 Other prognostic equations

Other prognostic equations include those for the following: tau, τ, the time elapsed for a process; the mean cloud water collected following the motion; the amount of rime collected by ice; the amount of ice from vapor deposition; the density of the rime ice collected; and the density of ice.

2.7.5.1 Lagrangian cloud exposure time

The Lagrangian equation for tau, τ, the time elapsed for a process is given by

$$\frac{\partial \tau_{\text{rime},x}}{\partial t} = -u_i\frac{\partial \tau_{\text{rime},x}}{\partial x_i} + \frac{\partial}{\partial x_i}\left(K_h\frac{\partial \tau_{\text{rime},x}}{\partial x_i}\right) + \delta_{i3}\frac{\partial\left(\bar{V}_{TQ}\tau_{\text{rime},x}\right)}{\partial x_i} + c\tau_{\text{rime},x}, \tag{2.224}$$

where c is a variable that is 1 if a process is occurring and 0 if it is not.

2.7.5.2 Lagrangian mean cloud mixing ratio

A variable $\bar{Q}_{cw}$ is the mean cloud water collected following the motion of the air and is defined by

$$\bar{Q}_{cw} = \frac{\int_0^{\tau} Q_{cw}(\tau')d\tau'}{\int_0^{\tau} d\tau'} = \frac{1}{\tau_{cw}} \int_0^{\tau} Q_{cw}(\tau')d\tau', \tag{2.225}$$

where τ' is a temporary variable for integration of the Leibnitz rule. The prognostic equation is given by

$$\begin{aligned} \frac{d\bar{Q}_{cw}}{dt} &= \frac{d}{dt}\left[\frac{1}{\tau_{cw}} \int_0^{\tau_{cw}} \bar{Q}_{cw}(\tau')d\tau'\right] = \frac{d\tau_{cw}}{dt}\left[\frac{1}{\tau_{cw}} \int_0^{\tau_{cw}} \bar{Q}_{cw}(\tau')d\tau'\right] \\ &= \frac{Q_{cw}}{\tau_{cw}} - \tau_{cw}^2 \int_0^{\tau_{cw}} \bar{Q}_{cw}(\tau')d\tau' = \frac{Q_{cw}}{\tau_{cw}} - \frac{\bar{Q}_{cw}}{\tau_{cw}}, \end{aligned} \tag{2.226}$$

which allows

$$\frac{\partial \bar{Q}_{cw}}{\partial t} = -u_i \frac{\partial \bar{Q}_{cw}}{\partial x_i} + \frac{\partial}{\partial x_i}\left(K_h \frac{\partial \bar{Q}_{cw}}{\partial x_i}\right) + \delta_{i3}\frac{1}{\rho}\frac{\partial\left(\bar{V}_T Q_{cw} \bar{Q}_{cw}\right)}{\partial x_i} + c\left(\frac{Q_{cw}}{\tau_{cw}} - \frac{\bar{Q}_{cw}}{\tau_{cw}}\right). \tag{2.227}$$

In the equation above c is the same as in the equation for τ, $c = 1$ for cloud-water mixing ratio present, and $c = 0$ for no cloud-water mixing ratio present.

2.7.5.3 Mixing ratio of deposition

A prognostic equation for the amount of vapor deposition/condensation mixing ratio that is collected on, or sublimed/evaporated from, a spectrum of particles is (Morrison and Grabowski 2008)

$$\begin{aligned} \frac{\partial Q_{dep,x}}{\partial t} &= -\frac{1}{\rho}\frac{\partial \rho u_i Q_{dep,x}}{\partial x_i} + \frac{Q_{dep,x}}{\rho}\frac{\partial \rho u_i}{\partial x_i} + \frac{\partial}{\partial x_i}\left(\rho K_h \frac{\partial Q_{dep,x}}{\partial x_i}\right) \\ &+ \delta_{i3}\frac{1}{\rho}\frac{\partial\left(\rho \bar{V}_{TQ} Q_{dep,x}\right)}{\partial x_i} + SQ_{dep,x}. \end{aligned} \tag{2.228}$$

2.7.5.4 Mixing ratio of rime

Similarly, an equation for the amount of rime mixing ratio that is collected on a spectrum of particles is given as (Morrison and Grabowski 2008)

$$\frac{\partial Q_{\text{rime},x}}{\partial t} = -\frac{1}{\rho}\frac{\partial \rho u_i Q_{\text{rime},x}}{\partial x_i} + \frac{Q_{\text{rime},x}}{\rho}\frac{\partial \rho u_i}{\partial x_i} + \frac{\partial}{\partial x_i}\left(\rho K_h \frac{\partial Q_{\text{rime},x}}{\partial x_i}\right)$$
$$+ \delta_{i3}\frac{1}{\rho}\frac{\partial\left(\rho \bar{V}_{\text{TQ}} Q_{\text{rime},x}\right)}{\partial x_i} + SQ_{\text{rime},x} \quad . \tag{2.229}$$

2.7.5.5 *Rime density*

An equation for the rime density follows as

$$\frac{\partial \rho_{\text{rime},x}}{\partial t} = -u_i \frac{\partial \rho_{\text{rime},x}}{\partial x_i} + \frac{\partial}{\partial x_i}\left(K_h \frac{\partial \rho_{\text{rime},x}}{\partial x_i}\right) + \delta_{i3}\frac{\partial\left(\bar{V}_{\text{TQ}} \rho_{\text{rime},x}\right)}{\partial x_i} + S\rho_{\text{rime},x}. \tag{2.230}$$

2.7.5.6 *Density*

Similarly an equation to predict density is given by,

$$\frac{\partial \rho_x}{\partial t} = -u_i \frac{\partial \rho_x}{\partial x_i} + \frac{\partial}{\partial x_i}\left(K_h \frac{\partial \rho_x}{\partial x_i}\right) + \delta_{i3}\frac{\partial\left(\bar{V}_{\text{TQ}} \rho_x\right)}{\partial x_i} + S\rho_x. \tag{2.231}$$

These previous six prognostic equations permit a means to parameterize a smooth transfer of particles from one density of species to another (e.g low-density graupel to medium-density graupel, etc.). This process will be discussed further in Chapter 9.

2.8 Bin microphysical parameterization spectra and moments

In bin microphysical parameterizations there is a need to find a way to represent the bin spectrum or number density function $n(x)$ reasonably, where x is mass. If a linear scale is used to represent the bin spectrum for a typical droplet and drop spectrum, which spans from 4 microns to 4 millimeters, to capture the spectrum reasonably, far too many bins would be required to be economical. Instead, a logarithmic scale can be incorporated, such as the one presented in Ogura and Takahashi (1973), with sizes closer together at small particle sizes and wider apart at large sizes. Many, such as Berry and Reinhardt (1974a), Farley and Orville (1986), and Farley (1987), use exponential functions for radius,

$$r = r_0 \exp\left(\frac{J-1}{J_0}\right), \tag{2.232}$$

where r_0 is the initial radius of the distribution and J is an integer that determines the spacing of the bins. The corresponding mass is

$$x = x_0 \exp\left(\frac{3(J-1)}{J_0}\right). \tag{2.233}$$

In this case for a spherical hydrometeor,

$$x_0 = \frac{4}{3}\pi r_0^3 \rho_L. \tag{2.234}$$

A mass scale that corresponds to the size distribution is given by defining $x(J)$,

$$x(J) = x_0 2^{\left(\frac{J-1}{2}\right)}, \tag{2.235}$$

where x_0 is the smallest mass (2.68×10^{-10} g), with 61 categories, each $2^{1/2}$, times the mass of the preceding category.

The transformation presents a new number density function given by $n(J)$, which is related to the original size distribution, $n(x)$, by the following,

$$n(J) = \left(\frac{\ln 2}{2}\right) n(x). \tag{2.236}$$

Others such as Tzivion *et al.* (1987), also use bins in the spectrum that increase in mass from one bin to the next by factors of $p = 2$, $2^{1/2}$, $2^{1/3}$, or $2^{1/4}$, etc., according to

$$x_{k+1} = p x_k, \tag{2.237}$$

where k is the bin index. Following Tzivion *et al.* (1987), the j-th moments, M, are given in terms of x for mass and $n(x,t)$ for the number density function,

$$M_k^j = \int_{x_k}^{x_{k+1}} x^j n_k(x,t)\,\mathrm{d}x. \tag{2.238}$$

Having laid a theoretical foundation for the bulk and bin parameterization of microphysical processes, the subject of nucleation of liquid-water droplets and ice crystals will be discussed in the next chapter.

3

Cloud-droplet and cloud-ice crystal nucleation

3.1 Introduction

In this chapter modes of cloud droplet nucleation and ice-crystal nucleation are examined as well as parameterizations of number concentrations of cloud condensation nuclei and ice nuclei. The nuclei in general are small aerosols of various sizes called Aitken aerosols $O(10^{-2}\ \mu m)$, large aerosols $O(10^{-1}\ \mu m)$, giant aerosols $O(10^{0}\ \mu m)$, and ultra-giant aerosols $O(10^{1}$ to $10^{2}\ \mu m)$. Nucleation by cloud condensation nuclei and ice nuclei is called heterogeneous nucleation as it involves a foreign substance on which cloud water and ice water can form, compared to homogeneous nucleation, for which no foreign substance is needed for nucleation. Supersaturations have to exceed values not found on Earth (e.g. 400%) for homogeneous nucleation of liquid droplets, which is discussed at length in Pruppacher and Klett (1997). An examination of the Kelvin curve described in the next section shows why this is so. As homogeneous nucleation does not occur on Earth for liquid particles, it is not parameterized in models. In general, cloud condensation nuclei made of some salt compound such as sodium chloride (table salt) are the most effective for heterogeneous nucleation of liquid droplets. Heterogeneous nucleation of liquids can be a function of several variables, such as temperature, vapor pressure or supersaturation, pressure, and factors or activation coefficients related to the composition of aerosols involved. As a result the means of expressing heterogeneous nucleation have become more complex over the years as a result of new observations and new techniques to represent nuclei numbers. One technique involves the incorporation of the nucleation activation coefficients in parameterizations.

Homogeneous nucleation of ice occurs in Earth's atmosphere when temperatures of cloud droplets or larger drops become low enough. In general, the smaller the droplet, the colder it has to be for homogeneous nucleation to

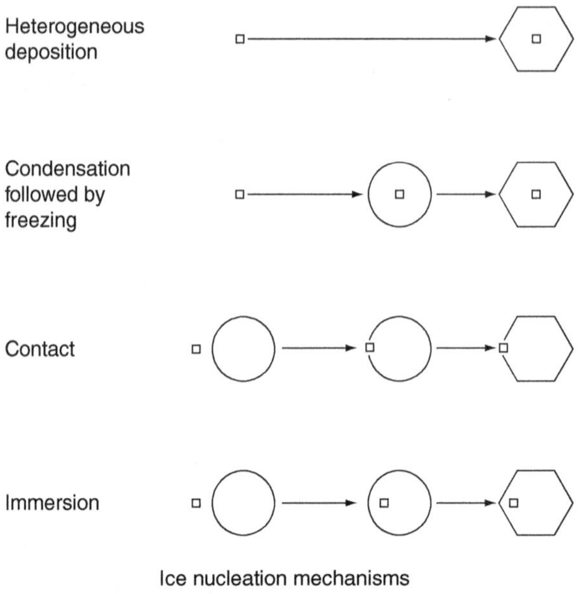

Fig. 3.1. Schematic picture of the ways atmospheric ice nucleation can account for ice formation. (From Rogers and Yau 1989; courtesy of Elsevier.)

occur. Heterogeneous nucleation of ice can occur in a variety of modes. The first mode is heterogeneous deposition nucleation (Fig. 3.1). This deposition mode requires that supersaturation with respect to ice be achieved. The second kind of nucleation that can occur is condensation-freezing ice nucleation (Fig. 3.1). The condensation-freezing mode requires that at temperatures below freezing the air becomes supersaturated with respect to liquid. Then one of the aerosols that makes up a water drop activates as an ice nucleus during condensation. In the third mode, the contact nuclei mode, the ice nucleation occurs with collision of a supercooled liquid-water droplet and an ice nucleus (Fig. 3.1), which is generally thought to be some form of clay, such as kaolinite. Recent evidence points to certain types of bacteria being very plentiful as nuclei. Last, there is the fourth mode called the immersion mode of ice nucleation. The immersion mode occurs when a droplet is nucleated on an aerosol particle at temperatures above freezing. Then, as the temperature of the droplet falls below freezing sufficiently, the aerosol activates as an ice nucleus (Fig. 3.1). The ratio of ice-forming nuclei to liquid-droplet-forming nuclei is usually very small and varies from near nil at temperatures near freezing to 1000 m^{-3} per 1×10^8 m^{-3} aerosol particles at $-20\,°C$. As the temperature gets colder, the number of ice nuclei approximately increases exponentially.

The heterogeneous nucleation of ice is a very complex physical process, and it requires a number of physical conditions to be met. These include first the insolubility requirement. Ice nuclei almost always are insoluble. For example, if the nuclei were soluble, the freezing point of the solute could decrease substantially and thereby prevent nucleation from occurring, moreover the aerosol could dissolve. Second, there is the size requirement. Large or giant nuclei make better ice nuclei than Aitken particles because once nucleated they are at a size at which they can immediately begin collecting much smaller particles. Third, there is a chemical-bond requirement. That is, if there is a hydrogen-bond site on the ice nucleus, it provides a location where embryonic ice crystals can grow. Fourth, there is the crystallographic requirement, which may be more important than some of the other requirements. This requirement, simply stated, is that the ice-nuclei aerosol must have a crystalline structure similar to that of ice water, which means that water molecules can align themselves in a structure similar to ice, readily permitting nucleation. Fifth, there is the activation-site requirement. This means that there must be a site in the ice-nuclei aerosol that is favorable for initiating ice. These five requirements and the models of ice nucleation are discussed in detail in Pruppacher and Klett (1997) and the reader is referred to their textbook for a comprehensive examination. Because these are not included directly in most parameterizations, they will not be addressed further here.

In the rest of this chapter, cloud condensation nuclei will be examined first in the context of heterogeneous nucleation of cloud droplets. Then ice crystal nucleation will be examined, including heterogeneous nucleation, homogeneous nucleation, and secondary nucleation of ice crystals.

3.2 Heterogeneous nucleation of liquid-water droplets for bulk model parameterizations

3.2.1 Nucleation rate as a function of S,w,T

With a detailed, high-resolution, bulk parameterization model it seems more appropriate to compute nucleation and condensation explicitly and to do away with the saturation adjustment that was used in Soong and Ogura (1973); Klemp and Wilhelmson (1978); Tao *et al.* (1989); Gilmore *et al.* (2004a); and Straka and Mansell (2005), as well as many other models. Instead, nucleation can be represented explicitly. For example, a modified method to that used by Ziegler *et al.* (1985) can be used. The parameterization is based on the activation of cloud condensation nuclei N_{CCN}, of which the number available is at present only based on a power law as a function of the saturation ratio (with respect to liquid), S_L, in percent. The equation for cloud condensation nuclei activated is given as

$$N_{CCN} = C_{CCN} S_L^k \tag{3.1}$$

where the constant $C_{CCN} = 1.26 \times 10^9$ m^{-3} and $k = 0.308$ for continental conditions (Seifert and Beheng 2005) and $C_{CCN} = 1.0 \times 10^8$ m^{-3} and $k = 0.462$ for maritime conditions (Seifert and Beheng 2005 and Khain *et al.* 2000). Pruppacher and Klett (1997) put the value of k somewhere between 0.2 and 0.6 with a median value of 0.5. Seifert and Beheng assume all cloud condensation nuclei are activated at a maximum S_L of 1.1% and do not allow any additional activation. It is not necessary to include this condition; the nucleation parameterization includes derivatives of S_L in the Z direction. For very high-resolution models derivatives can be computed in the X and Y direction as well. The derivatives can be found using the following equation (3.2) by incorporating centered finite difference schemes. However, one-sided differences should be used at cloud boundaries,

$$\left.\frac{\partial N_T}{\partial t}\right|_{nuc} = \left\{ \begin{matrix} C_{CCN} k S^{k-1} \left\{ \max\left[\left(u\frac{\partial S_L}{\partial x}\right), 0\right] + \max\left[\left(v\frac{\partial S_L}{\partial y}\right), 0\right] + \max\left[\left(w\frac{\partial S_L}{\partial z}\right), 0\right] \right\} > 0 \\ \text{otherwise zero} \end{matrix} \right\}, \tag{3.2}$$

where u is x-directional horizontal velocity, v is y-directional horizontal velocity, and w is z-directional vertical velocity.

Equations such as this for different aerosol sizes such as Aitken, large, giant, and ultra-giant cloud condensation nuclei can be developed. However, most of the time, only one aerosol size, generally unspecified, is used. Saleeby and Cotton (2008) have attempted to use two aerosol sizes, one of which is a giant nucleus; and Straka *et al.* (2009a) used four aerosol sizes in high-resolution simulations in order to increase the dependence of nucleation on aerosol size.

Note that the local tendency of S, $\partial S/\partial t$, is not explicitly predicted. At present various investigators assume that advection dominates nucleation (Ziegler 1985; Seifert and Beheng 2005). The maximum concentration permitted is 1.50×10^9 m^{-3} for continental clouds and 1.50×10^8 m^{-3} for maritime clouds (Seifert and Beheng 2005).

Next, a tendency equation for the mixing ratio Q_{cw} by cloud-water nucleation is given by

$$\left.\frac{\partial Q_{cw}}{\partial t}\right|_{nuc} = \frac{x_{L,nuc}}{\rho} \left.\frac{\partial N_{Tcw}}{\partial t}\right|_{nuc}, \tag{3.3}$$

where, $x_{L,nuc} = 5 \times 10^{-13}$ to 1×10^{-12} kg, which is the minimum mass of a cloud drop and corresponds to a diameter up to about 5×10^{-6} m.

3.2.2 Nucleation rate as a function of S,w,T,P, activation parameters

The process of nucleation of cloud condensation nuclei and their growth by condensation thereafter into cloud droplets is a particularly challenging problem. The future of bulk microphysics parameterizations of cloud droplets in cloud-resolving models in particular hinges on the ability to predict mixing ratio and number concentration of newly nucleated cloud droplets. The activation of cloud condensation nuclei and subsequent growth controls the initial production of cloud-droplet mixing ratio Q_{cw} and number concentration N_{T}. In some liquid-cloud nucleation schemes, a prognostic equation for activated cloud condensation nuclei is used to distinguish them from all of the cloud condensation nuclei available. One scheme, which will be the focus here, is that of Cohard *et al.* (1998, 2000), and Cohard and Pinty (2000). This scheme is based on cloud condensation nuclei concentrations that are prognosed with a source term that is a function of temperature T, pressure P, and vertical motion w.

Difficulties with these schemes exist as they are based on maximum (cloud condensation nuclei activation) and mean (condensation) local supersaturation S_{LV}, where the subscript LV means supersaturation of vapor with respect to liquid. The quantity S_{LV} is dependent on T, P, w, and Q_{SL}. The value of S_{LV} generally is not well captured in cloud models and can vary unphysically when there is non-homogeneous mixing in clouds, near the physical boundaries of the cloud (Stevens *et al.* 1996) and owing to its dependence on the numerical timestep. With these issues duly noted, Cohard and collaborators looked for a method of computing S_{LV} at the gridscale. The smallest cloud condensation nuclei are activated as a function of the Kohler curve (Pruppacher and Klett 1997) variables such as chemical, hygroscopic, size criteria, and thermodynamic variables. Owing to the influence on aerosol size, some bin models predict a bin spectrum of aerosol sizes. This will be discussed later.

In the years since the strict use of saturation adjustment schemes, several of the bulk microphysical models that have been developed with predictive equations for cloud-droplet mixing ratio and number concentration have followed these earlier approaches (Cohard and collaborators). It should be noted that even Twomey's (1959) full method shown below has received some criticism for not accommodating all of the most necessary factors in describing the situation when an aerosol particle will activate as a cloud condensation nucleus. The nucleation of water droplets in bulk parameterization models is a particularly difficult problem to manage as bulk parameterizations need to take into account factors such as temperature, supersaturation, activation size spectrum, and vertical velocity. As pointed out by Cohard *et al.* (1998) and Cohard and Pinty (2000), not only do these factor influence early cloud growth, they are important factors in radiative properties of clouds. As a result, it may be questionable to compute gridscale actions of nucleation and condensation owing to timesteps used in coarser-grid three-dimensional models such as Global Circulation Models (GCMs), but perhaps not in cloud-scale models. Pruppacher and Klett (1997) discuss the role of size (radius of curvature) and solute effects through the Kohler equation for critical radii, as well as critical supersaturations (Twomey 1959), laying the foundation for a powerful approach. This involves a power law with regard to supersaturation, and was derived from a simplified form of the Kohler equation,

$$S_{\max} = \left[\frac{1.63 \times 10^{-3} w^{\frac{3}{2}}}{ckB\left[\frac{3}{2}, \frac{k}{2}\right]} \right]^{\frac{1}{k+2}} \tag{3.4}$$

$$N_{\mathrm{CCN}} = cS_{\max}^{k} = c^{\frac{k}{k+2}} \left[\frac{1.63 \times 10^{-3} w^{\frac{3}{2}}}{kB\left[\frac{3}{2}, \frac{k}{2}\right]} \right]^{\frac{k}{k+2}} \tag{3.5}$$

where N_{CCN} is the number of activated condensation nuclei or number of cloud condensation nuclei, c is a constant, k is the exponent of $S_{\max}$, w is the vertical motion, and B is the Beta function. Note that c and k are parameters that fit each aerosol type. This equation has its limitations owing to the likely inability for this method or any similar method to capture all of the possible nucleation mechanisms owing to limitations mentioned above.

Cohard *et al.* (1998) devised a more elegant and general method of describing factors that influence activation of condensation nuclei, namely the saturation ratio (in percent), size distribution, and solubility of aerosols. The foundations of their parameterization are based in differences between maritime and continental sources of aerosols. They begin by exploring how a modified form of Twomey's equation performs, but with four activation spectrum coefficients. They also come to the conclusion that this method reduces to a simple power law at very large and very small supersaturations. They begin with the saturation development equation discussed in detail in Chapter 4, but written here as (3.6) and (3.7), which describe the change in diameter of a single droplet that includes Kelvin size effect and Raoult's solution effects, where both effects are represented by $y(T,D)$

$$\frac{dS_{LV}}{dt} = \psi_1(T,P)W - \psi_2 \frac{dQ_{cw}}{dt}, \tag{3.6}$$

and

$$D\frac{dD}{dt} = 4G(T,P)(S_{LV} - y(T,D)). \tag{3.7}$$

Using Pruppacher and Klett (1997) as a guide, the rate of change of Q_{cw}, the mixing ratio of liquid condensed during nucleation, is given by

$$\frac{dQ_{cw}}{dt} \approx 2\pi \frac{\rho_L}{\rho}(2G(T,P))^{\frac{3}{2}} S_{LV} \int_0^{S_{LV}} n(S') \left(\int_{\tau(S')}^{t} S dt' \right)^{\frac{1}{2}} dS', \tag{3.8}$$

where ρ_L is the density of liquid water. The concentration number of nuclei $n(S)$ active between S and $S + dS$ is

$$N_{CCN}(S_{LV}) = \int_0^{S_{LV}} n(S) dS. \tag{3.9}$$

The equation for Q_{cw} is very complex. Twomey (1959) suggested that it leads to a lower-bounds estimation of the integral in time of the saturation development equation (3.6) that can be used in the analytical derivation of S_{LVmax} (Cohard *et al.* 1998). Thus,

$$\int_{\tau(S')}^{t} S dt' > \frac{S_{LV}^2 - S'^2}{2\psi_1 w}. \tag{3.10}$$

Combining (3.8) and (3.10) gives a condensation-rate estimation,

$$\frac{dQ_{cw}}{dt} > 4\pi \frac{\rho_L}{\rho} \frac{G^{3/2}}{2(\psi_1 w)^{\frac{1}{2}}} S_{LV} \int n(S') (S_{LV}^2 - S'^2)^{\frac{1}{2}} dS'. \tag{3.11}$$

Twomey took

$$n(S') = kcS'^{k-1}, \tag{3.12}$$

which provides a concise expression for S_{LVmax}.

In the discussion by Cohard *et al.* (1998), an expression that retains the behavior of the four activation spectra coefficients is given by

$$n(S') = kcS^{k-1}(1 + \beta S'^2)^{-\mu}, \tag{3.13}$$

where β, μ, c, and k are the activation spectra coefficients.

Using the change in variable,

$$x = \left(\frac{S'}{S_{\mathrm{LV}}}\right)^2, \tag{3.14}$$

so that

$$S_{\mathrm{LV}}\frac{1}{2}x^{-\frac{1}{2}}\mathrm{d}x = \mathrm{d}S' \tag{3.15}$$

and integrating (3.11) results in

$$\frac{\mathrm{d}Q_{\mathrm{cw}}}{\mathrm{d}t} > 4\pi\frac{\rho_{\mathrm{L}}}{\rho}\frac{G^{\frac{3}{2}}}{(\psi_1 w)^{\frac{1}{2}}}kcS_{\mathrm{LV}}\int_0^1\left(x^{\frac{1}{2}}S_{\mathrm{LV}}\right)^{k-1}\left(1+\beta S_{\mathrm{LV}}^2 x\right)^{-\mu} \times \left(S_{\mathrm{LV}}^2 - xS_{\mathrm{LV}}^2\right)^{\frac{1}{2}}S_{\mathrm{LV}}\frac{1}{2}x^{-\frac{1}{2}}\mathrm{d}x. \tag{3.16}$$

Rearrangement gives

$$\frac{\mathrm{d}Q_{\mathrm{cw}}}{\mathrm{d}t} > 4\pi\frac{\rho_{\mathrm{L}}}{\rho}\frac{G^{\frac{3}{2}}}{(\psi_1 w)^{\frac{1}{2}}}kcS_{\mathrm{LV}}\int_0^1\left(x^{\frac{1}{2}}S_{\mathrm{LV}}\right)^{k-1}\left(1+\beta S_{\mathrm{LV}}^2 x\right)^{-\mu} \times S_{\mathrm{LV}}(1-x)^{\frac{1}{2}}S_{\mathrm{LV}}\frac{1}{2}x^{-\frac{1}{2}}\mathrm{d}x. \tag{3.17}$$

Combining the S_{LV} terms and bringing them outside the integral,

$$\frac{\mathrm{d}Q_{\mathrm{cw}}}{\mathrm{d}t} > 2\pi\frac{\rho_{\mathrm{L}}}{\rho}\frac{G^{\frac{3}{2}}}{(\psi_1 w)^{\frac{1}{2}}}kcS_{\mathrm{LV}}^{k+2}\int_0^1 x^{\frac{(k-2)}{2}}\left(1+\beta S_{\mathrm{LV}}^2 x\right)^{-\mu}(1-x)^{\frac{1}{2}}\mathrm{d}x, \tag{3.18}$$

and combining the x terms and integrating gives

$$\frac{\mathrm{d}Q_{\mathrm{cw}}}{\mathrm{d}t} > 2\pi\frac{\rho_{\mathrm{L}}}{\rho}\frac{G^{\frac{3}{2}}}{(\psi_1 w)^{\frac{1}{2}}}kcS_{\mathrm{LV}}^{k+2}B\left(\frac{k}{2},\frac{3}{2}\right)\ {}_2F_1\left(\mu,\frac{k}{2},\frac{k}{2}+\frac{3}{2};-\beta S_{\mathrm{LV}}^2\right). \tag{3.19}$$

Following Cohard *et al.* (1998), the maximum supersaturation S_{LVmax}, is given by the saturation development equation (3.6) by setting $\mathrm{d}S_{\mathrm{LV}}/\mathrm{d}t = 0$. Thus,

$$\psi_2(T,P)\frac{\mathrm{d}Q_{\mathrm{cw}}}{\mathrm{d}t} = \psi_1(T,P)w. \tag{3.20}$$

Solving for $\mathrm{d}Q_{\mathrm{cw}}/\mathrm{d}t$ gives

$$\frac{\mathrm{d}Q_{\mathrm{cw}}}{\mathrm{d}t} = \frac{\psi_1(T,P)w}{\psi_2(T,P)}. \tag{3.21}$$

Substituting (3.19) for $\mathrm{d}Q_{\mathrm{cw}}/\mathrm{d}t$ in (3.21) gives

$$\frac{\psi_1(T,P)w}{\psi_2(T,P)} > 2\pi \frac{\rho_{\mathrm{L}}}{\rho} \frac{G^{\frac{3}{2}}}{(\psi_1 w)^{\frac{1}{2}}} kc S_{\mathrm{LV\,max}}^{k+2} B\left(\frac{k}{2}, \frac{3}{2}\right) {}_2F_1\left(\mu, \frac{k}{2}, \frac{k}{2} + \frac{3}{2}; -\beta S_{\mathrm{LV\,max}}^2\right). \quad (3.22)$$

The final expression has a form (Cohard *et al.* 1998) given by

$$S_{\mathrm{LV\,max}}^{k+2}\, {}_2F_1\left(\mu, \frac{k}{2}, \frac{k}{2} + \frac{3}{2}; -\beta S_{\mathrm{LV\,max}}^2\right) = \rho \frac{(\psi_1 w)^{3/2}}{2kc\pi\rho_{\mathrm{L}}\psi_2 G^{3/2} B\left(\frac{k}{2}, \frac{3}{2}\right)}, \quad (3.23)$$

where the following are defined as

$$\psi_1(T,P) = \frac{g}{R_{\mathrm{d}}T}\left(\frac{\epsilon L_{\mathrm{V}}}{c_p T} - 1\right); \quad (3.24)$$

$$\psi_2(T,P) = \left(\frac{p}{\epsilon e_{\mathrm{s}}(T)} + \frac{\epsilon L_{\mathrm{V}}^2}{R_{\mathrm{d}}T^2 c_p}\right); \quad (3.25)$$

and

$$G(T,P) = \frac{1}{\rho_{\mathrm{L}}}\left(\frac{R_{\mathrm{v}}T}{\psi e_{\mathrm{s}}(T)} + \frac{L_{\mathrm{V}}}{k_{\mathrm{a}}T}\left(\frac{L_{\mathrm{V}}}{R_{\mathrm{v}}T} - 1\right)\right)^{-1}. \quad (3.26)$$

Here c, k, β and μ are the four activation spectrum coefficients (Cohard *et al.* 1998) as mentioned earlier; ${}_2F_1$ and B are the Gauss hypergeometric function and the Beta function, respectively; see Cohard *et al.* (1998) or Appendix.

The value for activated N_{CCN} is solved by Cohard *et al.* (1998) using (3.9) and (3.13) to obtain,

$$N_{\mathrm{CCN}} = c S_{\mathrm{LV\,max}}^{k}\, {}_2F_1\left(\mu, \frac{k}{2}, \frac{k}{2} + 1; -\beta S_{\mathrm{LV\,max}}^2\right). \quad (3.27)$$

This equation has different values of c and k than Twomey's expression. Equation (3.27) has been discussed by Cohard *et al.* (1998) as having four previously unused activation coefficients, which can express various aspects of aerosols involved in nucleation. This makes it possible parameterically to include the aspects of activation size spectrum, chemical composition, and solubility into the equation for heterogeneous nucleation (3.27).

An estimate of the maximum number of N_{CCN} that might be activated is given by (3.27). With this estimate of N_{CCN}, the production rate of nucleated droplets is given as a comparison to the number of aerosols already activated, N_{a}. The source term for activated aerosols is in simplified form, with a centered timestep,

$$\max(0, N_{\rm CCN}(S_{\rm LV\,max}) - N_{\rm a}(t - \delta t)). \quad (3.28)$$

Equation (3.28) also dictates whether cloud-drop concentration is permitted to increase.

Employing the equation from the Kohler curve, $(S - 1) - a/r + b/r^3$, where r is radius, a and b are constants, the critical diameter for nucleation with unstable growth rate is then given by

$$D_{\rm crit} = \frac{4A}{3} S_{\rm LV\,max}. \quad (3.29)$$

Lastly, the change of mixing ratio of cloud droplets is given as

$$\max(0, N_{\rm CCN}(S_{\rm LV\,max}) - N_{\rm a}(t - \delta t)) \frac{\rho_L}{\rho} \frac{\pi}{6} D^3_{\rm crit}. \quad (3.30)$$

3.3 Heterogeneous liquid-water drop nucleation for bin model parameterizations

The initiation of liquid water in bin parameterization models can be accomplished with any number of functions to describe the distribution of newly nucleated cloud-water droplets in a cloud given that there is supersaturation at a grid point. Some parameterizations predict cloud condensation nuclei, whilst others do not. There are normalized functional forms for the distribution of droplets and usually a prognostic equation of total cloud condensation nuclei. An approach like this bypasses the need to know any explicit information about the makeup of aerosols that are cloud condensation nuclei in the atmosphere, even if solute effects and curvature effects are ignored. This was the methodology in many early models, such as those proposed by Ogura and Takahashi (1973) and Soong (1974) to initialize cloud droplets in supersaturated regions of the model domain.

In model studies of warm-rain growth, Ogura and Takahashi (1973) compared three different initial droplet distributions in different cloud simulations. Note that it was found by these authors that the distribution choice was not of major significance (Soong 1974). Soong (1974) found that the choice of the initial droplet spectra for bin models as given by Ogura and Takahashi (1973) was not that important in the final solutions. Two of the types of Ogura and Takahashi's distributions are discussed below. For example, they parameterized the condensation process of the initial droplet distribution with some prescribed form $f(x)$, which is normalized and where x is mass. This is multiplied by the number of nuclei available $\xi(z, t)$ for a one-dimensional model as a function of height z and time t. The rate of change of the number concentration of cloud droplets $N_{cw}(x)$ due to the formation of cloud droplets about nuclei is written simply as

$$\left[\frac{\partial N_{cw}(x)}{\partial(t)}\right]_{init} = \frac{\xi(z,t)f(x)}{\delta t}. \tag{3.31}$$

The different normalized forms of $f(x)$ can potentially control different precipitation evolutions. The distribution called type (1) was used by Twomey (1966), Warshaw (1967) and Kovetz and Olund (1969). Its form is given by

$$f(x) = \frac{1}{x_f}\left(\frac{x}{x_f}\right)\exp\left(-\frac{x}{x_f}\right), \tag{3.32}$$

where x_f is a constant in the function defining the spectrum of CCN activated.

According to Scott (1968), it is close to a Gaussian distribution with respect to radius and has a relative variance of 0.25. Scott also states this form is easier to work with than a pure Gaussian distribution. In this distribution, a maritime spectrum results with $x_f = 1.029 \times 10^{-10}$ kg.

From (3.32), $f(r)$ is related to $f(x)$ by

$$f(r) = 4\pi r^2 f(x). \tag{3.33}$$

The so-called type (2) distribution was used by Golovin (1963), Berry (1967) and Soong (1974) and is given as

$$f(x) = \frac{x}{\bar{x}}\exp\left(-\frac{x}{\bar{x}}\right), \tag{3.34}$$

where for the mean mass $\bar{x} = 2 \times 10^{-10}$ kg, the liquid content should be 1 kg kg^{-1}, and the number of nucleated droplets should be 5×10^7 m^{-3}.

In general, most models make the number concentration of cloud condensation nuclei decrease as nucleation of cloud droplets takes place, and the concentration of cloud condensation nuclei increases as cloud-droplet-sized particles evaporate below a certain size, nominally a radius of 4 μm. These can be reactivated if recycled into the cloud's supersaturated regions. As given by Ogura and Takahashi (1973) the increase of cloud condensation nuclei by evaporation is just

$$\left[\frac{\partial \xi}{\partial t}\right]_{evap} = -N_{cw}(x)\left.\frac{dx}{dt}\right|_{x=x_0}, \tag{3.35}$$

or the total change in nuclei owing to initiation of cloud droplets and evaporation of cloud droplets is

$$\frac{\partial \xi}{\partial t} = \left.\frac{\partial \xi}{\partial t}\right|_{\text{evap}} + \left.\frac{\partial \xi}{\partial t}\right|_{\text{init}}. \tag{3.36}$$

Note there are advection and diffusion terms as well for the cloud condensation nuclei conservation equation (3.36). One problem with this equation is that while one cloud condensation nucleus is needed to produce a cloud droplet, and orders of magnitude more cloud droplets are required to make precipitation-sized particles, when these precipitation particles evaporate, they reintroduce only one cloud condensation nucleus, which is larger than the original cloud condensation nucleus. Thus the conservation equation is somewhat flawed. To overcome this deficiency efforts began to predict a range of cloud condensation nuclei sizes, as discussed next.

3.3.1 Aerosol size distributions and nucleation for initiation of liquid-water droplets in bin model parameterizations

The initiation of liquid water in bin parameterization models can be accomplished using a spectrum of cloud condensation nuclei, which are in bins similar to those used for cloud droplets and drops. Usually the range is smaller than that from the smallest to the largest liquid drop. The conservation equation for including nuclei is given, following Kogan (1991), quite simply as the advection and diffusion tendencies plus a sink term on cloud condensation nuclei that results from nucleation. Computing nucleation with a spectrum of cloud condensation nuclei and a spectrum of liquid droplets allows a modeler to avoid having two-dimensional cloud-droplet spectra. The two-dimensional cloud-droplet spectrum has one dimension as the cloud mass and the other dimension as the salt mass in the case of soluble cloud condensation nuclei.

When drops evaporate, the cloud condensation nuclei spectrum is returned to its original size, though washout by drizzle or rain is allowed. More realistic cloud condensation nuclei redistribution after evaporation is a very difficult problem and how cloud condensation nuclei actually redistribute in the size spectrum after evaporation is not known (Khairoutdinov and Kogan 1999).

3.4 Homogeneous ice-crystal nucleation parameterizations

Homogeneous freezing occurs when air temperatures are colder than –40 °C (233.15 K) and liquid drops freeze instantaneously. Some use –30 °C as the demarcation temperature for homogeneous freezing, whilst others use values as cold as −50 °C. Newer parameterizations have been developed that take into account not only the temperature, but also the size of the liquid droplet. At −40 °C clusters of about 200 to 300 water molecules will freeze spontaneously. At warmer temperatures the clusters of water molecules have to be larger for homogeneous freezing, whilst at colder temperatures clusters of fewer than 200 or so water molecules need to come together for freezing to take place without the presence of an ice nucleus.

The most recent studies and parameterization of homogeneous freezing of cloud droplets suggest that this occurs approximately between −30 °C and −50 °C (DeMott *et al.* 1994). The authors give a number of droplets that freeze ΔN_{freeze} due to homogeneous freezing by the following

$$\Delta N_{\text{freeze}} = \int_0^\infty [1 - \exp(-JV\Delta t)] N_{\text{Tcw}}(D) \text{d}D. \tag{3.37}$$

In this equation, J, the homogeneous freezing rate of cloud drops to frozen cloud drops, is given by the following, where T_c is temperature in °C,

$$\begin{aligned} \log_{10} J = & - 603.952 - 52.6611 T_c - 1.7439 T_c^2 - 2.65 \times 10^{-2} T_c^3 \\ & - 1.536 \times 10^{-4} T_c^4. \end{aligned} \tag{3.38}$$

In (3.37), the volume V is approximated by the mean-droplet diameter in units of cm by Milbrant and Yau (2005b). Therefore, a fraction of freezing in one timestep may be written as

$$F_{\text{freeze}} = \frac{\Delta N_{\text{freeze}}}{N_{\text{Tcw}}} = \left[1 - \exp\left(-J \frac{\pi}{6} D_{\text{cwmv}}^3 \Delta t\right)\right], \tag{3.39}$$

where D_{cwmv} is the mean volume diameter of cloud droplets.

Based on this rate equation (3.39), equations for mixing ratio and number concentration are simply

$$Q_{\text{freeze}} = \frac{F_{\text{freeze}} Q_{\text{cw}}}{\Delta t} \tag{3.40}$$

and

$$N_{\text{freeze}} = \frac{F_{\text{freeze}} N_{\text{Tcw}}}{\Delta t}. \tag{3.41}$$

As described by DeMott (1994), F_{freeze} is 0 at −30 °C and 1 at −50 °C. This means that many supercooled liquid-cloud drops freeze at temperatures slightly warmer than the standard homogeneous freezing temperature for supercooled liquid-cloud drops of −40 °C, but allows some supercooled liquid-cloud droplets to exist at temperatures as low as −50 °C.

3.5 Heterogeneous ice-crystal nucleation parameterizations

3.5.1 Early parameterizations

The Fletcher (1962) ice-nucleus curve fit is perhaps one of the most primitive curves still used in cloud models today. The formula is

$$N_{\mathrm{id}} = n_0 \exp(-a[T - T_0]), \tag{3.42}$$

where N_{id} is the ice deposition number concentration, n_0, the number of ice nuclei that are active $= 10^{-2}\,\mathrm{m}^{-3}$, $a = 0.6\,^{\circ}\mathrm{C}^{-1}$, and $T_0 = 273.15$ K. There has been much discussion about this parameterization in the literature concerning the over-production of ice nuclei by deposition at very cold temperatures ($T <$ 245 K). Also this parameterization does not take into account the degree of supersaturation over ice.

An alternative to this was proposed by Cotton *et al.* (1986) by including Huffman and Vali's (1973) equation for relative supersaturation dependence on ice nucleation, and is given by

$$N_{\mathrm{id}} = \left[\frac{(S_{\mathrm{i}} - 1)}{(S_0 - 1)}\right]^b, \tag{3.43}$$

where $S_{\mathrm{i}} - 1$ is the fractional supersaturation with respect to ice and $S_0 - 1$ is the fractional ice supersaturation at water saturation, where $b = 4.5$.

A hybrid parameterization was produced by Cotton *et al.* (1986) by combining Huffman and Vali's (1973) equation with the Fletcher (1962) parameterization

$$N_{\mathrm{id}} = n_0 \left[\frac{(S_{\mathrm{i}} - 1)}{(S_0 - 1)}\right]^b \exp(a[T - T_0]), \tag{3.44}$$

where a and b are constants given above. This parameterization underestimates ice nuclei at warmer temperatures.

3.5.2 Explicit cloud ice-crystal nucleation

As carried out by Seifert and Beheng (2005), an ice nucleation mechanism, following Reisner *et al.* (1998) and various other authors, is used to make an ice-crystal number-concentration nucleation source as follows (Meyers *et al.* 1992),

$$\frac{\partial N_{\mathrm{Tid}}}{\partial t} = \left\{ \begin{matrix} \max\left[\dfrac{(N_{\mathrm{Tid}} - N_{\mathrm{TI}})}{\Delta t}\right] \\ \text{otherwise zero} \end{matrix} \right\}, \tag{3.45}$$

where

$$N_{\mathrm{Tid}} = 0.001 \exp[-0.639 + 12.96 S_{\mathrm{I}}]. \tag{3.46}$$

The subscripts id and I refer to ice deposition and ice, respectively. Meyers *et al.* (1992) strictly developed this equation from data between temperatures of –7 to –20 °C and between ice saturations of 2 to 25% or –5 to 4.5% water supersaturation. In practice, the use of this equation is usually arbitrarily limited to temperatures colder than –5 °C. In addition, the value of N_{Tid} is bounded by 10 times and 0.1 times the result from the following equation (3.47) from Reisner *et al.* (1998),

$$N_{\mathrm{Tid}} = 0.01 \exp(-\min(T, 246.15) - 273.15). \tag{3.47}$$

Seifert and Beheng (2005) include this as they claim there is an instability with the Meyers *et al.* (1992) scheme at very cold temperatures, though this needs to be investigated further. The maximum number of ice-crystal concentration is arbitrarily limited to the same number as the maximum number of cloud drops permitted, which is stated above as 1.5×10^9 m^{-3}.

The nucleation of ice-crystal water and cloud water is integrated using time-splitting, with small timesteps of between 0.4 and 0.6 s. Thus, for a model timestep of 5 s, the number of small steps is set to 10 assuming a small timestep of 0.5 s. This causes some computation increase, but considering the complexity of some models, it is only a small fractional increase.

3.5.3 Contact nucleation

Next contact nucleation, studied by Young (1974b), is considered and is governed by

$$N_{\mathrm{ic}} = N_{\mathrm{a0}}(270.15 - T_{\mathrm{c}})^{1.3}, \tag{3.48}$$

where N_{ic} is contact nucleation and $N_{\mathrm{a0}} = 2 \times 10^5$ m^{-3} at all levels (Cotton *et al.* 1986). However, Young proposed that N_{a0} varies from 2×10^5 m^{-3} at MSL (mean sea level) to 10^6 m^{-3} at 5000 m MSL. For reference, N_{a0} is the aerosol population that can activate to make ice nuclei. Later experiments showed this relationship to be not very accurate. Therefore, Meyers *et al.* (1992) designed a new relationship that was exponential in nature and given by

$$N_{\mathrm{ic}} = \exp[-2.8 + 0.262(273.15 - T_{\mathrm{c}})], \tag{3.49}$$

where N_{ic} is in number per liter. Ice is not permitted by this method at temperatures warmer than –2 °C. The main mechanisms of contact nucleation are by Brownian, thermophoretic, and diffusiophoretic forcing following Young (1974a, b). Contact freezing nuclei are assumed to be 0.2 microns in diameter.

3.5.3.1 Brownian motion

Brownian motion causes contact nucleation of supercooled cloud droplets from random collisions of cloud drops with aerosols. A highly detailed discussion of aerosol physics and Brownian motion is found in Pruppacher and Klett (1997). Brownian-motion-induced contact nucleation has been parameterized by Young (1974a, b) and others for use in models with complex ice-nucleation mechanisms using

$$\left.\frac{1}{N_{\mathrm{a}}}\frac{\mathrm{d}N_{\mathrm{a}}}{\mathrm{d}t}\right|_{Br} = 4\pi Q_{\mathrm{cw}}\psi_{\mathrm{a}}\left(1 + 0.3N_{\mathrm{re}}^{1/2}N_{\mathrm{sc}}^{1/3}\right), \tag{3.50}$$

where N_{a} is the number of aerosols, Q_{cw} is the cloud-water mixing ratio, and ψ_{a} is aerosol diffusivity given as

$$\psi_{\mathrm{a}} = \frac{kT_{\infty}}{6\pi r_{\mathrm{a}}\eta_{\infty}}(1 + N_{\mathrm{kn}}). \tag{3.51}$$

In (3.51), $k = 1.38047 \times 10^{-23}\ \mathrm{J\,K^{-1}}$ is the Boltzmann constant, r_{a} is the aerosol radius, η_{∞} is the viscosity of air, and N_{kn} is the Knudsen number, which is defined as,

$$N_{\mathrm{kn}} = \frac{7.37T}{288pr_{\mathrm{a}}} = \frac{\lambda}{r_{\mathrm{a}}}, \tag{3.52}$$

where λ is the mean free path of air and p is the pressure.

A scaled parameterization of contact nucleation owing to Brownian motion has also been developed by Cotton *et al.* (1986).

3.5.3.2 Thermophoresis effects

Thermophoresis effects are explained by a dependence on the Knudsen number N_{kn}, which is defined by (3.52). The thermophoresis effect is the motion of an aerosol caused by a radiometric or thermally induced force. Details of this effect are discussed by Pruppacher and Klett (1997), with key points repeated here for completeness. This force comes from non-uniform heating of particles owing to temperature gradients in an aerosol's suspending gas. When $N_{\mathrm{kn}} \gg 1$, temperature gradients induce gas molecules to deliver a greater net impulse on the warm side of a particle than on the cold side, thus driving the particle in the direction of the cold side of the particle. For $N_{\mathrm{kn}} \ll 1$, the problem is more complex. Consider a region on the surface layer around an aerosol which has a larger wavelength than the thermal gradient.

According to Pruppacher and Klett, the layer of this gas closest to this surface will acquire a temperature gradient that conforms approximately to that on the surface of the aerosol. Therefore, gas molecules on the warmer direction impart a greater impulse force to the surface of the aerosol than those on the cooler direction. This allows a situation where the particle as a whole can experience a force along the temperature gradient of the gas. Put plainly, contact nucleation by thermophoresis effects occurs due to the attraction or repulsion of aerosol particles to the droplet along the thermal gradients (Cotton *et al.* 1986).

A parameterization for contact nucleation by thermophoresis is given by Young (1974a) as

$$\frac{1}{N_{\mathrm{a}}}\frac{\mathrm{d}N_{\mathrm{a}}}{\mathrm{d}t}\bigg|_{\mathrm{Th}} = 4\pi r_{\mathrm{cw}} f_t (T_\infty - T_{\mathrm{w}})\frac{K_g^*}{p}\, 4.185\times 10^4\,\mathrm{J\,k\,cal^{-1}}, \tag{3.53}$$

where K_g^* is the thermal conductivity of air, with influences of convection on heat diffusion f_1^* and molecular boundary layer considerations on heat diffusion included in f_2^*. These are related by

$$K_g^* = K_g f_1^* f_2^*, \tag{3.54}$$

where

$$f_1^* = 1 + \frac{P^*}{4\pi C}\left(0.56 N_{\mathrm{re}}^{1/2} N_{\mathrm{pr}}^{1/3}\right) \tag{3.55}$$

and

$$f_2^* \approx 1. \tag{3.56}$$

In f_1^*, P^* is the semi-perimeter normal to fall axis, C is the shape parameter for diffusional processes for ice crystals following McDonald (1963), and N_{pr} is the Prandtl number. The thermophoretic factor f_t is complicated and given by

$$f_t = \frac{0.4[1 + 1.45N_{\mathrm{kn}} + 0.4N_{\mathrm{kn}}\exp(-1/N_{\mathrm{kn}})]\left(K_g + 2.5N_{\mathrm{kn}}K_{\mathrm{a}}\right)}{(1 + 3N_{\mathrm{kn}})\left(2K_g + 5K_{\mathrm{a}}N_{\mathrm{kn}} + K_{\mathrm{a}}\right)} \tag{3.57}$$

where K_{a} is the aerosol conductivity.

3.5.3.3 *Diffusiophoresis effects*

Following Pruppacher and Klett (1997), diffusiophoretic effects on aerosol particle motions refer to concentration gradients in a gaseous medium or mixture. The motion of aerosol particles is in the direction of the gradient. For the interested reader Pruppacher and Klett (1997) provide a comprehensive discussion of this effect. To make diffusiophoretic effects simpler to understand, a description following Cotton *et al.* (1986) follows. Diffusiophoresis is due to attraction and repulsion of aerosol particles to a droplet along gradients of water vapor. Thermophoresis effects dominate so that the net effect of thermophoresis and diffusiophoresis is to inhibit contact nucleation of cloud droplets during supersaturation and enhance contact nucleation during subsaturation as described by Cotton *et al.* (1986). The parameterization of diffusiophoresis contact nucleation effects is described by Young (1974a) as follows,

$$\left.\frac{1}{N_{\mathrm{a}}}\frac{\mathrm{d}N_{\mathrm{a}}}{\mathrm{d}t}\right|_{\mathrm{Th}} = 4\pi r_{\mathrm{cw}} g_{\mathrm{d}}(\rho_{\mathrm{v}\infty} - \rho_{\mathrm{vSL}})\psi_{\mathrm{v}}^{*}(3.34\times10^{22}\ \text{molecules g}^{-1}), \tag{3.58}$$

where $\rho_{\mathrm{v}\infty}$ is vapor density at infinity and ρ_{vSL} is the vapor density over the liquid water droplet's surface, ψ_{v}^{*} is vapor diffusivity influences of convection on heat diffusion f_1^* and molecular boundary layer considerations on heat diffusion included in f_2^*.

The last three variables are defined as

$$\psi_{\mathrm{v}}^{*} = \psi_{\mathrm{v}} f_1^* f_2^*, \tag{3.59}$$

where

$$f_1 = 1 + \frac{P^*}{4\pi C}\left(0.56 N_{\mathrm{re}}^{1/2} N_{\mathrm{sc}}^{1/3}\right), \tag{3.60}$$

and

$$f_2 = \frac{r_{\mathrm{cw}}}{r_{\mathrm{cw}} + \frac{\psi_{\mathrm{v}}}{\beta}\left[\frac{2\pi}{R_{\mathrm{v}} T_{\infty}}\right]^{1/2}}, \tag{3.61}$$

where R_{v} is the gas constant for water vapor and $\beta = 0.4$ is the condensation or deposition coefficient. Also in (3.58), g_{d} is given as

$$g_{\mathrm{d}} = g_{\mathrm{d}}'\left[\frac{m_{\mathrm{v}}^{1/2}}{N_{\mathrm{v}} m_{\mathrm{v}}^{1/2} + N_{\mathrm{a}} m_{\mathrm{a}}^{1/2}}\right], \tag{3.62}$$

where $g_{\mathrm{d}}' = 0.8$ to 1.0, m is mass of a molecule or aerosol, and N is number concentrations of molecules or aerosols.

3.5.4 Secondary ice nucleation

There are two parameterizations for the ice multiplication hypothesis given by Hallet and Mossop (1974) and Mossop (1976). The first is the most commonly used; in this, approximately 350 splinters are produced for every milligram of rime collected onto each graupel particle at –5 °C (Hallet and Mossop 1974). The formulation for this processes is temperature dependent and given by,

$$\frac{dN_{\mathrm{Tisp}}}{dt} = \rho 3.5 \times 10^{8} f_1(T_{\mathrm{cw}})\left(Q_{\mathrm{gw}} AC_{\mathrm{cw}} + Q_{\mathrm{hw}} AC_{\mathrm{cw}}\right), \tag{3.63}$$

where the subscripts gw and hw are for graupel and hail, and cw is for cloud water. The subscript isp stands for ice splintering. The term $f_1(T)$ is defined by,

$$f_1(T) = \begin{matrix} 0 & T > 270.15 \\ [(T - 268.15)/2] & 270.15 > T > 268.15 \\ [(T - 268.15)/3] & 268.15 > T > 265.15 \\ 0 & 265.15 > T \end{matrix}. \tag{3.64}$$

A source term in the prognostic equation for the mixing ratio of ice splinters is

$$\frac{dQ_{\mathrm{isp}}}{dt} = \frac{m_{\mathrm{i0}}}{\rho}\frac{dN_{\mathrm{Tisp}}}{dt}, \tag{3.65}$$

where m_{i0} is the minimum ice crystal mass.

Now that nucleation has been presented, it is possible to explore condensation/evaporation, and deposition/sublimation processes of newly activated cloud and ice crystals, respectively. Next saturation adjustment schemes will be discussed, followed by descriptions of explicit condensation/evaporation and deposition/sublimation by vapor diffusion.

4

Saturation adjustment

4.1 Introduction

Saturation adjustment schemes are usually designed to bring the relative humidity back to exactly 100% when supersaturation occurs. In doing so, the enthalpy of condensation or deposition is released, the temperature is increased just the right amount for 100% humidity, and the air becomes laden with condensate in the form of cloud droplets at temperatures warmer than 273.15 K. At temperatures colder than freezing, in order to adjust the relative humidity to 100% with respect to ice, a mixture of cloud droplets and ice crystals may be found, and finally at temperatures colder than 233.15 K, only ice crystals are generally produced. For the case of a mixture of cloud droplets and ice crystals, the adjustment is made such that the saturation mixing ratio of each phase, liquid and ice, is weighted in the calculation of relative humidity (Tao *et al.* 1989). Some of the earliest adjustment schemes were described by McDonald (1963), for example, to simulate fog formation. The adjustment process can be prescribed for a single step as in Rutledge and Hobbs (1983; 1984), or an iteration process such as that in Bryan and Fritsch (2002), using potential-temperature, vapor, and mixing ratios. In Tripoli and Cotton (1981), an ice-liquid potential temperature and vapor are used to diagnose quickly the cloud-water mixing ratio required to bring a parcel to 100% humidity with an appropriate associated temperature increase (condensation) or temperature decrease (evaporation).

Alternatively, schemes have been developed by Asai (1965), Langlois (1973), and Soong and Ogura (1973) to adjust potential-temperature fields, vapor fields, and condensate fields with a single non-iterative step when supersaturation exists. In addition, a single-step adjustment to capture the evaporative cooling and loss of cloud particles at subsaturation is built into these systems of equations. Moreover, equations for change in temperature and pressure owing to phase change of water can be computed with the Soong and Ogura (1973) scheme. Few models actually consider pressure change following Wilhelmson and Ogura (1972), though Bryan and Fritsch (2002) claimed notable differences through the inclusion of pressure changes. Finally, some models predict saturation ratio and this is used to determine how much condensation/evaporation, and heating/cooling should occur (e.g. Hall 1980). In addition, the saturation ratio can be expressed as a function of vertical motion.

Many bulk parameterization models, and some bin microphysical models, use saturation adjustment schemes that exactly eliminate supersaturation after a timestep to account for some of the nucleation and condensation. Some bulk parameterization models use both an explicit nucleation scheme, followed by a saturation adjustment scheme (Seifert and Beheng 2005) to bring about saturation. Still some bulk and bin microphysical parameterizations only use explicit nucleation and condensation schemes (Ziegler 1985) to bring about near saturation in a timestep. In the former of these models, where exact saturation is brought about, the advection and diffusion of some measure of temperature, water vapor, cloud water and perhaps cloud ice are computed in the dynamical part of a model first and then the saturation adjustment scheme is applied. Supersaturation occurs as the mixing ratio Q_0 exceeds $Q_s(T)$ at some pressure. Kogan and Martin (1994) refer to this as dynamical supersaturation given by the equation,

$$S_0 = \frac{Q_0 - Q_s(T)}{Q_s(T)}, \tag{4.1}$$

where Q_0 and S_0 are the actual mixing and saturation ratios; the subscript s indicates the saturation value.

Then an adjustment is applied to reach zero supersaturation, as shown at point P_b in Fig. 4.1. In Seifert and Beheng (2005) the advection and diffusion of some measure of temperature and water vapor are also computed first in the dynamical part of a model. With both bulk and bin microphysical models, some form of explicit nucleation and condensation parameterization is used to determine cloud-droplet or ice-crystal growth (subscript m). As the model is advanced forward, the final state is rarely ever characterized by zero. This can be seen at P_m on Fig. 4.1. Here it is seen that the model solution represents a supersaturated state given by

$$S_m = \frac{Q_m - Q_s(T_m)}{Q_s(T_m)}, \tag{4.2}$$

which is rarely at saturation, but rather it gets close to saturation. The explicit nucleation schemes were discussed in Chapter 3 and the explicit condensation

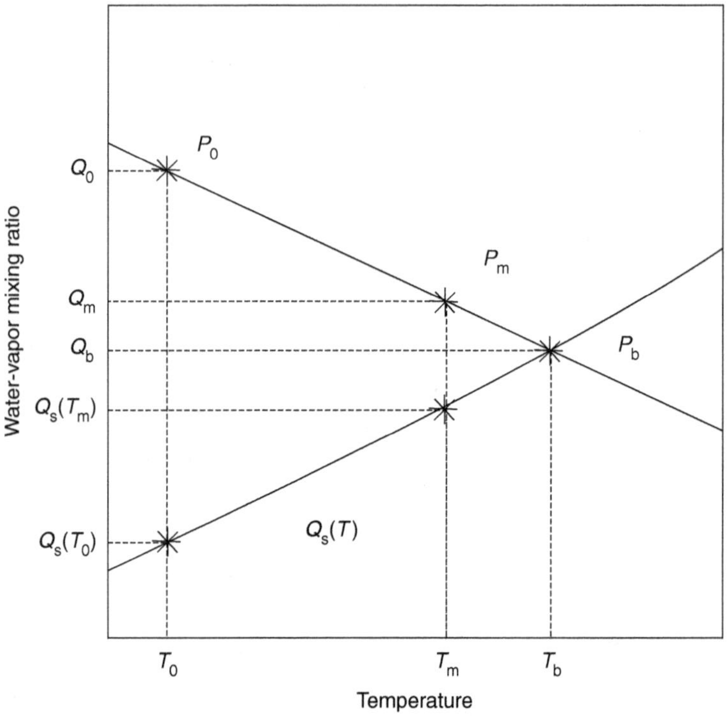

Fig. 4.1. Conceptual model of the moist saturation adjustment process; explained in text. (From Kogan and Martin 1994; courtesy of the American Meteorological Society.)

equation is described in Chapter 5. It should be noted that typically the explicit condensation is solved with a smaller timestep, in time-splitting fashion, than with the dynamical timestep. Usually the small timestep is on the order of 0.5 s as described by Clark (1973), Kogan (1991), and Kogan and Martin (1994), but can range from 0.1 s to 0.5 s depending on vigor of the updrafts simulated.

The parameters on which the degree of supersaturation depends include the cloud-drop number concentration N_{Tcw}, the average radius r_{a} of cloud drops in the spectrum, and the diffusivity of water vapor ψ as discussed by Kogan and Martin (1994). With these parameters, a phase relaxation time-scale to zero supersaturation can be written as

$$\tau_{\mathrm{r}} = \frac{1}{4\pi\psi N_{\mathrm{Tcw}} r_{\mathrm{a}}}, \tag{4.3}$$

with cloud supersaturations S_m typically ranging from a few tenths of a percent to several percent in actuality. The error given by S_m/S_0 is stated to be large for weak updrafts or updrafts with small numbers of large cloud condensation nuclei, but is small for large updrafts on the order of 15 m s^{-1} and for large cloud condensation nuclei concentrations.

4.2 Liquid bulk saturation adjustments schemes

In general, saturation adjustment schemes adjust the potential temperature or temperature and water-vapor mixing ratio isobarically to near-perfect saturation. If the air is subsaturated, cloud water is evaporated and that can continue until saturation is reached or all the cloud water is evaporated. It was shown by Wilhemson and Ogura (1972) that pressure influences could be neglected. But Bryan and Fritsh (2002) more recently found that perhaps they should be included for deep vigorous thunderstorms.

4.2.1 A simple liquid saturation adjustment

Perhaps the simplest saturation adjustment strategy is the non-iterative/iterative scheme used by many including Rutledge and Hobbs (1983; 1984) and Bryan and Fritsch (2002) and given by

$$\frac{\mathrm{d}Q_\mathrm{v}}{\mathrm{d}t} = \frac{Q_\mathrm{v} - Q_\mathrm{SL}}{\Delta t\left(1 + \frac{L_\mathrm{v}^2 Q_\mathrm{SL}}{c_p R_\mathrm{v} T^2}\right)}, \tag{4.4}$$

where Q_v is the vapor mixing ratio, and Q_SL is the saturation mixing ratio with respect to the liquid. This equation can be iterated once (Rutledge and Hobbs 1983; 1984) or several times, typically five or six times, in more vigorous weather systems such as strong thunderstorms (Bryan and Fritsch 2002) until the newest potential-temperature value converges to the previous one. The equation for potential temperature follows as

$$\frac{\mathrm{d}\theta}{\mathrm{d}t} = \gamma_L \frac{\mathrm{d}Q_\mathrm{v}}{\mathrm{d}t}, \tag{4.5}$$

where γ_L is

$$\gamma_L = \frac{L_\mathrm{v}}{c_p \pi}, \tag{4.6}$$

and L_v is the enthalpy of vaporisation. Care must be taken with this parameterization because the timestep Δt appears in the denominator of (4.4) for $\mathrm{d}Q_\mathrm{v}/\mathrm{d}t$. If the timestep is very large (> 5 s) and no iteration is done, the scheme will artificially overshoot or undershoot saturation. However, iteration usually solves this problem.

4.2.2 Soong and Ogura liquid-water saturation adjustment

Perhaps the most popular saturation adjustment scheme as of this writing is one that has been around for more than thirty years. This scheme is one proposed by Soong and Ogura (1973). Soong and Ogura start with Teten's formula, which is

$$Q_{\mathrm{SL}} = a \exp\left(\frac{b[T - T_0]}{[T - c]}\right) \tag{4.7}$$

where $a = 380/p$ (Pa), $b = 17.269\,3882$, $c = 35.86$, and $T = 273.15$ K.

Next, we let Q^*, and θ^* be intermediate values of Q and θ that exist after all other forcing besides saturation adjustment (including advection, diffusion, source and sink terms, etc.) is applied. Then, an expression for $\mathrm{d}\theta$ can be written,

$$\mathrm{d}\theta = \theta^{\tau+1} - \theta^* = -\frac{L_{\mathrm{v}}}{c_p \pi} \mathrm{d}Q_{\mathrm{SL}} = -\frac{L_{\mathrm{v}}}{c_p \pi}\left(Q_{\mathrm{v}}^{\tau+1} - Q_{\mathrm{v}}^*\right), \tag{4.8}$$

where π is the Exner function,

$$\pi = c_p \left(\frac{p}{p_{00}}\right)^{R_{\mathrm{d}}/c_p}, \tag{4.9}$$

where R_{d} is the dry gas constant ($= 287.04$ J kg^{-1} K^{-1}) and P_{00} is the reference pressure equal to 100 000 Pa. Now with

$$\theta^{\tau+1} = \Delta\theta + \theta^*, \tag{4.10}$$

and (4.7) and (4.8) an equation can be written,

$$Q_{\mathrm{SL}}^{\tau+1} = Q_{\mathrm{v}}^{\tau+1} = a \exp\left(\frac{b[\Delta T + T^* - T_0]}{[\Delta T + T^* - c]}\right). \tag{4.11}$$

The following detailed steps are used to arrive at the final equations for the saturation adjustment, ignoring pressure adjustments as was shown to be acceptable by Wilhelmson and Ogura (1972). Multiplying the top and bottom by $T^* - c - \Delta T$ inside the exponential gives

$$Q_{\mathrm{SL}}^{\tau+1} = a \exp\left(\frac{b[\Delta T + T^* - T_0]}{[\Delta T + T^* - c]}\frac{[T^* - c - \Delta T]}{[T^* - c - \Delta T]}\right). \tag{4.12}$$

Letting $\chi = T^* - c$,

$$Q_{\mathrm{SL}}^{\tau+1} = a \exp\left(\frac{b[\Delta T + T^* - T_0][\chi - \Delta T]}{[\chi - \Delta T]^2}\right). \tag{4.13}$$

Now simplifying and neglecting terms of higher order than ΔT gives

$$Q_{\mathrm{SL}}^{\tau+1} = a \exp\left(\frac{b[T^* - T_0]}{[T^* - c]}\right) \exp\left(\frac{b\chi\Delta T - bT^*\Delta T + bT_0\Delta T}{[T^* - c]^2}\right). \tag{4.14}$$

Next, χ is expanded and the terms on the right-hand side inside the exponential are canceled,

$$Q_{\mathrm{SL}}^{\tau+1} = Q_{\mathrm{SL}}^{*} \exp\left(\frac{\cancel{bT^*\Delta T} - bc\Delta T - \cancel{bT^*\Delta T} + bT_0\Delta T}{[T^* - c]^2}\right). \tag{4.15}$$

Now, using a series expansion to express the exponential, and eliminating all higher-order terms, (4.15) may be written as

$$Q_{\mathrm{SL}}^{\tau+1} = Q_{\mathrm{SL}}^{*} \exp\left(\frac{b\Delta T(T_0 - c)}{[T^* - c]^2}\right) = Q_{\mathrm{SL}}^{*}\left(1 + \frac{b\Delta T[T_0 - c]}{[T^* - c]^2}\right). \tag{4.16}$$

Now R_1 and $\theta^{\tau+1}$ are defined, which are

$$R_1 = \left(1 + \frac{b\Delta T[T_0 - c]}{[T^* - c]^2}\right)^{-1}, \tag{4.17}$$

and

$$\theta^{\tau+1} = \theta^* + \frac{R_1 L_{\mathrm{v}}}{c_p \pi}\left(Q_{\mathrm{v}}^* - Q_{\mathrm{SL}}^*\right). \tag{4.18}$$

Now from (4.10) it can be written

$$\Delta\theta = -\Delta Q_{\mathrm{SL}} \frac{L_{\mathrm{v}} c_p}{\pi}. \tag{4.19}$$

Thus, the following can now be written

$$Q_{\mathrm{v}}^{\tau+1} = Q_{\mathrm{v}}^* - R_1\left(Q_{\mathrm{v}}^* - Q_{\mathrm{SL}}^*\right). \tag{4.20}$$

A pictorial diagram of this parameterization (Fig. 4.2; reproduced from Soong and Ogura 1973) shows that the parcel is lifted to saturation and along the path PSG, which is dry adiabatic to S and moist adiabatic to G. Alternatively Asai (1965) lifts a parcel along the path, PSR, which is dry adiabatic to R, and then the parcel is adjusted isobarically to G.

4.2.3 The Langlois saturation adjustment scheme

Another liquid-only scheme has been used by Langlois (1973) and adopted by Cohard and Pinty (2000) as a non-iterative adjustment for liquid-water

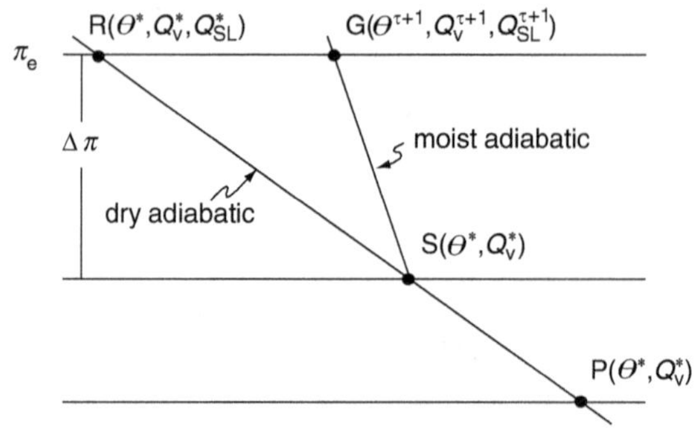

Fig. 4.2. Schematic diagram of the saturation adjustment technique. Consider the following. Lift an air parcel from point P up to point G during time step τ to $\tau + 1$. As the parcel is saturated at G, it must become saturated at a level (denoted by S) between P and G, inclusive. Along the path PS on the chart θ and Q_v take the values θ^* and Q_v^*, respectively, as they are conserved. Values $\theta^{\tau+1}$ and $Q_{SL}^{\tau+1}$ are reached directly by going through PSG rather than PRG as in Asai's (1965) approach. (From Soong and Ogura 1973; courtesy of the American Meteorological Society.)

saturation. The scheme is repeated here for completeness following closely the presentation in Cohard and Pinty (2000). As they pointed out from the first law of thermodynamics, an estimation of the condensation rate is given by

$$(T - T^*) + \frac{L_v(T)}{c_p}\left(Q_{SL}(T) - Q_v^*\right) = 0. \tag{4.21}$$

The variables T^* and Q_v^* are intermediate values obtained after integrating all other processes and source and sink terms. They also make the condensation rate equal to

$$\left.\frac{dQ}{dt}\right|_{cond} = \max\left[-Q_{cw}, Q_v^* - Q_{SL}(T)\right], \tag{4.22}$$

where Q_{cw} is the mixing ratio for cloud water. The parameterization details for evaporation of cloud are discussed in Cohard and Pinty (2000). The equation,

$$F(T) = (T - T^*) + \frac{L_v(T)}{c_p}\left(Q_{SL}(T) - Q_v^*\right), \tag{4.23}$$

is solved with the exception of condensation or evaporation of cloud drops initially. To solve the above equation, Langlois' (1973) approach is employed with a quasi-second-order expansion of $F(T) = 0$ about T^*. The approach begins using the following equation,

$$T = T^* - \frac{F(T^*)}{F'(T^*)}\left\{1 + \frac{1}{2}\frac{F(T^*)}{F'(T^*)}\frac{F''(T^*)}{F'(T^*)}\right\}, \quad (4.24)$$

where superscript primes denote first and second derivatives. Next the saturation mixing ratio at temperature T is defined as

$$Q_{\mathrm{SL}} = \frac{\epsilon e_{\mathrm{SL}}}{P - e_{\mathrm{SL}}}, \quad (4.25)$$

where $\epsilon = R_{\mathrm{d}}/R_{\mathrm{v}}$ and e_{SL} is the saturation vapor pressure over liquid water given by the expression

$$e_{\mathrm{SL}}(T) = \exp(\alpha_{\mathrm{v}} - \beta_{\mathrm{v}}/T - \gamma_{\mathrm{v}}\ln(T)), \quad (4.26)$$

which is more complicated than most approaches that are used. Now the values of α_{v}, β_{v}, and γ_{v} are defined as

$$\alpha_{\mathrm{v}} = \ln(e_{\mathrm{SL}}(T_{00})) + \beta_{\mathrm{v}}/T_{00} - \gamma_{\mathrm{v}}\ln(T_{00}), \quad (4.27)$$

$$\beta_{\mathrm{v}} = \frac{L_{\mathrm{v}}(T_{00})\gamma_{\mathrm{v}}T_{00}}{R_{\mathrm{v}}}, \quad (4.28)$$

and

$$\gamma_{\mathrm{v}} = \frac{c_{v\mathrm{v}} - c_{p\mathrm{v}}}{R_{\mathrm{v}}}, \quad (4.29)$$

where $c_{v\mathrm{v}}$ and $c_{p\mathrm{v}}$ are specific heats at constant volume and pressure, respectively, for vapor v and T_{00} is the temperature at the freezing of water.

Using these expressions, derivatives $Q'_{\mathrm{SL}}(T^*)$ and $Q''_{\mathrm{SL}}(T^*)$ are expressed as

$$Q'_{\mathrm{SL}}(T^*) = A_{\mathrm{w}}Q_{\mathrm{SL}}(T^*)\left\{1 + \frac{Q_{\mathrm{SL}}(T^*)}{\epsilon}\right\} \quad (4.30)$$

and

$$Q''_{\mathrm{SL}}(T^*) = Q'_{\mathrm{SL}}(T^*)\left\{\frac{A'_{\mathrm{w}}(T^*)}{A_{\mathrm{w}}(T^*)} + A'_{\mathrm{w}}(T^*)\left(1 + 2\frac{Q_{\mathrm{SL}}(T^*)}{\epsilon}\right)\right\}, \quad (4.31)$$

where

$$A_{\mathrm{w}}(T) = \frac{\beta_{\mathrm{v}}}{T^2} - \frac{\gamma_{\mathrm{v}}}{T} \quad (4.32)$$

and

$$A'_{\mathrm{w}}(T) = -2\frac{\beta_{\mathrm{v}}}{T^3} + \frac{\gamma_{\mathrm{v}}}{T^2}. \quad (4.33)$$

Using the equations above gives

$$T = T^{*} - \Delta_1\left(1 + \frac{1}{2}\Delta_1\Delta_2\right), \tag{4.34}$$

where Δ_1 and Δ_2 are given as

$$\Delta_1 = \frac{F(T^{*})}{F'(T^{*})} = \frac{L_v(T^{*})}{c_p + L_v(T^{*})Q'_{SL}(T^{*})}\left(Q_{SL}(T^{*}) - Q_v^{*}\right), \tag{4.35}$$

and

$$\Delta_2 = \frac{F''(T^{*})}{F'(T^{*})} = \frac{L_v(T^{*})}{c_p + L_v(T^{*})Q'_{SL}(T^{*})}Q''_{SL}(T^{*}). \tag{4.36}$$

This ends the saturation adjustment methods presented by Cohard and Pinty (2000) and Langlois (1973). Compared to the Soong and Ogura (1973) scheme, which is also non-iterative, this scheme appears to require somewhat more calculations to be done to obtain approximately the same solution.

4.3 Ice and mixed-phase bulk saturation adjustments schemes

4.3.1 A simple ice saturation adjustment scheme

As with the liquid schemes, the simplest ice scheme is probably similar to that for liquid by Rutledge and Hobbs (1983) and Bryan and Fritsch (2002) and is given for ice initiation as

$$\frac{dQ_v}{dt} = \frac{Q_v - Q_{SI}}{\Delta t\left(1 + \dfrac{L_s^2 Q_{SI}}{c_p R_v T^2}\right)}, \tag{4.37}$$

where Q_{SI} is the saturation mixing ratio with respect to ice. This equation can be iterated once (Rutledge and Hobbs 1983) or several times, typically five or six times in more vigorous systems (Bryan and Fritsch 2002) until the newest potential-temperature value converges to the previous one. The equation for potential temperature follows as

$$\frac{d\theta}{dt} = \gamma_{ice}\frac{dQ_v}{dt}, \tag{4.38}$$

where γ_{ice} is

$$\gamma_{ice} = \frac{L_s}{c_p \pi}, \tag{4.39}$$

and L_s is the enthalpy of sublimation.

4.3.2 Soong and Ogura-type ice-water saturation adjustment

An ice-only version of the Soong and Ogura (1973) scheme can be devised by first defining the saturation vapor mixing ratio for ice using Teten's formula for ice,

$$Q_{\mathrm{SI}} = a \exp\left(\frac{b_{\mathrm{ice}}[T - T_0]}{[T - c_{\mathrm{ice}}]}\right) \tag{4.40}$$

where $a = 380/p$ (Pa), $b_{\mathrm{ice}} = 21.87455$, $c_{\mathrm{ice}} = 7.66$, and $T = 273.15$ K.

Following the same general steps that were used for the liquid version of the Soong and Ogura liquid saturation adjustment parameterization, one arrives at the ice-only saturation adjustment equation.

First $R_{1,\mathrm{ice}}$ and $\theta^{\tau+1}$ are defined:

$$R_{1,\mathrm{ice}} = \left(1 + \frac{b_{\mathrm{ice}} \Delta T [T_0 - c_{\mathrm{ice}}]}{[T^* - c_{\mathrm{ice}}]^2}\right)^{-1} \tag{4.41}$$

and

$$\theta^{\tau+1} = \theta^* + \frac{R_{1,\mathrm{ice}} L_{\mathrm{s}}}{c_p \pi}\left(Q_{\mathrm{v}}^* - Q_{\mathrm{SI}}^*\right). \tag{4.42}$$

Now using (4.10),

$$\Delta\theta = -\Delta Q_{\mathrm{SI}} \frac{L_{\mathrm{s}} c_p}{\pi}, \tag{4.43}$$

and the following can be written,

$$Q_{\mathrm{v}}^{\tau+1} = Q_{\mathrm{v}}^* - R_{1,\mathrm{ice}}\left(Q_{\mathrm{v}}^* - Q_{\mathrm{SI}}^*\right). \tag{4.44}$$

4.3.3 *Tao* et al. *saturation adjustment for liquid and ice mixtures*

A mixed-phase saturation adjustment scheme was proposed by Tao *et al.* (1989) that adjusts the potential-temperature and water-vapor mixing ratio in saturation conditions isobarically to 0% supersaturation for ice, liquid, or mixed-phase clouds. Tao *et al.* begin with two assumptions. The first assumption is that the saturation mixing ratio with respect to ice and liquid, Q_{SS}, is given as a liquid-cloud mixing-ratio and ice-cloud mixing-ratio-weighted mean of ice and liquid water saturation values,

$$Q_{\mathrm{SS}} = \frac{(Q_{\mathrm{cw}} Q_{\mathrm{SL}} + Q_i Q_{\mathrm{SI}})}{(Q_{\mathrm{cw}} + Q_i)}. \tag{4.45}$$

At temperatures warmer than T_{frz}, only liquid water is permitted, and at T_{hom} only ice water is permitted.

The second assumption is that under super- or sub-saturation conditions condensation and deposition occur such that they are linearly dependent on $T_{\text{frz}} = 273.15$ K and $T_{\text{hom}} = 233.15$ K. Excess vapor goes into liquid, ice, or a liquid–ice mix for cloud particles. Cloud water (represented by subscript cw) and cloud ice (subscript ci) evaporate or sublime immediately when sub-saturation conditions exist. Evaporation or sublimation will continue to occur to the point of exhaustion of cloud droplets or when enough cloud drops evaporate such that saturation conditions exist.

With these two assumptions Tao *et al.* (1989) write that

$$dQ_{\text{v}} = Q_{\text{v}} - Q_{\text{SS}}, \tag{4.46}$$

$$dQ = dQ_{\text{v}}\, CND, \tag{4.47}$$

and

$$dQ_{\text{ice}} = dQ_{\text{v}}\, DEP; \tag{4.48}$$

where *CND* and *DEP* are given by

$$CND = \frac{(T - T_{\text{hom}})}{(T_{\text{frz}} - T_{\text{hom}})}, \tag{4.49}$$

and

$$DEP = \frac{(T_{\text{frz}} - T)}{(T_{\text{frz}} - T_{\text{hom}})}, \tag{4.50}$$

where dQ_{v}, dQ_{cw}, and dQ_{ci} are the changes in Q_{v}, Q_{cw}, and Q_{ci}, respectively.

Following Tao *et al.* (1989), the procedure for the adjustment is to compute all sources and sinks of θ, Q_{v}, Q_{cw}, and Q_{ci} and label them at time $t + \Delta t$ as q^*, Q_{v}^*, Q_{cw}^*, and Q_{ci}^*. Then the saturation mixing ratios for Q_{SL}^* and Q_{SI}^* are given using Teten's formula as,

$$Q_{\text{SL}}^* = \frac{380}{p(\text{Pa})} \exp\left(\frac{a_{\text{liq}}(T^* - T_{\text{frz}})}{(T^* - b_{\text{liq}})}\right), \tag{4.51}$$

and

$$Q_{\text{SI}}^* = \frac{380}{p(\text{Pa})} \exp\left(\frac{a_{\text{ice}}(T^* - T_{\text{frz}})}{(T^* - b_{\text{ice}})}\right), \tag{4.52}$$

where $a_{\text{liq}} = 17.2693882$, $b_{\text{liq}} = 35.86$, $a_{\text{ice}} = 21.8735584$, and $b_{\text{ice}} = 7.66$.

The adjustment is toward a moist adiabatic condition, under isobaric (constant pressure) processes. With liquid only present, the parameterization adjusts to a moist adiabat for liquid processes only. For ice-only processes, the parameterization adjusts to a moist adiabat for ice processes only. For mixed phases, when both liquid and ice are present, the parameterization adjusts to a moist adiabat for mixed ice and liquid processes. The representation of this process is not a trivial task and estimations of the amount of ice and cloud produced during the saturation adjustment unfortunately are based on inadequate information. The potential temperature is found from

$$\mathrm{d}\theta = \theta^{t+1} - \theta^{*} = \frac{(L_{\mathrm{v}}\mathrm{d}Q_{\mathrm{cw}} + L_{\mathrm{s}}\mathrm{d}Q_{\mathrm{ci}})}{c_p \pi}, \tag{4.53}$$

and the vapor mixing ratio is

$$Q_{\mathrm{v}}^{t+\Delta t} = \frac{\left(Q_{\mathrm{cw}}^{*} Q_{\mathrm{SL}}^{t+\Delta t} + Q_{\mathrm{ice}}^{*} Q_{\mathrm{SI}}^{t+\Delta t}\right)}{\left(Q_{\mathrm{cw}}^{*} + Q_{\mathrm{ci}}^{*}\right)}. \tag{4.54}$$

Now

$$\theta^{t+\Delta t} = \theta^{*} + \mathrm{d}\theta \tag{4.55}$$

is substituted into the Teten's formula for Q_{SL}^{*} (4.51) and Q_{SI}^{*} (4.52). The calculation is made simpler by converting all T^{*} variables into $\pi\theta^{*}$ or simply θ^{*}. Then following the method of Soong and Ogura (1973) demonstrated above, the first-order terms in $\mathrm{d}\theta$ are used to write

$$Q_{\mathrm{v}}^{t+\Delta t} = Q_{\mathrm{v}}^{*} - R_1 + R_2 \mathrm{d}\theta, \tag{4.56}$$

where, according to Tao *et al.* (1989),

$$R_1 = Q_{\mathrm{v}}^{*} - \frac{\left(Q_{\mathrm{SL}}^{*} Q_{\mathrm{cw}}^{*} + Q_{\mathrm{ci}}^{*} Q_{\mathrm{SI}}^{*}\right)}{\left(Q_{\mathrm{cw}}^{*} + Q_{\mathrm{ci}}^{*}\right)} \tag{4.57}$$

$$R_2 = \frac{\left(A_1 Q_{\mathrm{SL}}^{*} Q_{\mathrm{cw}}^{*} + A_2 Q_{\mathrm{ci}}^{*} Q_{\mathrm{SI}}^{*}\right)}{\left(Q_{\mathrm{cw}}^{*} + Q_{\mathrm{i}}^{*}\right)}. \tag{4.58}$$

We now let

$$A_1 = \frac{(237.3 a_{\mathrm{liq}} \pi)}{(T^{*} - 35.86)^2}, \tag{4.59}$$

$$A_2 = \frac{(237.3 a_{\mathrm{ice}} \pi)}{(T^{*} - 7.66)^2}, \tag{4.60}$$

$$A_3 = \frac{(L_v\, CND + L_s\, DEP)}{c_p \pi}. \tag{4.61}$$

Next, using A_1, A_2, A_3, R_1, and R_2, the changes in θ and Q_v can be found by the adjustment,

$$\theta^{t+\Delta t} = \theta^* + \frac{R_1 A_3}{(1 + R_2 A_3)}, \tag{4.62}$$

$$Q_v^{t+\Delta t} = Q_v^* + \frac{R_1}{(1 + R_2 A_3)}. \tag{4.63}$$

An interesting concept about this saturation adjustment for mixed-phase cloud particles is that supersaturation with respect to ice is permitted to occur. This can happen because the saturation with respect to liquid water is larger than that with respect to ice water. This allows, in more sophisticated models, the nucleation of different ice habits that depend on ice- or liquid-water sub- or super-saturation. It also permits the depositional growth of ice crystals by explicit means as well as by the adjustment procedure. There have been a few models that use nucleation methods discussed previously for liquid water and ice water, and use the saturation adjustment as a proxy for deposition growth on already nucleated ice particles. It should be noted that the scheme above does not predict the number concentration of ice- or liquid-water particles nucleated. Particles nucleated have to be supplied as above; or by some means that specifies ice concentration by temperature (e.g. Fletcher's curve); or some other parameterization based upon temperature and supersaturation; or constants for liquid-water drop concentrations.

4.3.4 Ice–liquid-water potential-temperature iteration

In this scheme the cloud-water mixing ratio is diagnosed, and if temperatures are below the homogeneous freezing temperature, the ice-crystal water mixing ratio is computed. Closely following Flatau *et al.*'s (1989) explanation of Cotton and Tripoli's (1980) and Tripoli and Cotton's (1981) approach, two variables are first taken from Chapter 1, including the Exner function (4.9) and the ice–liquid-water potential temperature θ_{il},

$$\theta = \theta_{il}\left(1 + \frac{L_v Q_{liq} + L_s Q_{ci}}{c_p \max(T, 253.15)}\right). \tag{4.64}$$

In the first step, the Exner function and potential temperature are computed. Next the supersaturation is computed to see if any liquid should exist at a location. The procedure continues by determining if the vapor and cloud-water mixing ratios exist from the total water mixing ratio (cloud water plus water vapor) Q_{T}. Cloud water exists if the atmosphere is supersaturated and the amount is the excess of Q_{T} over saturation mixing ratio Q_{VL}. To make this scheme work Q_{liq} needs to be defined,

$$Q_{\mathrm{liq}} = Q_{\mathrm{cw}} + Q_{\mathrm{rw}}, \tag{4.65}$$

where Q_{cw} is the cloud-water mixing ratio and Q_{rw} is rain-water mixing ratio. If there is ice present, the total ice is computed as the sum of all of the ice mixing ratios ($Q_{\mathrm{ice-species}}$),

$$Q_{\mathrm{ci}} = \sum Q_{\mathrm{ice-species}}. \tag{4.66}$$

The water-vapor and cloud-water mixing ratios are computed as follows. First, the water-vapor mixing ratio is computed from

$$Q_{\mathrm{v}} = \max\left(0, Q_{\mathrm{T}} - Q_{\mathrm{liq}} - Q_{\mathrm{ci}}\right). \tag{4.67}$$

Then the cloud-water mixing ratio is computed using the above variables and the saturation mixing ratio over liquid Q_{SL},

$$Q_{\mathrm{cw}} = \max\left(0, Q_{\mathrm{T}} - Q_{\mathrm{liq}} - Q_{\mathrm{ci}} - Q_{\mathrm{SL}}\right). \tag{4.68}$$

The above system of equations can be iterated to diagnose T, θ, Q_{SL}, Q_{v}, and Q_{cw}, from the predictive equations for θ_{il}, π, Q_{T}, Q_{rw}, and all the ice mixing ratio species.

When temperatures are below the homogeneous temperature, the following procedure is added to the iteration instead of the steps immediately above. First $Q_{\mathrm{cw}} = 0$, and $Q^*_{\mathrm{rw}} = 0$, where the asterisk denotes the intermediate value during iteration. Next, Q^*_{ci}, the temporary ice-crystal mixing ratio is defined as

$$Q^*_{\mathrm{ci}} = Q_{\mathrm{ci}} + Q_{\mathrm{rw}} + \max(0, Q_{\mathrm{T}} - Q_{\mathrm{ci}} - Q_{\mathrm{SL}}). \tag{4.69}$$

When the iteration is done, the temporary values are set to be the new permanent values. That completes the iteration.

4.4 A saturation adjustment used in bin microphysical parameterizations

Droplet growth by water-vapor diffusion condensation occurs with supersaturation, and evaporation with subsaturation. When drops gain mass or lose mass, they move to larger or smaller sizes, respectively, with the constraint that

$$\int_0^\infty n(x)\mathrm{d}x = C, \tag{4.70}$$

where C is a constant. An equation that describes the change in the size distribution, $n(x)$, with time by condensation and evaporation is given as

$$\left[\frac{\partial n(x)}{\partial t}\right]_{\text{cond,evap}} = -\frac{\partial}{\partial x}\left[n(x)\frac{\mathrm{d}x}{\mathrm{d}t}\right]. \tag{4.71}$$

Following Ogura and Takahashi (1973), first the model variables are all updated by advection, diffusion, filtering, etc., and advanced to the intermediate values, T^*, Q_v^*, n^*, and the number of aerosols ξ^* at $t = \tau + 1$. Then the saturation mixing ratio is computed with Teten's formula,

$$Q_\mathrm{SL}^* = \frac{380}{p}\exp\left(a_\mathrm{liq}\frac{[T^* - 273.15]}{[T^* - b_\mathrm{liq}]}\right), \tag{4.72}$$

where p is the environmental pressure. No supersaturation is allowed at $t = \tau + 1$. Then, following Asai (1965), if $\delta M > 0$,

$$\delta M = Q_\mathrm{v}^* - Q_\mathrm{SL}^* > 0. \tag{4.73}$$

Then δM_1 is computed by

$$\delta M_1 = \delta M\left(1 + \frac{L_\mathrm{v}^2 Q_\mathrm{SL}}{c_p R_\mathrm{d} T^2}\right)^{-1}, \tag{4.74}$$

such that δM_1 is condensed, so that air is brought to exact supersaturation with the water-vapor mixing ratio $Q_\mathrm{v}^* - \delta M_1$ at temperature $T^* + (L_\mathrm{v}/c_p)\,\delta M_1$.

First δM_1 is allowed to condense on nuclei. The total mass of vapor condensed per unit mass of air during the time increment Δt is given by

$$S(t) = \frac{\xi}{\rho}\sum_{J=1}^{J\max} x(J)f(J), \tag{4.75}$$

where $f(J)$ is described as one of the function types for nucleation of cloud condensation nuclei given by expressions in Chapter 3. When $S > \delta M_1$, only a fraction of ξ^* given by $\xi^*(\delta M_1/S)$ is activated. So ξ at $t = \tau + 1$ is provided by the following,

$$\xi = \xi^* - \left(\frac{\delta M_1}{S}\right)\xi^*. \tag{4.76}$$

In addition, in step 2 of the computation $n^{**}(J)$ is computed from $n^*(J)$ by

$$n^{**}(J) = n^*(J) + \left(\frac{\delta M_1}{S}\right)\xi^* f(J). \tag{4.77}$$

Now, if $S < \delta M_1$, the remainder of $(\delta M_1 - S)$ will be exhausted by allowing existing droplets and drops to grow by condensation. The change in mass of a droplet or drop is given by

$$x'(J) = x(J) + \left(\frac{\mathrm{d}x}{\mathrm{d}t}\right)_J \Delta t, \tag{4.78}$$

where x is mass. The rate of mass growth is given in the next chapter. The final value of $n^{**}(J)$ can be computed using the method of Kovetz and Olund (1969) that is also described in Chapter 5.

Owing to the fact that no supersaturation is allowed, the total condensed water vapor for each Δt is given by

$$G(t) = \frac{\Delta t}{\rho} \sum_{J=1}^{J\max} n^{**}(J) \left(\frac{\mathrm{d}x}{\mathrm{d}t}\right)_J, \tag{4.79}$$

and the growth per category by condensation therefore is just

$$\left[\frac{(\delta M_1 - S)}{G}\right] \left(\frac{\mathrm{d}x}{\mathrm{d}t}\right). \tag{4.80}$$

For the case of evaporation, when $\delta M < 0$, the evaporation rate is computed by the same equation used to compute condensation growth. The change in $n(J)$ for evaporation also is similar to that used for condensation, except that $J' = J$ to $J_{\max}$. According to this method, the number of droplets less than 4 μm are computed and evaporated completely and their cloud condensation nuclei are added to the number of nuclei ξ. There are problems with this, in that a drop that evaporates is made up of many droplets, and thus contains many cloud condensation nuclei. Therefore the actual number of cloud condensation nuclei is not conserved.

4.5 Bulk model parameterization of condensation from a bin model with explicit condensation

The effects of bulk parameterization saturation adjustments versus bin models with explicit nucleation, which allow supersaturation to exist (Fig. 4.3) show that errors from the bin model are most significant for the small cloud condensation nucleation number-concentration case (maritime environments), but they improve as cloud condensation nuclei numbers approach values that would be considered average or large (continental environments).

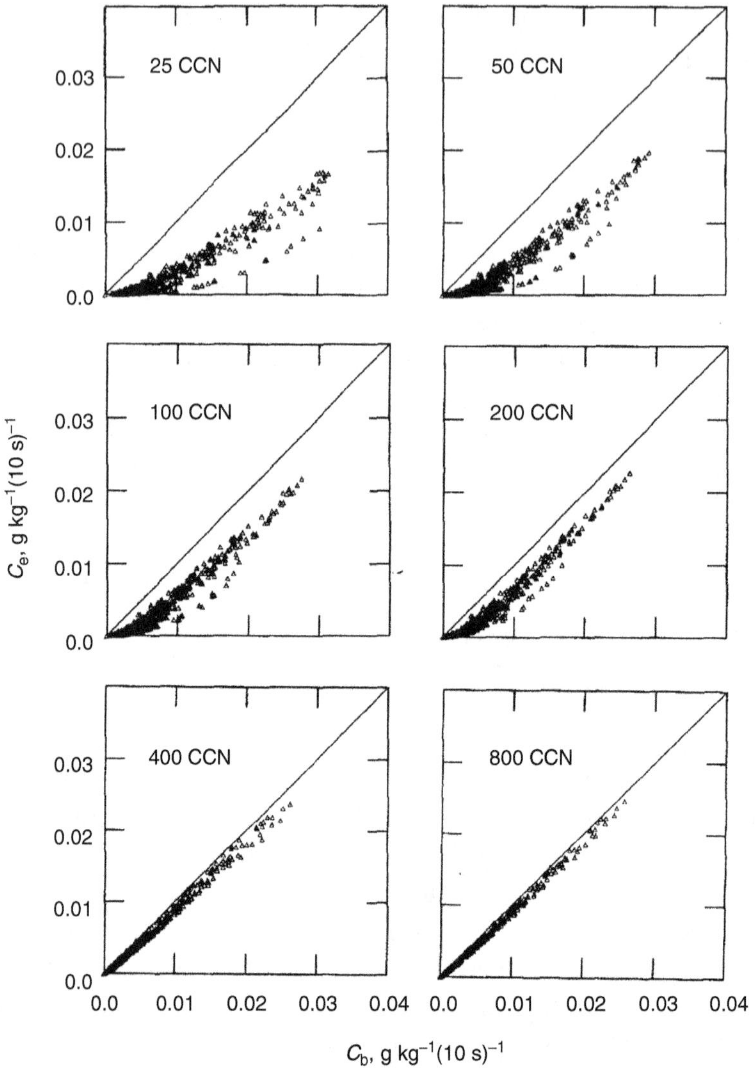

Fig. 4.3. Scatterplots of explicit, C_e, versus bulk, C_b, condensation rates at points in the model with non-zero explicit condensation rates, obtained every two minutes for cases with cloud condensation nucleation concentrations from 25 to 800 cm^{-3}. The difference between this figure and Fig. 4.4 is that the higher-order equation was used to parameterize the bulk condensation. (From Kogan and Martin 1994; courtesy of the American Meteorological Society.)

Kogan and Martin (1994) did multiple regression analyses on the predicted variables in a bin microphysical model with explicit condensation to derive two new bulk microphysical models with bulk condensation parameterizations. The more accurate of the two formulations is

$$C_{\rm rb} = \left(\beta_1 + 2S_0\beta_2 + \beta_3 Q_{\rm cw} + \frac{Q_{\rm cw}\beta_4}{S_0}\right)C_{\rm b} + S_0\beta_5. \tag{4.81}$$

The variables of interest include $C_{\rm rb}$, which is the revised bulk condensation rate found from the regression coefficients β_1, β_2, β_3, β_4 and β_5 and $C_{\rm b}$, which is defined as a first-guess bulk condensation rate. The bulk microphysical model formulation for first-guess bulk condensation rate was computed following McDonald (1963). The cloud droplet and cloud condensation nuclei numbers are not usually known in bulk models. In McDonald's formulation, exact saturation is achieved using

$$Q_{\rm v} = Q_{\rm v}^0 + \delta Q = Q_{\rm SL}, \tag{4.82}$$

$$-L\delta Q = c_p \delta T, \tag{4.83}$$

and

$$T = T^0 + \delta T. \tag{4.84}$$

In the above, $Q_{\rm v}^0$ and T^0 are the values of vapor mixing ratio and temperature before the adjustment. The saturation mixing ratio is $Q_{\rm SL}$. In addition, $Q_{\rm v}$ and T are the values of vapor mixing ratio and temperature after the adjustment. Finally δQ and δT are the changes in $Q_{\rm v}^0$ and T^0 that are needed to reach perfect saturation.

The bin microphysical model with explicit condensation that was used is approximately the same as that given in Chapter 5.

Empirical regression coefficients for various initial total numbers of cloud condensation nuclei (in cm^{-3}) are given in Table 4.1. The residual error in Table 4.1 is calculated using

$$R = \frac{\sum(\text{Exact condensation} - C_{\rm rb})^2}{\text{total points}}. \tag{4.85}$$

Table 4.1

Initial CCN	β_1	β_2	β_3	β_4	β_5	Residual error R
25	0.32	5.2	−0.24	0.028	−0.12	2.7×10^{-4}
50	0.45	8.4	−0.36	0.028	−0.18	2.4×10^{-4}
100	0.66	11.0	−0.45	0.027	−0.25	2.0×10^{-4}
200	0.88	13.0	−0.45	0.022	−0.32	1.5×10^{-4}
400	2.10	3.9	−0.12	0.0037	−0.57	0.30×10^{-4}
800	2.00	1.2	−0.016	0.00082	−0.52	0.038×10^{-4}

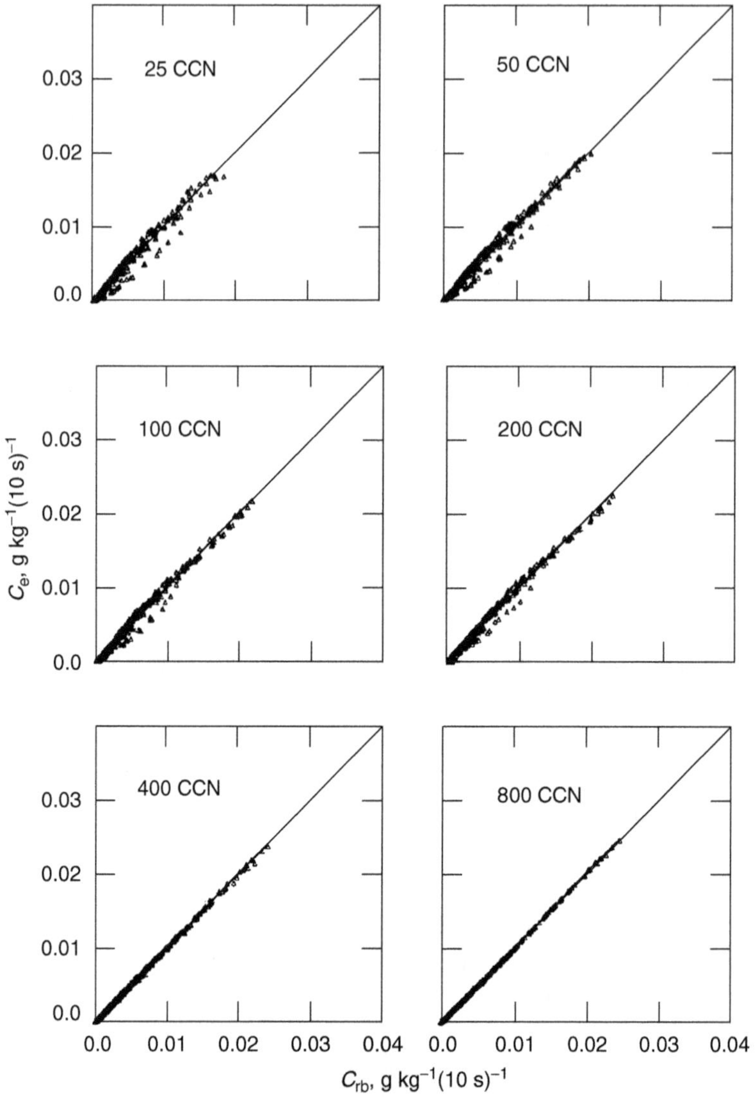

Fig. 4.4. Scatterplots of explicit, C_e, versus bulk, C_b, condensation rates at points in the model with non-zero explicit condensation rates, obtained every two minutes for cases with cloud condensation nucleation concentrations from 25 to 800 cm^{-3}. (From Kogan and Martin 1994; courtesy of the American Meteorological Society.)

Results from the use of (4.81) to compute bulk microphysical parameterizations of condensation are shown in Fig. (4.4) as compared with bin model solutions. The solutions are remarkably good at reproducing the bin microphysical parameterization condensation rates, and improve as large concentrations are used.

4.6 The saturation ratio prognostic equation

Often modelers find convenience in predicting saturation ratio as one of the prognostic equations they employ. For liquid-phase-only clouds, the saturation development equation is given as

$$\frac{dS}{dt} = Q_1 \frac{dz}{dt} - Q_2 \frac{d\chi_{liq}}{dt}. \tag{4.86}$$

In Rogers and Yau (1989) and Pruppacher and Klett (1997) it is explained that the first term on the right of (4.86) is the change in saturation ratio by adiabatic ascent or descent. The second term in (4.86) is the change in saturation ratio by condensation or evaporation of vapor onto or from droplets. The terms Q_1 and Q_2 are given following Rogers and Yau (1989), Pruppacher and Klett (1997) and others as

$$Q_1 = \frac{1}{T}\left[\frac{\epsilon L_v g}{R_d c_p T} - \frac{g}{R_d}\right], \tag{4.87}$$

and

$$Q_2 = \rho\left[\frac{R_d T}{\epsilon e_s} - \frac{\epsilon L_v^2}{PTc_p}\right], \tag{4.88}$$

where g is the acceleration due to gravity, 9.8 m s^{-2}. Whilst it may appear easy to obtain Q_1 and Q_2, the procedure is sketched out for completeness for those readers not accustomed to working with these equations. Let us start with the assumption of no condensation under adiabatic ascent,

$$\frac{dS}{dt} = Q_1 \frac{dz}{dt}. \tag{4.89}$$

Let us assume that we can write dS/dt as

$$\frac{dS}{dt} = \frac{d}{dt}\left(\frac{e}{e_s}\right) = \left(\frac{1}{e_s}\frac{de}{dt} - \frac{e}{e_s^2}\frac{de_s}{dt}\right). \tag{4.90}$$

Defining the approximate mixing ratio,

$$Q = \epsilon \frac{e}{p}, \tag{4.91}$$

and differentiating with respect to time, we obtain

$$\frac{dQ}{dt} = 0 = \frac{1}{p}\frac{de}{dt} = \frac{e}{p^2}\frac{dp}{dt}, \tag{4.92}$$

then

$$\frac{1}{p}\frac{\mathrm{d}e}{\mathrm{d}t} = -\frac{e}{p^2}\frac{pg}{R_\mathrm{d}T}\frac{\mathrm{d}z}{\mathrm{d}t}, \tag{4.93}$$

or

$$\frac{\mathrm{d}e}{\mathrm{d}t} = -e\frac{g}{R_\mathrm{d}T}\frac{\mathrm{d}z}{\mathrm{d}t}. \tag{4.94}$$

The next term in Q_1 in (4.89) can be found by manipulating the Clausius–Clapeyron equation,

$$\frac{\mathrm{d}e_\mathrm{s}}{\mathrm{d}t} = -\frac{e_\mathrm{s}L_\mathrm{v}}{R_\mathrm{v}T^2}\frac{\mathrm{d}T}{\mathrm{d}z}\frac{\mathrm{d}z}{\mathrm{d}t} = -\frac{e_\mathrm{s}L_\mathrm{v}}{R_\mathrm{v}T^2}\frac{g}{c_p}\frac{\mathrm{d}z}{\mathrm{d}t}. \tag{4.95}$$

Substituting the definition of $\mathrm{d}e_\mathrm{s}/\mathrm{d}t$ into (4.89), results in the following equation if $S = e/e_\mathrm{s}$ is about O(1) on the right-hand side,

$$\frac{\mathrm{d}S}{\mathrm{d}t} = \left(\frac{L_\mathrm{v}}{R_\mathrm{v}T^2}\frac{g}{c_p}\frac{\mathrm{d}z}{\mathrm{d}t} - \frac{g}{R_\mathrm{d}T}\frac{\mathrm{d}z}{\mathrm{d}t}\right), \tag{4.96}$$

or

$$\frac{\mathrm{d}S}{\mathrm{d}t} = \left(\frac{L_\mathrm{v}}{R_\mathrm{v}T^2}\frac{g}{c_p}W - \frac{g}{R_\mathrm{d}T}W\right). \tag{4.97}$$

Next a solution for Q_2 is sought. This solution is a bit more complex. Starting with the assumption that there is no vertical motion, (4.86) becomes

$$\frac{\mathrm{d}S}{\mathrm{d}t} = Q_2\frac{\mathrm{d}\chi}{\mathrm{d}t} \tag{4.98}$$

and we know what $\mathrm{d}S/\mathrm{d}t$ is from above. Next, as before, solving for the relationship between Q, e, and p,

$$e = \frac{Q_\mathrm{s}p}{\epsilon} = e_\mathrm{s} \tag{4.99}$$

and differentiating gives

$$\frac{\mathrm{d}e_\mathrm{s}}{\mathrm{d}t} = \frac{p}{\epsilon}\frac{\mathrm{d}Q_\mathrm{s}}{\mathrm{d}t} + \frac{Q_\mathrm{s}}{\epsilon}\frac{\mathrm{d}p}{\mathrm{d}t}. \tag{4.100}$$

Rearranging,

$$\frac{1}{e_\mathrm{s}}\frac{\mathrm{d}e}{\mathrm{d}t} = \frac{\rho R_\mathrm{d}T}{\epsilon e_\mathrm{s}}\frac{\mathrm{d}Q_\mathrm{s}}{\mathrm{d}t} + \frac{Q_\mathrm{s}}{\epsilon}\frac{\mathrm{d}p}{\mathrm{d}z}\frac{\mathrm{d}z}{\mathrm{d}t}, \tag{4.101}$$

where the last term on the right-hand side is assumed to be zero. There is a somewhat unexpected step that needs to be done to get the solution to $\mathrm{d}e_{\mathrm{s}}/\mathrm{d}t$,

$$\frac{\mathrm{d}e_{\mathrm{s}}}{\mathrm{d}t} = \frac{\mathrm{d}e_{\mathrm{s}}}{\mathrm{d}T}\frac{\mathrm{d}T}{\mathrm{d}z}\frac{\mathrm{d}z}{\mathrm{d}Q_{\mathrm{s}}}\frac{\mathrm{d}Q_{\mathrm{s}}}{\mathrm{d}t} = \frac{\mathrm{d}e_{\mathrm{s}}}{\mathrm{d}T}\frac{\mathrm{d}T}{\mathrm{d}Q_{\mathrm{s}}}\frac{\mathrm{d}Q_{\mathrm{s}}}{\mathrm{d}t}; \tag{4.102}$$

or, since $\mathrm{d}z/\mathrm{d}t = 0$, then

$$\begin{aligned} \frac{\mathrm{d}e_{\mathrm{s}}}{\mathrm{d}t} &= \frac{\mathrm{d}e_{\mathrm{s}}}{\mathrm{d}T}\frac{\mathrm{d}Q_{\mathrm{s}}}{\mathrm{d}t}\left(-\frac{\mathrm{d}T}{\mathrm{d}z}\frac{\mathrm{d}z}{\mathrm{d}Q_{\mathrm{s}}}\right) + \frac{\mathrm{d}e_{\mathrm{s}}}{\mathrm{d}T}\frac{\mathrm{d}T}{\mathrm{d}Q_{\mathrm{s}}}\frac{\mathrm{d}Q_{\mathrm{s}}}{\mathrm{d}t} \\ &= \frac{\mathrm{d}e_{\mathrm{s}}}{\mathrm{d}T}\left(\frac{\mathrm{d}T}{\mathrm{d}Q_{s}} - \frac{\mathrm{d}T}{\mathrm{d}z}\frac{\mathrm{d}z}{\mathrm{d}Q_{\mathrm{s}}}\right)\frac{\mathrm{d}Q_{\mathrm{s}}}{\mathrm{d}t}. \end{aligned} \tag{4.103}$$

Then, the following can be written,

$$\frac{\mathrm{d}e_{\mathrm{s}}}{\mathrm{d}t} = \frac{\mathrm{d}e_{\mathrm{s}}}{\mathrm{d}T}\left(\frac{\mathrm{d}T}{\mathrm{d}Q_{\mathrm{s}}} + \frac{g}{c_p}\frac{\mathrm{d}z}{\mathrm{d}Q_{\mathrm{s}}}\right)\frac{\mathrm{d}Q_{\mathrm{s}}}{\mathrm{d}t}. \tag{4.104}$$

Now from the definition of a moist adiabat,

$$\frac{\mathrm{d}z}{\mathrm{d}Q_{\mathrm{s}}} = \frac{L_{\mathrm{v}}}{g} + \frac{\mathrm{d}T}{\mathrm{d}Q_{\mathrm{s}}}\frac{c_p}{g}, \tag{4.105}$$

the following can be written,

$$-\frac{L_{\mathrm{v}}}{c_p} = \frac{\mathrm{d}T}{\mathrm{d}Q_{\mathrm{s}}} + \frac{g}{c_p}\frac{\mathrm{d}z}{\mathrm{d}Q_{\mathrm{s}}}. \tag{4.106}$$

Rewriting the expression for $\mathrm{d}e_{\mathrm{s}}/\mathrm{d}t$, and noting that for an isobaric process $e \sim e_{\mathrm{s}}$ during condensation, the following is obtained,

$$\frac{e}{e_{\mathrm{s}}^2}\frac{\mathrm{d}e_{\mathrm{s}}}{\mathrm{d}t} = \frac{e}{e_{\mathrm{s}}^2}\frac{\mathrm{d}e_{\mathrm{s}}}{\mathrm{d}T}\left(\frac{L_{\mathrm{v}}}{c_p}\right)\frac{\mathrm{d}Q_{\mathrm{s}}}{\mathrm{d}t} \approx \frac{1}{e_{\mathrm{s}}}\frac{\mathrm{d}e_{\mathrm{s}}}{\mathrm{d}T}\left(\frac{L_{\mathrm{v}}}{c_p}\right)\frac{\mathrm{d}Q_{\mathrm{s}}}{\mathrm{d}t}. \tag{4.107}$$

Substituting (4.107) into the Clausius–Clapyeron equation (4.95) and remembering that $R_{\mathrm{v}} = R_{\mathrm{d}}/\epsilon$ (and using the equation of state for dry air),

$$\frac{\mathrm{d}e_{\mathrm{s}}}{\mathrm{d}T} = \frac{L_{\mathrm{v}}\,e_{\mathrm{s}}}{R_{\mathrm{v}}T^2}. \tag{4.108}$$

From (4.107),

$$\frac{e}{e_{\mathrm{s}}^2}\frac{\mathrm{d}e_{\mathrm{s}}}{\mathrm{d}t} = \frac{\epsilon\rho}{R_{\mathrm{d}}pT}\left(\frac{L^2}{c_p}\right)\frac{\mathrm{d}Q_{\mathrm{s}}}{\mathrm{d}t}. \tag{4.109}$$

Now Q_2 can be written as

$$Q_2 = \rho \left[\frac{R_\mathrm{d} T}{\epsilon e_\mathrm{s}} - \frac{\epsilon L_\mathrm{v}^2}{P T c_p} \right]. \tag{4.110}$$

This can be substituted into (4.98) and an equation for liquid-phase saturation development is found.

Alternatively, Chen (1994) writes the saturation development equation for mixed-phase processes slightly differently as

$$\frac{\mathrm{d}S}{\mathrm{d}t} = Q_1' \frac{\mathrm{d}z}{\mathrm{d}t} - Q_2' \frac{\mathrm{d}\chi_\mathrm{liq}}{\mathrm{d}t} - Q_3' \frac{\mathrm{d}\chi_\mathrm{ice}}{\mathrm{d}t}, \tag{4.111}$$

where,

$$Q_1' = S \frac{g M_\mathrm{a}}{R_\mathrm{d} T} \left[\frac{L_\mathrm{v}}{c_p T} - 1 \right], \tag{4.112}$$

$$Q_2' = \left[S \frac{L_\mathrm{v}^2}{c_p R_\mathrm{d} T^2} + \frac{P}{e_\mathrm{s}} \right], \tag{4.113}$$

and

$$Q_3' = \left[S \frac{L_\mathrm{v} L_\mathrm{s}}{c_p R_\mathrm{d} T^2} + \frac{P}{e_\mathrm{s}} \right]. \tag{4.114}$$

5

Vapor diffusion growth of liquid-water drops

5.1 Introduction

Once a cloud droplet is nucleated it can continue to grow by water-vapor diffusion or condensation, at first rapidly, then slowly as diameter increases, if supersaturation conditions with respect to liquid water continue to occur around the droplet or drop. Conversely, a cloud droplet or raindrop will decrease in diameter by water-vapor diffusion or evaporation, first slowly when large, then rapidly when small, as diameter decreases, assuming sub-saturation conditions with respect to liquid water continue to occur around the cloud droplet or raindrop.

Condensation and evaporation are governed by the same equation, the water-vapor diffusion equation. To understand condensation and evaporation of some particle, two diffusive processes must be considered. The first of these includes water-vapor transfer to or from a particle by steady-state water-vapor diffusion. It is a result of vapor gradients that form around a particle; thus the particle is not in equilibrium with its environment. The second of these processes is conduction owing to thermal diffusion of temperature gradients around a particle that is growing or decreasing in size. Fick's law of diffusion describes these diffusion processes. In summary, consideration must be made for mass and heat flux to and away from particles. These steady-state diffusion processes are derived independently and then a net mass change is obtained iteratively, or by a direct method, by combining the equations with the help of the Clausius–Clapyeron equation.

There are several ways to solve the steady-state equations, and two will be presented. One method includes kinetic effects and one does not. Of these basic approaches, the one that includes kinetic effects is perhaps 10 percent more accurate than the other in the early stages of growth. At later times, the simpler of these methods, which includes just a basic growth equation, approximately gives the same basic result as the more accurate method, which includes several second-order effects that are not in a basic growth equation. The more accurate equation is primarily valid during the first 5 to 10 seconds after nucleation. After that time, the two equations basically give the same answers as other methods. Thus, it probably is not necessary to have this sort of accuracy for either bulk or bin model parameterizations. Nevertheless, these second-order effects will be reviewed for completeness as they do become important if growth after nucleation of aerosols or cloud condensation nuclei is studied.

The basic assumptions that need to be made for vapor diffusion mass change to a liquid particle include the following:

- the particle is larger than the critical radius with regard to the Kohler curve
- the particle is stationary
- the particle is isolated
- the particle is stationary with surface area $4\pi r^2$ where r is radius of the drop
- the vapor field is steady state with infinite extent and supply.

Other first- and second-order effects that can be included in the basic growth equation are:

- ventilation (advective effects)
- kinetic effects
- competitive effects among particles
- radius and solution effects for very small particles.

This provides a basis to start examinating the basics of vapor diffusional growth (condensation and evaporation) of liquid-water cloud droplets, drizzle, and raindrops.

5.2 Mass flux of water vapor during diffusional growth of liquid-water drops

The diffusional change in mass of liquid-water drops owing to subsaturation or supersaturation with respect to liquid water primarily depends on thermal and vapor diffusion. In addition, for larger particles, advective processes are important, and the influences of advective processes have to be approximated using data from laboratory experiments. In the following pages, equations will be derived to arrive at a parameterization equation for diffusional growth changes in a spherical liquid-water particle that is large enough, on the order of a few microns in diameter, so that surface curvature effects can be ignored. Moreover, the liquid-water drops will be assumed to be pure. Later in the chapter the influence of liquid-water drop size and solutes will be considered.

First, from the continuity equation for density of water-vapor molecules the following equation can be written,

$$\frac{\partial \rho_v}{\partial t} + u \bullet \nabla \rho_v = \psi \nabla^2 \rho_v. \tag{5.1}$$

The flow is assumed to be non-divergent, and ρ_v is the water-vapor density given by $\rho_v = nm$. In this definition, n is the number of molecules and m is the mass of a water molecule. In addition, ψ is the vapor diffusivity, given more precisely as,

$$\psi = 2.11 \times 10^{-5} (T/T_0)^{1.94} (p/p_{00}) \,\mathrm{m^2\,s^{-1}}, \tag{5.2}$$

where, $T_0 = 273.15$ K and $p_{00} = 101325$ Pa.

If it is assumed that u is zero so the flow is zero or stationary flow (sum of the air-flow velocity and the vapor-flow velocity) is zero, (5.1) becomes

$$\frac{\partial \rho_v}{\partial t} = \psi \nabla^2 \rho_v. \tag{5.3}$$

With the steady-state assumption, a basic form of Fick's first law of diffusion results for the number of molecules, n, where m is a constant (similar to Rogers and Yau, 1989),

$$\nabla^2 \rho_v = \nabla^2 nm = \nabla^2 n = 0. \tag{5.4}$$

Assuming isotropy, which permits the use of spherical coordinates for this problem, (5.4) becomes

$$\nabla^2 n = \frac{1}{R^2} \frac{\partial}{\partial R} \left(R^2 \frac{\partial n}{\partial R} \right) = 0, \tag{5.5}$$

where R is the distance from the center of the drop.

The product rule is applied to (5.5),

$$\frac{R^2}{R^2} \frac{\partial}{\partial R} \left(\frac{\partial n}{\partial R} \right) + \frac{1}{R^2} \frac{\partial (R^2)}{\partial R} \left(\frac{\partial n}{\partial R} \right) = 0, \tag{5.6}$$

which is written more precisely as

$$\frac{\partial}{\partial R} \left(\frac{\partial n}{\partial R} \right) + \frac{2}{R} \frac{\partial n}{\partial R} = 0. \tag{5.7}$$

Now letting,

$$x = \left(\frac{\partial n}{\partial R} \right), \tag{5.8}$$

substitution of (5.8) into (5.7) results in

$$\frac{\partial x}{\partial R} = -\frac{2}{R}x. \tag{5.9}$$

Integration of (5.9) over R gives,

$$\int \frac{\partial x}{\partial R}\mathrm{d}R = -2\int \frac{x}{R}\mathrm{d}R. \tag{5.10}$$

Rearranging,

$$\int \mathrm{d}\ln x = -2\int \mathrm{d}\ln R; \tag{5.11}$$

and finally integration gives

$$\ln(x) = -2\ln(R) + c', \tag{5.12}$$

where c' is a constant of integration. Taking the exponential of both sides of (5.12) gives,

$$x = c''R^{-2}. \tag{5.13}$$

Now, substituting (5.8) back into (5.13) results in

$$\frac{\partial n}{\partial R} = c''R^{-2}. \tag{5.14}$$

Integrating (5.14) over dR,

$$\int \frac{\partial n}{\partial R}\mathrm{d}R = \int c''R^{-2}\mathrm{d}R, \tag{5.15}$$

results in

$$n(R) = -\frac{c''}{R} + c''', \tag{5.16}$$

where c''' is another constant of integration.

Now, the constants of integration can be determined from the boundary conditions, which are: as R approaches R_∞, n approaches n_∞; and when R equals the drop radius, R_r, n is equal to n_r. Application of these boundary conditions to (5.16) gives

$$n_\infty = -\frac{c''}{R_\infty} + c''' = c''', \tag{5.17}$$

where $c''/R_\infty \ll c'''$, so that $n_\infty = c'''$.

Next, using the boundary conditions and (5.17), (5.16) is then

$$n_r = -\frac{c''}{R_r} + c''' = -\frac{c''}{R_r} + n_\infty. \tag{5.18}$$

Simplifying,

$$n_r = -\frac{c''}{R_r} + n_\infty. \tag{5.19}$$

Solving (5.19) for c'',

$$c'' = a(n_{r=a} - n_\infty). \tag{5.20}$$

From (5.17) c''' was given by

$$c''' = n_\infty. \tag{5.21}$$

Thus (5.16) becomes

$$n(R) = \frac{R_r(n_r - n_\infty)}{R} + n_\infty. \tag{5.22}$$

Now the rate of mass increase or decrease at the drop's surface by way of a flux of droplets toward or away from the drop can be written as $\mathrm{d}M/\mathrm{d}t$, where M is mass,

$$\frac{\mathrm{d}M}{\mathrm{d}t} = \psi 4\pi R_r^2 m \left(\frac{\partial n}{\partial R}\right)_{R=R_r}. \tag{5.23}$$

Equation (5.22) is used to find the derivative of n with respect to R while holding $R = R_r$,

$$\frac{\mathrm{d}M}{\mathrm{d}t} = \psi 4\pi R^2 m \left(\frac{\partial[R_r(n_r - n_\infty)/R + n_\infty]}{\partial R}\right)_{R=R_r}; \tag{5.24}$$

rearranging,

$$\frac{\mathrm{d}M}{\mathrm{d}t} = \psi 4\pi R^2 m \left\{ \frac{\partial}{\partial R}\left[\frac{1}{R} R_r(n_r - n_\infty)\right]_{R=R_r} + \left(\frac{\partial n_\infty}{\partial R}\right)_{R=R_r} \right\}. \tag{5.25}$$

Now by using

$$\frac{\partial n_\infty}{\partial R} = 0 \tag{5.26}$$

in (5.25), the following is found,

$$\frac{\mathrm{d}M}{\mathrm{d}t} = \psi 4\pi R^2 m \frac{\partial}{\partial R}\left[\frac{1}{R} R_r(n_r - n_\infty)\right]_{R=R_r}. \tag{5.27}$$

Taking the derivative gives

$$\frac{\mathrm{d}M}{\mathrm{d}t} = \psi 4\pi R_r (n_\infty - n_r) m. \tag{5.28}$$

Now we note that

$$\rho_{\mathrm{v},r} = n_r m, \tag{5.29}$$

and

$$\rho_{\mathrm{v},\infty} = n_\infty m, \tag{5.30}$$

where $\rho_{\mathrm{v},r}$ is the vapor density at the drop's surface, and $\rho_{\mathrm{v},\infty}$ is the uniform density.

Substitution of (5.29) and (5.30) into (5.28) gives

$$\frac{\mathrm{d}M}{\mathrm{d}t} = \psi 4\pi R_r \left(\rho_{\mathrm{v},\infty} - \rho_{\mathrm{v},r}\right), \tag{5.31}$$

which is the mass change owing to vapor gradients.

5.3 Heat flux during vapor diffusional growth of liquid water

An analogous procedure can be followed to get a relationship $\mathrm{d}q/\mathrm{d}t$, which is the heat flux owing to temperature gradients. Based on $\mathrm{d}q/\mathrm{d}t$, another equation for $\mathrm{d}M/\mathrm{d}t$, different from (5.31), may be written.

From the continuity equation for temperature T the following equation can be written, where K is thermal diffusivity,

$$\frac{\partial T}{\partial t} + u\bullet\nabla T = K\nabla^2 T. \tag{5.32}$$

Assuming again that the flow is non-divergent, and that u is zero, so the flow is zero or stationary flow (sum of the air-flow velocity and the vapor-flow velocity) is zero. Note the value for thermal conductivity κ is given as

$$\kappa = 2.43 \times 10^{-2} \frac{1.832 \times 10^{-5}}{1.718 \times 10^{-5}} \left(\frac{T}{296.0}\right)^{1.5} \left(\frac{416.0}{[T - 120.0]}\right) \mathrm{J\,m^{-1}\,s^{-1}\,K^{-1}}. \tag{5.33}$$

Next, applying $u = 0$ in (5.32),

$$\frac{\partial T}{\partial t} = K\nabla^2 T. \tag{5.34}$$

With the steady-state assumption a basic Fick's law of diffusion for T results,

$$\nabla^2 T = 0. \tag{5.35}$$

The assumption of isotropy permits the use of spherical coordinates for this problem, so that (5.35) becomes

$$\nabla^2 T = \frac{1}{R^2}\frac{\partial}{\partial R}\left(R^2\frac{\partial T}{\partial R}\right) = 0. \tag{5.36}$$

Expanding and simplifying gives

$$\frac{R^2}{R^2}\frac{\partial}{\partial R}\left(\frac{\partial T}{\partial R}\right) + \frac{1}{R^2}\frac{\partial(R^2)}{\partial R}\left(\frac{\partial T}{\partial R}\right) = \frac{\partial}{\partial R}\left(\frac{\partial T}{\partial R}\right) + \frac{2}{R}\frac{\partial T}{\partial R} = 0. \tag{5.37}$$

Now letting

$$y = \left(\frac{\partial T}{\partial R}\right), \tag{5.38}$$

and substituting (5.38) into (5.37) gives

$$\frac{\partial y}{\partial R} = -\frac{2y}{R}. \tag{5.39}$$

Integration of (5.39) over R is,

$$\int \frac{\partial y}{\partial R}\mathrm{d}R = -2\int \frac{y}{R}\mathrm{d}R. \tag{5.40}$$

Then rewriting gives

$$\int \mathrm{d}\ln y = -2\int \mathrm{d}\ln R; \tag{5.41}$$

and finally integration results in

$$\ln(y) = -2\ln(R) + c', \tag{5.42}$$

where c' is a constant of integration.

Taking the exponential of both sides of (5.42) gives

$$y = c''R^{-2}, \tag{5.43}$$

where c'' is a constant.

Substituting (5.38) back into (5.43) results in,

$$\frac{\partial T}{\partial R} = c''R^{-2}. \tag{5.44}$$

Integration of the (5.44) expression with respect to R gives,

$$\int \frac{\partial T}{\partial R}\mathrm{d}R = \int c''R^{-2}\mathrm{d}R; \tag{5.45}$$

and finally,

$$T(R) = -\frac{c''}{R} + c''', \tag{5.46}$$

where c''' is another constant of integration.

Now, the constants of integration can be determined again from the boundary conditions, which are: as R approaches R_∞, T approaches T_∞; and when R equals the drop radius, R_r, T is equal to T_r. Application of these boundary conditions to (5.46) gives

$$T_\infty = -\frac{c''}{R_\infty} + c''' = c''', \tag{5.47}$$

where $c''/R_\infty \ll c'''$, so that $c''' = T_\infty$.

Next, using the boundary conditions and (5.47), (5.46) becomes

$$T_r = -\frac{c''}{R_r} + c''' = -\frac{c''}{R_r} + T_\infty. \tag{5.48}$$

Simplifying,

$$T_r = -\frac{c''}{R_r} + T_\infty. \tag{5.49}$$

Now solving for c'',

$$c'' = R_r(T_\infty - T_r), \tag{5.50}$$

and using (5.47) gives c''' as

$$c''' = T_\infty, \tag{5.51}$$

and (5.46) becomes

$$T(R) = \frac{R_r(T_r - T_\infty)}{R} + T_\infty. \tag{5.52}$$

Now we can write an expression for the energy change $\mathrm{d}q/\mathrm{d}t$ which takes place at the drop's surface,

$$\frac{\mathrm{d}q}{\mathrm{d}t} = -4\pi R_r^2 \rho K c_p \left(\frac{\partial T}{\partial R}\right)_{R=R_r}. \tag{5.53}$$

Equation (5.52) is used to find the derivative of T with respect to R while holding $R = R_r$,

$$\frac{\mathrm{d}q}{\mathrm{d}t} = -4\pi R_r^2 \rho K c_p \left(\frac{\partial[R_r(T_r - T_\infty)/R + T_\infty]}{\partial R}\right)_{R=R_r}. \tag{5.54}$$

Rearranging gives

$$\frac{\mathrm{d}q}{\mathrm{d}t} = -4\pi R_r^2 \rho K c_p \left\{ \frac{\partial}{\partial R} \left[\frac{1}{R} R_r (T_r - T_\infty) \right]_{R=R_r} + \left(\frac{\partial T_\infty}{\partial R} \right)_{R=R_r} \right\}. \tag{5.55}$$

Now using

$$\frac{\partial T_\infty}{\partial R} = 0, \tag{5.56}$$

(5.55) becomes,

$$\frac{\mathrm{d}q}{\mathrm{d}t} = -4\pi R_r^2 \rho K c_p \frac{\partial}{\partial R} \left[\frac{1}{R} R_r (T_r - T_\infty) \right]_{R=R_r}. \tag{5.57}$$

Taking the derivative results in

$$\frac{\mathrm{d}q}{\mathrm{d}t} = -4\pi R_r^2 \rho K c_p \left[-\frac{1}{R^2} R_r (T_r - T_\infty) \right]_{R=R_r}. \tag{5.58}$$

Now (5.58) is applied at $R = R_r$, and simplifing,

$$\frac{\mathrm{d}q}{\mathrm{d}t} = 4\pi R_r^2 \rho K c_p \left[\frac{R_r}{R_r^2} (T_r - T_\infty) \right]. \tag{5.59}$$

Lastly,

$$\frac{\mathrm{d}q}{\mathrm{d}t} = 4\pi R_r \rho K c_p (T_r - T_\infty), \tag{5.60}$$

which is the energy change owing to temperature gradients.

5.4 Plane, pure, liquid-water surfaces

The diffusional change in mass of a liquid-water drop owing to sub- or super-saturation depends on thermal and vapor diffusion along with advective processes. In the following, these will be employed to arrive at a parameterization equation for diffusional growth changes in a liquid-water drop, that are large enough, on the order of a few microns in diameter, so that surface curvature effects can be ignored. Moreover, the liquid drop is assumed to be pure. The following derivation closely follows Byers (1965).

The mass flux of vapor to or from a droplet can be written as

$$\frac{1}{A} \frac{\mathrm{d}M}{\mathrm{d}t} = \psi \left(\frac{\mathrm{d}\rho_\mathrm{v}}{\mathrm{d}R} \right), \tag{5.61}$$

where A is the surface area of the droplet, M is the mass of the droplet, R at R_r is radius of the droplet, t is time, ψ is diffusivity of water vapor in air, and ρ_{v} is the water-vapor density.

The following relationship can be written that represents condensation where the surface area of a sphere has been used,

$$\frac{\mathrm{d}M}{\mathrm{d}t} = A\psi\left(\frac{\mathrm{d}\rho_{\mathrm{v}}}{\mathrm{d}R}\right) = 4\pi R^2\psi\left(\frac{\mathrm{d}\rho_{\mathrm{v}}}{\mathrm{d}R}\right). \tag{5.62}$$

Now consider a spherical droplet as a discontinuity between two phases, such as a liquid-water droplet that has a vapor density at its surface $\rho_{\mathrm{v},r}$ and a water-vapor field of uniform density, $\rho_{\mathrm{v},\infty}$. The continuous gradient $\mathrm{d}\rho_{\mathrm{v},\infty}/\mathrm{d}t$ is now replaced by the gradient of these two values. It is desirable to have an expression for the growth rate in terms of radius. The following equation is obtained by first rearranging (5.62) and then integrating such that the transport of the vapor to the droplet is

$$\frac{\mathrm{d}M}{\mathrm{d}t}\int_{R_r}^{R_\infty}\frac{\mathrm{d}R}{R^2} = 4\pi\psi\int_{\rho_{\mathrm{v},r}}^{\rho_{\mathrm{v},\infty}}\mathrm{d}\rho_{\mathrm{v}}. \tag{5.63}$$

Integration gives

$$-\frac{\mathrm{d}M}{\mathrm{d}t}\left(\frac{1}{R_\infty} - \frac{1}{R_r}\right) = 4\pi\psi\left(\rho_{\mathrm{v},\infty} - \rho_{\mathrm{v},r}\right), \tag{5.64}$$

where,

$$\frac{1}{R_\infty} \to 0. \tag{5.65}$$

Thus,

$$\frac{\mathrm{d}M}{\mathrm{d}t} = 4\pi R_r\psi\left(\rho_{\mathrm{v},\infty} - \rho_{\mathrm{v},r}\right), \tag{5.66}$$

and R_∞ can be considered to be some distance from the droplet, such as one-half the distance to the next droplet, which is probably 1×10^2 to 1×10^3 of radii away.

Now the equation in terms of the rate change of the radius of a sphere in time can be written using the fact that $M = \rho V$, where V is the volume of a sphere, $V = 4/3\pi R_r^3$,

$$\frac{\mathrm{d}M}{\mathrm{d}t} = \rho_{\mathrm{L}}\frac{\mathrm{d}V}{\mathrm{d}t} = \rho_{\mathrm{L}}4\pi R_r^2\frac{\mathrm{d}R_r}{\mathrm{d}t} = \rho_{\mathrm{L}}4\pi R_r\left(R_r\frac{\mathrm{d}R_r}{\mathrm{d}t}\right), \tag{5.67}$$

where ρ_{L}, the density of water, is assumed to be constant.

Substitution of (5.66) into (5.67) and solving for $R_r \mathrm{d}R_r/\mathrm{d}t$ gives

$$R_r \frac{\mathrm{d}R_r}{\mathrm{d}t} = \frac{\psi}{\rho_\mathrm{L}}\left(\rho_{\mathrm{V},\infty} - \rho_{\mathrm{V},r}\right). \tag{5.68}$$

Now from the equation of state for water vapor, the vapor pressure, e, is

$$e = R_\mathrm{V} T \rho_\mathrm{V}, \tag{5.69}$$

and assuming the temperature T_r of the droplet at its surface and the temperature of the air next to the drop T_∞ are equal, $R_r \mathrm{d}R_r/\mathrm{d}t$ in (5.68) is written as

$$R_r \frac{\mathrm{d}R_r}{\mathrm{d}t} = \frac{\psi}{\rho_\mathrm{L} R_\mathrm{V} T}(e_\infty - e_r). \tag{5.70}$$

Recall that enthalpy is the energy transferred between two phases with no temperature change occurring in the two phases. Therefore, a balanced state requires that the enthalpy resulting from condensation must be liberated to the environment; the opposite is true for evaporation. That is, the enthalpy must be absorbed from the environment. Now the enthalpy associated with the phase change (5.66) and (5.68) is L_V. Next, multiplying both left and right hand sides of (5.66) by the L_V results in

$$L_\mathrm{V} \frac{\mathrm{d}M}{\mathrm{d}t} = L_\mathrm{V} 4\pi R_r \psi \left(\rho_{\mathrm{V},\infty} - \rho_{\mathrm{V},r}\right). \tag{5.71}$$

It must be remembered that this is a constant-pressure or isobaric process.

An expression of the diffusion of heat energy away from droplets during condensation is arrived at analogously to that for water vapor toward a drop. To see this, we first start with

$$\frac{1}{A}\frac{\mathrm{d}q}{\mathrm{d}t} = \rho K c_p \frac{\mathrm{d}T}{\mathrm{d}R}, \tag{5.72}$$

where q is heat, c_p is specific heat for an isobaric process, K is thermal diffusivity, and ρ is the density of air. From this, following the steps above,

$$\frac{\mathrm{d}q}{\mathrm{d}t} = 4\pi R^2 \rho K c_p \frac{\mathrm{d}T}{\mathrm{d}R}. \tag{5.73}$$

At this point, a discontinuity is assumed for the temperature field between the drop and its environment. To obtain the expression for the diffusion of heat away from the drop, rearrange (5.73) and integrate both sides,

$$\frac{\mathrm{d}q}{\mathrm{d}t} \int_{R_r}^{R_\infty} \frac{\mathrm{d}R}{R^2} = -4\pi \rho K c_p \int_{T_r}^{T_\infty} \mathrm{d}T, \tag{5.74}$$

where T_r is temperature at the drop's surface. Upon integration, the result is,

$$-\frac{\mathrm{d}q}{\mathrm{d}t}\left(\frac{1}{R_\infty}-\frac{1}{R_r}\right)=4\pi\rho K c_p(T_\infty - T_r). \tag{5.75}$$

Assuming $1/R_\infty$ is small,

$$\frac{\mathrm{d}q}{\mathrm{d}t}=4\pi R_r \rho K c_p(T_r - T_\infty). \tag{5.76}$$

Often thermal conductivity is used instead of thermal diffusivity. The relationship between these two is

$$\kappa = \rho K c_p, \tag{5.77}$$

which makes (5.76) equal to

$$\frac{\mathrm{d}q}{\mathrm{d}t}=4\pi R_r \kappa (T_r - T_\infty). \tag{5.78}$$

Now to balance the heat diffusion associated with enthalpy and that with temperature differences between the drop surface and at some distance from the drop (where influence of the drop is not felt) the following is written,

$$\frac{\mathrm{d}q}{\mathrm{d}t}=L_\mathrm{V}\frac{\mathrm{d}M}{\mathrm{d}t}. \tag{5.79}$$

Using (5.70) in (5.67) and (5.78) in (5.79), the following can be written,

$$4\pi R_r \kappa (T_r - T_\infty) = 4\pi R_r L_\mathrm{V} \psi (e_\infty - e_r)/R_\mathrm{V} T. \tag{5.80}$$

This can be simplified to be

$$\frac{(e_\infty - e_r)}{T(T_r - T_\infty)} = \frac{R_\mathrm{V}\kappa}{L_\mathrm{V}\psi}, \tag{5.81}$$

which at an equilibrium state is the same as the wet-bulb relationship.

Now substituting (5.70) into (5.81) we obtain the following,

$$R_r\frac{\mathrm{d}R_r}{\mathrm{d}t} = \frac{\kappa(T_r - T_\infty)}{\rho_\mathrm{L} L_\mathrm{V}}. \tag{5.82}$$

So, to summarize so far, when the air is saturated with respect to the droplet, vapor diffuses toward the droplet surface and heat diffuses away from the droplet surface. Now if the air is unsaturated with respect to vapor pressure over the droplet surface, then vapor diffuses away from the droplet surface and heat diffuses toward the droplet surface.

The goal is to represent both of these processes in one rate equation. Specifically, all terms are desired to be cast as observables.

Now the rate equation for diffusion without considering enthalpy is (5.70). Dividing this equation by $e_{s,\infty}$, which gives

$$\frac{(e_\infty - e_r)}{e_{s,\infty}} = \frac{1}{\psi}\frac{\rho_L R_v T}{e_{s,\infty}}\left(R_r \frac{dR_r}{dt}\right). \tag{5.83}$$

Now, following Mason (1957) and Byers (1965), the Clausius–Clapeyron equation is integrated for an ideal vapor between $e_{s,\infty}$ and e_r, and T_∞ to T_r, resulting in

$$\ln\left(\frac{e_{s,r}}{e_{s,\infty}}\right) = \frac{L_v}{R_v}\left(\frac{1}{T_\infty} - \frac{1}{T_r}\right) = \frac{L_v}{R_v T_r T_\infty}(T_r - T_\infty) = \frac{L_v}{R_v T_\infty^2}(T_r - T_\infty), \tag{5.84}$$

where for multiplication purposes, in the denominator, the temperatures at $R = \infty$ and $R = R_r$ can be assumed to be given by temperature at infinity when multiplied together, but not when differences are taken. Next (5.82) is rearranged,

$$(T_r - T_\infty) = \frac{\rho_L L_v}{\kappa}\left(R_r \frac{dR_r}{dt}\right), \tag{5.85}$$

and (5.85) is substituted into (5.84) so that

$$\ln\left(\frac{e_{s,r}}{e_{s,\infty}}\right) = \frac{L_v^2 \rho_L}{\kappa R_v T^2}\left(R_r \frac{dR_r}{dt}\right). \tag{5.86}$$

Now we want to replace the quantity in the natural logarithm in (5.86). Rewriting (5.83),

$$\frac{e_\infty}{e_{s,\infty}} = \frac{e_{s,r}}{e_{s,\infty}} + \frac{\rho_L R_v T}{\psi e_{s,\infty}}\left(R_r \frac{dR_r}{dt}\right). \tag{5.87}$$

The exponential of both sides of (5.86) is taken,

$$\left(\frac{e_{s,r}}{e_{s,\infty}}\right) = \exp\left(\frac{L_v^2 \rho_L}{\kappa R_v T^2} R_r \frac{dR_r}{dt}\right), \tag{5.88}$$

and substituting (5.88) into the first term on the right-hand side of (5.87) and letting $S_L = (e_\infty / e_{s,\infty})$ be the ambient saturation ratio,

$$S_L = \frac{e_\infty}{e_{s,\infty}} = \exp\left(\frac{L_v^2 \rho_L}{\kappa R_v T^2} R_r \frac{dR_r}{dt}\right) + \frac{\rho_L R_v T}{\psi e_{s,\infty}}\left(R_r \frac{dR_r}{dt}\right). \tag{5.89}$$

Now following Byers (1965) we let

$$x = R_r \frac{dR_r}{dt}, \tag{5.90}$$

so that (5.89) can now be written as

$$S_{\mathrm{L}} = \exp(a''_{\mathrm{L}}x) + b''_{\mathrm{L}}x, \tag{5.91}$$

where, $a''_{\mathrm{L}} = L_{\mathrm{v}}^2\rho_{\mathrm{L}}/(\kappa R_{\mathrm{v}}T^2)$, and $b''_{\mathrm{L}} = \rho_{\mathrm{L}}R_{\mathrm{v}}T/(\psi e_{\mathrm{s},\infty})$.

When $a''_{\mathrm{L}}x \ll 1$, (5.91) can be written approximately as, using the expansion of an exponential,

$$S_{\mathrm{L}} = 1 + a''_{\mathrm{L}}x + b''_{\mathrm{L}}x = 1 + (a''_{\mathrm{L}} + b''_{\mathrm{L}})x. \tag{5.92}$$

Using (5.90) and rearranging (5.92) gives a rate equation,

$$R_r\frac{\mathrm{d}R_r}{\mathrm{d}t} = \frac{(S_{\mathrm{L}}-1)}{a''_{\mathrm{L}} + b''_{\mathrm{L}}} = (S_{\mathrm{L}}-1)G_{\mathrm{L}}(T,P), \tag{5.93}$$

where the function $G_{\mathrm{L}}(T, P)$ is

$$G_{\mathrm{L}}(T,P) = \frac{1}{\dfrac{\rho_{\mathrm{L}}L_{\mathrm{v}}^2}{R_{\mathrm{v}}\kappa T^2} + \dfrac{\rho_{\mathrm{L}}R_{\mathrm{v}}T}{e_{\mathrm{s},\infty}\psi}}. \tag{5.94}$$

Now (5.66), (5.68) and (5.93) are used to write a mass change rate as

$$\frac{\mathrm{d}M}{\mathrm{d}t} = \rho_{\mathrm{L}}4\pi R_r(S_{\mathrm{L}}-1)G_{\mathrm{L}}(T,P) = \rho_{\mathrm{L}}2\pi D_r(S_{\mathrm{L}}-1)G_{\mathrm{L}}(T,P), \tag{5.95}$$

where D_r is the diameter of the drop in (5.95).

The vapor diffusion, mass growth equation also can be derived following Rogers and Yau (1989), using a linear function for vapor density put forth by Mason (1971). We start by writing the steady-state-diffusion mass rate equation as

$$\frac{\mathrm{d}M}{\mathrm{d}t} = 4\pi R_r\psi(\rho_{\mathrm{v},\infty} - \rho_{\mathrm{v},r}), \tag{5.96}$$

where $\rho_{\mathrm{v},\infty}$ is the vapor density at the ambient temperature and $\rho_{\mathrm{v},r}$ is the vapor density over the drop. The steady-state diffusion of heat toward a particle is given by

$$\frac{\mathrm{d}q}{\mathrm{d}t} = 4\pi R_r\kappa(T_r - T_\infty), \tag{5.97}$$

where T_r is the temperature at the drop's radius or surface, and T_∞ is the ambient temperature. From (5.96) and (5.97) a rate of change of temperature at the drop's surface can be given as

$$\frac{4}{3}\pi R_r^3\rho_{\mathrm{L}}c_{pw}\frac{\mathrm{d}T_r}{\mathrm{d}t} = L_{\mathrm{v}}\frac{\mathrm{d}M}{\mathrm{d}t} - \frac{\mathrm{d}q}{\mathrm{d}t}, \tag{5.98}$$

where c_{pw} is the specific heat of water at constant pressure.

But if a steady-state-diffusion process is assumed, $dT_r/dt = 0$, the result is the following balance relation between the temperature and density fields,

$$\frac{(\rho_{v,\infty} - \rho_{v,r})}{(T_r - T_\infty)} = \frac{\kappa}{L_v \psi}. \tag{5.99}$$

At almost all times, the values of T_r and $\rho_{v,r}$ are unknown. Incorporating the solute and surface-tension (curvature) terms and the equation of state for water vapor, it is found that

$$\rho_{v,r} = \frac{e'_{s,r}}{R_v T_r} = \left(1 + \frac{a}{R_r} - \frac{b}{R_r^3}\right) \frac{e_{s,r}}{R_v T_r}. \tag{5.100}$$

In (5.100) it is important to remember that $e_{s,r}$ is the equilibrium vapor pressure over a pure, plane surface. The value of $e_{s,r}$ at T_r then can be found by the Clausius–Clapeyron equation. Using (5.99) and (5.100), which are an implicit simultaneous system of equations, numerical iterative techniques can be used to find an exact solution to the mass growth rate of the vapor-diffusion equation.

Alternatively to the numerical method of solution, following Mason (1971) and, closely, Rogers and Yau (1989), in a field of saturated vapor, changes in vapor density can be related to temperature differences,

$$\frac{d\rho_v}{\rho_v} = \frac{L_v}{R_v} \frac{dT_\infty}{T_\infty^2} - \frac{dT_\infty}{T_\infty}. \tag{5.101}$$

Now integrating this equation from temperature T_r to T_∞ and, assuming T_∞/T_r is approximately unity, gives

$$\ln\left(\frac{\rho_{v,s,\infty}}{\rho_{v,s,r}}\right) = (T_\infty - T_r)\left(\frac{L_v}{R_v T_\infty T_r} - \frac{1}{T_r}\right), \tag{5.102}$$

where the "s" subscript denotes saturation vapor density.

As the ratio of vapor densities is near unity, an approximation can be made such that

$$\frac{\rho_{v,s,\infty} - \rho_{v,s,r}}{\rho_{v,s,r}} = \left(\frac{T_\infty - T_r}{T_\infty}\right)\left(\frac{L_v}{R_v T_\infty} - 1\right), \tag{5.103}$$

where the approximation that $T_\infty T_r = T_\infty^2$ is also used. Now using (5.97) and (5.98) and substituting for $(T_\infty - T_r)$ in (5.103) the following results,

$$\frac{\rho_{v,s,\infty} - \rho_{v,s,r}}{\rho_{v,s,r}} = \left(1 - \frac{L_v}{R_v T_\infty}\right)\left(\frac{L_v}{4\pi R_r \kappa T_\infty}\right)\frac{dM}{dt}; \tag{5.104}$$

and using

$$\frac{\rho_{\mathrm{v},\infty} - \rho_{\mathrm{v},r}}{\rho_{\mathrm{v},r}} = \left(\frac{1}{4\pi R_r \psi \rho_{\mathrm{v},r}} \right) \frac{\mathrm{d}M}{\mathrm{d}t}, \tag{5.105}$$

along with the result of subtracting (5.104) from (5.105), the following approximate relation can be found, after a bit of algebra, assuming that $\rho_{\mathrm{v},r} = \rho_{\mathrm{v,s},r}$,

$$R_r \frac{\mathrm{d}R_r}{\mathrm{d}t} = \frac{(S_\mathrm{L} - 1)}{\left(\frac{L_\mathrm{v}}{R_\mathrm{v}T} - 1 \right) \frac{L_\mathrm{v}\rho_\mathrm{L}}{\kappa T} + \frac{\rho_\mathrm{L} R_\mathrm{v} T}{\psi e_\mathrm{s}}}. \tag{5.106}$$

Note that compared to the rate equation derived using Byers' approximation (5.93) and (5.94), this equation has a correction term in the denominator, which is small compared to the other two terms, and can be retained or neglected. The mass growth rate approximation, which is needed to produce a parameterization, is given by

$$\frac{\mathrm{d}M}{\mathrm{d}t} = f_\mathrm{v} 2\pi D_r (S_\mathrm{L} - 1) G'_\mathrm{L}(T, P), \tag{5.107}$$

where D_r denotes diameter here, and $G'_\mathrm{L}(T,P)$ is similar to $G_\mathrm{L}(T, P)$ except $G'_\mathrm{L}(T,P)$ has the correction term,

$$G'(T,P) = \frac{1}{\left(\frac{L_\mathrm{v}}{R_\mathrm{v}T} - 1 \right) \frac{L_\mathrm{v}\rho_\mathrm{L}}{\kappa T} + \frac{\rho_\mathrm{L} R_\mathrm{v} T}{\psi e_\mathrm{s}}}. \tag{5.108}$$

5.5 Ventilation effects

In the derivation of a vapor diffusion equation for a liquid sphere, it was assumed that the drop was stationary, and that the vapor and thermal gradients around the sphere were symmetric. This is only accurate for a drop at rest. For falling drops, or drops moving relative to the flow, the vapor and temperature gradients around the drop are distorted with steeper gradients in front of a drop falling vertically relative to the flow and weaker gradients behind the drop. During condensation, energy is convected away from the drop more efficiently for a drop in motion relative to the flow, and the vapor supply is enhanced more efficiently than if the drop were stationary. During evaporation, energy is convected toward the drop more efficiently and vapor is removed away from the drop more efficiently, when the drop is in motion.

The influences of the flow relative to the drops for steady-state diffusion, are represented in the mass-growth and energy-flux equations by modifying them using empirical formulas to adjust the equations with coefficients. These coefficients in the empirical formulas are called the ventilation coefficients for heat, for the heat-flux equation, and for vapor, for the vapor-flux equation. These coefficients actually arise from including the influences of advection that were ignored in deriving the vapor-diffusion equation, e.g.

$$\nabla^2 \rho_v = \vec{u} \cdot \vec{\nabla} \rho_v. \tag{5.109}$$

The vapor-diffusion rate equation is modified by a vapor ventilation coefficient by computing the growth of a droplet at rest and the growth of a droplet in freefall (flow relative to the drop makes the drop appear to be in freefall); and taking the ratio of the two with the growth rate of the stationary drop in the denominator and the rate of the falling drop in the numerator. For the vapor mass equation, the following mass ventilation coefficient is computed using

$$f_v = \frac{dM/dt}{dM_0/dt} = \frac{\text{mass rate for a falling drop}}{\text{mass rate for a stationary drop}}. \tag{5.110}$$

The same is done for the heat-flux equation to arrive at a heat ventilation coefficient,

$$f_h = \frac{dq/dt}{dq_0/dt} = \frac{\text{heat flux for a falling drop}}{\text{heat flux for a stationary drop}}. \tag{5.111}$$

The ventilation coefficients are usually parameterized in terms of the Reynolds number, the Schmidt number, and the Prandtl number, which are all dimensionless numbers. The Reynolds number is the ratio of the inertial to viscous terms in the velocity equations. The Reynolds number thus is given as

$$N_{re} = \frac{U_\infty D}{\nu}, \tag{5.112}$$

where U_∞ is the terminal velocity of the drop, D is the characteristic diameter of the drop and ν is the kinematic viscosity of air. The Schmidt number is used in the equation for the vapor ventilation coefficient and is the ratio of kinematic viscosity to vapor diffusivity and can be written as

$$N_{sc} = \frac{\nu}{\psi}, \tag{5.113}$$

where ψ is vapor diffusivity. The Prandtl number is used in the equation for the heat ventilation,

$$N_{pr} = \frac{\nu}{K}, \tag{5.114}$$

where K is the thermal diffusivity for dry air.

From these non-dimensional numbers, the following can be derived for raindrops and similar-sized and larger-sized spheroidal ice particles (Pruppacher and Klett 1997). For vapor ventilation equations, there are the following two conditions,

$$\begin{cases} f_{\mathrm{v}} = 1 + 0.108\left(N_{\mathrm{sc}}^{1/3}N_{\mathrm{re}}^{1/2}\right)^2 & \text{for } N_{\mathrm{sc}}^{1/3}N_{\mathrm{re}}^{1/2} < 1.4 \\ f_{\mathrm{v}} = 0.78 + 0.308N_{\mathrm{sc}}^{1/3}N_{\mathrm{re}}^{1/2} & \text{for } N_{\mathrm{sc}}^{1/3}N_{\mathrm{re}}^{1/2} > 1.4. \end{cases} \tag{5.115}$$

For heat ventilation equations, there are also the following two conditions,

$$\begin{cases} f_{\mathrm{h}} = 1 + 0.108\left(N_{\mathrm{pr}}^{1/3}N_{\mathrm{re}}^{1/2}\right)^2 & \text{for } N_{\mathrm{pr}}^{1/3}N_{\mathrm{re}}^{1/2} < 1.4 \\ f_{\mathrm{h}} = 0.78 + 0.308N_{\mathrm{pr}}^{1/3}N_{\mathrm{re}}^{1/2} & \text{for } N_{\mathrm{pr}}^{1/3}N_{\mathrm{re}}^{1/2} > 1.4. \end{cases} \tag{5.116}$$

When solving for the diffusion-growth equation (5.95) or (5.107), it is assumed that

$$f_{\mathrm{v}} \approx f_{\mathrm{h}}, \tag{5.117}$$

which is a reasonably good first guess.

With the ventilation coefficient, the mass growth equations (5.95) or (5.107), respectively, are written as,

$$\frac{\mathrm{d}M}{\mathrm{d}t} = \rho_{\mathrm{L}} 2\pi D_r (S_{\mathrm{L}} - 1) G_{\mathrm{L}}(T, P) f_{\mathrm{v}}, \tag{5.118}$$

or

$$\frac{\mathrm{d}M}{\mathrm{d}t} = \rho_{\mathrm{L}} 2\pi D_r (S_{\mathrm{L}} - 1) G'_{L}(T, P) f_{\mathrm{v}}. \tag{5.119}$$

5.6 Curvature effects on vapor diffusion and Kelvin's law

In this section, curvature effects are considered. Embryonic droplets nucleated on very small aerosols or nucleated homogeneously are small enough that curvature effects related to the radii of the droplet, and surface tension, which is a function of temperature, must be included. Forces that bind water molecules together [O(10–1000) molecules] of newly nucleated embryonic droplets require higher vapor pressure and thus greater supersaturation to grow (supersaturation $S > 1.5$ to 5); they also lose water molecules more easily than droplets with larger radii owing to weaker net forces of hydrogen bonds holding the water embryo together. In addition, the higher the surface tension the more easily molecules may desorb from droplets. Thus, it is found that the equilibrium saturation vapor pressure over a very small drop is much larger than that for bigger droplets for a given surface tension. Droplets smaller than a critical radius will evaporate or break apart owing to collisions by other water molecules or clusters of molecules, whilst larger droplets than this critical radius will continue to grow as the internal forces holding the droplet together are strong enough to withstand collisions with clusters of other water molecules.

To account for curvature effects, the surface tension of the particle must be considered. The rate of growth depends on partial pressure of vapor in the ambient, which determines the impact rate of vapor molecules on the droplets. The decay rate is controlled by surface tension and to a greater degree by the temperature. This is because embryonic droplets that make up these droplets must contain enough binding energy through hydrogen bonds to withstand breakup by thermal agitation against the surface tension at the droplet's surface. The surface tension is the free energy per unit area of some substance, such as water, and is also defined as the work per unit area required to extend the surface of the droplet. Work per unit area is force times distance per unit area with units of J m^{-1}. The surface tension of droplets, as given between a water–air interface, $\sigma_{\mathrm{L/a}}$, is temperature dependent, and decreasing with increasing temperature according to

$$\sigma_{\mathrm{L/a}} = 0.0761 - 0.00155(T - 273.15)\,\mathrm{N\,m^{-1}}. \tag{5.120}$$

Other values of surface tension for solutes can be found in Pruppacher and Klett (1997). Molecules can leave the surface of smaller droplets much more easily than from larger droplets with smaller droplets thus requiring larger equilibrium supersaturations with respect to planar surfaces. When a droplet is in equilibrium with its environment, it is losing as many water molecules as it is gaining. Critical-sized droplets are formed by random collisions of vapor molecules, and if they become supercritical they will continue to grow spontaneously. If they do not reach a critical size, then they fall apart through collisions by other vapor molecules (clusters) as the free energy needed from excess water vapor in the environment is not sufficient to expand the surface of the droplets. The equilibrium vapor pressure for this state is given by,

$$e_{\mathrm{s},r} = e_{\mathrm{s},\infty} \exp\left(\frac{2\sigma_{\mathrm{L/a}}}{RR_{\mathrm{v}}\rho_{\mathrm{L}}T_r}\right), \tag{5.121}$$

where $e_{\mathrm{s},r}$ is the saturation vapor pressure over a spherical droplet with radius r, temperature T_r, density ρ_{L} and surface tension $\sigma_{\mathrm{L/a}}$. The variable $e_{\mathrm{s},\infty}$ is the saturation vapor pressure over a bulk planar surface of infinite length and width. As the droplet size decreases, the supersaturation or vapor pressure for equilibrium increases and vice versa. From (5.121) a critical radius r_{critical} for equilibrium can be derived for any supersaturation. With $S_t = e_{s,r}/e_{s,\infty}$, the critical radius is given by Kelvin's law, which includes curvature effects,

$$r_{\text{critical}} = \left(\frac{2\sigma_{L/a}}{R_v \rho_L T_r \ln S_L}\right). \tag{5.122}$$

When r_{critical} is exceeded, then the droplet will grow, and when r is less than r_{critical}, the droplet will evaporate as molecules desorb from the surface.

5.7 Solute effects on vapor diffusion and Raoult's law

The results in previous subsections of this chapter were obtained with the assumption that planar surfaces were free of solutes, which are substances dissolved in water.

The presence of solutes generally lowers the equilibrium vapor pressure over a droplet. The equilibrium vapor pressure occurs in part as a result of water molecules at the surface of the drop being replaced by molecules of the solute. For planar water surfaces, we start with the following equation to estimate the reduction in equilibrium vapor pressure for dilute solutions,

$$\frac{e'_r}{e_{s,r}} = \frac{n_0}{n_0 + n}, \tag{5.123}$$

where e'_r is the equilibrium vapor pressure over a solution with n_0 molecules of water and n molecules of a solute (Rogers and Yau 1989). For $n \ll n_0$, or for dilute solutions, (5.123) reduces to the following,

$$\frac{e'_r}{e_{s,\infty}} = 1 - \frac{n}{n_0}. \tag{5.124}$$

Some solutions break up into ionic components by ionic dissociation. The effect is that there are more ions than molecules in the solute. As a result, n in (5.124) needs to be modified by i, the poorly understood van't Hoff factor, the number of effective ions, which approximates ionic availability. For sodium chloride, for example, i is approximately 2 (Low 1969; Rogers and Yau 1989). For ammonium sulfate, i is approximately 3. For solutes that do not dissociate, $i = 1$. So the more a substance dissociates, the lower the equilibrium vapor pressure for a given droplet size. With i, using Avogadro's number N_0 (molecules per mole), and solute of mass M and molecular weight m_s, n becomes

$$n = \frac{iN_0 M}{m_s}, \tag{5.125}$$

and with water spheres with molecular weight m_w, and mass M_w, the number of water molecules n_0 can be expressed as

$$n_0 = \frac{N_0 M_w}{m_w}. \tag{5.126}$$

Now a factor b' can be written where,

$$b' = \frac{3 i m_v M}{4\pi \rho_L m_s}. \tag{5.127}$$

An expression can also be written for solute effects using (5.124)–(5.127) as

$$\frac{e'_r}{e_{s,\infty}} = 1 - \frac{b'}{R_r^3}, \tag{5.128}$$

which is Raoult's law.

5.8 Combined curvature and solute effects and the Kohler curves

Both the curvature and solute effects can be combined as the following families of curves called the Kohler curves,

$$\frac{e'_{s,r}}{e_{s,\infty}} = \left(1 - \frac{b'}{R_r^3}\right) \exp\left(\frac{a'}{R_r}\right), \tag{5.129}$$

with a' given as

$$a' = \frac{2\sigma_{L/a}}{\rho_L R_v T}. \tag{5.130}$$

Using the series approximation for an exponential and neglecting higher-order terms, the expression given by (5.129) can be reduced to

$$\frac{e'_{s,r}}{e_{s,\infty}} = \left(1 + \frac{a'}{R_r} - \frac{b'}{R_r^3}\right). \tag{5.131}$$

The curves of this equation show that at small radii, solute effects dominate, sometimes so much so that equilibrium occurs at $S < 1$. As the radius increases, new equilibrium values will be established. This continues until the critical radius is reached, if it ever is. The critical radius occurs at

$$r_{\text{critical}} = \sqrt{\frac{3b'}{a'}}. \tag{5.132}$$

This value is achieved by taking the derivative of (5.131), setting it equal to zero, and solving for $R_r = r_{\text{critical}}$. Once the critical radius is reached, the smallest amount of increase in S will lead to unstable growth of the droplet solute. This critical saturation S_{critical} value also can be written as follows,

$$S_{\mathrm{L}_{\text{critical}}} = 1 + \sqrt{\frac{4a'^3}{27b'}}. \tag{5.133}$$

The vapor-diffusion equations with saturation vapor equations adjusted for curvature and solute effects for liquid water are now given. The vapor-diffusion equations (5.118) and (5.119) including the Kohler-curve effects and ventilation effects are

$$\frac{\mathrm{d}M}{\mathrm{d}t} = \rho_{\mathrm{L}} 2\pi D_r \left[(S_{\mathrm{L}} - 1) - \frac{a'}{R_r} + \frac{b'}{R_r^3} \right] G_{\mathrm{L}}(T,P) f_{\mathrm{v}}, \tag{5.134}$$

or, after Mason (1971) and Rogers and Yau (1989),

$$\frac{\mathrm{d}M}{\mathrm{d}t} = \rho_{\mathrm{L}} 2\pi D_r \left[(S_{\mathrm{L}} - 1) - \frac{a'}{R_r} + \frac{b'}{R_r^3} \right] G'_{\mathrm{L}}(T,P) f_{\mathrm{v}}. \tag{5.135}$$

5.9 Kinetic effects

Only the very basics of kinetic effects are covered here. A more complete discussion would require a more formal analysis in the physics and energetics of condensation and evaporation (Pruppacher and Klett 1997; Young 1993). Heat, mass, and momentum transfer between small aerosols, droplets, or small pristine crystals, and molecules of water vapor in the environment, depend on the Knudsen number, N_{kn}; this was defined earlier as the ratio of the mean free path, λ, in the atmosphere, which is about 6×10^{-8} m at sea-level pressure and temperature, to the radius of the particle. The value of λ is the distance a molecule will travel on average before colliding with another molecule and exchanging momentum. It also can be written as being dependent upon temperature T and inversely on atmospheric pressure P. For very small particles such as small aerosols, where, $N_{\mathrm{kn}} \gg 1$, the theory of molecular collisions holds; for larger droplets, when the $N_{\mathrm{kn}} \ll 1$ and the continuum approximations of Maxwell hold, the heat- and mass-transfer equations developed earlier for diffusional growth are valid. At the size at which water droplets are just nucleated, typically radii of 0.1 to 1.0 micron, N_{kn} is O(1) for which both previously mentioned theories break down. As a result, approximate relationships were developed to account for kinetic effects for very small droplets to be incorporated into the diffusional growth equations.

For mass growth, a factor called the condensation coefficient, β, for vapor was found experimentally (see Pruppacher and Klett 1997). With a water drop in an air and vapor mixture interface environment, a fraction of impinging molecules actually are incorporated into the drop (condensed). The condensation coefficient is approximately 0.01 to 0.07 with an average of 0.026 to 0.035; a list of experimental results for this coefficient can be found in Pruppacher and Klett (1997). A length scale λ_β as a function of β (Rogers and Yau 1989) was found experimentally to be

$$\lambda_\beta = \frac{\psi}{\beta}\left(\frac{2\pi}{R_\mathrm{v}T}\right)^{1/2}. \tag{5.136}$$

This normalization factor can be used to include kinetic effects of vapor within a distance approximated by the mean free path of air from the droplet,

$$g(\beta) = \frac{R_r}{R_r + \lambda_\beta}, \tag{5.137}$$

so that the mass-transfer equation including kinetic effects, which holds for very small drops is given by

$$\frac{\mathrm{d}M}{\mathrm{d}t} = 4\pi R_r \psi g(\beta)\left(\rho_\mathrm{v} - \rho_{\mathrm{v},r}\right). \tag{5.138}$$

Notice that this equation reverts back to (5.96) when R_r becomes large compared to λ_β. Similarly an accommodation coefficient, α, was developed and described by Rogers and Yau (1989) and Pruppacher and Klett (1997) as approximating the fraction of molecules bouncing off the drop and acquiring the temperature of the drop. The definition of α is given below, and is found to be approximately unity (or near 0.96),

$$\alpha = \frac{T_2' - T_1}{T_2 - T_1}, \tag{5.139}$$

where T_2' is approximately the temperature of vapor molecules leaving the surface of the liquid, T_2 is the temperature of the liquid, and T_1 is the temperature of the vapor. Similar to the mass-growth rate equation, the heat-transfer rate can be written in a form such that in the limit when larger droplets exist the coefficient converges to 1, and the Maxwell continuum theory prevails. The factor is defined as

$$f(\alpha) = \frac{R_r}{R_r + \lambda_\alpha}, \tag{5.140}$$

and λ_α is given by

$$\lambda_\alpha = \left(\frac{\kappa}{\alpha p}\right)\frac{(2\pi R_\mathrm{d}T)^{1/2}}{(c_\mathrm{v} + R_\mathrm{d}/2)}. \tag{5.141}$$

In general, most bulk parameterizations do not include kinetic effects; these are only important for very small droplets, which grow within a very short period of time, on O(10) s, by diffusional growth, to sizes where the kinetic effects are very small. If kinetic effects are included, the $G_L(T, P)$ and $G'_L(T, P)$ are given as,

$$G_L(T,P) = \frac{1}{\dfrac{\rho_L L_v^2}{R_v \kappa T_\infty^2 f(\alpha)} + \dfrac{\rho_L R_v T_\infty}{e_s \psi g(\beta)}} = \frac{1}{\dfrac{\rho_L L_v^2}{R_v \kappa T_\infty^2 f(\alpha)} + \dfrac{\rho_L}{\rho Q_{SL} \psi g(\beta)}} \tag{5.142}$$

$$G'_L(T,P) = \frac{1}{\left(\dfrac{L_v}{R_v T} - 1\right)\dfrac{L_v \rho_L}{\kappa T f(\alpha)} + \dfrac{\rho_L R_v T}{\psi e_s g(\beta)}} = \frac{1}{\left(\dfrac{L_v}{R_v T} - 1\right)\dfrac{\rho_L L_v}{\kappa T f(\alpha)} + \dfrac{\rho_L R_v T}{\psi e_s g(\beta)}}. \tag{5.143}$$

Normally $G_L(T, P)$ and $G'_L(T, P)$ are multiplied through by ρ_L eliminating it in the denominator.

5.10 Higher-order approximations to the mass tendency equation

Higher than first-order, linear-function approximations for saturation vapor density were first used by Srivastava and Coen (1992) under the assumption of steady-state diffusion conditions, where the rate increase of mass M for a particle of radius r is

$$\frac{dM}{dt} = 4\pi R_r \psi f_v \left(\rho_{v,\infty} - \rho_{v,r}\right). \tag{5.144}$$

In (5.144), $\rho_{v,\infty}$ is the ambient vapor density and $\rho_{v,r}$ is the vapor density at the particle surface. With steady-state conditions assumed, the rate at which energy is released or absorbed by condensation/deposition or evaporation/sublimation is equal to the rate at which it is conducted away from/toward a particle, which is written as

$$L_v \frac{dM}{dt} = 4\pi R_r \kappa f_h (T_r - T_\infty) = 4\pi R_r \kappa f_h \Delta T, \tag{5.145}$$

where f_h is the ventilation for heat.

Similarly to the solution method presented previously, to solve (5.144) and (5.145), ψ and κ need to be evaluated at appropriate temperatures and pressures; for simplicity this has been done at ambient temperatures (Srivastava and Coen (1992) and others). Neglecting kinetic effects results in minor shortcomings compared to the improvement of using the higher-order approximation approach to reduce the error over the traditional linear-difference approximations. Heat storage and radiative effects are ignored here as well

(Srivastava and Coen 1992). Some debate exists in the literature concerning the use of simple expressions for thermodynamic functions such as K, ψ, L_v, etc., but Srivastava and Coen (1992) showed these errors to be rather insignificant compared to the improvements made by their higher-order approximations.

As given in Srivastava and Coen (1992), the vapor density at the surface of a particle is

$$\rho_r = \rho_{s,r}[1 + s_r], \tag{5.146}$$

which can be written for s_r in terms of the vapor density ratio minus one, i.e.

$$s_r = \left[\frac{\rho_r}{\rho_{s,r}} - 1\right] \tag{5.147}$$

where s_r is the equilibrium supersaturation over a particle with surface tension and solute effects included. With iterative numerical techniques the above three implicit equations (5.144), (5.145), and (5.146) can be solved for an exact solution. This is an unattractive approach for numerical models, so modelers usually make vapor-density difference a linear function of temperature following Fletcher (1962), Mason (1971), Pruppacher and Klett (1978, 1997), and Rogers and Yau (1993),

$$\rho_{s,r} \approx \rho_{s,\infty} + \frac{\partial\left(\rho_{s,\infty}\right)}{\partial T}\Delta T, \tag{5.148}$$

in a similar manner to the method presented above.

Using (5.146) and (5.148), the following can be solved for temperature difference,

$$(\Delta T)_1 = \frac{\rho_s}{\dfrac{\partial\left(\rho_{s,\infty}\right)}{\partial T}}\frac{\gamma}{1+\gamma}(s - s_r). \tag{5.149}$$

The mass-growth rate (5.144) becomes

$$\left(\frac{dM}{dt}\right)_1 = \frac{4\pi R_r \psi f_v \rho_{s,\infty}}{1+\gamma}(s - s_r), \tag{5.150}$$

where the following dimensionless parameter γ is defined as

$$\gamma \equiv \frac{L_v \psi}{\kappa}\frac{f_v}{f_h}\frac{\partial\left(\rho_{s,\infty}\right)}{\partial T}. \tag{5.151}$$

In the equations above, the variable "s" is the ambient supersaturation given as

$$s = \left[\frac{\rho_\infty}{\rho_{s,\infty}} - 1\right], \tag{5.152}$$

or

$$\rho_\infty = \rho_{s,\infty}(1 + s). \tag{5.153}$$

In the case of (5.150), "1" denotes a linear function of temperature difference and is used in approximating the saturation vapor density at a drop's surface. The accuracy of (5.150) depends on the accuracy of (5.148). Srivastava and Coen (1992) show that for very warm and dry conditions, such as those with dry microbursts, tremendous temperature deficits can occur between particle and the air around the droplet (Fig. 5.1); the linear function approximation (5.150) will begin to fail, and will result in even worse errors from the exact iterative solution (5.144)–(5.146) where $s_r = 0$. The curves in Fig. 5.1 are for the evaporation of raindrops for two ambient pressures. Equation (5.150) always results in an underestimate of evaporation rate.

To relax this error and obtain a better solution, an additional term is added to the saturation vapor-density function approximation (5.148) to make it second order,

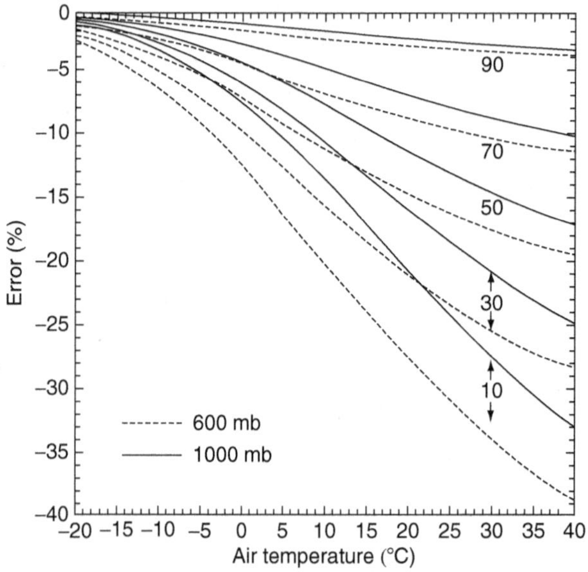

Fig. 5.1. Error in the rate of evaporation of raindrops from the exact solution as a function of air temperature for selected relative humidities using the traditional equations (solid lines 1000 mb; dashed lines 600 mb). (From Srivastava and Coen 1992; courtesy of the American Meterological Society.)

$$\rho_{s,r} \approx \rho_{s,\infty} + \frac{\partial(\rho_{s,\infty})}{\partial T}\Delta T + \frac{1}{2!}\frac{\partial^2(\rho_{s,\infty})}{\partial T^2}(\Delta T)^2 + \ldots \tag{5.154}$$

Now, as described by Srivastava and Coen (1992) the $\mathrm{d}M/\mathrm{d}t$ is eliminated between (5.144) and (5.145) and (5.146) and (5.154) are used to obtain the following quadratic for the temperature difference,

$$\frac{1}{2\rho_s}\frac{\partial^2(\rho_{s,\infty})}{\partial T^2}(\Delta T)^2 + \frac{1}{\rho_s}\left(\frac{1+\gamma}{\gamma}\right)\frac{\partial(\rho_{s,\infty})}{\partial T} - (s - s_r) = 0. \tag{5.155}$$

The solution for the temperature difference is given after some algebra as

$$\Delta T = \frac{\frac{\partial(\rho_{s,\infty})}{\partial T}}{\frac{\partial^2(\rho_{s,\infty})}{\partial T^2}}\left(\frac{1+\gamma}{\gamma}\right)\left\{-1 \pm \left[1 + \left(\frac{\gamma}{1+\gamma}\right)^2 \frac{\frac{\partial^2(\rho_{s,\infty})}{\partial T^2}}{\frac{\partial(\rho_{s,\infty})}{\partial T}}\frac{(\rho_{s,\infty})}{\frac{\partial(\rho_{s,\infty})}{\partial T}}(s - s_r)\right]^{1/2}\right\}. \tag{5.156}$$

Now, with a lot of algebra, and expanding the square root to terms of the order of $(s - s_r)^2$, the second-order temperature difference equation is

$$\begin{aligned}(\Delta T)_2 &= \frac{\rho_s}{\frac{\partial(\rho_{s,\infty})}{\partial T}}\left(\frac{\gamma}{1+\gamma}\right)(s - s_r)[1 - \alpha(s - s_r)] \\ &= (\Delta T)_1[1 - \alpha(s - s_r)];\end{aligned} \tag{5.157}$$

and the second-order mass rate equation is

$$\begin{aligned}\left(\frac{\mathrm{d}M}{\mathrm{d}t}\right)_2 &= \frac{4\pi R_r \psi f_v(\rho_{s,\infty})}{1+\gamma}(s - s_r)[1 - \alpha(s - s_r)] \\ &= \left(\frac{\mathrm{d}M}{\mathrm{d}t}\right)_1[1 - \alpha(s - s_r)].\end{aligned} \tag{5.158}$$

In (5.157) and (5.158), α is given as

$$\alpha \equiv \frac{1}{2}\left(\frac{\gamma}{1+\gamma}\right)^2 \frac{\frac{\partial^2(\rho_{s,\infty})}{\partial T^2}}{\frac{\partial(\rho_{s,\infty})}{\partial T}}\frac{(\rho_{s,\infty})}{\frac{\partial(\rho_{s,\infty})}{\partial T}}. \tag{5.159}$$

Even higher-order equations can be found that are more accurate than $(\mathrm{d}M/\mathrm{d}t)_2$ given in (5.158) by expanding the square root to terms of $(s - s_r)^3$ and $(s - s_r)^4$. These are given by Srivastava and Coen (1992) as

$$\left(\frac{\mathrm{d}M}{\mathrm{d}t}\right)_3 = \left(\frac{\mathrm{d}M}{\mathrm{d}t}\right)_1\left[1 - \alpha(s - s_r) + 2\alpha^2(s - s_r)^2\right] \tag{5.160}$$

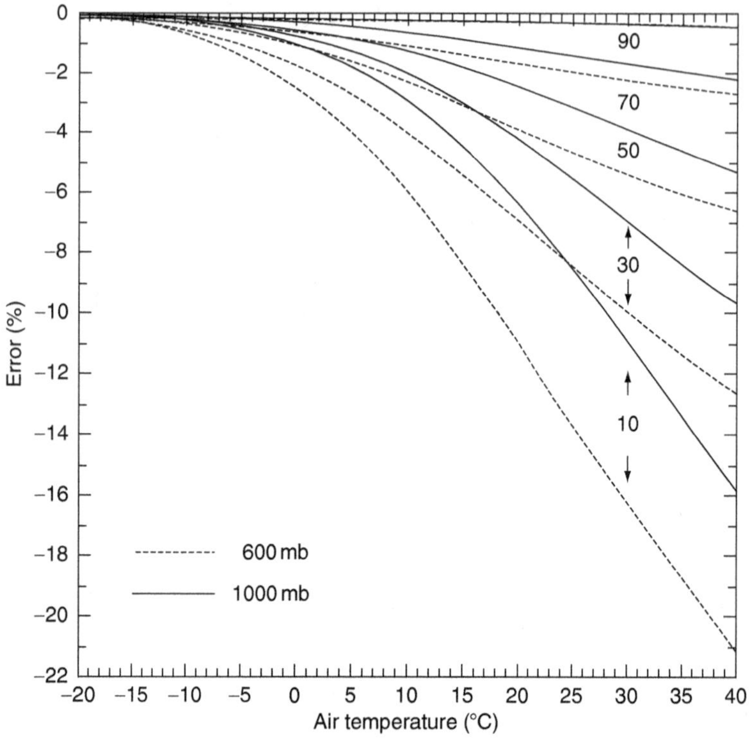

Fig. 5.2. Error in the rate of evaporation of raindrops from the exact solution as a function of air temperature for selected relative humidities using the second-order equations (solid lines 1000 mb; dashed lines 600 mb). (From Srivastava and Coen 1992; courtesy of the American Meteorological Society.)

and

$$\left(\frac{\mathrm{d}M}{\mathrm{d}t}\right)_4 = \left(\frac{\mathrm{d}M}{\mathrm{d}t}\right)_1 \left[1 - \alpha(s - s_r) + 2\alpha^2(s - s_r)^2 - 5\alpha^3(s - s_r)^3\right], \tag{5.161}$$

for third- and fourth-order expansions. Users of these higher-order expansions should refer to Srivastava and Coen (1992, pp. 1645–1646) for further information about solutions to the positive root versus the negative root in (5.156). Under certain conditions (very warm and very dry) the argument of the square root may be negative.

Figure 5.2 shows the temperature difference between equation (5.158) and the exact solution (5.144)–(5.146) as a function of air temperature for certain relative humidities and two pressures, 1000 mb and 600 mb. Here the percent errors are generally much lower than the case in Fig. 5.1 (notice the change in scale in the ordinate). Again in extremely warm and dry conditions, errors become large and unacceptable.

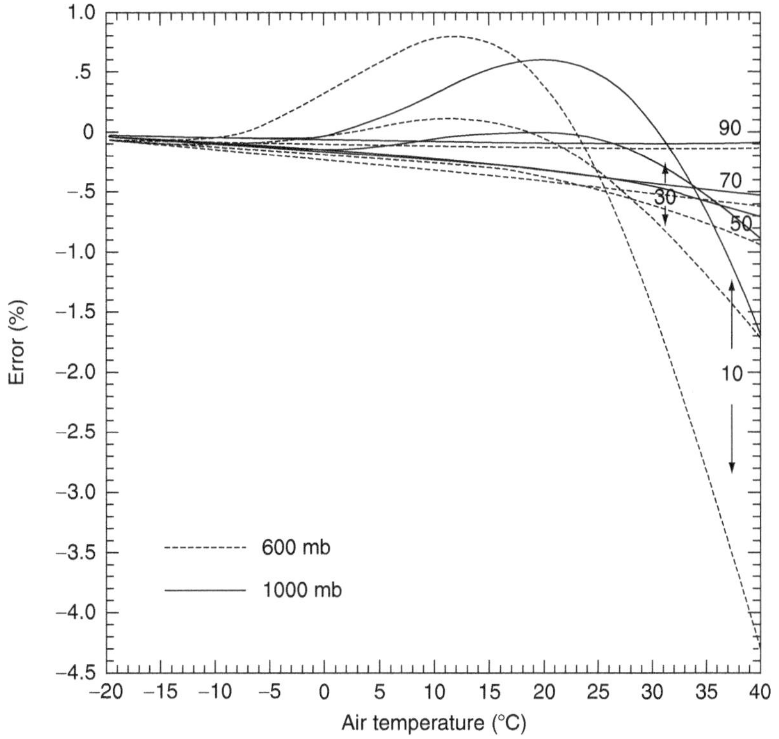

Fig. 5.3. Error in the rate of evaporation of raindrops from the exact solution as a function of air temperature for selected relative humidities using the third-order equations (solid lines 1000 mb; dashed lines 600 mb). (From Srivastava and Coen 1992; courtesy of the American Meteorological Society.)

The use of equation (5.160) results in the errors shown in Fig. 5.3 (note that fourth order doesn't gain much accuracy over the third order). In this case, clearly the use of higher-order equations has a significant impact on the percent errors. Even under the most extreme conditions, errors are 5% or less. Therefore, the use of equation (5.160) over the equation (5.150) is indicated since it is only slightly more complex and has significantly more accuracy.

5.11 Parameterizations

5.11.1 Gamma distribution

For the vapor diffusion of liquid water particles, the basic equation that is solved is

$$\frac{\mathrm{d}M}{\mathrm{d}t} = \frac{4\pi D_r (S_\mathrm{L} - 1) f_\mathrm{v}}{\left[\frac{L_\mathrm{v}^2}{\kappa_r R_\mathrm{v} T^2} + \frac{1}{\rho \psi Q_\mathrm{SL}}\right]}, \tag{5.162}$$

where Q_{SL} is the saturation ratio with respect to the liquid. To parameterize this equation for say, a modified gamma distribution (2.26), the mass change equation owing to vapor diffusion is written as

$$\frac{1}{\rho}\int_0^\infty \frac{dM(D_x)n(D_x)}{dt}dD_x = \frac{1}{\rho}\int_0^\infty 2\pi D_x(S_L-1)G_L(T,P) \times \left(0.78+0.308N_{sc}^{1/3}N_{re}^{1/2}\right)n(D_x)dD_x, \tag{5.163}$$

where D_x is the diameter of some hydrometeor species x. Substitution of (2.25), the complete gamma distribution for $n(D_x)$ into (5.163) gives

$$Q_xCE_v = \frac{1}{\rho_0}\int_0^\infty 2\pi D_x(S_L-1)G_L(T,P) \times \left[0.78+0.308N_{sc}^{1/3}\left(\frac{D_xV_{Tx}}{\nu_x}\right)^{1/2}\left(\frac{\rho_0}{\rho}\right)^{1/4}\right] \times \left[\frac{N_{Tx}\mu_x\alpha_x^{\nu_x}}{\Gamma(\nu_x)}\left(\frac{D_x}{D_{nx}}\right)^{\nu_x\mu_x-1}\exp\left(-\alpha_x\left[\frac{D_x}{D_{nx}}\right]^{\mu_x}\right)\right]d\left(\frac{D_x}{D_{nx}}\right). \tag{5.164}$$

Integrating gives the desired generalized gamma distribution parameterization equation for vapor-diffusion growth, assuming that terminal velocity is given by the following power law,

$$V_T(D_x) = c_xD^{d_x}\left(\frac{\rho_0}{\rho}\right)^{1/2}. \tag{5.165}$$

The complete gamma distribution solution is given by

$$Q_xCE_v = \frac{1}{\rho_0}2\pi(S_L-1)G_L(T,P)\frac{N_{Tx}\alpha_x^{\nu_x}}{\Gamma(\nu_x)} \times \left[0.78\frac{\Gamma\left(\frac{1+\nu_x\mu_x}{\mu_x}\right)}{\alpha_x^{\left(\frac{1+\nu_x\mu_x}{\mu_x}\right)}}D_{nx}+0.308\Gamma\frac{\left(\frac{3+d_x}{2\mu_x}+\frac{\mu_x\nu_x}{\mu_x}\right)}{\alpha_x^{\left(\frac{3+d_x}{2\mu_x}+\frac{\mu_x\nu_x}{\mu_x}\right)}}N_{sc}^{1/3}\nu_x^{-1/2}c_x^{1/2}D_{nx}^{\frac{3+d_x}{2}}\left(\frac{\rho_0}{\rho}\right)^{1/4}\right], \tag{5.166}$$

whilst the modified gamma distribution solution is

$$Q_xCE_v = \frac{1}{\rho_0}2\pi(S_L-1)G_L(T,P)\frac{N_{Tx}}{\Gamma(\nu_x)} \times \left[0.78\Gamma\left(\frac{1+\nu_x\mu_x}{\mu_x}\right)D_{nx}+0.308\Gamma\left(\frac{3+d_x}{2\mu_x}+\frac{\mu_x\nu_x}{\mu_x}\right)N_{sc}^{1/3}\nu_x^{-1/2}c_x^{1/2}D_{nx}^{\frac{3+d_x}{2}}\left(\frac{\rho_0}{\rho}\right)^{1/4}\right], \tag{5.167}$$

and the gamma distribution solution is

$$Q_xCE_v = \frac{1}{\rho}2\pi(S_L - 1)G_L(T,P)\frac{N_{Tx}}{\Gamma(v_x)} \times \left[0.78\Gamma(1+v_x)D_{nx} + 0.308\Gamma\left(\frac{3+d_x}{2}+v_x\right)N_{sc}^{1/3}v_x^{-1/2}c_x^{1/2}D_{nx}^{\frac{3+d_x}{2}}\left(\frac{\rho_0}{\rho}\right)^{1/4}\right], \tag{5.168}$$

where CE_v is the condensation or evaporation of vapor. After a cloud droplet is nucleated, it grows by vapor diffusion until it is large enough to grow mostly by coalescence. For cloud drops with diameters less than 120 microns, the following parameterized diffusion growth equations can be used (Pruppacher and Klett 1997) assuming a ventilation coefficient of

$$f_v = 1 + 0.108\left(N_{sc}^{1/3}N_{re}^{1/2}\right)^2. \tag{5.169}$$

Thus, the complete gamma distribution solution is

$$Q_xCE_v = \frac{1}{\rho}2\pi(S_L - 1)G_L(T,P)\frac{N_{Tx}\alpha_x^{v_x}}{\Gamma(v_x)} \times \left[1.0\frac{\Gamma\left(\frac{1+v_x\mu_x}{\mu_x}\right)}{\alpha_x^{\left(\frac{1+v_x\mu_x}{\mu_x}\right)}}D_{nx} + 0.108\frac{\Gamma\left(\frac{3+d_x}{\mu_x}+\frac{v_x\mu_x}{\mu_x}\right)}{\alpha_x^{\left(\frac{3+d_x}{\mu_x}+\frac{v_x\mu_x}{\mu_x}\right)}}N_{sc}^{2/3}v_x^{-1}c_xD_{nx}^{3+d_x}\left(\frac{\rho_0}{\rho}\right)^{1/2}\right], \tag{5.170}$$

whilst the modified gamma distribution solution is given by,

$$Q_xCE_v = \frac{1}{\rho}2\pi(S_L - 1)G_L(T,P)\frac{N_{Tx}}{\Gamma(v_x)} \times \left[1.0\Gamma\left(\frac{1+v_x\mu_x}{\mu_x}\right)D_{nx} + 0.108\Gamma\left(\frac{3+d_x}{\mu_x}+\frac{v_x\mu_x}{\mu_x}\right)N_{sc}^{2/3}v_x^{-1}c_xD_{nx}^{3+d_x}\left(\frac{\rho_0}{\rho}\right)^{1/2}\right], \tag{5.171}$$

and the gamma distribution solution is,

$$Q_xCE_v = \frac{1}{\rho}2\pi(S_L - 1)G_L(T,P)\frac{N_{Tx}}{\Gamma(v_x)} \times \left[1.0\Gamma(1+v_x)D_{nx} + 0.108\Gamma(3+d_x+v_x)N_{sc}^{2/3}v_x^{-1}c_xD_{nx}^{3+d_x}\left(\frac{\rho_0}{\rho}\right)^{1/2}\right]. \tag{5.172}$$

The change in number concentration during evaporation is a complicated issue in many regards. There is no change in number concentration during condensation. For simplicity, many assume that the number concentration change is related to the mixing ratio change as follows,

$$N_{Tx}CE_v = \frac{N_{Tx}}{Q_x}(Q_x CE_v), \tag{5.173}$$

where $(Q_x CE)$ is found from the above equations.

There are problems with this formulation in that it really does not capture the nature of the number of droplets that evaporate. An alternative is presented below. Starting with the equation for rate of change of radius, the ventilation coefficient is set to one, and it is assumed that only the very smallest drops are fully evaporating,

$$R_r \frac{dR_r}{dt} = (S_L - 1)G_L(T, P). \tag{5.174}$$

From (5.174), it can be written that

$$\int_{R_{max}}^{0} R_r \frac{dR_r}{dt} dt = \int_{t=0}^{t=\Delta t} (S_L - 1)G_L(T, P)dt. \tag{5.175}$$

Now R_{max} is the largest remaining drop after t seconds of evaporation and is

$$\frac{R_{max}^2}{2} = -(S_L - 1)G_L(T, P)\Delta t, \tag{5.176}$$

or

$$D_{max} = [-8(S_L - 1)G_L(T, P)\Delta t]^{1/2}. \tag{5.177}$$

With this D_{max}, one can integrate the number of particles in the distribution that will evaporate completely so that a distribution of sizes from 0 to ∞ is recovered,

$$\frac{dN_{Tx,evap}}{dt} = -\frac{N_{Tx}}{\Delta t \Gamma(\nu_x)} \Gamma\left(\nu_x, \frac{D_{x,max}}{D_{nx}}\right). \tag{5.178}$$

5.11.2 Log-normal distribution

Start with the vapor diffusion equation for liquid,

$$\frac{1}{\rho}\int_0^\infty \frac{dM(D_x)n(D_x)}{dt} dD_x = \frac{1}{\rho}\int_0^\infty 2\pi D_x(S_L - 1)G_L(T, P) f_v n(D_x) dD_x. \tag{5.179}$$

The log-normal distribution spectrum is defined as

$$n(D_x) = \frac{N_{Tx}}{\sqrt{2\pi}\sigma_x D_x} \exp\left(-\frac{[\ln(D_x/D_{nx})]^2}{2\sigma_x^2}\right). \tag{5.180}$$

The prognostic equation for the mixing ratio, Q_x, for vapor diffusion can be written as

$$Q_vCE_x = \frac{dQ_x}{dt} = \frac{1}{\rho}\int_0^\infty 2\pi(S_L - 1)G_L(T,P) \times \left\{0.78 + 0.308N_{sc}^{1/3}\left[\frac{c_x}{\nu_x}\left(\frac{\rho_0}{\rho}\right)^{1/2} D_x^{d_x+1}\right]^{1/2}\right\} D_x n(D_x)dD_x, \tag{5.181}$$

where N_{sc} is the Schmidt number, ν_x is the viscosity of air, ρ is the density of air, and ρ_0 is the mean density at sea level for a standard atmosphere.

Expanding (5.181) results in two integrals,

$$Q_vCE_x = \frac{2\pi(S_L - 1)G_L(T,P)}{\rho} \times \left\{\int_0^\infty 0.78D_x n(D_x)dD_x + \int_0^\infty 0.308N_{sc}^{1/3}\left[\frac{c_x}{\nu_x}\left(\frac{\rho_0}{\rho}\right)^{1/2} D_x^{d_x+1}\right]^{1/2} D_x n(D_x)dD_x\right\}. \tag{5.182}$$

Substituting (5.180) into (5.182) gives

$$Q_vCE_x = \frac{2\pi(S_L - 1)G_L(T,P)}{\rho}\left\{\frac{0.78N_{Tx}}{\sqrt{2\pi}\sigma_x}\int_0^\infty \exp\left(-\frac{[\ln(D_x/D_{nx})]^2}{2\sigma_x^2}\right)dD_x + \left(\frac{c_x}{\nu_x}\right)^{1/2}\left(\frac{\rho_0}{\rho}\right)^{1/4} 0.308N_{sc}^{1/3}D_{nx}^{\left(\frac{d_x+3}{2}\right)}\int_0^\infty \left(\frac{D_x}{D_{nx}}\right)^{\frac{d_x+1}{2}}\exp\left(-\frac{[\ln(D_x/D_{nx})]^2}{2\sigma_x^2}\right)dD_x\right\}. \tag{5.183}$$

All D_x terms are divided by D_{nx} for each of the two integrals,

$$Q_vCE_x = \frac{2\pi(S_L - 1)G_L(T,P)}{\rho}\frac{N_{Tx}}{\sqrt{2\pi}\sigma_x}\left\{0.78D_{nx}\int_0^\infty \exp\left(-\frac{[\ln(D_x/D_{nx})]^2}{2\sigma_x^2}\right)d\left(\frac{D_x}{D_{nx}}\right) + \left(\frac{c_x}{\nu_x}\right)^{1/2}\left(\frac{\rho_0}{\rho}\right)^{1/4} 0.308N_{sc}^{1/3}D_{nx}^{\left(\frac{d_x+3}{2}\right)}\int_0^\infty \left(\frac{D_x}{D_{nx}}\right)^{\frac{d_x+1}{2}}\exp\left(-\frac{[\ln(D_x/D_{nx})]^2}{2\sigma_x^2}\right)d\left(\frac{D_x}{D_{nx}}\right)\right\}. \tag{5.184}$$

We now let $u = D_x/D_{nx}$,

$$Q_{\mathrm{v}}CE_x = \frac{2\pi(S_{\mathrm{L}}-1)G_{\mathrm{L}}(T,P)}{\rho}\frac{N_{\mathrm{T}x}}{\sqrt{2\pi}\sigma_x}\left\{0.78D_{nx}\int_0^\infty \exp\left(-\frac{[\ln u]^2}{2\sigma_{x2}^2}\right)du\right.$$
$$\left.+\left(\frac{c_x}{v_x}\right)^{1/2}\left(\frac{\rho_0}{\rho}\right)^{1/4}0.308N_{\mathrm{sc}}^{1/3}D_{nx}^{\left(\frac{d_x+3}{2}\right)}\int_0^\infty u^{\left(\frac{d_x+1}{2}\right)}\exp\left(-\frac{[\ln u]^2}{2\sigma_x^2}\right)du\right\}. \tag{5.185}$$

By, letting $y = \ln(u)$, $u = \exp(y)$, $du/u = dy$, so,

$$Q_{\mathrm{v}}CE_x = \frac{2\pi(S_{\mathrm{L}}-1)G_{\mathrm{L}}(T,P)}{\rho}\frac{N_{\mathrm{T}x}}{\sqrt{2\pi}\sigma_x}\left\{0.78D_{nx}\int_{-\infty}^\infty \exp(y)\exp\left(-\frac{y^2}{2\sigma_x^2}\right)dy\right.$$
$$\left.+\left(\frac{c_x}{v_x}\right)^{1/2}\left(\frac{\rho_0}{\rho}\right)^{1/4}0.308N_{\mathrm{sc}}^{1/3}D_{nx}^{\frac{d_x+3}{2}}\int_{-\infty}^\infty \exp\left(\frac{d_x+3}{2}y\right)\exp\left(-\frac{y^2}{2\sigma_x^2}\right)dy\right\}, \tag{5.186}$$

where the limits of the integral change as u approaches zero from positive values, $\ln(u)$ approaches negative infinity. Likewise, for the upper limit, as u approaches positive infinity, $\ln(u)$ approaches positive infinity.

Now the following integral definition is applied:

$$\int_{-\infty}^\infty \exp(2b'x)\exp(-a'x^2)dx = \sqrt{\frac{\pi}{a'}}\exp\left(\frac{b'^2}{a'}\right), \tag{5.187}$$

by allowing $y = x$, and for the first integral, $a' = 1/(2\sigma^2)$, $b' = 1/2$, and for the second integral, $a' = 1/(2\sigma^2)$, $b' = (d + 3)/4$. Therefore, (5.186) becomes the prognostic equation for Q_x for the vapor-diffusion process assuming a log-normal distribution,

$$Q_{\mathrm{v}}CE_x = \frac{2\pi(S_{\mathrm{L}}-1)G_{\mathrm{L}}(T,P)N_{\mathrm{T}x}}{\rho}\left\{0.78D_{nx}\exp\left(\frac{\sigma_x^2}{2}\right)\right.$$
$$\left.+\left(\frac{c_x}{v_x}\right)^{1/2}\left(\frac{\rho_0}{\rho}\right)^{1/4}0.308N_{\mathrm{sc}}^{1/3}D_{nx}^{\left(\frac{d_x+3}{2}\right)}\exp\left(\frac{(d_x+3)^2\sigma_x^2}{8}\right)\right\}. \tag{5.188}$$

5.12 Bin model methods to vapor-diffusion mass gain and loss

5.12.1 Kovetz and Olund method

To accommodate the mass transfer with mass gain and loss owing to vapor-diffusion processes, the constraint is that the mass must be conserved, as expressed by

$$\int_0^\infty n(M)\mathrm{d}M = \text{constant}, \tag{5.189}$$

where accommodations need to be made for complete evaporation.

The general form of the vapor-diffusion gain and loss transfer equation is given as

$$\frac{\partial n(M)}{\partial t} = -\frac{\partial}{\partial M}\left[n(M)\frac{\mathrm{d}M}{\mathrm{d}t}\right], \tag{5.190}$$

where, $n(M)$ is the number of particles of mass M. One of the most commonly used schemes in the 1970s was the Kovetz and Olund (1969) scheme. Generally, the starting place is to compute the diffusion growth $\mathrm{d}M/\mathrm{d}t$ as in (5.190). Then, the following can be written using index J to indicate the bin to which a droplet belongs, and the mass of droplets $M(J)$ within that bin, to predict an intermediate value of M,

$$M'(J) = M(J) + \Delta t\left(\frac{\mathrm{d}M}{\mathrm{d}t}\right)_J. \tag{5.191}$$

The new $n(J)$ at $t = \tau + 1$ is computed from the latest $n^*(J)$ by

$$n^{**}(J) = \sum_{J'}^{J} R(J,J')n^*(J'), \tag{5.192}$$

with $R(J,J')$ defined by

$$R(J,J') = \begin{cases} \dfrac{M'(J) - M(J-1)}{M(J) - M(J-1)} & \text{for } M(J-1) < M'(J') < M(J) \\[2ex] \dfrac{M(J+1) - M'(J)}{M(J+1) - M(J)} & \text{for } M(J) < M'(J') < M(J+1). \\[2ex] 0 & \text{for all other cases} \end{cases} \tag{5.193}$$

This scheme by Kovetz and Olund satisfies the constraint (5.189).

5.12.2 The Tzivion et al. method

The procedure of Tzivion *et al.* (1989) begins in a similar manner to the Kovetz and Olund (1969) method with the general form of the vapor-diffusion mass gain and loss transfer equation given as

$$\left.\frac{\partial n(M)}{\partial t}\right|_{\text{evap,cond}} = -\frac{\partial}{\partial M}\left[n(M)\frac{\mathrm{d}M}{\mathrm{d}t}\right], \tag{5.194}$$

for which there is an analytical solution (see Appendix B of Tzivion *et al.* 1989). In (5.194), the following describes the change in mass with time for a single drop,

$$\left.\frac{dM}{dt}\right|_{\text{evap,cond}} = C(P,T)\Delta S M^{1/3}, \tag{5.195}$$

where the specific humidity surplus is denoted by ΔS, and $C(P, T)$ is a known function of pressure and temperature.

If $\Delta S < 0$ then evaporation occurs, and if $\Delta S > 0$, condensation occurs. The term $C(P, T)$ is similar in many ways to $G(T, P)$ or $G'(T, P)$ in the evaporation term for bulk parameterization. The analytical solution, presented in Tzivion *et al.* (1989) is given without derivation here as

$$n(M,t) = M^{-1/3}\left[\left(M^{2/3} - \frac{2}{3}\tau\right)^{1/2}\right] \times n_{t=0}\left[\left(M^{2/3} - \frac{2}{3}\tau\right)^{3/2}\right], \tag{5.196}$$

where $n_{t=0}$ is the initial drop distribution.

The variables τ and τ^* are given by

$$\tau = \int_0^t C(P,T)\Delta S(t)dt \tag{5.197}$$

and

$$\tau^* = \int_t^{t+\Delta} C(P,T)\Delta S(t)dt. \tag{5.198}$$

Now following Tzivion *et al.* (1989) and integrating $n(M, t)$ on the interval given by $[x_k; x_{k+1}]$, the following equations for the zeroth and first moment can be found,

$$N_k(t+\Delta t) = \int_{y_k}^{y_{k+1}} n_k(M,t)dM \tag{5.199}$$

and

$$I_k(t+\Delta t) = \int_{y_k}^{y_{k+1}} \left[\left(M^{2/3} + \frac{2}{3}\tau^*\right)^{3/2}\right] n_k(M,t)dM. \tag{5.200}$$

The variables y_k and y_{k+1} (limits of integration) are defined as

$$y_k = \left(x_k^{2/3} - \frac{2}{3}\tau^*\right)^{3/2} \tag{5.201}$$

and

$$y_{k+1} = \left(x_{k+1}^{2/3} - \frac{2}{3}\tau^* \right)^{3/2}. \tag{5.202}$$

Tzivion *et al.* describe the physical interpretation as the following. Start with a category (k) with bounds given by $[x_k; x_{k+1}]$. Next determine the region $[y_k; y_{k+1}]$ that encompasses all of these particle sizes, which will get smaller by evaporation or get larger by condensation in time increment Δt and fall into the region $[x_k; x_{k+1}]$.

As the integrals for the zeroth and first moments do not span entire categories $[x_k; x_{k+1}]$, the method used for collection is employed. This follows as

$$M^j n_k(M,t) = x_k^j f_k(t) \left(\frac{x_{k+1} - M}{x_{k+1} - x_k} \right) + x_{k+1}^j g_k(t) \left(\frac{M - x_k}{x_{k+1} - x_k} \right), \tag{5.203}$$

where $f_k(t)$ and $g_k(t)$ are values of n_k at $M = x_k$ and $M = x_{k+1}$, respectively. This can be substituted directly into the zeroth-moment equation, (5.199). The first-moment equation above can be solved by multiplying and dividing the integrand in (5.200) by M, which gives

$$I_k(t + \Delta t) = \int_{y_k}^{y_{k+1}} \left(M^{2/3} + \frac{2}{3}\tau^* \right)^{3/2} \frac{M n_k(M,t)}{M} \mathrm{d}M. \tag{5.204}$$

To get the solution, realize that $Mn_k(M, t)$ is linear in M, so the following integrals have to be solved to complete the system of solutions for the first moment

$$\int_{y_k}^{y_{k+1}} \frac{\left(M^{2/3} + \frac{2}{3}\tau^* \right)^{3/2}}{M} \mathrm{d}M \tag{5.205}$$

and

$$\int_{y_k}^{y_{k+1}} \left(M^{2/3} + \frac{2}{3}\tau^* \right)^{3/2} \mathrm{d}M. \tag{5.206}$$

These integrals have analytical solutions given in Tzivion *et al.* (1989). The performance of the two-moment scheme in finite difference form (5.199), (5.203) and (5.204) against the analytical solution (5.196) for an initial gamma distribution shows excellent agreement (Fig. 5.4) for an environment characterized by 50% relative humidity after 20 minutes.

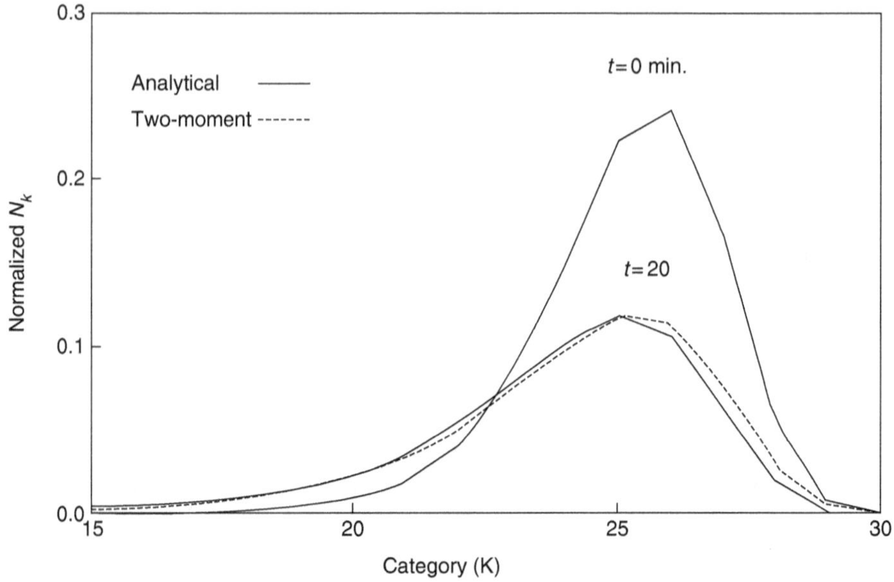

Fig. 5.4. An analytical solution to the evaporation equation as compared with the proposed approximation after 20 minutes of evaporation in a subsaturated environment of 50% relative humidity. The category drop concentration, N_k, is normalized by the initial drop concentration. (From Tzivion *et al.* 1989; courtesy of the American Meteorological Society.)

5.13 Perspective

Nearly two decades ago Srivastava (1989) made an argument for not using macroscale supersaturation as in traditional diffusion theories, such as the ones explained in this chapter (as in Byers 1965; and Rogers and Yau 1989), and even in Srivastava and Coen's (1992) own work. Rather he advocated using microscale approximations to supersaturations. These are of a form such that turbulent fluctuations are taken into account. Srivastava (1989) did realize in the end that his approach of attempting to include microscopic turbulence would be exceptionally complex to derive and program, but also exceptionally computationally intensive. It was noted that equations could be developed for bin microphysical parameterizations and bulk microphysical parameterizations. He hoped that one day a simpler representation of the concept of including microscopic supersaturation influences on droplet growth might be found for both bulk and bin microphysical parameterizations.

6

Vapor diffusion growth of ice-water crystals and particles

6.1 Introduction

After the nucleation of an ice-water particle or crystal, the addition of ice mass to the particle or crystal owing to supersaturation with respect to ice is called deposition. Furthermore, the loss of ice mass from an ice-water particle or crystal owing to subsaturation with respect to ice is called sublimation. Together these are called vapor diffusion of ice-water particles and crystals that are both governed by the same equation, which is nearly identical in form to that for diffusion of liquid-water particles except for some constants and shape parameters. The derivation for the vapor diffusion equation follows much the same path as that for deriving a basic equation to represent vapor diffusion for liquid-water particles. The main differences are related to the enthalpies of heat (enthalpy of sublimation instead of enthalpy of evaporation), and the particular shape factors for ice crystals, which include, for example: spheres; plates; needles; dendrites; sectors; stellars; and bullets and columns that can be either solid or hollow, etc. (see Pruppacher and Klett 1997 for habits at temperatures between 273.15 and 253.15 K, and Bailey and Hallet 2004 for habits at temperatures colder than 253.15 K).

Typically, for diffusion growth of ice-water particles and crystals, the electrostatic analog is invoked. This is similar to stating that the vapor diffusion growth of the various ice-water crystal shapes is related to the capacitance of the various shapes. The main shapes that are representative of the various ice-water particles include spheres, thin plates, oblates, and prolates. Kelvin's equation is useful for predicting the nucleation of pristine ice crystals; however, solute effects are usually not considered, as solutes often do not freeze until the solute reaches rather cold temperatures. Finally, ventilation effects are included even at small particle sizes just as ventilation effects are included for small cloud droplets. Admittedly, though, the ventilation effects are nearly negligible at the smallest ice-water particle and crystal sizes.

6.2 Mass flux of water vapor during diffusional growth of ice water

The diffusional change in mass of ice-water particles owing to subsaturation or supersaturation with respect to ice water primarily depends on thermal and vapor diffusion. In addition, for larger particles, advective processes are important, and have to be approximated from laboratory experiments. In the following pages equations will be developed to arrive at a parameterization equation for diffusional growth changes in a spherical ice-water particle that is large enough, on the order of a few microns in diameter, that surface curvature effects can be ignored. Moreover, the ice-water particles will be assumed to be pure (non-solutes). Other shapes besides spheres will be considered later in this chapter.

Following the same steps as with liquid-water drops, except replacing variables associated with liquid water with those associated with ice water, an equation for vapor diffusion growth of ice water can be obtained.

For ice-water particles, the same continuity equation for vapor molecules can be used as was used for liquid-water particles,

$$\frac{\partial \rho_{\mathrm{v}}}{\partial t} + u \bullet \nabla \rho_{\mathrm{v}} = \psi \nabla^2 \rho_{\mathrm{v}}. \tag{6.1}$$

The vapor density is given by $\rho_{\mathrm{v}} = nm$. In this definition n is the number of water-vapor molecules and m is the mass of a water molecule. Assume that the flow is non-divergent, and that u is zero, or stationary flow exists (sum of the air-flow velocity and the vapor-flow velocity is zero). When the steady-state assumption is used, as for liquid-water drops, Fick's first law of diffusion for n results. The variable ψ is the vapor diffusivity as given in Chapter 5. With these assumptions and definitions we can easily arrive at an expression for $\mathrm{d}M/\mathrm{d}t$ for ice particles in a similar manner as was done for liquid-water drops,

$$\frac{\mathrm{d}M}{\mathrm{d}t} = 4\pi R_r^2 m\psi \frac{\partial n}{\partial R}, \tag{6.2}$$

where R is the distance from the droplet center, and R_r is the radius. The boundary conditions are as before, as n approaches n_∞, R approaches infinity

R_∞ and as n approaches n_r, R approaches R_r, then the following can be written,

$$\frac{\mathrm{d}M}{\mathrm{d}t} = 4\pi R_r m\psi(n_\infty - n_r), \tag{6.3}$$

or,

$$\frac{\mathrm{d}M}{\mathrm{d}t} = 4\pi R_r \psi(\rho_{\mathrm{v},\infty} - \rho_{\mathrm{v},r}), \tag{6.4}$$

which is the mass change of ice particles owing to vapor density gradients.

6.3 Heat flux during vapor diffusional growth of ice water

An analogous procedure to that used to obtain an expression for $\mathrm{d}M/\mathrm{d}t$, can be followed to obtain a relationship $\mathrm{d}q/\mathrm{d}t$, which is the heat flux owing to temperature. From the continuity equation for temperature T, the following equation can be written,

$$\frac{\partial T}{\partial t} + u \bullet \nabla T = \kappa \nabla^2 T. \tag{6.5}$$

Again assume that the flow is non-divergent, and that u is zero or stationary flow exists (sum of the air-flow velocity and the vapor-flow velocity is zero). When the steady-state assumption is used as with liquid-water drops, Fick's law of diffusion for T results. The variable κ is the thermal conductivity as given in Chapter 5. With these assumptions and definitions we can easily arrive at an expression for $\mathrm{d}q/\mathrm{d}t$ for vapor diffusion of ice.

Following the method for deriving an equation for diffusion for a liquid-water drop, an expression for $\mathrm{d}q/\mathrm{d}t$ results,

$$\frac{\mathrm{d}q}{\mathrm{d}t} = 4\pi R_r^2 \rho K c_p \left(\frac{\partial T}{\partial r}\right)_{R=R_r}, \tag{6.6}$$

where the following can be written for $R = R_r$,

$$\frac{\mathrm{d}q}{\mathrm{d}t} = 4\pi R_r \rho K c_p (T_r - T_\infty); \tag{6.7}$$

this is the energy exchange owing to thermal gradients.

6.4 Plane, pure, ice-water surfaces

The diffusional change in mass of an ice-water sphere owing to sub- or super-saturation depends on thermal and vapor diffusion. Next an equation is found for diffusional growth changes in an ice-water sphere that is large enough, on the order of a few microns in diameter, that surface curvature effects can be ignored. Moreover the ice-water sphere is assumed to be pure.

Following steps used in deriving the vapor diffusion of a liquid-water drop, but assuming an ice-water sphere, the following can be found as

$$R_r \frac{dR_r}{dt} = (S_i - 1)G_i(T, P), \tag{6.8}$$

where $G_i(T, P)$, a thermodynamic variable that is a function of T and P, is

$$G_i(T, P) = \frac{1}{\frac{\rho_i L_s^2}{R_v K T_\infty^2} + \frac{\rho_i R_v T_\infty}{e_{SI}\psi}} = \frac{1}{\frac{\rho_i L_s^2}{R_v K T_\infty^2} + \frac{\rho_i}{\rho Q_{SI}\psi}}, \tag{6.9}$$

where L_s is the enthalpy of sublimation, e_{SI} is the vapor pressure over ice, Q_{SI} is the saturation mixing ratio over ice, and ρ_i is the density of ice. The mass growth rate approximation, which is needed to develop a parameterization, is given by the following for an ice sphere as

$$\frac{dM}{dt} = \rho_i 2\pi D(S_i - 1)G_i(T, P). \tag{6.10}$$

The vapor diffusion, mass growth equation for ice-water spheres can also be derived following Rogers and Yau (1989), using the methods for liquid-water drops, but replacing variables for the liquid phase with the ice phase,

$$R_r \frac{dR_r}{dt} = \frac{(S_i - 1)}{\left(\frac{L_s}{R_v T} - 1\right)\frac{L_s \rho_i}{KT} + \frac{\rho_i R_v T}{\psi e_{SI}(T)}}. \tag{6.11}$$

Note that this equation has a correction term in the denominator, which is small compared to the other two terms, and can be retained or neglected. The mass growth rate approximation, which is needed to develop a parameterization, is given by

$$\frac{dM}{dt} = \rho_i 2\pi D(S_i - 1)G_i'(T, P), \tag{6.12}$$

where $G_i'(T, P)$ is similar to $G_i(T, P)$, except $G_i'(T, P)$ has the correction term as in the liquid vapor diffusion growth equation,

$$G_i'(T, P) = \frac{1}{\left(\frac{L_s}{R_v T} - 1\right)\frac{L_s \rho_i}{KT} + \frac{\rho_i R_v T}{\psi e_{SI}(T)}}. \tag{6.13}$$

6.5 Ventilation effects for larger ice spheres

One aspect we have not considered is ventilation effects. The equation for vapor diffusion growth derived so far is based on a stationary drop with no relative flow past it; that is, advective effects on the vapor density gradients have been ignored. These effects in general are very difficult to include explicitly, so they are parameterized in terms of the Schmidt number ($N_{sc} = \nu/\psi \approx 0.71$) and the Reynolds number ($DV_T(D)/\nu$) where V_T is terminal velocity. The ventilation factor used for ice particles is written in the form of

$$f_v = 0.78 + 0.308 N_{sc}^{1/3} N_{re}^{1/2}. \tag{6.14}$$

With (6.14), the mass growth equation is written as

$$\frac{dM}{dt} = 2\pi D(S_i - 1)G_i(T, P)f_v, \tag{6.15}$$

or

$$\frac{dM}{dt} = 2\pi D(S_i - 1)G_i'(T, P)f_v. \tag{6.16}$$

Again $G_i(T, P)$ and $G_i'(T, P)$ are multiplied through by ρ_i as in Chapter 5.

6.6 Parameterizations

6.6.1 Generalized gamma distribution

The vapor diffusion of ice-water particles, which can include sublimation and deposition, can only occur at temperatures below 273.15 K. The basic equation that is solved is

$$\frac{dM(D)}{dt} = \frac{4\pi D(S_i - 1)f_v}{\left[\frac{L_s^2}{KR_vT^2} + \frac{1}{\rho\psi Q_{SI}}\right]}, \tag{6.17}$$

where D is the capacitance using the electrostatic analog (Pruppacher and Klett 1997), Q_{SI} is the saturation mixing ratio with respect to ice, and the ventilation coefficient for snow aggregates (Rutledge and Hobbs 1984) is

$$f_v = \left(0.65 + 0.44 N_{sc}^{1/3} N_{re}^{1/2}\right). \tag{6.18}$$

For graupel, frozen drops and hail with which $D > 120$ mm, the ventilation coefficient is given by

$$f_v = \left(0.78 + 0.308 N_{sc}^{1/3} N_{re}^{1/2}\right), \tag{6.19}$$

and for ice particles with $D < 120$ mm, the ventilation coefficient is given by

$$f_v = \left(1.00 + \left[0.108 N_{sc}^{1/3} N_{re}^{1/2}\right]^2\right). \tag{6.20}$$

The form of the equation integrated for larger spherical particles is

$$\frac{1}{\rho}\int_0^\infty \frac{\mathrm{d}M(D_x)n(D_x)}{\mathrm{d}t}\mathrm{d}D_x = \frac{1}{\rho}\int_0^\infty 2\pi D_x(S_\mathrm{i}-1)G_\mathrm{i}(T,P) \times \left(0.78+0.308N_\mathrm{sc}^{1/3}N_\mathrm{re}^{1/2}\right)n(D_x)\mathrm{d}D_x, \tag{6.21}$$

where the subscript x denotes any ice habit x. This is then rewritten as

$$Q_xDS_\mathrm{L} = \frac{1}{\rho_0}\int_0^\infty 2\pi D_x(S_\mathrm{i}-1)G_\mathrm{i}(T,P)\left[0.78+0.308N_\mathrm{sc}^{1/3}\left(\frac{D_xV_{\mathrm{T}x}}{\nu_x}\right)^{1/2}\left(\frac{\rho_0}{\rho}\right)^{1/4}\right] \times \left[\frac{N_{\mathrm{T}x}\mu_x}{\Gamma(\nu_x)}\left(\frac{D_x}{D_{nx}}\right)^{\nu_x\mu_x-1}\exp\left(-\left[\frac{D_x}{D_{nx}}\right]^{\mu_x}\right)\right]\mathrm{d}\left(\frac{D_x}{D_{nx}}\right), \tag{6.22}$$

and

$$Q_xDS_\mathrm{L} = \frac{1}{\rho_0}\int_0^\infty 2\pi D_x(S_\mathrm{i}-1)G_\mathrm{i}(T,P)\left[0.78+0.308N_\mathrm{sc}^{1/3}\left(\frac{D_xV_{\mathrm{T}x}}{\nu_x}\right)^{1/2}\left(\frac{\rho_0}{\rho}\right)^{1/4}\right] \times \left[\frac{N_{\mathrm{T}x}\mu_x\alpha_x^{\nu_x}}{\Gamma(\nu_x)}\left(\frac{D_x}{D_{nx}}\right)^{\nu_x\mu_x-1}\exp\left(-\alpha_x\left[\frac{D_x}{D_{nx}}\right]^{\mu_x}\right)\right]\mathrm{d}\left(\frac{D_x}{D_{nx}}\right), \tag{6.23}$$

where S_L is the saturation ratio with respect to liquid.

For larger ice particles ($D > 120$ microns) that are nearly spherical the following equation can be derived for a modified gamma distribution of deposition/sublimation. Integrating gives the desired parameterization equation for vapor diffusion growth, assuming terminal velocity, is

$$V_{\mathrm{T}x}(D_x) = c_xD_x^{d_x}\left(\frac{\rho_0}{\rho}\right)^{1/2}. \tag{6.24}$$

The complete gamma distribution solution is then given as

$$Q_xDS_\mathrm{L} = \frac{1}{\rho}2\pi(S_\mathrm{i}-1)G_\mathrm{i}(T,P)\frac{N_{\mathrm{T}x}\alpha_x^{\nu_x}}{\Gamma(\nu_x)} \times \left[0.78\frac{\Gamma\left(\frac{1+\nu_x\mu_x}{\mu_x}\right)}{\alpha_x^{\left(\frac{1+\nu_x\mu_x}{\mu_x}\right)}}D_{nx}+0.308\Gamma\frac{\left(\frac{3+d_x}{2\mu_x}+\frac{\mu_x\nu_x}{\mu_x}\right)}{\alpha_x^{\left(\frac{3+d_x}{2\mu_x}+\frac{\mu_x\nu_x}{\mu_x}\right)}}N_\mathrm{sc}^{1/3}\nu_x^{-1/2}c_x^{1/2}D_{nx}^{\frac{3+d_x}{2}}\left(\frac{\rho_0}{\rho}\right)^{1/4}\right], \tag{6.25}$$

whereas, the modified gamma distribution solution is

$$Q_xDS_L = \frac{1}{\rho}2\pi(S_i - 1)G_i(T,P)\frac{N_{Tx}}{\Gamma(\nu_x)} \times \left[0.78\Gamma\left(\frac{1+\nu_x\mu_x}{\mu_x}\right)D_{nx} + 0.308\Gamma\left(\frac{3+d_x}{2\mu_x} + \frac{\nu_x\mu_x}{\mu_x}\right)N_{sc}^{1/3}\nu_x^{-1/2}c_x^{1/2}D_{nx}^{\frac{3+d_x}{2}}\left(\frac{\rho_0}{\rho}\right)^{1/4}\right]. \quad (6.26)$$

The gamma distribution solution is

$$Q_xDS_L = \frac{1}{\rho}2\pi(S_i - 1)G_i(T,P)\frac{N_{Tx}}{\Gamma(\nu_x)} \times \left[0.78\Gamma(1+\nu_x)D_{nx} + 0.308\Gamma\left(\frac{3+d_x}{2} + \nu_x\right)N_{sc}^{1/3}\nu_x^{-1/2}c_x^{1/2}D_{nx}^{\frac{3+d_x}{2}}\left(\frac{\rho_0}{\rho}\right)^{1/4}\right]. \quad (6.27)$$

For smaller particles ($D < 120$ mm) that are nearly spherical, the following equation has a more appropriate ventilation coefficient. The complete gamma distribution solution is given by

$$Q_xDS_L = \frac{1}{\rho}2\pi(S_i - 1)G_i(T,P)\frac{N_{Tx}\alpha_x^{\nu_x}}{\Gamma(\nu_x)} \times \left[1.0\frac{\Gamma\left(\frac{1+\nu_x\mu_x}{\mu_x}\right)}{\alpha_x^{\left(\frac{1+\nu_x\mu_x}{\mu_x}\right)}}D_{nx} + 0.108\frac{\Gamma\left(\frac{3+d_x}{\mu_x} + \frac{\nu_x\mu_x}{\mu_x}\right)}{\alpha_x^{\left(\frac{3+d_x}{\mu_x} + \frac{\nu_x\mu_x}{\mu_x}\right)}}N_{sc}^{2/3}\nu_x^{-1}c_xD_{nx}^{3+d_x}\left(\frac{\rho_0}{\rho}\right)^{1/2}\right], \quad (6.28)$$

whereas the modified gamma distribution solution is

$$Q_xDS_L = \frac{1}{\rho}2\pi(S_i - 1)G_i(T,P)\frac{N_{Tx}}{\Gamma(\nu_x)} \times \left[1.0\Gamma\left(\frac{1+\nu_x\mu_x}{\mu_x}\right)D_{nx} + 0.108\Gamma\left(\frac{3+d_x}{\mu_x} + \frac{\nu_x\mu_x}{\mu_x}\right)N_{sc}^{2/3}\nu_x^{-1}c_xD_{nx}^{3+d_x}\left(\frac{\rho_0}{\rho}\right)^{1/2}\right]. \quad (6.29)$$

The gamma distribution solution is given as

$$Q_xDS_L = \frac{1}{\rho}2\pi(S_i - 1)G_i(T,P)\frac{N_{Tx}}{\Gamma(\nu_x)} \times \left[1.0\Gamma(1+\nu_x)D_{nx} + 0.108\Gamma(3+d_x+\nu_x)N_{sc}^{2/3}\nu_x^{-1}c_xD_{nx}^{3+d_x}\left(\frac{\rho_0}{\rho}\right)^{1/2}\right]. \quad (6.30)$$

The change in number concentration during sublimation is a complicated issue in many regards. There is no change in number concentration during condensation. For simplicity, many assume that the number-concentration change is related to the mixing-ratio change as follows,

$$N_{Tx}SB_V = Q_xSB_V\frac{N_{Tx}}{Q_x}, \quad (6.31)$$

where mixing ratio tendencies for deposition and sublimation are written, respectively, as

$$Q_x DP_v = \max(Q_x DS_v, 0.0), \tag{6.32}$$

and

$$Q_x SB_v = \min(Q_x DS_v, 0.0). \tag{6.33}$$

There are problems with this formulation in that it really does not capture the nature of the number of droplets that evaporate. An alternative is to start with an equation for the rate of change of the radii of the particles; the ventilation coefficient is set as equal to one, assuming it is only the very smallest ice particles that are fully subliming,

$$R_r \frac{dR_r}{dt} = (S_i - 1)G_i(T, P). \tag{6.34}$$

From (6.34) it can be written that

$$\int_{R_{max}}^{0} R_r \frac{dR_r}{dt} dt = \int_{t=0}^{t=\Delta t} (S_i - 1)G_i(T, P) dt. \tag{6.35}$$

Now R_{max}, the largest remaining drop after Δt seconds of sublimation, is

$$\frac{R_{max}^2}{2} = -(S_i - 1)G_i(T, P)\Delta t \tag{6.36}$$

or

$$D_{max} = [-8(S_i - 1)G_i(T, P)\Delta t]^{1/2}. \tag{6.37}$$

With this D_{max} one can integrate the number of particles in the distribution that will evaporate completely so that you return to a distribution of sizes from 0 to ∞,

$$N_x SB_v = -\frac{N_{Tx}}{\Delta t \Gamma(\nu_x)} \Gamma\left(\nu_x, \frac{D_{max}}{D_{nx}}\right). \tag{6.38}$$

6.6.2 Log-normal distribution

The log-normal distribution parameterization for a spherical piece of ice is now given. We start with the vapor diffusion equation for ice,

$$\frac{1}{\rho}\int_0^{\infty} \frac{dM(D_x)n(D_x)}{dt} dD_x = \frac{1}{\rho}\int_0^{\infty} 2\pi D_x (S_i - 1)G_i(T, P) f_v n(D_x) dD_x. \tag{6.39}$$

The log-normal distribution spectrum is defined as

$$n(D_x) = \frac{N_{Tx}}{\sqrt{2\pi}\sigma_x D_x} \exp\left(-\frac{[\ln(D_x/D_{nx})]^2}{2\sigma_x^2}\right), \tag{6.40}$$

where σ is a distribution parameter. The prognostic equation for Q_x for vapor diffusion of ice can be written as

$$\begin{aligned} Q_v DS_x = \frac{dQ_x}{dt} = \frac{1}{\rho} \int_0^\infty 2\pi G_i(T,P)(S_i - 1) \\ \times \left\{ 0.78 + 0.308 N_{sc}^{1/3} \left[\frac{c_x}{\nu_x} \left(\frac{\rho_0}{\rho} \right)^{1/2} D_x^{d_x+1} \right]^{1/2} \right\} D_x n(D_x) dD_x. \end{aligned} \tag{6.41}$$

Expanding results in two integrals,

$$\begin{aligned} Q_v DS_x = \frac{2\pi G_i(T,P)(S_i - 1)}{\rho} \left\{ \int_0^\infty 0.78 D_x n(D_x) dD_x \right. \\ \left. + \int_0^\infty 0.308 N_{sc}^{1/3} \left[\frac{c_x}{\nu_x} \left(\frac{\rho_0}{\rho} \right)^{1/2} D_x^{d_x+1} \right]^{1/2} D_x n(D_x) dD_x \right\}. \end{aligned} \tag{6.42}$$

Substituting (6.40) into (6.42) gives

$$\begin{aligned} Q_v DS_x = \frac{2\pi G_i(T,P)(S_i - 1)}{\rho} \left\{ \frac{0.78 N_{Tx}}{\sqrt{2\pi}\sigma_x} \int_0^\infty \exp\left(-\frac{[\ln(D_x/D_{nx})]^2}{2\sigma_x^2}\right) dD_x \right. \\ \left. + \frac{N_{Tx}}{\sqrt{2\pi}\sigma_x} \left(\frac{c_x}{\nu_x}\right)^{1/2} \left(\frac{\rho_0}{\rho}\right)^{1/4} 0.308 N_{sc}^{1/3} \int_0^\infty D_x^{\frac{d_x+1}{2}} \exp\left(-\frac{[\ln(D_x/D_{nx})]^2}{2\sigma_x^2}\right) dD_x \right\}. \end{aligned} \tag{6.43}$$

Dividing all D_x terms by D_{nx} for each of the two integrals gives

$$\begin{aligned} Q_v DS_x = \frac{2\pi G_i(T,P)(S_i - 1)}{\rho} \frac{N_{Tx}}{\sqrt{2\pi}\sigma_x} \left\{ 0.78 D_{nx} \int_0^\infty \exp\left(-\frac{[\ln(D_x/D_{nx})]^2}{2\sigma_x^2}\right) d\left(\frac{D_x}{D_{nx}}\right) \right. \\ \left. + \left(\frac{c_x}{\nu_x}\right)^{1/2} \left(\frac{\rho_0}{\rho}\right)^{1/4} 0.308 N_{sc}^{1/3} D_{nx}^{\left(\frac{d_x+3}{2}\right)} \int_0^\infty \left(\frac{D_x}{D_{nx}}\right)^{\frac{d_x+1}{2}} \exp\left(-\frac{[\ln(D_x/D_{nx})]^2}{2\sigma_x^2}\right) d\left(\frac{D_x}{D_{nx}}\right) \right\}. \end{aligned} \tag{6.44}$$

We now let $u = D_x/D_{nx}$,

$$Q_\mathrm{v}DS_x = \frac{2\pi G_\mathrm{i}(T,P)(S_\mathrm{i}-1)}{\rho}\frac{N_{\mathrm{T}x}}{\sqrt{2\pi}\sigma_x}\left\{0.78D_{nx}\int_0^\infty \exp\left(-\frac{[\ln u]^2}{2\sigma_x^2}\right)\mathrm{d}u\right.$$
$$\left.+\left(\frac{c_x}{\nu_x}\right)^{1/2}\left(\frac{\rho_0}{\rho}\right)^{1/4} 0.308N_\mathrm{sc}^{1/3}D_{nx}^{\left(\frac{d_x+3}{2}\right)}\int_0^\infty u^{\left(\frac{d_x+1}{2}\right)}\exp\left(-\frac{[\ln u]^2}{2\sigma_x^2}\right)\mathrm{d}u\right\}, \tag{6.45}$$

and letting $y = \ln(u)$, $u = \exp(y)$, $\mathrm{d}u/u = \mathrm{d}y$, so,

$$Q_\mathrm{v}DS_x = \frac{2\pi G_\mathrm{i}(T,P)(S_\mathrm{i}-1)}{\rho}\frac{N_{\mathrm{T}x}}{\sqrt{2\pi}\sigma_x}\left\{0.78D_{nx}\int_{-\infty}^{\infty}\exp(y)\exp\left(-\frac{y^2}{2\sigma_x^2}\right)\mathrm{d}y\right.$$
$$\left.+\left(\frac{\alpha}{\nu_x}\right)^{1/2}\left(\frac{\rho_0}{\rho}\right)^{1/4} 0.308N_\mathrm{sc}^{1/3}D_{nx}^{\frac{d_x+3}{2}}\int_{-\infty}^{\infty}\exp\left(\frac{d_x+3}{2}y\right)\exp\left(-\frac{y^2}{2\sigma_x^2}\right)\mathrm{d}y\right\}, \tag{6.46}$$

where the limits of the integral change as u approaches zero from positive values and, $\ln(u)$ approaches negative infinity. Likewise, for the upper limit, as u approaches positive infinity, $\ln(u)$ approaches positive infinity.

Now the following integral definition is applied,

$$\int_{-\infty}^{\infty}\exp(2b'x)\exp(-a'x^2)\mathrm{d}x = \sqrt{\frac{\pi}{a'}}\exp\left(\frac{b'^2}{a'}\right), \tag{6.47}$$

by allowing $y = x$, and for the first integral, $a' = 1/(2\sigma^2)$, $b' = 1/2$; and for the second integral, $a' = 1/(2\sigma^2)$, $b' = (d+3)/4$. Therefore (6.46) becomes the prognostic equation for Q_x for the vapor diffusion process,

$$Q_\mathrm{v}DS_x = \frac{2\pi G_\mathrm{i}(T,P)(S_\mathrm{i}-1)N_{\mathrm{T}x}}{\rho}\left\{0.78D_{nx}\exp\left(\frac{\sigma_x^2}{2}\right)\right.$$
$$\left.+\left(\frac{c_x}{\nu_x}\right)^{1/2}\left(\frac{\rho_0}{\rho}\right)^{1/4} 0.308N_\mathrm{sc}^{1/3}D_{nx}^{\left(\frac{d_x+3}{2}\right)}\exp\left(\frac{(d_x+3)^2\sigma_x^2}{8}\right)\right\}. \tag{6.48}$$

6.7 Effect of shape on ice-particle growth

The electrostatic analog is employed with ice crystals (Pruppacher and Klett 1997) and the following in the subsections are used to represent capacitance analogs. Hall and Pruppacher (1976) provide the following ventilation expressions for ice crystals,

$$\begin{cases} f_v = 1 + 0.14X^2 & \text{for } X < 1 \\ f_v = 0.86 + 0.28X & \text{for } X > 1, \end{cases} \tag{6.49}$$

where X is given by

$$X = N_{sc}^{1/3} N_{re}^{1/2}, \tag{6.50}$$

$$N_{sc} = \frac{\nu}{\psi}, \tag{6.51}$$

and

$$N_{re} = \frac{V_T(D)L^*}{\nu}. \tag{6.52}$$

In (6.52), L^* is the ratio of the total surface area, W, to the perimeter, P, of the crystal.

Wang (1985) was decidedly convincing that simple shape approximations for crystals are not sufficient for computing ventilation coefficients. Thus Wang and Ji (1992) derived ventilation coefficients assuming Stokes flow for columns, plates, and broad branched crystals. These are included below as they represent the newest values available for research.

6.7.1 Sphere

A sphere is represented by the radius r with a capacitance of

$$C_0 = r \tag{6.53}$$

and the above ventilation coefficient should be valid, or $L^* = W/P = 2a$, with $P = 2\pi r$ and $W = 4\pi r^2$.

6.7.2 Hexagonal-shaped plate

A hexagonal plate's capacitance can be well represented by a circular disk and is the easiest to parameterize,

$$C_0 = \frac{2r}{\pi}, \tag{6.54}$$

where C_0 is the capacitance, and r is the radius of the disk that describes the hexagonal plate.

The P for a disk is $P = 2\pi r$ for the major axis of the disk falling perpendicular to the relative flow. The surface area is given by the volume divided by the height, with values tabulated by Pruppacher and Klett (1997), e.g.

$$h = 1.41 \times 10^{-2} D^{0.474} \tag{6.55}$$

with height h and circumscribed diameter D in cm. The volume V in cm^3 is

$$V = 9.17 \times 10^{-3} D^{2.475}. \tag{6.56}$$

Alternatively, following Wang and Ji (1992), the value of f_v is given as

$$\begin{aligned} f_v = 1 &- 0.06042(X/10) + 2.79820(X/10)^2 \\ &- 0.31933(X/10)^3 + 0.06247(X/10)^4, \end{aligned} \tag{6.57}$$

where X is as defined in (6.50) and is valid for $1 < N_{re} < 120$. For broad branched crystals, a similar relation can be derived from numerical results as

$$f_v = 1 + 0.35463(X/10) + 3.55333(X/10)^2, \tag{6.58}$$

and is valid for $1 < N_{re} < 120$.

6.7.3 Simple ice plate shapes of various thickness

Simple plates of various thickness are approximated by oblate spheroids with major and minor axes lengths of $2a$ and $2b$, respectively. Following Pruppacher and Klett (1997) and McDonald (1963), the equation used for plates of varying thickness is

$$C_0 = \frac{a\epsilon}{\sin^{-1}(\epsilon)}, \tag{6.59}$$

with ϵ given as

$$\epsilon = \left(1 - \frac{b^2}{a^2}\right)^{1/2}. \tag{6.60}$$

The P for an oblate is $P = 2\pi a$ for the major axis falling perpendicular to the relative flow. Alternatively, P can be computed as

$$P = 2\pi a^2 + \pi(b^2/\epsilon)\ln[(1+\epsilon)/(1-\epsilon)], \tag{6.61}$$

where a and b are minor and major axes of an oblate spheroid. The surface area is given by dividing V by h,

$$h = 0.138\ D^{0.778}, \tag{6.62}$$

with height h and circumscribed diameter D in cm. The volume V in cm^3 is,

$$V = 8.97 \times 10^{-2}\ D^{2.778}. \tag{6.63}$$

6.7.4 Columnar-shaped ice crystals

McDonald (1963) and Pruppacher and Klett (1997) also proposed the following prolate-spheroid capacitance equation for the capacitance of columnar crystals,

$$C_0 = \frac{A}{\ln[(a+A)/b]}, \tag{6.64}$$

where A is given by

$$A = \left(a^2 - b^2\right)^{1/2}. \tag{6.65}$$

For prolates the value of P is computed as

$$P = \pi 2a\left(1 - 0.25\epsilon^2 - 0.0469\epsilon^4 - 0.0195\epsilon^6 - 0.0107\epsilon^8 - 0.0067\epsilon^{10} - \cdots\right). \tag{6.66}$$

The surface area is approximated following that for a bullet rosette where w is the width of the crystal and L is the length of an individual bullet. The surface area of a column is given by

$$\Omega = \left(\pi w^2/4\right) + \pi w L, \tag{6.67}$$

where $L = 2a$. Following Wang and Ji (1992) again, the value of the ventilation coefficient is

$$f_v = 1 - 0.00668(X/4) + 2.79402(X/4)^2 - 0.73409(X/4)^3 + 0.73911(X/4)^4, \tag{6.68}$$

which is valid for $0.2 < N_{re} < 20$.

6.7.5 Needle-like ice-crystal shapes

Needle-like crystals are treated like extreme prolates, where $b \ll a$; thus the equation for a prolate given above can be written as

$$C_0 = \frac{a}{\ln(2a/b)}. \tag{6.69}$$

The values of P and L for needle-like crystals could be approximated with that of extreme prolates.

7
Collection growth

7.1 Introduction

An issue that has perplexed the minds of great meteorologists for many years now, and still does, is the determination of the length of time that it takes for rain to form and fall to the ground (Knight and Miller 1993). This problem has been the center of much past and present research. First, nucleation occurs, followed by condensation growth, and finally drops begin to grow to a size that is large enough that the probability of a collision becomes non-negligible. This size seems to be around diameters of about 41 μm. Until drops grow to this size by vapor diffusion and collection from very small droplets, or if aerosols of the size of ultra-giant cloud condensation nuclei are available, droplets may not grow to the size necessary for rapid coalescence. If they do, then rapid coalescence or collection growth begins to dominate. In general, it takes some time for a few particles finally to reach about $D = 82$ mm, a size where more rapid coalescence can take place.

Collection growth can be presented as a relatively straightforward two-body collection continuous growth problem or a complex, statistical collection problem. Both of these are included in the discussion that follows. A primary mode by which hydrometeors come together is by differing fall-speeds such that particles of different sizes, densities, or shapes fall at differing speeds, which allows collisions to occur. Furthermore, electrical forces can act if particles are differentially charged, which may enhance collection; or collection may decrease if the particles have the same charge sign. Finally, though there is much debate surrounding this, turbulent forces may play a role in the collection of droplets. This latter issue will be left to discussion in periodicals for now. In general, gravitation effects that result in relative-fallspeed differences between particles dominate over electrical and turbulent effects.

When particles begin to collide there are at least two factors that need to be considered. These are the probability that droplets will collide; and the probability that collisions will result in coalescence (that particles will stick together). The product of these two probabilities is called the collection efficiency.

When two droplets collide, the following are possible outcomes. First, when particles collide they may coalesce. Second, the particles may collide and bounce apart (rebound). Third, particles may coalesce and then separate with original sizes preserved. Fourth, particles coalesce and then separate with resulting different sizes and may possibly produce additional droplets.

The important variables in collisional growth include; the size of the particles involved; the fall velocities of the particles; the trajectory of the particles; the number of collisions; the number of coalescing collisions (collections); electrical effects; and turbulent effects. Finally it is important to understand that collisional growth goes by several names including collision growth, accretion growth, coalescence growth, riming growth, and aggregation growth.

7.2 Various forms of the collection equation

There are three models of the collection equations following Gillespie (1975) and Young (1975 and 1993): the continuous growth model; the quasi-stochastic or discrete model; and the pure-stochastic, probabilistic, statistical, or Poisson model. Figure 7.1, reproduced from Young (1975) illustrates these models well. According to Pruppacher and Klett (1997), the goal of collection is to describe the growth of N drops in a spectrum of drops. What is needed to describe this is the collection kernel, which is described by only the drop mass where A_{ik} approaches $A(m)$. It is important to realize that A_{ik} represents collection in a well-mixed cloud model. This is really describing collection as a whole in a very simple, idealized, whole cloud. N_k is the number of droplets in bin k; $N_k = n_k$ times V (volume); K_{ik} is the collection kernel between drops in bins i and k – this describes the rate of i collecting k size droplets; and A_{ik} is K_{ik} divided by V. The parameter A can be thought of as the probability per unit time of collections between any pair of i and k size drops or drops of bin size i and k. This is the so-called "well-mixed" cloud model assumption. The model for growth is that there are initially N drops in a cloud with mass M, N' droplets having mass m, that $N \ll N'$; coalescences only are possible between drops and droplets (Pruppacher and Klett 1997). Growth of droplets by the three models of collection is summarized below.

The continuous growth model predicts that all collector drops of a given size will collect the same number of smaller droplets and grow to the same size

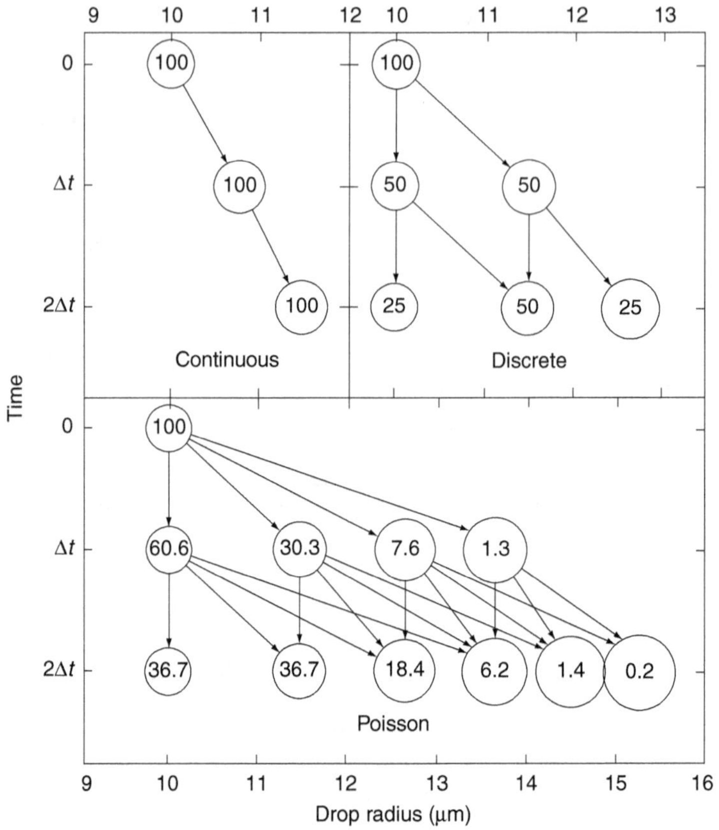

Fig. 7.1. The growth of 10 mm-radius drops collecting 8 mm drops for the continuous, discrete, and Poisson collection models. The expected number of collection events within a given timestep 0.5 s. Numbers within the circles reflect the percentage of drops of that size; arrows show growth paths. (From Young 1975; courtesy of the American Meteorological Society.)

as each other after time interval dt (Fig. 7.1). This implies that each collector drop grows at the same rate. Mathematically, the number dN_S (change in small droplets collected) is interpreted as the fractional number of small droplets collected by every collector droplet of radius R_L in time interval dt. Or, as described by Gillespie (1975) and Pruppacher and Klett (1997), $A(m)N'\mathrm{d}t$ is the number of droplets of mass m, which any drop of mass M will collect in time dt.

The quasi-stochastic or discrete model predicts that all clouds will have the same size distributions after time dt. Mathematically, the number of small collected droplets dN_S is interpreted as the fraction of drops that collect a droplet in the interval dt. As the collection is a discrete process, drops do not collect fractional droplets. Thus, this is interpreted as saying that a fraction f of drops will grow collecting droplets of a given size, and a fraction $(1-f)$ will not. Therefore, drops that are initially the same size may have different growth histories, which allows a spectrum of drops to develop (Fig. 7.1). However, only one outcome is permitted for each initial condition. There are two interpretations that can be made. Droplets may be distributed uniformly. A drop then either collects a droplet, or it does not collect a droplet. Secondly, droplets may be randomly distributed. Some droplets may collect one, two, or more droplets; while others may collect none. As in the continuous model, in this quasi-stochastic model, $A(m)N'\mathrm{d}t$ is the number of drops of mass M, which will collect a droplet of mass m in time dt.

The purely stochastic, probabilistic, or Poisson, model predicts that all clouds will have a unique distribution after a time interval dt. Mathematically, the number of $\mathrm{d}N_S$ is interpreted as the probability that a collector drop will collect a droplet of some size in a time interval dt. In this model it is assumed that all droplets have positions that are probabilistic in nature (Fig. 7.1). Moreover, as before, it can be stated that $A(m)N'\mathrm{d}t$ is the probability that any drop of mass M will collect a droplet of mass m in time dt.

Ironically, considering the mathematical differences between the quasi-stochastic model and the pure-stochastic model (Pruppacher and Klett 1997), they are believed to produce essentially the same result after some sufficient time interval dt, though this contention is not uniformly accepted by the community.

7.3 Analysis of continuous, quasi-stochastic, and pure-stochastic growth models

The purpose here is to explore the theoretical bases and analyses presented by Gillespie (1975) and summarized by Pruppacher and Klett (1997) of the continuous, quasi-stochastic, and probabilistic growth models. All three will be examined in some detail. A discussion will be made of the probabilistic growth model, though it is not used in bulk or in many bin model parameterizations very often, not even very simple bin models. This section hopefully will provide a background for understanding the three possible considerations of drop–droplet collection modes, and how to develop parameterizations for the continuous growth and quasi-stochastic growth models possible. The models are analyzed assuming that $A(M) = A =$ constant. Also, it is assumed that a cloud is well mixed at time $t = 0$.

7.3.1 The continuous growth collection equation for a model cloud

With the continuous growth model, all drops start at time $t = 0$ with the same mass m_0. Each drop of mass m_0 also grows at the same rate. Therefore the state of drops of mass m can be described by $M(t)$ = mass of any drop at time t. Also μ is the size of the drops collected. Thus, for continuous growth, it can be written

$$\frac{dM(t)}{dt} = \mu AN', \tag{7.1}$$

and integrating over time with the initial condition $M(0) = m_0$, is the initial mass, results in

$$\int_{M}^{M'(t)} \frac{dM(t)}{dt} dt = \int_{t=0}^{t} \mu AN' dt, \tag{7.2}$$

so that,

$$M(t) = m_0 + \mu AN't. \tag{7.3}$$

This model requires that every drop of mass m_0 collect a certain number of droplet(s) continuously (linearly, in this case) of mass μ in time interval dt. Except for very large drops and hailstones, the restrictive nature of the continuous growth model that requires that all drops of a given mass grow at the same continuous rate is unrealistic.

7.3.2 The quasi-stochastic collection equation for a model cloud

When using the quasi-stochastic model, it must be remembered that only a fraction of the drops of mass m_0 will collect a droplet of mass μ in interval dt. This can be justified by the fact that there are random positions of drops m_0 and droplets m; some drops will collect one or more droplets, whilst others will collect none. From Pruppacher and Klett (1997) and Gillespie (1975) this means that not all drops of mass m_0 will grow at the same rate at the same time. Now the problem changes from the continuous growth model. The quasi-stochastic model described here follows very closely the presentation put forth by Gillespie (1975). The following can be stated as $N(m,t)$ is defined as the number of drops of mass m_0 (or fraction of drops of mass m_0) that grow at time t by collecting a droplet of mass μ. This results in

$$m(t) = m_0, m_0 + \mu, m_0 + 2\mu \ldots. \tag{7.4}$$

Note that this requires drops to be described in terms of discrete sizes, rather than drops becoming random sizes.

Now the governing equations for $N(m,t)$ become the following. In time $(t, t + dt)$ it is found that $N(m - \mu, t)AN'dt$ drops of mass $m - \mu$ will each collect with a droplet to become drops of mass m, and $N(m, t)AN'dt$ drops of mass m will each collect a droplet to reach drops of mass $m + \mu$. Thus, according to Gillespie (1975), the net increase in number of drops of mass m in a time interval $(t, t + dt)$ can be written as

$$dN(m, t) = AN'[N(m - \mu; t) - N(m; t)]dt, \tag{7.5}$$

or expressing this as a growth rate equation as given by,

$$\frac{\partial N(m, t)}{\partial t} = AN'[N(m - \mu; t) - N(m; t)]. \tag{7.6}$$

Telford (1955) was the first to consider this model as presented in Gillespie (1972, 1975) and Pruppacher and Klett (1997). This equation is a set of coupled, linear, first-order differential equations, and can be solved with the initial conditions,

$$N(m, 0) = \begin{cases} N', & \text{for } m = m_0 \\ 0, & \text{for } m \neq m_0 \end{cases}. \tag{7.7}$$

The equation (7.6) is what Gillespie (1975) chooses to call the stochastic collection equation. It is valid for the simple cloud model he prescribed.

In solving this coupled, linear first-order differential equation, there is no need to be concerned at the initial time with $N(m_0 - \mu, 0)$ as it is zero, because m_0 is the defined smallest drop size (7.4). So, the first three equations describing initial and subsequent growth of the initial droplet are given as

$$\frac{dN(m_0, t)}{dt} = AN'[-N(m_0; t)], \tag{7.8}$$

$$\frac{dN(m_0 + \mu, t)}{dt} = AN'[N(m_0; t) - N(m_0 + \mu; t)], \tag{7.9}$$

and

$$\frac{dN(m_0 + 2\mu, t)}{dt} = AN'[N(m_0 + \mu; t) - N(m_0 + 2\mu; t)], \tag{7.10}$$

and so on, to as many steps as are needed.

Equation (7.8) is solved with the initial conditions (7.7). Its solution can be substituted into (7.9), which can be used in (7.10) and so on. This procedure results in

$$N(m + k\mu, t) = \frac{N(AN't)^k \exp(-AN't)}{k!} \quad \text{for } k = 0, 1, 2, \ldots. \tag{7.11}$$

Equation (7.11) will give the number of drops of mass $m = m_0 + k\mu$, that can be found at any time t. Also, at $t > 0$ there is a spectrum of drops of various masses (i.e. a mass spectrum; Gillespie 1975). This is in contrast to the continuous model where all droplets have the same mass; there is no spectrum of drops.

The total number of drops at time t is found by writing $x = AN't$, using (7.7), and then by inspection,

$$\begin{aligned}\sum_{m=m_0}^{\infty} N(m, t) &= \sum_{k=0}^{\infty} \frac{Nx^k \exp(-x)}{k!} = N \exp(-x) \sum_{k=0}^{\infty} \frac{x^k}{k!} \\ &= N \exp(-x) \exp(x) = N(m_0, t = 0).\end{aligned} \tag{7.12}$$

The normalized power moments are given as the following in order to reveal features of the mass spectrum (Gillespie 1975),

$$M_j(t) \equiv \frac{1}{N} \sum_{m=m_0}^{\infty} m^j N(m, t) \quad j = 1, 2, 3, \ldots. \tag{7.13}$$

Gillespie notes that $M_j(t)$ is just the average $(\text{mass})^j$ of drops at time t. The average drop mass would be helpful to know. This is found by

$$M_1(t) \equiv \frac{1}{N} \sum_{m=m_0}^{\infty} mN(m, t) = \frac{1}{N} \sum_{k=0}^{\infty} (m_0 + k\mu) \frac{Nx^k \exp(-x)}{k!}, \tag{7.14}$$

$$M_1(t) = \exp(-x) \left[m_0 \sum_{k=0}^{\infty} \frac{x^k}{k!} + \mu x \sum_{k=0}^{\infty} \frac{x^{k-1}}{(k-1)!} \right], \tag{7.15}$$

$$M_1(t) = \exp(-x)[m_0 \exp(x) + \mu AN't \exp(x)], \tag{7.16}$$

and

$$M_1(t) = m_0 + \mu AN't. \tag{7.17}$$

The quantity, $M_1(t)$, is the average drop mass at time t, which is the center of the mass spectrum or center of the graph of N versus m. This model differs from the continuous model in that $M(t)$ represents the mass of all of the drops of mass m_0 that collect droplets of mass μ at a certain rate, in other words, the mass spectrum.

To add more information about the mass spectrum to the average drop mass, $M_1(t)$, the root-mean-square deviation of the spectrum gives the width of the spectrum of drops (remember there is no width with the continuous model with cloud included). The root-mean-squared deviation is provided by Gillespie (1975) as

$$\Delta t = \left[M_2(t) + M_1(t)^2\right]^{\frac{1}{2}}. \tag{7.18}$$

$M_2(t)$ is calculated from (7.11) and (7.13) the same way that $M_1(t)$ was calculated using (7.14). Gillespie also writes that the calculation of $M_2(t)$ is facilitated by writing $k^2 = [k(k-1)+k]$. It is found that the width of the spectrum is

$$\Delta(t) = \mu(ANt)^{\frac{1}{2}}. \tag{7.19}$$

Equation (7.19) shows that while the center of a graph of N versus m grows linearly with t, the width is proportional to $t^{1/2}$. There are two important points to be made from the analyses above. First, using (7.3) and (7.17), it can be written that $M_1(t) = M(t)$, which means that the drops together collect droplets at the same rate in the continuous growth and quasi-stochastic model for the cloud model specified. But as $\Delta(t)$ increases with time, the drops do not all grow at the same rate in the quasi-stochastic model. As pointed out by Gillespie (1975), some grow slower, and some faster than in the continuous growth model.

Pruppacher and Klett (1997) hint that the quasi-stochastic model is also too restrictive. They state that some drops of mass m will collect droplets of mass m at different rates; i.e., that drops of mass m collect droplets independent of other drops of mass m, and that some drop of mass m at the same starting time as another drop of mass m may collect a different number of droplets of mass m. From this point onward the probabilistic model is discussed.

7.3.3 The pure-stochastic collection equation for a model cloud

It has been noted that the quasi-stochastic collection equation is too restrictive by Gillespie (1975) and Pruppacher and Klett (1997) as it has the requirement that all m-drops together collect other drops and droplets at a definite rate in an idealized cloud. There is no fluctuation in the number of drops and droplets collected by a given drop as there would be in the real atmosphere. However, in a pure-stochastic model, such fluctuations in collection are permitted. As a result, there is no longer a number to associate with the number of drops and droplets collected by a drop at any time t. However, it is possible to predict the probability of finding a given number of m-drops of a particular size at time t with the pure-stochastic model. With this noted, there is a state function in the pure-stochastic model given by the following statement,

"$P(n, m, t) \equiv$ probability that exactly n drops have mass m at time t", (7.20)

where, $n = 0, 1, 2, 3, \ldots, N$, and $m = m_0, m_0 + \mu, m_0 + 2\mu \ldots$. Next we want to compute $P(n, m, t)$, which also requires considering the probability $P(k,t)$ that any drop will collect exactly k drops and droplets in time t. Pruppacher and Klett (1997) and Gillespie (1975) note this is just the expression for the common Poisson distribution,

$$\Pi(k,t) = \frac{(AN't)^k \exp(-AN't)}{k!}. \tag{7.21}$$

This is derived by starting with the definition of the probabilistic modeled cloud $AN'\mathrm{d}t$, which is the probability a drop will collect another drop or droplet in time $\mathrm{d}t$ (Gillespie 1975). For $k = 0$,

$$\Pi(0, t + \mathrm{d}t) = \Pi(0,t)(1 - AN'\mathrm{d}t), \tag{7.22}$$

or this can be written as

$$\frac{\mathrm{d}\Pi(0,t)}{\Pi(0,t)} = -AN'\mathrm{d}t. \tag{7.23}$$

Next, integrating with the initial condition, $\Pi(0,0) = 1$, gives

$$\Pi(0,t) = \exp(-AN't). \tag{7.24}$$

Therefore, for any $k \geq 1$ there is a probability that a drop will collect exactly k drops or droplets in time $(0, t)$ as denoted by

$$\Pi(k,t) = \int_0^t \Pi(k-1;t')\Pi(0,t-t')AN'\mathrm{d}t. \tag{7.25}$$

Gillespie (1975) interprets (7.25) as the product of the

$$\begin{aligned}&[\text{probability that the drop will collect exactly } k-1 \text{ drops in } (0,t')]\\ &\quad\times [\text{probability that the drop will collect one more drop in } \mathrm{d}t' \text{ at } t']\\ &\quad\times [\text{probability that the drop will collect no more drops in } (t,t')]\end{aligned} \tag{7.26}$$

and summed over all t' from 0 to t.

The equation (7.25) is a recursion relation for $\Pi(k,t)$. Using (7.24), formulas for $\Pi(k, 1)$, $\Pi(k, 2)$, $\Pi(k, t)$, and so on can be calculated, so that the following can be written,

$$\Pi(k,t) = \frac{(AN't)^k \exp(-AN't)}{k!}, k = 0, 1, 2, 3, \ldots. \tag{7.27}$$

Following Pruppacher and Klett (1997) and Gillespie (1975) closely, this means that as each drop collects droplets independently of other drops, and the probability is that some of the n drops will all collect exactly k droplets in time t whilst the other $N - n$ drops will not, it can be written

$$\Pi^n(k,t)[1 - \Pi(k,t)]^{N-n}. \tag{7.28}$$

The number of ways of arranging two groups of n drops and $N - 1$ drops from a set of N drops is given by,

$$\binom{N}{n} = \frac{N!}{[n!(N-n)!]}. \tag{7.29}$$

The probability that precisely n of the N drops will collect precisely k droplets in time t is given by the product of (7.28) and (7.29),

$$\binom{N}{n}\Pi^n(k,t)[1 - \Pi(k,t)]^{N-n}. \tag{7.30}$$

Using (7.30), it is possible to write

$$P(n, m_0 + k\mu, t) = \frac{N!}{n!(N-n)!}\Pi^n(k,t)[1 - \Pi(k,t)]^{N-n}$$
$$n = 0, 1, 2, 3 \ldots \text{ and } m = 0, 1, 2, 3 \ldots. \tag{7.31}$$

Equations (7.27) and (7.31) comprise the pure-stochastic solution for $P(n, m, t)$ for the simple drop and droplet cloud. At any time t, it is possible for there to be zero drops to N drops of mass m. This is in contrast to the quasi-stochastic model in which there will be exactly $N(m, t)$ drops of mass m.

The equation for the moments of $P(m, n, t)$ with respect to n, can be written

$$N_j(m,t) = \sum_{n=0}^{N} n^j P(n,m,t), j = 1, 2, \ldots. \tag{7.32}$$

Next, the first moment of $P(m, n, t)$ with respect to n has the following physical meaning. The value of $P(m, n, t)$ is the average of number of $N_1(m, t)$ of m-drops in the cloud at time t (average spectrum) and is given by

$$N_1(m,t) = \sum_{n=0}^{N} nP(n,m,t) = \sum_{n=0}^{N} \frac{nN!}{n!(N-n)!}\Pi^n(k,t)[1 - \Pi(k,t)]^{N-n}$$
$$= N\Pi(k,t) = N_m(t). \tag{7.33}$$

For the simple cloud, the average spectrum (7.33) is the solution for the stochastic collection equation. Equation (7.33) is equivalent mathematically to quasi-stochastic equation (7.11). However, $N(m, t)$ is the number of drops of mass m at time t, whereas, $N_1(m, t)$ is only the average number of drops of mass m at time t, which is the center of a graph of $P(n, m, t)$ versus n.

Following Pruppacher and Klett (1997), in analogy with the previous decision to choose $\Delta(t)$ as the width of the quasi-stochastic collection equation spectrum, the root mean square deviation of stochastic $P(m, n, t)$ is chosen with respect to n; or in other words, the following is true,

$$\Delta(m,t) \equiv \left[N_2(m,t) - N_1^2(m,t)\right]^{1/2}, \tag{7.34}$$

with $N_2(m, t)$ given as the second moment of $P(m, n, t)$ with respect to n. Using (7.31) and (7.32) to calculate $N_2(m, t)$, the following can be written,

$$\Delta(m,t) \equiv [N_1(m,t)]^{1/2}\left[1 - \frac{N_1^2(m,t)}{N}\right]^{1/2}. \tag{7.35}$$

Note that similarly to the quasi-stochastic case and according to Gillespie (1975), the calculation of $N_2(m, t)$, is made easier by using $n^2 = [n(n-1)+1]$.

The second factor on the right-hand side of (7.35) approaches unity as $t \to \infty$, which can be proved by (7.33). Somewhere between about

$$N_1(m,t) - [N_1(m,t)]^{1/2} \text{ and } N_1(m,t) + [N_1(m,t)]^{1/2} \tag{7.36}$$

drops of mass m in the cloud should be expected to be found at time t.

Gillespie (1972) described an analysis of a cloud strictly from the point of view of the Poisson model or pure-stochastic model. It was shown in Gillespie (1975) that if (i) certain correlations between drops and droplets can be neglected and (ii) coalescences of drops the same size will not occur, then we can state something about the standard stochastic collection equation, i.e. that it describes the mean drop-size spectrum $N_1(m, t)$. In this case, it can be shown as in Gillespie (1972) that the function $P(n, m, t)$ tends to the Poisson form (Gillespie 1975),

$$\lim_{t\to\infty} P(n,m;t) = \frac{N_1^n(m,t)\exp[-N_1(m,t)]}{n!}, \tag{7.37}$$

and the spectrum width approaches the following,

$$\Delta(m,t) \to [N_1(m,t)]^{1/2}. \tag{7.38}$$

If the simplifying conditions above, i.e. (i) and (ii) are not met, the situation becomes unclear. This is a result of the necessity of conditional probabilities given by: $P^{(1)}(n,m|n',m';t)$ = probability of having n drops of mass m at time t,

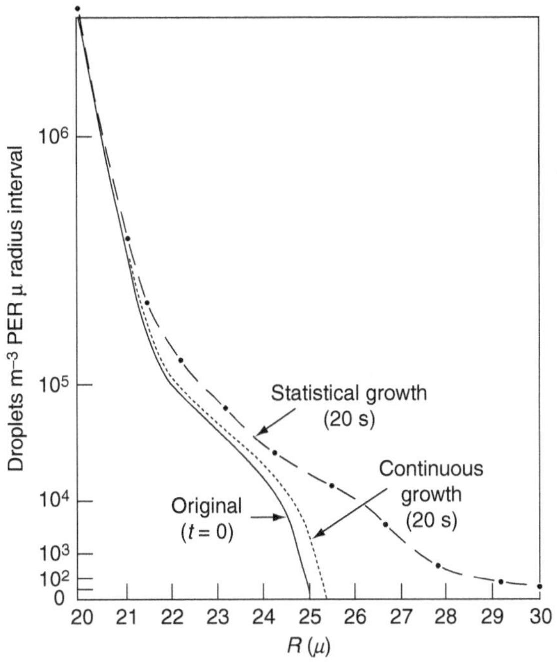

Fig. 7.2. The initial distribution and distribution after 20 s (on a cube root scale for number concentration per micron). (From Twomey 1964; courtesy of the American Meteorological Society.)

given that there are n' drops of mass m'; and $P^{(2)}(n, m|n', m'; n'', m''; t)$ = probability of having n drops of mass m at time t, given that there are n' drops of mass m' and n'' drops of mass m'', and so on.

For the drop–droplet cloud, $P^{(1)}(n, m|n', m'; t)$ is given by

$$P^{(1)}(n, m|n', m'; t) = \frac{(N-n')!}{n!(N-n'-n)!}\Pi^n(k,t)[1-\Pi(k,t)]^{N-n'-n} \tag{7.39}$$

$$n = 0, 1, 2, 3, \ldots, N \text{ and } m = 0, 1, 2, 3 \ldots,$$

$$P^{(2)}(n, m|n', m'; n'', m''; t) = \frac{(N-n'-n'')!}{n!(N-n'-n''-n)!}\Pi^n(k,t)[1-\Pi(k,t)]^{N-n'-n''-n} \tag{7.40}$$

$$n = 0, 1, 2, 3, \ldots, N \text{ and } m = 0, 1, 2, 3 \ldots,$$

and $P^{(3)} = \cdots$ etc., and so on.

The no-correlation approximation (i) leads only to an approximation of the conditional probabilities by $P(n, m; t)$. Unfortunately, there are no ways to evaluate exactly these probabilities, P, $P^{(1)}$, $P^{(2)}$, $P^{(3)}$, ... Gillespie (1972) points to Bayewitz *et al.* (1974) for a possible answer based on the extent of correlations in a stochastic coalescence process. Some slight amount of information is provided in Pruppacher and Klett (1997), but perhaps not enough to evaluate the probabilities fully.

A comparison of the continuous and statistical (Poisson) growth models is shown in Fig. 7.2, reproduced here from Twomey (1964). This figure shows rapid broadening of the drop spectrum after 20 s for the probabilistic model whereas the continuous model shows only very slow growth of the spectrum.

7.4 Terminal velocity

Essential to understanding the collection growth of a particle is the understanding of the terminal velocities of the particles involved in the process. There are many empirical and derived values for terminal velocity for varying types of particles. The easiest and simplest description of terminal velocity is the terminal velocity of a sphere.

The terminal velocity of an object is the maximum speed to which an object will accelerate in freefall resulting from a balance of drag and gravity forces. The gravity force accelerates a particle downward, whereas the drag force is a result of resistance of air molecules to the motion of an object.

The gravity force is

$$F_{\mathrm{g}} = gm = gV(\rho_x - \rho), \tag{7.41}$$

where V is volume (for a sphere, $V = D^3\pi/6$), m is mass, $g = 9.8$ m s^{-2} is the acceleration owing to gravity, ρ_x is the density of the particle, and ρ is the density of the air. Typically, the density of the particle is 100 to 1000 times greater than the density of air at mean sea level, so the density of air can be neglected in this equation. Using this approximation and the definition of volume for a sphere, the gravitational force is,

$$F_{\mathrm{g}} = gV\rho_x = g(\pi/6)D^3\rho_x. \tag{7.42}$$

The drag force is

$$F_{\mathrm{d}} = Ac_{\mathrm{d}}\rho u_\infty^2 = (\pi/8)D^2 c_{\mathrm{d}}\rho u_\infty^2, \tag{7.43}$$

where $c_{\mathrm{d}} = 0.6$ is the drag coefficient, u_∞ is the terminal velocity, A is the area (for a sphere, $A = (\pi/8)D^2$).

The terminal velocity is found by equating the gravity force (7.42) and the drag force (7.43). Thus, $F_{\mathrm{g}} = F_{\mathrm{d}}$ or,

$$(\pi/8)D^2 c_{\mathrm{d}}\rho u_\infty^2 = g(\pi/6)D^3\rho_x, \tag{7.44}$$

which can be solved for u_∞,

$$u_\infty = \left(\frac{4}{3}\frac{gD}{c_d}\frac{\rho_x}{\rho}\right)^{\frac{1}{2}}. \tag{7.45}$$

7.5 Geometric sweep-out area and gravitational sweep-out volume per unit time

Traditionally the gravitational collection kernel for hydrodynamical capture (Rogers and Yau 1989; Pruppacher and Klett 1997) is defined as being related to the geometric sweep-out area, which is given as a function of the downward projected area or footprint of the two colliding particles of diameters D_1 and D_2. This can be written as,

$$\text{geometric sweep-out area} \equiv \frac{\pi}{4}(D_1 + D_2)^2, \tag{7.46}$$

where $D_1 > D_2$ is assumed. This equation does not take into account the deviation of small droplets approaching a larger drop and being pushed away from the larger drop by dynamic pressure forces such that the droplet is not captured by or sticks to the larger drop without tearing away. These influences are usually incorporated in the collision efficiencies (E_{coll}) and coalescence (E_{coal}) efficiencies. These are probabilities of a collision, and the probability of sticking after collision.

Now the collision efficiency and coalescence efficiency together when multiplied give the collection efficiency (E_{collect}). Next the geometric sweep-out area is multiplied by these efficiencies (E_{coll} and E_{coal}) or the collection efficiency (E_{collect}) as well as the difference of particle D_1's terminal velocity $V_T(D_1)$ and particle D_2's terminal velocity $V_T(D_2)$ to get the geometric sweep-out volume per second or the gravitational collection kernel, K,

$$K(D_1, D_2) = E_{\text{coll}}E_{\text{coal}}\frac{\pi}{4}(D_1 + D_2)^2[V_T(D_1) - V_T(D_2)]. \tag{7.47}$$

This general form (Rogers and Yau 1989; Pruppacher and Klett 1997) is used in nearly all models now, with only a few exceptions (e.g. Cohard and Pinty 2000; see next section).

7.6 Approximate polynomials to the gravitational collection kernel

Other models use polynomials for the collection kernel. Long (1974) devised a number of different polynomials to represent the gravitational collection kernel in search of analytical solutions to the quasi-stochastic collection equation. These have limited usefulness in general. The first step to generating these polynomials is to calculate the gravitational collection kernel accurately. Then the second step is to fit an approximation to it. Long used the collision efficiencies of Shafrir and Gal-Chen (1971) to provide the rationale for developing these polynomials.

Two of Long's gravitational collection kernels given by Pruppacher and Klett (1981) are the following in the case where $v > u$ where v and u are dimensions of D^3,

$$P(v,u) = \begin{cases} 1.10 \times 10^9 (v^2) & \text{for } 20 \leq D \leq 100 \text{ microns} \\ 6.33 \times 10^3 (v) & \text{for } D > 100 \text{ microns} \end{cases} \tag{7.48}$$

and

$$P(v,u) = \begin{cases} 9.44 \times 10^9 (v^2 + u^2) & \text{for } 20 \leq D \leq 100 \text{ microns} \\ 5.78 \times 10^3 (v + u) & \text{for } D > 100 \text{ microns.} \end{cases} \tag{7.49}$$

The leading coefficients are derived from minimizing the root mean square of the logarithm of the approximating polynomial and the collection kernel. The logarithm is used because of the wide range in values of the gravitational kernel, and the need for accurate solution over this wide spectrum. Note that these approximating polynomials do not include the influence of turbulence or other possible influences. Cohard and Pinty (2000) are the latest group of investigators to use these polynomials (at least the second set of these polynomials).

7.7 The continuous collection growth equation as a two-body problem

In general, collection is a many-body collection problem. However, for larger drops in a population of much smaller drops, arguments can be made that collection growth can be simplified to a two-body problem. The justification for the two-body collection problem is provided by example. If there are 1000 droplets per cm^3 (1×10^9 m^{-3}) and the average droplet size is about 10 microns in diameter, then the average spacing of the droplets is on the order of about 100 droplet diameters. In this case, the droplets can be said to be relatively sparse in the cloud. Larger drops are found on the order of about 1000 to 10 000 m^{-3}.

For two-body problems it is assumed then that there are two spherical drops of different sizes and that shape effects are considered to be of secondary importance for the basic problem. A larger drop collecting a smaller droplet is considered. The drop and droplet are relatively distant from each

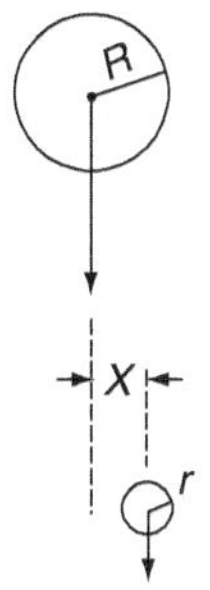

Fig 7.3. Particle collision geometry, where R is the radius of the collector, r is the radius of the collected particle, and x is the distance of the collected particle from the fall line of the larger particle. (From Rogers and Yau 1989; courtesy of Elsevier.)

other initially. The droplet and drop are assumed to fall at their terminal velocities and are widely spaced in the vertical. The air and internal motions of the drop and droplet are assumed to be calm.

The basic physical model is one of finding the sweep-out area of the two-body problem. First the effective cross-sectional area for the drop and droplet must be found. In the model R_L is the radius of the large drop and R_S is the radius of the smaller droplet. The geometric sweep-out area for the drop and droplet is

$$\text{geometric sweep-out area} = \pi(R_L + R_S)^2. \tag{7.50}$$

The simple geometric sweep-out volume per second swept out in this two-body system is given as function of radius R as

$$\text{geometric sweep-out volume per second} = \pi(R_L + R_S)^2(V_{TL} - V_{TS}), \tag{7.51}$$

which is shown in Fig. 7.3; or the geometric sweep-out volume per second can be given as a function of diameter D as

$$\text{geometric sweep-out volume per second} = \frac{\pi}{4}(D_L + D_S)^2(V_{TL} - V_{TS}), \tag{7.52}$$

where V_{TL} and V_{TS} are the terminal velocities of the larger and smaller drop and droplet, respectively. The volume increase per unit time of the larger drop is related to the geometric sweep-out volume per unit time and the number and size of the smaller droplets in the sweep-out volume (now no longer a two-body problem),

$$\begin{aligned}\frac{dV_L}{dt} &= \pi(R_L + R_S)^2(V_{TL} - V_{TS})V_S N_S \\ &= \pi(R_L + R_S)^2(V_{TL} - V_{TS})\pi\frac{4}{3}R_S^3 N_{TS}.\end{aligned} \tag{7.53}$$

Now, if the increase in the mass of the large drop can also be written by including the density of the drop the following can be written,

$$\frac{\mathrm{d}M_\mathrm{L}}{\mathrm{d}t} = \rho\pi(R_\mathrm{L} + R_\mathrm{S})^2(V_\mathrm{TL} - V_\mathrm{TS})\pi\frac{4}{3}R_\mathrm{S}^3 N_\mathrm{TS}. \tag{7.54}$$

This is a highly simplified expression of the two-body formulation. In reality, the geometric sweep-out volume should take into account the aerodynamic pressure forces that can push a droplet out of the way of a collector drop. This is done by defining grazing trajectories, which are the trajectories of the droplets furthest from the fall line of the collector drops that just make contact with the collector drops. These are found at radius R_G from the fall line of the larger drop.

This growth is often called the continuous growth equation in terms of change of R_L. With some simple algebra, the rate of mass increase of the drop (7.54) can be written in terms of the rate of increase of the radius of the drop,

$$\frac{\mathrm{d}R_\mathrm{L}}{\mathrm{d}t} = \pi\frac{N_\mathrm{S}}{3}\left(\frac{R_\mathrm{S}^3}{R_\mathrm{L}^2}\right)(R_\mathrm{L} + R_\mathrm{S})^2(V_\mathrm{TL} - V_\mathrm{TS}). \tag{7.55}$$

For a sphere, the mass of the larger drop can be written as

$$M_\mathrm{L} = \rho_\mathrm{L} V_\mathrm{L} = \rho_\mathrm{L}\pi\frac{4}{3}R_\mathrm{L}^3. \tag{7.56}$$

Now using (7.56), (7.54) becomes

$$\frac{\mathrm{d}M_\mathrm{L}}{\mathrm{d}t} = \rho_\mathrm{L} 4\pi R_\mathrm{L}^2 \frac{\mathrm{d}R_\mathrm{L}}{\mathrm{d}t}, \tag{7.57}$$

which completes the simple two-body problem, except there has been no consideration for collection efficiencies. The collection efficiency is incorporated as a factor $E(R_\mathrm{L}, R_\mathrm{S})$, such that

$$\frac{\mathrm{d}R_\mathrm{L}}{\mathrm{d}t} = \pi\frac{N_\mathrm{S}}{3}\left(\frac{R_\mathrm{S}^3}{R_\mathrm{L}^2}\right)(R_\mathrm{L} + R_\mathrm{S})^2(V_\mathrm{TL} - V_\mathrm{TS})E(R_\mathrm{L}, R_\mathrm{S}). \tag{7.58}$$

The collection efficiencies are computed with the collision efficiencies multiplied by the coalescence efficiencies, and are smaller than either of the components by up to 30–90 or so percent.

7.8 The basic form of an approximate stochastic collection equation

The basic form of the approximate stochastic collection equation for say, liquid drops is given as

$$\frac{\partial N_k}{\partial t} = \frac{1}{2}\sum_{i=1}^{k-1} K_{i,k-i} N_i N_{k-i} - N_k \sum_{i=1}^{\infty} K_{i,k} N_i, \tag{7.59}$$

where N is number concentration times volume and K is related to the collection kernel. The first sum is called the gain sum for mass bin k. Droplets of mass k are produced by collisions between droplets of mass bins i, and $k - i$. The factor of 1/2 prevents double counting. For example if mass bin $k = 5$ is in consideration, then droplets in bin $k - i = 3$ ($i = 2$) plus droplets in bin 2 make drops of the mass of those in bin $k = 5$. Now, depending on how this is programmed, droplets in mass bin $k - i = 2$ ($i = 3$) plus droplets in mass bin 3 also make droplets of the same mass as those in bin 5. Therefore double counting occurs. The second sum is a loss term for mass bin k. Droplets of size k are lost by coalescence with drops of all other sizes.

7.9 Quasi-stochastic growth interpreted by Berry and Reinhardt

In this section the results of Berry and Rienhardt's (1974a–d) studies on collection are briefly examined. First, their basic definitions must be defined to understand the processes they explain. A drop spectrum is denoted by the density function $f(x)$, where $f(x)$ is, as described earlier, the number of drops per unit volume of air in the size interval x, $x + dx$, where x here is the mass of a drop of any size. Quite simply, it is known that mass and volume are related by $x = \rho_L V$ where V is volume and ρ_L is the density of water. Moreover the density functions $f(x)$ and $n(V)$ are related by $f(x)dx = n(V)dV$. In their work, a mass density function $g(x) = xf(x)$ was defined. The concentration of droplets N and liquid-water content L are defined as integrals over the density functions $f(x)$ and $g(x)$ as follows,

$$N = \int f(x)\mathrm{d}x \tag{7.60}$$

and

$$L = \int g(x)\mathrm{d}x. \tag{7.61}$$

Now the mean mass of a droplet is given simply by $x_f = \frac{L}{N}$ or

$$x_f = \frac{L}{N} = \frac{\int xf(x)\mathrm{d}x}{\int f(x)}. \tag{7.62}$$

Drops with mean mass x_f have mean radius given simply by

$$r_f = \left(\frac{3\pi\rho_L}{4}\right)^{1/3} x_f^{1/3}. \tag{7.63}$$

The mean mass of the density function is given as

$$x_g = \frac{\int xg(x)\mathrm{d}x}{\int g(x)} = \frac{\int x^2 f(x)\mathrm{d}x}{\int xf(x)}. \tag{7.64}$$

The value of the equivalent radius of x_g is given as

$$r_g = \left(\frac{3\pi\rho_L}{4}\right)^{1/3} x_g^{1/3}. \tag{7.65}$$

Berry and Reinhardt's (1974a–d) simulations start with an initial distribution of drops given by a gamma distribution. The relative variance is given by

$$var\ x = \frac{x_g}{x_f} - 1. \tag{7.66}$$

In the problem at hand, the given initial conditions make *var x* equal to 1. That is with $r_f = 12$ microns, water content equal to 1 g m^{-3}, and a concentration of 166 drops per cm^3. In displaying plots, they present a function best described as a log-increment mass density function $g(\ln r)$, with a design so that $g(\ln r)\mathrm{d}(\ln r)$ is the mass of cloud drops per unit volume of air with radii in $\mathrm{d}(\ln r)$. The reason that this is done is that it de-emphasizes the smallest drops and emphasizes the drops of interest, which are the large drops. This results in a density function of $g(\ln r) = 3x^2 f(x)$ (see Berry and Reinhardt 1974a–d). Of interest in the results is that if there are drops with initial size greater than 20 mm in radius, two modes develop from a single mode: the mode of small drops shortens and the distribution locally widens and follows r_f of the solution closely; whereas large drops increase and follow r_g closely.

Various physical processes lead to broadening of the droplet solution, which are essential to understanding the growth of initial distributions. These are carefully summarized by Berry and Reinhardt (1974a–d) and are reproduced here. For brevity's sake and for clarity, their discussion is followed closely. Two modes are possible, one called spectrum S1 centered at 10 microns with a water content of 0.8 g m^{-3}, and the spectrum S2 centered at 20 microns with a water content of 0.2 g m^{-3}. The different physical processes that need to be parameterized in bulk parameterization models are shown in Figs. 7.4–7.7. Four assumptions are examined:

(a) The first assumption is that collisions between all drop pairs are allowed (Fig. 7.4).
(b) The second assumption is that collisions are only permitted for drops in S1 (Fig. 7.5).
(c) The third assumption is that collisions can only take place between a drop in S2 and one in S1 (Fig. 7.6).
(d) The fourth assumption is that collisions can only take place between drops in S2 (Fig. 7.7).

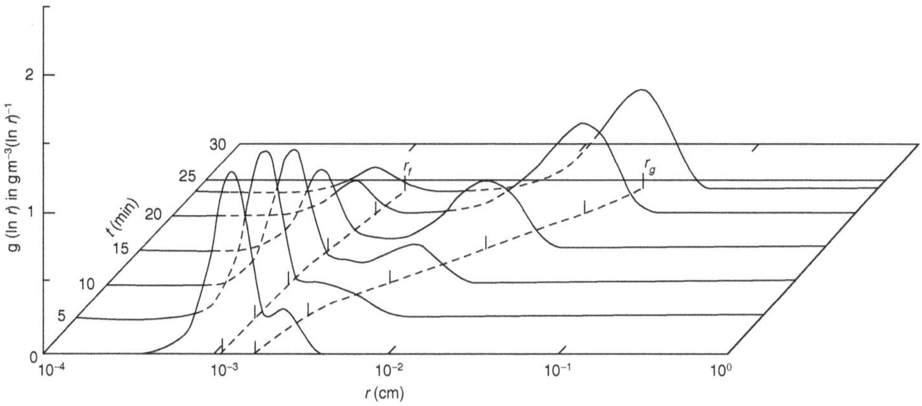

Fig. 7.4. Time evolution of the initial spectrum composed of 0.8 g m^{-3} with small drops centered at radius $r_f^0 = 10\,\mu m$, and 0.2 g m^{-3} with larger drops centered at radius $r_f^0 = 50\,\mu m$, both with *var x* = 1. The variables r_f and r_g are the final mean values of the smaller drop and larger drop spectra, respectively. (From Berry and Reinhardt 1974a; courtesy of the American Meteorological Society.)

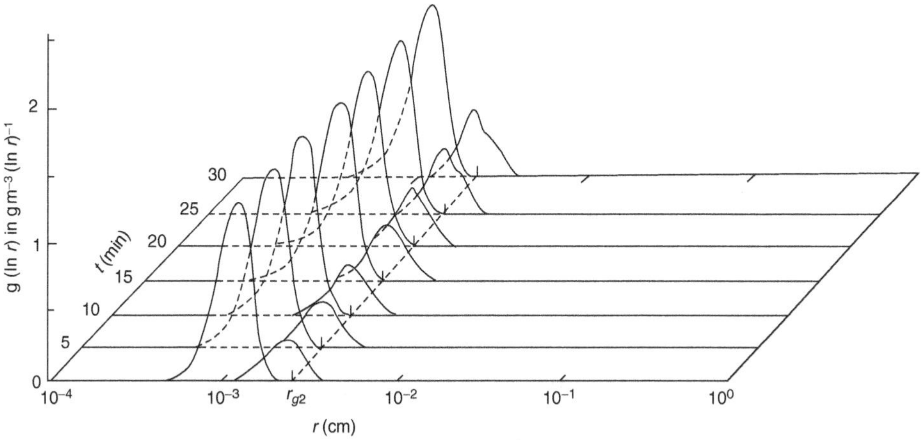

Fig. 7.5. Time evolution of the initial spectrum given in Fig. 7.4 with only S1–S1 interactions allowed. The interactions S1–S1 are called autoconversion and are interactions between small droplets to produce S2 droplets. Only a small amount of liquid water is converted to S2 droplets. With no growth of S2 droplets allowed no further growth of the S2 spectrum occurs. (From Berry and Reinhardt 1974a; courtesy of the American Meteorological Society.)

In summary, there are three modes in general that lead to large drops or change the large-drop distribution. These are those processes described by considering each assumption individually, starting with (b) which is called autoconversion (care must be taken in understanding what is really

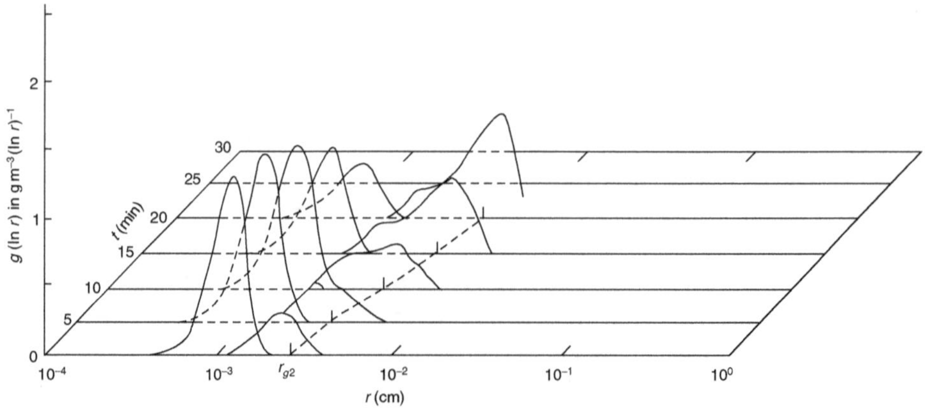

Fig. 7.6. Time evolution of the initial spectrum given in Fig. 7.4 with only S2–S1 interactions allowed. The S2–S1 interactions are called accretion and involve the collection of small S1 droplets by larger S2 drops. This mode adds liquid water to S2 but at a rate much larger than the S1–S1 autoconversion rates. The mode acts quickly at first but then depletes the S1 droplets and the growth of the S2 drops stabilizes. Still some reason must be found to explain the production of much larger drops. (From Berry and Reinhardt 1974a; courtesy of the American Meteorological Society.)

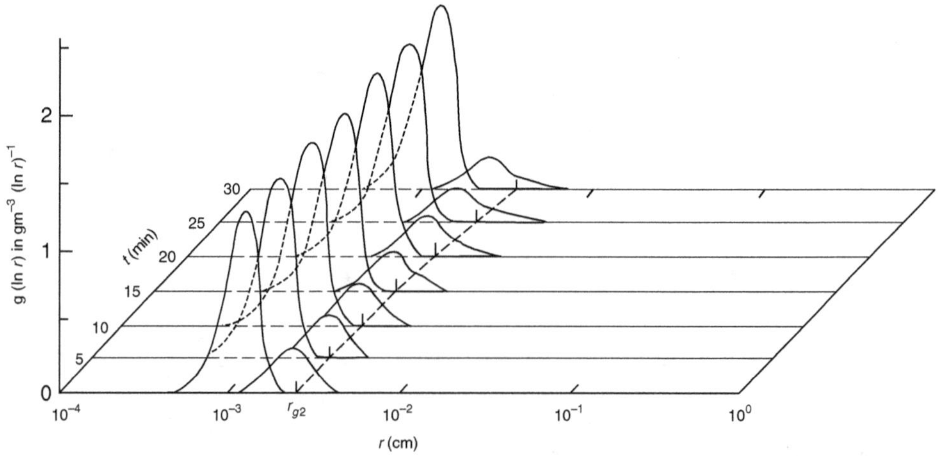

Fig. 7.7. Time evolution of the initial spectrum given in Fig. 7.4 with only S2–S2 interactions allowed. The S2–S2 interactions are called large-hydrometeor self-collection and involve the collection of large S2 drops by large S2 drops. With S2–S2 self-collection interactions only the S2 drops grow. The process here is slow as the liquid water in S2 is small. Ultimately it is this mode that produces large drops quickly. (From Berry and Reinhardt 1974a; courtesy of the American Meteorological Society.)

autoconversion); (c) which is called accretion; and (d) which is called large-hydrometeor self-collection.

The result from case (a) includes all of these possibilities. In (a) S1 is depleted but the S1 mode does not change significantly, whilst the S2 mode changes shape substantially. In case (b), S2 gains only by interaction of S1 drops with S1 drops, and the influence is marginal in producing S2 drops only on the left side of the S2 distribution. With no S2-drop interactions there is no significant growth of the S2 spectrum. In case (c) the rate transfer of water to S2 drops is much faster than the previous cases showing that accretion between S2 collecting S1 drops is more important than the process called autoconversion, here in transferring water to S2 drops. Finally, in case (d) it is seen that the interaction amongst the large S2 drops, leads to distribution flattening, though this progression is slow.

In Berry and Reinhardt (1974a–d) it is stated that it is autoconversion that begins the process of transferring water from S1 to S2 drops and allows other mechanisms to operate. Its rate is almost always slower than other rates; whilst small, it is essential to droplet distribution evolution. It is accretion that is the primary mechanism for transferring drops from S1 to S2. Though S1 generally maintains its shape and position during accretion, S1 loses water to S2. Growth is initially quick for the S2 spectrum; then it slows as all of S1 is consumed by S2. The large-drop self-collection S2 with S2 interactions makes S2 distribution flatter, though growth of the spectrum by S2–S2 self-collections tends to be relatively slow. This process is increased as water mass is added to S2 from S1 drops by accretion, which is the main mechanism of the growth of the spectrum in general. Without water from S1 drops the S2 distribution just becomes flatter and eventually growth slows. The rate of growth by S2–S2-drop interactions is initially slow, but as S1 is added to S2 primarily by accretion, it is the S2–S2-drop interactions that cause rapid growth of the large-drop tail of the spectrum, which means an increasing x_g.

7.10 Continuous collection growth equation parameterizations

7.10.1 Gamma distribution function for continuous collection equation

The continuous collection equation is still used in many models in cases where larger particles such as rain, snow, graupel, frozen drops, and hail collect cloud water and cloud ice. In addition, it is used to make quick calculations to determine if an autoconversion should be activated. This continuous collection growth equation takes its simple form from roots described in the earlier part of this chapter, Rogers and Yau (1989), and Pruppacher and Klett

(1997), among others. The basic equation that is solved is the following where particle x is collecting particle y,

$$\frac{1}{\rho}\int_0^\infty \frac{\mathrm{d}M(D_x)n(D_x)}{\mathrm{d}t} = \frac{\mathrm{d}Q_x}{\mathrm{d}t} = Q_xAC_y = \int_0^\infty \frac{\pi}{4}D_x^2 E_{xy}Q_y V_{\mathrm{T}x}\left(\frac{\rho_0}{\rho}\right)^{1/2} n(D_x)\mathrm{d}D_x, \quad (7.67)$$

which can be written for the complete gamma distribution as

$$Q_xAC_y = \int_0^\infty \frac{\pi}{4}D_x^2 E_{xy}Q_y\mu_x\alpha_x^{\nu_x}c_xD_x^{d_x}\left(\frac{\rho_0}{\rho}\right)^{1/2}\frac{N_{\mathrm{T}x}}{\Gamma(\nu_x)}\left(\frac{D_x}{D_{nx}}\right)^{\nu_x\mu_x-1} \times \exp\left(-\alpha_x\left[\frac{D_x}{D_{nx}}\right]^{\mu_x}\right)\mathrm{d}\left(\frac{D_x}{D_{nx}}\right), \quad (7.68)$$

or next the modified gamma distribution

$$Q_xAC_y = \int_0^\infty \frac{\pi}{4}D_x^2 E_{xy}Q_y\mu_x c_xD_x^{d_x}\left(\frac{\rho_0}{\rho}\right)^{1/2}\frac{N_{\mathrm{T}x}}{\Gamma(\nu_x)}\left(\frac{D_x}{D_{nx}}\right)^{\nu_x\mu_x-1} \times \exp\left(-\left[\frac{D_x}{D_{nx}}\right]^{\mu_x}\right)\mathrm{d}\left(\frac{D_x}{D_{nx}}\right). \quad (7.69)$$

Then the gamma distribution is

$$Q_xAC_y = \int_0^\infty \frac{\pi}{4}D_x^2 E_{xy}Q_y c_xD_x^{d_x}\left(\frac{\rho_0}{\rho}\right)^{1/2}\frac{N_{\mathrm{T}x}}{\Gamma(\nu_x)}\left(\frac{D_x}{D_{nx}}\right)^{\nu_x-1} \times \exp\left(-\left[\frac{D_x}{D_{nx}}\right]\right)\mathrm{d}\left(\frac{D_x}{D_{nx}}\right). \quad (7.70)$$

Each can be rewritten; (7.68) using the complete gamma distribution by choosing appropriate values for α, ν, μ:

$$Q_xAC_y = D_{nx}^{2+d_x}\int_0^\infty \frac{\pi}{4}c_x\left(\frac{D_x}{D_{nx}}\right)^{2+d_x} E_{xy}Q_y\mu_x\alpha_x\left(\frac{\rho_0}{\rho}\right)^{1/2}\frac{N_{\mathrm{T}x}}{\Gamma(\nu_x)}\left(\frac{D_x}{D_{nx}}\right)^{\mu_x\nu_x-1} \times \exp\left(-\alpha_x\left[\frac{D_x}{D_{nx}}\right]^{\mu_x}\right)\mathrm{d}\left(\frac{D_x}{D_{nx}}\right); \quad (7.71)$$

(7.69) with the modified gamma distribution,

$$Q_xAC_y = D_{nx}^{2+d_x} \int_0^\infty \frac{\pi}{4} c_x \left(\frac{D_x}{D_{nx}}\right)^{2+d_x} E_{xy}\mu_x Q_y \left(\frac{\rho_0}{\rho}\right)^{1/2} \frac{N_{Tx}}{\Gamma(\nu_x)} \left(\frac{D_x}{D_{nx}}\right)^{\nu_x\mu_x - 1} \times \exp\left(-\left[\frac{D_x}{D_{nx}}\right]^{\mu_x}\right) d\left(\frac{D_x}{D_{nx}}\right); \tag{7.72}$$

and (7.70) with the gamma distribution,

$$Q_xAC_y = D_{nx}^{2+d_x} \int_0^\infty \frac{\pi}{4} c_x \left(\frac{D_x}{D_{nx}}\right)^{2+d_x} E_{xy} Q_y \left(\frac{\rho_0}{\rho}\right)^{1/2} \frac{N_{Tx}}{\Gamma(\nu_x)} \left(\frac{D_x}{D_{nx}}\right)^{\nu_x - 1} \times \exp\left(-\left[\frac{D_x}{D_{nx}}\right]\right) d\left(\frac{D_x}{D_{nx}}\right). \tag{7.73}$$

Then each can be simplified to the following final expressions: the continuous collection equation for the complete gamma distribution,

$$Q_xAC_y = \frac{\pi}{4} c_x D_{nx}^{2+d_x} E_{xy} Q_y \frac{N_{Tx}\alpha_x^{\nu_x}}{\Gamma(\nu_x)} \left(\frac{\rho_0}{\rho}\right)^{1/2} \frac{\Gamma\left(\frac{2+d_x+\mu_x\nu_x}{\mu_x}\right)}{\alpha_x^{\left(\frac{2+d_x+\mu_x\nu_x}{\mu_x}\right)}}; \tag{7.74}$$

for the modified gamma distribution,

$$Q_xAC_y = \frac{\pi}{4} c_x D_{nx}^{2+d_x} E_{xy} Q_y \frac{N_{Tx}}{\Gamma(\nu_x)} \left(\frac{\rho_0}{\rho}\right)^{1/2} \Gamma\left(\frac{2 + d_x + \nu_x\mu_x}{\mu_x}\right); \tag{7.75}$$

and for the gamma distribution,

$$Q_xAC_y = \frac{\pi}{4} c_x D_{nx}^{2+d_x} E_{xy} Q_y \frac{N_{Tx}}{\Gamma(\nu_x)} \left(\frac{\rho_0}{\rho}\right)^{1/2} \Gamma(2 + d_x + \nu_x). \tag{7.76}$$

Verlinde *et al.* (1990) studied the integration of the continuous growth equation directly and analytically. A figure of terminal velocities used is shown in Verlinde *et al.* (1990, their Figure 1). Also in Verlinde *et al.* (1990, their Figure 2) errors are plotted on a graph of the diameter of the collectee drop versus the diameter of the collector drop. For the case when a faster moving particle is collecting a smaller, much slower moving particle, it is seen that the smallest errors are in the middle ranges of sizes, the second largest errors are at all collector sizes for medium-sized collected particles, and the largest errors ($>$ O[10]), surprisingly, are at the large collector sizes. As discussed by Twomey (1964) and Verlinde *et al.* (1990), the largest collectors have the largest errors collecting smaller particles as collection efficiencies are not prescribed, but rather are set to unity. When collector and collectee particles fall at about the same speed, then significant (> O[0]) errors are incurred at the smaller collector sizes (< 4 mm) that collect particles of 0.05 to 3 mm. A region of 40% errors is found diagonally through the mid section of the plot. The rest of the plot shows generally good agreement with larger-sized collector particles collecting smaller-sized particles.

7.10.2 Log-normal distribution for continuous collection

The equation for continuous collection, where particle x is collecting particle y is

$$\begin{aligned}\frac{1}{\rho_0}\int_0^\infty \frac{\mathrm{d}M(D_x)n(D_x)}{\mathrm{d}t} = \frac{\mathrm{d}Q_x}{\mathrm{d}t} &= Q_xAC_y \\ &= \int_0^\infty \frac{\pi}{4}D_x^2 E_{xy}Q_y c_x D_x^{d_x}\left(\frac{\rho_0}{\rho}\right)^{1/2} n(D_x)\mathrm{d}D_x.\end{aligned} \tag{7.77}$$

The log-normal number spectral distribution function is defined as

$$n(D_x) = \frac{N_{\mathrm{T}x}}{\sqrt{2\pi}\sigma_x D_x}\exp\left(-\frac{[\ln(D_x/D_{nx})]^2}{2\sigma_x^2}\right). \tag{7.78}$$

Substitution of (7.78) into (7.77) results in

$$\begin{aligned}\frac{1}{\rho_0}\int_0^\infty \frac{\mathrm{d}M(D_x)n(D_x)}{\mathrm{d}t} = \frac{\mathrm{d}Q_x}{\mathrm{d}t} &= Q_xAC_y \\ &= \int_0^\infty \frac{\pi}{4}D_x^2 E_{xy}Q_y c_x D_x^{d_x}\left(\frac{\rho_0}{\rho}\right)^{1/2} n(D_x)\mathrm{d}D_x.\end{aligned} \tag{7.79}$$

Dividing all D_x terms by D_{nx} gives

$$\begin{aligned}Q_xAC_y = \frac{\pi}{4}\left(\frac{\rho_0}{\rho}\right)^{1/2} &\frac{E_{xy}Q_y c_x N_{\mathrm{T}x} D_{nx}^{d_x+2}}{\sqrt{2\pi}\sigma_x} \\ \times &\int_0^\infty \left(\frac{D_x}{D_{nx}}\right)^{d_x+1} \exp\left(-\frac{[\ln(D_x/D_{nx})]^2}{2\sigma_x^2}\right)\mathrm{d}\left(\frac{D_x}{D_{nx}}\right).\end{aligned} \tag{7.80}$$

Now letting $u = D_x/D_{nx}$,

$$Q_xAC_y = \frac{\pi}{4}\left(\frac{\rho_0}{\rho}\right)^{1/2}\frac{E_{xy}Q_yc_xN_{Tx}D_{nx}^{d_x+2}}{\sqrt{2\pi}\sigma_x}\int_0^\infty u^{d_x+1}\exp\left(-\frac{[\ln u]^2}{2\sigma_x^2}\right)du. \qquad (7.81)$$

By letting $y = \ln(u)$, $u = \exp(y)$, $du/u = dy$, so,

$$Q_xAC_y = \frac{\pi}{4}\left(\frac{\rho_0}{\rho}\right)^{1/2}\frac{E_{xy}Q_yc_xN_{Tx}D_{nx}^{d_x+2}}{\sqrt{2\pi}\sigma_x}\int_{-\infty}^\infty \exp[(d_x+2)y]\exp\left(-\frac{y^2}{2\sigma_x^2}\right)dy, \qquad (7.82)$$

where the limits of the integral change as u approaches zero from positive values, and $\ln(u)$ approaches negative infinity. Likewise, for the upper limit, as u approaches positive infinity, $\ln(u)$ approaches positive infinity.

Now the following integral definition is applied:

$$\int_{-\infty}^\infty \exp(2b'x)\exp(-a'x^2)dx = \sqrt{\frac{\pi}{a'}}\exp\left(\frac{b'^2}{a'}\right) \qquad (7.83)$$

by allowing $y = x$, $a' = 1/(2\sigma^2)$, $b' = (d_x+2)/2$. Therefore (7.82) becomes the expression for the continuous collection equation with a log-normal distribution,

$$Q_xAC_y = \frac{\pi}{4}\left(\frac{\rho_0}{\rho}\right)^{1/2} E_{xy}Q_yc_xN_{Tx}D_{nx}^{d_x+2}\exp\left(\frac{\sigma_x^2(d_x+2)^2}{2}\right). \qquad (7.84)$$

7.11 Gamma distributions for the general collection equations

A more general collection equation used in models can be represented in several ways. First the equations can be integrated using the Wisner *et al.* (1972) approximation, where $[V_{Tx}(D_{nx}) - V_{Ty}(D_{ny})]$ is moved outside of the double integral in the collection equation (7.85) below and approximated by $\Delta\bar{V}_{Txy} = |V_T(D_{nx}) - V_T(D_{ny})|$. Verlinde *et al.* (1990) have already shown the importance of the errors in using the Wisner *et al.* (1972) approximation, especially when terminal velocities are nearly similar; the errors can be several orders of magnitude. Mizuno (1990) and Murakami (1990) also showed this, and then made adjustments to the terminal-velocity differences to minimize errors. Nevertheless, the collection equation for the mixing ratio that is solved regardless of the approach is given for the complete gamma equation,

$$Q_xAC_y = \frac{1}{\rho}\int_0^\infty\int_0^\infty \frac{0.25\pi(D_x+D_y)^2(V_{TQx}-V_{TQy})\left(\frac{\rho_0}{\rho}\right)^{1/2} M(D_y)E_{xy}N_{Tx}N_{Ty}\mu_x\mu_y}{\Gamma(\nu_x)\Gamma(\nu_y)}$$
$$\times \frac{\alpha_x^{\nu_y}}{D_{nx}}\frac{\alpha_y^{\nu_y}}{D_{ny}}\left(\frac{D_x}{D_{nx}}\right)^{\nu_x\mu_x-1}\left(\frac{D_y}{D_{ny}}\right)^{\nu_y\mu_y-1}\exp\left(-\alpha_x\left[\frac{D_x}{D_{nx}}\right]^{\mu_x}\right)\exp\left(-\alpha_y\left[\frac{D_y}{D_{ny}}\right]^{\mu_y}\right)\mathrm{d}D_x\mathrm{d}D_y; \quad (7.85)$$

or next for the modified gamma distribution,

$$Q_xAC_y = \frac{1}{\rho}\int_0^\infty\int_0^\infty \frac{0.25\pi(D_x+D_y)^2(V_{TQx}-V_{TQy})\left(\frac{\rho_0}{\rho}\right)^{1/2} M(D_y)E_{xy}N_{Tx}N_{Ty}\mu_x\mu_y}{\Gamma(\nu_x)\Gamma(\nu_y)}$$
$$\times \frac{1}{D_{nx}}\frac{1}{D_{ny}}\left(\frac{D_x}{D_{nx}}\right)^{\nu_x\mu_x-1}\left(\frac{D_y}{D_{ny}}\right)^{\nu_y\mu_y-1}\exp\left(-\left[\frac{D_x}{D_{nx}}\right]^{\mu_x}\right)\exp\left(-\left[\frac{D_y}{D_{ny}}\right]^{\mu_y}\right)\mathrm{d}D_x\mathrm{d}D_y; \quad (7.86)$$

and lastly, for the gamma distribution,

$$Q_xAC_y = \frac{1}{\rho}\int_0^\infty\int_0^\infty \frac{0.25\pi(D_x+D_y)^2(V_{TQx}-V_{TQy})\left(\frac{\rho_0}{\rho}\right)^{1/2} M(D_y)E_{xy}N_{Tx}N_{Ty}}{\Gamma(\nu_x)\Gamma(\nu_y)}$$
$$\times \frac{1}{D_{nx}}\frac{1}{D_{ny}}\left(\frac{D_x}{D_{nx}}\right)^{\nu_x-1}\left(\frac{D_y}{D_{ny}}\right)^{\nu_y-1}\exp\left(-\left[\frac{D_x}{D_{nx}}\right]\right)\exp\left(-\left[\frac{D_y}{D_{ny}}\right]\right)\mathrm{d}D_x\mathrm{d}D_y. \quad (7.87)$$

Now an examination is made of the collection equation in terms of number concentration for the complete gamma distribution,

$$N_{Tx}AC_y = \int_0^\infty\int_0^\infty \frac{0.25\pi(D_x+D_y)^2(V_{TNx}-V_{TNy})\left(\frac{\rho_0}{\rho}\right)^{1/2} E_{xy}N_{Tx}N_{Ty}\mu_x\mu_y}{\Gamma(\nu_x)\Gamma(\nu_y)}$$
$$\times \frac{\alpha_x^{\nu_x}}{D_{nx}}\frac{\alpha_y^{\nu_y}}{D_{ny}}\left(\frac{D_x}{D_{nx}}\right)^{\nu_x\mu_x-1}\left(\frac{D_y}{D_{ny}}\right)^{\nu_y\mu_y-1}\exp\left(-\alpha_x\left[\frac{D_x}{D_{nx}}\right]^{\mu_x}\right)\exp\left(-\alpha_y\left[\frac{D_y}{D_{ny}}\right]^{\mu_y}\right)\mathrm{d}D_x\mathrm{d}D_y; \quad (7.88)$$

the modified gamma distribution,

$$N_{Tx}AC_y = \int_0^\infty\int_0^\infty \frac{0.25\pi(D_x+D_y)^2(V_{TNx}-V_{TNy})\left(\frac{\rho_0}{\rho}\right)^{1/2} E_{xy}N_{Tx}N_{Ty}\mu_x\mu_y}{\Gamma(\nu_x)\Gamma(\nu_y)}$$
$$\times \frac{1}{D_{nx}}\frac{1}{D_{ny}}\left(\frac{D_x}{D_{nx}}\right)^{\nu_x\mu_y-1}\left(\frac{D_y}{D_{ny}}\right)^{\nu_y\mu_y-1}\exp\left(-\left[\frac{D_x}{D_{nx}}\right]^{\mu_x}\right)\exp\left(-\left[\frac{D_y}{D_{ny}}\right]^{\mu_y}\right)\mathrm{d}D_x\mathrm{d}D_y; \quad (7.89)$$

and lastly the gamma distribution,

$$N_{\mathrm{T}x}AC_y = \int_0^\infty \int_0^\infty \frac{0.25\pi(D_x+D_y)^2(V_{\mathrm{T}Nx}-V_{\mathrm{T}Ny})\left(\frac{\rho_0}{\rho}\right)^{1/2} E_{xy}N_{\mathrm{T}x}N_{\mathrm{T}y}}{\Gamma(\nu_x)\Gamma(\nu_y)}$$
$$\times \frac{1}{D_{nx}}\frac{1}{D_{ny}}\left(\frac{D_x}{D_{nx}}\right)^{\nu_x-1}\left(\frac{D_y}{D_{ny}}\right)^{\nu_y-1}\exp\left(-\left[\frac{D_x}{D_{nx}}\right]\right)\exp\left(-\left[\frac{D_y}{D_{ny}}\right]\right)\mathrm{d}D_x\mathrm{d}D_y. \tag{7.90}$$

The equations (7.85)–(7.87) can be written with $M(D_y) = a_y D^{by}$ as the following, where M is mass. For the collection equation in terms of the mixing ratio using the complete gamma distribution the following is found,

$$Q_xAC_y = \frac{1}{\rho_0}\int_0^\infty \int_0^\infty \frac{0.25\pi a_y D_{ny}^{b_y}\left(\frac{D_y}{D_{ny}}\right)^{b_y} E_{xy}N_{\mathrm{T}x}N_{\mathrm{T}y}\mu_x\mu_y}{\Gamma(\nu_x)\Gamma(\nu_y)}$$
$$\times\left(D_{nx}^2\left(\frac{D_x}{D_{nx}}\right)^2+2D_{nx}D_{ny}\frac{D_x}{D_{nx}}\frac{D_y}{D_{ny}}+D_{ny}^2\left(\frac{D_y}{D_{ny}}\right)^2\right)(V_{\mathrm{T}Qx}-V_{\mathrm{T}Qy})\left(\frac{\rho_0}{\rho}\right)^{1/2} \tag{7.91}$$
$$\times\frac{\alpha_x^{\nu_x}}{D_{nx}}\frac{\alpha_y^{\nu_y}}{D_{ny}}\left(\frac{D_x}{D_{nx}}\right)^{\nu_x\mu_x-1}\left(\frac{D_y}{D_{ny}}\right)^{\nu_y\mu_y-1}\exp\left(-\alpha_x\left[\frac{D_x}{D_{nx}}\right]^{\mu_x}\right)\exp\left(-\alpha_y\left[\frac{D_y}{D_{ny}}\right]^{\mu_y}\right)\mathrm{d}D_x\mathrm{d}D_y;$$

or with the modified gamma distribution,

$$Q_xAC_y = \frac{1}{\rho_0}\int_0^\infty \int_0^\infty \frac{0.25\pi a_y D_{ny}^{b_y}\left(\frac{D_y}{D_{ny}}\right)^{b_y} E_{xy}N_{\mathrm{T}x}N_{\mathrm{T}y}\mu_x\mu_y}{\Gamma(\nu_x)\Gamma(\nu_y)}$$
$$\times\left(D_{nx}^2\left(\frac{D_x}{D_{nx}}\right)^2+2D_{nx}D_{ny}\frac{D_x}{D_{nx}}\frac{D_y}{D_{ny}}+D_{ny}^2\left(\frac{D_y}{D_{ny}}\right)^2\right)(V_{\mathrm{T}Qx}-V_{\mathrm{T}Qy})\left(\frac{\rho_0}{\rho}\right)^{1/2} \tag{7.92}$$
$$\times\frac{1}{D_{nx}}\frac{1}{D_{ny}}\left(\frac{D_x}{D_{nx}}\right)^{\nu_x\mu_x-1}\left(\frac{D_y}{D_{ny}}\right)^{\nu_y\mu_y-1}\exp\left(-\left[\frac{D_x}{D_{nx}}\right]^{\mu_x}\right)\exp\left(-\left[\frac{D_y}{D_{ny}}\right]^{\mu_y}\right)\mathrm{d}D_x\mathrm{d}D_y;$$

and finally the gamma distribution,

$$Q_xAC_y = \frac{1}{\rho_0}\int_0^\infty \int_0^\infty \frac{0.25\pi a_y D_{ny}^{b_y}\left(\frac{D_y}{D_{ny}}\right)^{b_y} E_{xy}N_{\mathrm{T}x}N_{\mathrm{T}y}}{\Gamma(\nu_x)\Gamma(\nu_y)}$$
$$\times\left(D_{nx}^2\left(\frac{D_x}{D_{nx}}\right)^2+2D_{nx}D_{ny}\frac{D_x}{D_{nx}}\frac{D_y}{D_{ny}}+D_{ny}^2\left(\frac{D_y}{D_{ny}}\right)^2\right)(V_{\mathrm{T}Qx}-V_{\mathrm{T}Qy})\left(\frac{\rho_0}{\rho}\right)^{1/2} \tag{7.93}$$
$$\times\frac{1}{D_{nx}}\frac{1}{D_{ny}}\left(\frac{D_x}{D_{nx}}\right)^{\nu_x-1}\left(\frac{D_y}{D_{ny}}\right)^{\nu_y-1}\exp\left(-\left[\frac{D_x}{D_{nx}}\right]\right)\exp\left(-\left[\frac{D_y}{D_{ny}}\right]\right)\mathrm{d}D_x\mathrm{d}D_y.$$

Then, the collection equation in terms of number concentration (7.88)–(7.90) can be rewritten using a complete gamma distribution,

$$N_{\mathrm{T}x}AC_y = \int_0^\infty \int_0^\infty \frac{0.25\pi E_{xy} N_{\mathrm{T}x} N_{\mathrm{T}y} \mu_x \mu_y}{\Gamma(\nu_x)\Gamma(\nu_y)}$$

$$\times \left(D_{nx}^2 \left(\frac{D_x}{D_{nx}}\right)^2 + 2D_{nx}D_{ny}\frac{D_x}{D_{nx}}\frac{D_y}{D_{ny}} + D_{ny}^2\left(\frac{D_y}{D_{ny}}\right)^2 \right) \left(V_{\mathrm{T}Qx} - V_{\mathrm{T}Qy}\right) \left(\frac{\rho_0}{\rho}\right)^{1/2} \quad (7.94)$$

$$\times \frac{\alpha_x^{\nu_x}}{D_{nx}} \frac{\alpha_y^{\nu_y}}{D_{ny}} \left(\frac{D_x}{D_{nx}}\right)^{\nu_x \mu_x - 1} \left(\frac{D_y}{D_{ny}}\right)^{\nu_y \mu_y - 1} \exp\left(-\alpha_x \left[\frac{D_x}{D_{nx}}\right]^{\mu_x}\right) \exp\left(-\alpha_y \left[\frac{D_y}{D_{ny}}\right]^{\mu_y}\right) \mathrm{d}D_x \mathrm{d}D_y;$$

the modified gamma distribution,

$$N_{\mathrm{T}x}AC_y = \int_0^\infty \int_0^\infty \frac{0.25\pi E_{xy} N_{\mathrm{T}x} N_{\mathrm{T}y} \mu_x \mu_y}{\Gamma(\nu_x)\Gamma(\nu_y)}$$

$$\times \left(D_{nx}^2 \left(\frac{D_x}{D_{nx}}\right)^2 + 2D_{nx}D_{ny}\frac{D_x}{D_{nx}}\frac{D_y}{D_{ny}} + D_{ny}^2\left(\frac{D_y}{D_{ny}}\right)^2 \right) \left(V_{\mathrm{T}Qx} - V_{\mathrm{T}Qy}\right) \left(\frac{\rho_0}{\rho}\right)^{1/2} \quad (7.95)$$

$$\times \frac{1}{D_{nx}} \frac{1}{D_{ny}} \left(\frac{D_x}{D_{nx}}\right)^{\nu_x \mu_x - 1} \left(\frac{D_y}{D_{ny}}\right)^{\nu_y \mu_y - 1} \exp\left(-\left[\frac{D_x}{D_{nx}}\right]^{\mu_x}\right) \exp\left(-\left[\frac{D_y}{D_{ny}}\right]^{\mu_y}\right) \mathrm{d}D_x \mathrm{d}D_y;$$

and finally the gamma distribution,

$$N_{\mathrm{T}x}AC_y = \int_0^\infty \int_0^\infty \frac{0.25\pi E_{xy} N_{\mathrm{T}x} N_{\mathrm{T}y}}{\Gamma(\nu_x)\Gamma(\nu_y)}$$

$$\times \left(D_{nx}^2 \left(\frac{D_x}{D_{nx}}\right)^2 + 2D_{nx}D_{ny}\frac{D_x}{D_{nx}}\frac{D_y}{D_{ny}} + D_{ny}^2\left(\frac{D_y}{D_{ny}}\right)^2 \right) \left(V_{\mathrm{T}Qx} - V_{\mathrm{T}Qy}\right) \left(\frac{\rho_0}{\rho}\right)^{1/2} \quad (7.96)$$

$$\times \frac{1}{D_{nx}} \frac{1}{D_{ny}} \left(\frac{D_x}{D_{nx}}\right)^{\nu_x - 1} \left(\frac{D_y}{D_{ny}}\right)^{\nu_y - 1} \exp\left(-\left[\frac{D_x}{D_{nx}}\right]\right) \exp\left(-\left[\frac{D_y}{D_{ny}}\right]\right) \mathrm{d}D_x \mathrm{d}D_y.$$

Integration of (7.91)–(7.96) provides expressions directly to compute approximations to the collection equations in terms of mixing ratio and number concentration using the gamma distributions.

The parameterization for (7.91) using a complete gamma distribution function is the following for mixing ratio tendency where Wisner's approximation has been used,

$$Q_xAC_y = \frac{0.25\pi\alpha_x^{\nu_x}\alpha_y^{\nu_y}E_{xy}N_{Tx}Q_y\Delta\bar{V}_{TQxy}\left(\frac{\rho_0}{\rho}\right)^{1/2}\alpha_y^{\left(\frac{b_y+\nu_y\mu_y}{\mu_y}\right)}}{\Gamma(\nu_x)\Gamma\left(\frac{b_y+\nu_y\mu_y}{\mu_y}\right)} \times \left[\begin{array}{l} \dfrac{\Gamma\left(\frac{2+\nu_x\mu_x}{\mu_x}\right)}{\alpha_x^{\left(\frac{2+\nu_x\mu_x}{\mu_x}\right)}}\dfrac{\Gamma\left(\frac{0+b_y+\nu_y\mu_y}{\mu_y}\right)}{\alpha_y^{\left(\frac{b_y+\nu_y\mu_y}{\mu_y}\right)}}D_{nx}^2 \\ +2\dfrac{\Gamma\left(\frac{1+\nu_x\mu_x}{\mu_x}\right)}{\alpha_x^{\left(\frac{1+\nu_x\mu_x}{\mu_x}\right)}}\dfrac{\Gamma\left(\frac{1+b_y+\nu_y\mu_y}{\mu_y}\right)}{\alpha_y^{\left(\frac{1+b_y+\nu_y\mu_y}{\mu_y}\right)}}D_{nx}D_{ny} \\ +\dfrac{\Gamma\left(\frac{0+\mu_x\nu_x}{\mu_x}\right)}{\alpha_x^{\left(\frac{0+\nu_x\mu_x}{\mu_x}\right)}}\dfrac{\Gamma\left(\frac{2+b_y+\mu_y\nu_y}{\mu_y}\right)}{\alpha_y^{\left(\frac{2+b_y+\nu_y\mu_y}{\mu_y}\right)}}D_{ny}^2 \end{array}\right]. \tag{7.97}$$

Next the modified gamma distribution is given by

$$Q_xAC_y = \frac{0.25\pi E_{xy}N_{Tx}Q_y\Delta\bar{V}_{TQxy}\left(\frac{\rho_0}{\rho}\right)^{1/2}}{\Gamma(\nu_x)\Gamma(b_y+\nu_y)} \times \left[\begin{array}{l} \Gamma\left(\frac{2+\nu_x\mu_x}{\mu_x}\right)\Gamma\left(\frac{0+b_y+\nu_y\mu_y}{\mu_y}\right)D_{nx}^2 \\ +2\Gamma\left(\frac{1+\nu_x\mu_x}{\mu_x}\right)\Gamma\left(\frac{1+b_y+\nu_y\mu_y}{\mu_y}\right)D_{nx}D_{ny} \\ +\ \Gamma\left(\frac{0+\nu_x\mu_x}{\mu_x}\right)\Gamma\left(\frac{2+b_y+\nu_y\mu_y}{\mu_y}\right)D_{ny}^2 \end{array}\right]. \tag{7.98}$$

Then for the gamma distribution, the following,

$$Q_xAC_y = \frac{0.25\pi E_{xy}N_{Tx}Q_y\Delta\bar{V}_{TQxy}\left(\frac{\rho_0}{\rho}\right)^{1/2}}{\Gamma(\nu_x)\Gamma(b_y+\nu_y)} \times \left[\begin{array}{l} \Gamma(2+\nu_x)\Gamma(0+b_y+\nu_y)D_{nx}^2 \\ +2\Gamma(1+\nu_x)\Gamma(1+b_y+\nu_y)D_{nx}D_{ny} \\ +\ \Gamma(0+\nu_x)\Gamma(2+b_y+\nu_y)D_{ny}^2 \end{array}\right]. \tag{7.99}$$

Next the final solution for the growth in terms of number concentration (7.94)–(7.96) using the complete gamma distribution is

$$N_{Tx}AC_y = \frac{0.25\pi\alpha_x^{\nu_x}\alpha_y^{\nu_y}E_{xy}N_{Tx}N_{Ty}\Delta\bar{V}_{TNxy}\left(\frac{\rho_0}{\rho}\right)^{1/2}}{\Gamma(\nu_x)\Gamma(\nu_y)} \times \left[\begin{array}{l} \dfrac{\Gamma\left(\frac{2+\mu_x\nu_x}{\mu_x}\right)}{\alpha_x^{\left(\frac{2+\nu_x\mu_x}{\mu_x}\right)}}\dfrac{\Gamma\left(\frac{2+\nu_y\mu_y}{\mu_y}\right)}{\alpha_y^{\left(\frac{2+\nu_y\mu_y}{\mu_y}\right)}}D_{nx}^2 \\ +2\Gamma\dfrac{\left(\frac{1+\mu_x\nu_x}{\mu_x}\right)}{\alpha_x^{\left(\frac{1+\nu_x\mu_x}{\mu_x}\right)}}\dfrac{\Gamma\left(\frac{1+\nu_y\mu_y}{\mu_y}\right)}{\alpha_y^{\left(\frac{1+\nu_y\mu_y}{\mu_y}\right)}}D_{nx}D_{ny} \\ +\Gamma\dfrac{\left(\frac{0+\mu_x\nu_x}{\mu_x}\right)}{\alpha_x^{\left(\frac{0+\nu_x\mu_x}{\mu_x}\right)}}\dfrac{\Gamma\left(\frac{2+\nu_y\mu_y}{\mu_y}\right)}{\alpha_y^{\left(\frac{0+\nu_y\mu_y}{\mu_y}\right)}}D_{ny}^2 \end{array}\right]; \quad (7.100)$$

the modified gamma distribution solution is

$$N_{Tx}AC_y = \frac{0.25\pi E_{xy}N_{Tx}N_{Ty}\Delta\bar{V}_{TNxy}\left(\frac{\rho_0}{\rho}\right)^{1/2}}{\Gamma(\nu_x)\Gamma(\nu_y)} \times \left[\begin{array}{l} \Gamma\left(\frac{2+\nu_x\mu_x}{\mu_x}\right)\Gamma\left(\frac{0+\nu_y\mu_y}{\mu_y}\right)D_{nx}^2 \\ +2\Gamma\left(\frac{1+\nu_x\mu_x}{\mu_x}\right)\Gamma\left(\frac{1+\nu_y\mu_y}{\mu_y}\right)D_{nx}D_{ny} \\ +\Gamma\left(\frac{0+\nu_x\mu_x}{\mu_x}\right)\Gamma\left(\frac{2+\nu_y\mu_y}{\mu_y}\right)D_{ny}^2 \end{array}\right]; \quad (7.101)$$

and finally the gamma function solution is

$$N_{Tx}AC_y = \frac{0.25\pi E_{xy}N_{Tx}N_{Ty}\Delta\bar{V}_{TNxy}\left(\frac{\rho_0}{\rho}\right)^{1/2}}{\Gamma(\nu_x)\Gamma(\nu_y)} \times \left[\begin{array}{l} \Gamma(2+\nu_x)\Gamma(0+\nu_y)D_{nx}^2 \\ +2\Gamma(1+\nu_x)\Gamma(1+\nu_y)D_{nx}D_{ny} \\ +\Gamma(0+\nu_x)\Gamma(2+\nu_y)D_{ny}^2 \end{array}\right]. \quad (7.102)$$

7.12 Log-normal general collection equations

The equation for the general collection is complicated and is given by

$$Q_xAC_y = \frac{1}{\rho_0}\int_0^\infty\int_0^\infty \left\{\frac{\pi}{4}E_{xy}\left|\bar{V}_{TQx} - \bar{V}_{TQy}\right|\left(\frac{\rho_0}{\rho}\right)^{1/2} \times m(D_y)(D_x + D_y)^2 n(D_x)n(D_y)\mathrm{d}D_x\mathrm{d}D_y\right\}, \tag{7.103}$$

where $m(D_y)$ is as before a_yD^{by}.

Substitution of (7.78) and $m(D_y)$ into (7.103) gives

$$Q_xAC_y = \frac{1}{\rho_0}\left(\frac{N_{Tx}N_{Ty}}{2\pi\sigma_x^2\sigma_y^2}\right)\frac{\pi}{4}E_{xy}\left|\bar{V}_{TQx} - \bar{V}_{TQy}\right|\left(\frac{\rho_0}{\rho}\right)^{1/2} \times a_y\int_0^\infty\int_0^\infty \left\{D_y^{b_y}(D_x + D_y)^2 D_x^{-1}D_y^{-1} \times \exp\left(-\frac{[\ln(D_x/D_{nx})]^2}{2\sigma_x^2}\right)\exp\left(-\frac{[\ln(D_y/D_{ny})]^2}{2\sigma_y^2}\right)\right\}\mathrm{d}D_x\mathrm{d}D_y. \tag{7.104}$$

Rearranging,

$$Q_xAC_y = \frac{1}{\rho_0}\left(\frac{N_{Tx}N_{Ty}}{2\pi\sigma_x^2\sigma_y^2}\right)\frac{\pi}{4}E_{xy}\left|\bar{V}_{TQx} - \bar{V}_{TQy}\right|\left(\frac{\rho_0}{\rho}\right)^{1/2} \times a_y\int_0^\infty\int_0^\infty \left\{(D_x^1D_y^{b_y-1} + 2D_x^0D_y^{b_y} + D_x^{-1}D_y^{b_y+1}) \times \exp\left(-\frac{[\ln(D_x/D_{nx})]^2}{2\sigma_x^2}\right)\exp\left(-\frac{[\ln(D_y/D_{ny})]^2}{2\sigma_y^2}\right)\mathrm{d}D_x\mathrm{d}D_y\right\}. \tag{7.105}$$

Expanding,

$$\begin{aligned} Q_xAC_y = {} & \frac{1}{\rho_0}\left(\frac{N_{Tx}N_{Ty}}{2\pi\sigma_x^2\sigma_y^2}\right)\frac{\pi}{4}E_{xy}\left|\bar{V}_{TQx} - \bar{V}_{TQy}\right|\left(\frac{\rho_0}{\rho}\right)^{1/2} a_y \\ & \times\left\{\int_0^\infty\int_0^\infty D_x^1D_y^{b_y-1}\exp\left(-\frac{[\ln(D_x/D_{nx})]^2}{2\sigma_x^2}\right)\exp\left(-\frac{[\ln(D_y/D_{ny})]^2}{2\sigma_y^2}\right)\mathrm{d}D_x\mathrm{d}D_y\right. \\ & + \int_0^\infty\int_0^\infty 2D_x^0D_y^{b_y}\exp\left(-\frac{[\ln(D_x/D_{nx})]^2}{2\sigma_x^2}\right)\exp\left(-\frac{[\ln(D_y/D_{ny})]^2}{2\sigma_y^2}\right)\mathrm{d}D_x\mathrm{d}D_y \\ & \left. + \int_0^\infty\int_0^\infty D_x^{-1}D_y^{b_y+1}\exp\left(-\frac{[\ln(D_x/D_{nx})]^2}{2\sigma_x^2}\right)\exp\left(-\frac{[\ln(D_y/D_{ny})]^2}{2\sigma_y^2}\right)\mathrm{d}D_x\mathrm{d}D_y\right\}. \end{aligned} \tag{7.106}$$

Dividing D_x terms by D_{nx} and D_y by D_{ny},

$$Q_xAC_y = \frac{1}{\rho_0}\left(\frac{N_{Tx}N_{Ty}}{2\pi\sigma_x^2\sigma_y^2}\right)\frac{\pi}{4}E_{xy}\left|\bar{V}_{TQx} - \bar{V}_{TQy}\right|\left(\frac{\rho_0}{\rho}\right)^{1/2} a_y$$
$$\times \left\{ D_{nx}^2 D_{ny}^{b_y} \int_0^\infty\int_0^\infty \left(\frac{D_x}{D_{nx}}\right)^1 \left(\frac{D_y}{D_{ny}}\right)^{b_y-1} \exp\left(-\frac{[\ln(D_x/D_{nx})]^2}{2\sigma_x^2}\right)\exp\left(-\frac{[\ln(D_y/D_{ny})]^2}{2\sigma_y^2}\right)\right.$$
$$\times\, \mathrm{d}\left(\frac{D_x}{D_{nx}}\right)\mathrm{d}\left(\frac{D_y}{D_{ny}}\right) + 2D_{nx}^1 D_{ny}^{b_y+1}\int_0^\infty\int_0^\infty \left(\frac{D_x}{D_{nx}}\right)^0\left(\frac{D_y}{D_{ny}}\right)^{b_y}\exp\left(-\frac{[\ln(D_x/D_{nx})]^2}{2\sigma_x^2}\right) \tag{7.107}$$
$$\times \exp\left(-\frac{[\ln(D_y/D_{ny})]^2}{2\sigma_y^2}\right)\mathrm{d}\left(\frac{D_x}{D_{nx}}\right)\mathrm{d}\left(\frac{D_y}{D_{ny}}\right) + D_{nx}^0 D_{ny}^{b_y+2}\int_0^\infty\int_0^\infty\left(\frac{D_x}{D_{nx}}\right)^{-1}\left(\frac{D_y}{D_{ny}}\right)^{b_y+1}$$
$$\left.\times \exp\left(-\frac{[\ln(D_x/D_{nx})]^2}{2\sigma_x^2}\right)\exp\left(-\frac{[\ln(D_y/D_{ny})]^2}{2\sigma_y^2}\right)\mathrm{d}\left(\frac{D_x}{D_{nx}}\right)\mathrm{d}\left(\frac{D_y}{D_{ny}}\right)\right\}.$$

Now letting $u = D_x/D_{nx}$ and $v = D_y/D_{ny}$ gives

$$Q_xAC_y = \frac{1}{\rho_0}\left(\frac{N_{Tx}N_{Ty}}{2\pi\sigma_x^2\sigma_y^2}\right)\frac{\pi}{4}E_{xy}\left|\bar{V}_{TQx} - \bar{V}_{TQy}\right|\left(\frac{\rho_0}{\rho}\right)^{1/2} a_y$$
$$\times\left\{D_{nx}^2 D_{ny}^{b_y}\int_0^\infty\int_0^\infty u^1 v^{b_y-1}\exp\left(-\frac{[\ln u]^2}{2\sigma_x^2}\right)\exp\left(-\frac{[\ln v]^2}{2\sigma_y^2}\right)\mathrm{d}u\mathrm{d}v\right.$$
$$+\,2D_{nx}^1 D_{ny}^{b_y+1}\int_0^\infty\int_0^\infty u^0 v^{b_y}\exp\left(-\frac{[\ln u]^2}{2\sigma_x^2}\right)\exp\left(-\frac{[\ln v]^2}{2\sigma_y^2}\right)\mathrm{d}u\mathrm{d}v \tag{7.108}$$
$$\left.+\,D_{nx}^0 D_{ny}^{b_y+2}\int_0^\infty\int_0^\infty u^{-1} v^{b_y+1}\exp\left(-\frac{[\ln u]^2}{2\sigma_x^2}\right)\exp\left(-\frac{[\ln v]^2}{2\sigma_y^2}\right)\mathrm{d}u\mathrm{d}v\right\}.$$

By letting $j = \ln(u)$, $u = \exp(j)$, $\mathrm{d}u/u = \mathrm{d}j$ and $k = \ln(v)$; thus $v = \exp(k)$, $\mathrm{d}v/v = \mathrm{d}k$, so,

$$Q_xAC_y = \frac{1}{\rho_0}\left(\frac{N_{Tx}N_{Ty}}{2\pi\sigma_x^2\sigma_y^2}\right)\frac{\pi}{4}E_{xy}\left|\bar{V}_{TQx} - \bar{V}_{TQy}\right|\left(\frac{\rho_0}{\rho}\right)^{1/2} a_y$$
$$\times\left\{D_{nx}^2 D_{ny}^{b_y}\int_{-\infty}^{\infty}\int_{-\infty}^{\infty}\exp(2j)\exp\left[(b_y+0)k\right]\exp\left(-\frac{j^2}{2\sigma_x^2}\right)\exp\left(-\frac{k^2}{2\sigma_y^2}\right)\mathrm{d}j\mathrm{d}k\right.$$

$$+\,2D_{nx}^{1}D_{ny}^{b_y+1}\int_{-\infty}^{\infty}\int_{-\infty}^{\infty}\exp(j)\exp\left[(b_y+1)k\right]\exp\left(-\frac{j^2}{2\sigma_x^2}\right)\exp\left(-\frac{k^2}{2\sigma_y^2}\right)\mathrm{d}j\mathrm{d}k$$

$$+\,D_{nx}^{0}D_{ny}^{b_y+2}\int_{-\infty}^{\infty}\int_{-\infty}^{\infty}\exp(0j)\exp\left[(b_y+2)k\right]\exp\left(-\frac{j^2}{2\sigma_x^2}\right)\exp\left(-\frac{k^2}{2\sigma_y^2}\right)\mathrm{d}j\mathrm{d}k\Bigg\}, \quad (7.109)$$

where the limits of the integral are that as u approaches zero from positive values, $\ln(u)$ approaches negative infinity; and as u approaches positive infinity, $\ln(u)$ approaches positive infinity. The same holds for the limits of v.

Now collecting like terms,

$$Q_xAC_y=\frac{1}{\rho_0}\left(\frac{N_{\mathrm{T}x}N_{\mathrm{T}y}}{2\pi\sigma_x^2\sigma_y^2}\right)\frac{\pi}{4}E_{xy}\left|\bar{V}_{\mathrm{T}Qx}-\bar{V}_{\mathrm{T}Qy}\right|\left(\frac{\rho_0}{\rho}\right)^{1/2}a_y$$
$$\times\Bigg\{D_{nx}^{2}D_{ny}^{b_y}\int_{-\infty}^{\infty}\exp(2j)\exp\left(-\frac{j^2}{2\sigma_x^2}\right)\mathrm{d}j\int_{-\infty}^{\infty}\exp\left[(b_y+0)k\right]\exp\left(-\frac{k^2}{2\sigma_y^2}\right)\mathrm{d}k$$
$$+\,2D_{nx}^{1}D_{ny}^{b_y+1}\int_{-\infty}^{\infty}\exp(j)\exp\left(-\frac{j^2}{2\sigma_x^2}\right)\mathrm{d}j\int_{-\infty}^{\infty}\exp\left[(b_y+1)k\right]\exp\left(-\frac{k^2}{2\sigma_y^2}\right)\mathrm{d}k \quad (7.110)$$
$$+\,D_{nx}^{0}D_{ny}^{b_y+2}\int_{-\infty}^{\infty}\exp(0j)\exp\left(-\frac{j^2}{2\sigma_x^2}\right)\mathrm{d}j\int_{-\infty}^{\infty}\exp\left[(b_y+2)k\right]\exp\left(-\frac{k^2}{2\sigma_y^2}\right)\mathrm{d}k\Bigg\}.$$

By Applying the integral definition (7.83), (7.110) becomes

$$Q_xAC_y=\frac{1}{\rho_0}\frac{\pi}{4}N_{\mathrm{T}x}N_{\mathrm{T}y}E_{xy}\left|\bar{V}_{\mathrm{T}Qx}-\bar{V}_{\mathrm{T}Qy}\right|\left(\frac{\rho_0}{\rho}\right)^{1/2}a_y$$
$$\times\Bigg\{D_{nx}^{2}D_{ny}^{b_y}\exp\left(2\sigma_x^2\right)\exp\left(\frac{\sigma_y^2b_y^2}{2}\right)$$
$$+\,2D_{nx}D_{ny}^{b_y+1}\exp\left(\frac{\sigma_x^2}{2}\right)\exp\left(\frac{\sigma_y^2(b_y+1)^2}{2}\right) \quad (7.111)$$
$$+\,D_{ny}^{b_y+2}\exp\left(\frac{\sigma_y^2(b_y+2)^2}{2}\right)\Bigg\}.$$

As described previously, the prognostic equation for N_{T} for the general collection equation is

$$N_{\mathrm{T}x}AC_y = \int_0^\infty \int_0^\infty \frac{\pi}{4} E_{xy} \left| \bar{V}_{\mathrm{TN}x} - \bar{V}_{\mathrm{TN}y} \right| (D_x + D_y)^2 n(D_x) n(D_y) \mathrm{d}D_x \mathrm{d}D_y. \tag{7.112}$$

Substituting (7.78) into (7.112) results in

$$\begin{aligned} N_{\mathrm{T}x}AC_y = & \frac{\pi}{4} E_{xy} \left| \bar{V}_{\mathrm{TN}x} - \bar{V}_{\mathrm{TN}y} \right| \frac{N_{\mathrm{TN}x} N_{\mathrm{TN}y}}{2\pi\sigma_x\sigma_y} \int_0^\infty \int_0^\infty (D_x + D_y)^2 D_x^{-1} D_y^{-1} \\ & \times \exp\left(-\frac{[\ln(D_x/D_{nx})]^2}{2\sigma_x^2} \right) \exp\left(-\frac{[\ln(D_y/D_{ny})]^2}{2\sigma_y^2} \right) \mathrm{d}D_x \mathrm{d}D_y. \end{aligned} \tag{7.113}$$

Expanding,

$$\begin{aligned} N_{\mathrm{T}x}AC_y = & \frac{\pi}{4} E_{xy} \left| \bar{V}_{\mathrm{TN}x} - \bar{V}_{\mathrm{TN}y} \right| \frac{N_{\mathrm{TN}x} N_{\mathrm{TN}y}}{2\pi\sigma_x\sigma_y} \\ & \times \left\{ \int_0^\infty \int_0^\infty D_y^{-1} D_x^{+1} \exp\left(-\frac{[\ln(D_x/D_{nx})]^2}{2\sigma_x^2} \right) \exp\left(-\frac{[\ln(D_y/D_{ny})]^2}{2\sigma_y^2} \right) \mathrm{d}D_x \mathrm{d}D_y \right. \\ & + 2 \int_0^\infty \int_0^\infty \exp\left(-\frac{[\ln(D_x/D_{nx})]^2}{2\sigma_x^2} \right) \exp\left(-\frac{[\ln(D_y/D_{ny})]^2}{2\sigma_y^2} \right) \mathrm{d}D_x \mathrm{d}D_y \\ & \left. + \int_0^\infty \int_0^\infty D_x^{-1} D_y^{1} \exp\left(-\frac{[\ln(D_x/D_{nx})]^2}{2\sigma_x^2} \right) \exp\left(-\frac{[\ln(D_y/D_{ny})]^2}{2\sigma_y^2} \right) \mathrm{d}D_x \mathrm{d}D_y \right\}. \end{aligned} \tag{7.114}$$

Dividing D_x by D_{nx} and D_y by D_{ny} gives

$$\begin{aligned} N_{\mathrm{T}x}AC_y = & \frac{\pi}{4} E_{xy} \left| \bar{V}_{\mathrm{TN}x} - \bar{V}_{\mathrm{TN}y} \right| \frac{N_{\mathrm{TN}x} N_{\mathrm{TN}y}}{2\pi\sigma_x\sigma_y} \\ & \times \left\{ D_{nx}^2 \int_0^\infty \int_0^\infty \left(\frac{D_y}{D_{ny}} \right)^{-1} \left(\frac{D_x}{D_{nx}} \right)^{1} \exp\left(-\frac{[\ln(D_x/D_{nx})]^2}{2\sigma_x^2} \right) \exp\left(-\frac{[\ln(D_y/D_{ny})]^2}{2\sigma_y^2} \right) \mathrm{d}\left(\frac{D_x}{D_{nx}} \right) \mathrm{d}\left(\frac{D_y}{D_{ny}} \right) \right. \\ & + 2 D_{nx}^1 D_{ny}^1 \int_0^\infty \int_0^\infty \exp\left(-\frac{[\ln(D_x/D_{nx})]^2}{2\sigma_x^2} \right) \exp\left(-\frac{[\ln(D_y/D_{ny})]^2}{2\sigma_y^2} \right) \mathrm{d}\left(\frac{D_x}{D_{nx}} \right) \mathrm{d}\left(\frac{D_y}{D_{ny}} \right) \\ & \left. + D_{ny}^2 \int_0^\infty \int_0^\infty \left(\frac{D_x}{D_{nx}} \right)^{-1} \left(\frac{D_y}{D_{ny}} \right)^{1} \exp\left(-\frac{[\ln(D_x/D_{nx})]^2}{2\sigma_x^2} \right) \exp\left(-\frac{[\ln(D_y/D_{ny})]^2}{2\sigma_y^2} \right) \mathrm{d}\left(\frac{D_x}{D_{nx}} \right) \mathrm{d}\left(\frac{D_y}{D_{ny}} \right) \right\}. \end{aligned} \tag{7.115}$$

Letting $u = D_x/D_{nx}$ and $v = D_y/D_{ny}$,

$$N_{\mathrm{T}x}AC_y = \frac{\pi}{4}E_{xy}\left|\bar{V}_{\mathrm{TN}x} - \bar{V}_{\mathrm{TN}y}\right| \frac{N_{\mathrm{TN}x}N_{\mathrm{TN}y}}{2\pi\sigma_x\sigma_y}$$

$$\times \left\{ D_{nx}^2 \int_0^\infty\int_0^\infty u^1 v^{-1} \exp\left(-\frac{[\ln u]^2}{2\sigma_x^2}\right)\exp\left(-\frac{[\ln v]^2}{2\sigma_y^2}\right)\mathrm{d}u\mathrm{d}v \right.$$

$$+ 2D_{nx}^1 D_{ny}^1 \int_0^\infty\int_0^\infty \exp\left(-\frac{[\ln u]^2}{2\sigma_x^2}\right)\exp\left(-\frac{[\ln v]^2}{2\sigma_y^2}\right)\mathrm{d}u\mathrm{d}v \tag{7.116}$$

$$\left. + D_{ny}^2 \int_0^\infty\int_0^\infty u^{-1} v^1 \exp\left(-\frac{[\ln u]^2}{2\sigma_x^2}\right)\exp\left(-\frac{[\ln v]^2}{2\sigma_y^2}\right)\mathrm{d}u\mathrm{d}v \right\}.$$

By letting $j = \ln(u)$, $u = \exp(j)$, $\mathrm{d}u/u = \mathrm{d}j$ and $k = \ln(v)$; thus $v = \exp(k)$, $\mathrm{d}v/v = \mathrm{d}k$, so,

$$N_{\mathrm{T}x}AC_y = \frac{\pi}{4}E_{xy}\left|\bar{V}_{\mathrm{TN}x} - \bar{V}_{\mathrm{TN}y}\right| \frac{N_{\mathrm{TN}x}N_{\mathrm{TN}y}}{2\pi\sigma_x\sigma_y}$$

$$\times \left\{ D_{nx}^2 \int_{-\infty}^{\infty}\int_{-\infty}^{\infty} \exp(2j)\exp\left(-\frac{j^2}{2\sigma_x^2}\right)\exp\left(-\frac{k^2}{2\sigma_y^2}\right)\mathrm{d}j\mathrm{d}k \right.$$

$$+ 2D_{nx}^1 D_{ny}^1 \int_{-\infty}^{\infty}\int_{-\infty}^{\infty} \exp(j)\exp(k)\exp\left(-\frac{j^2}{2\sigma_x^2}\right)\exp\left(-\frac{k^2}{2\sigma_y^2}\right)\mathrm{d}j\mathrm{d}k \tag{7.117}$$

$$\left. + D_{ny}^2 \int_{-\infty}^{\infty}\int_{-\infty}^{\infty} \exp(2k)\exp\left(-\frac{j^2}{2\sigma_x^2}\right)\exp\left(-\frac{k^2}{2\sigma_y^2}\right)\mathrm{d}j\mathrm{d}k \right\},$$

where the limits of the integral are that as u approaches zero from positive values, $\ln(u)$ approaches negative infinity; and as u approaches positive infinity, $\ln(u)$ approaches positive infinity. The same holds for v as well.

Collecting like terms,

$$N_{\mathrm{T}x}AC_y = \frac{\pi}{4}E_{xy}\left|\bar{V}_{\mathrm{TN}x} - \bar{V}_{\mathrm{TN}y}\right| \frac{N_{\mathrm{TN}x}N_{\mathrm{TN}y}}{2\pi\sigma_x\sigma_y}$$

$$\times \left\{ D_{nx}^2 \int_{-\infty}^{\infty} \exp(2j)\exp\left(-\frac{j^2}{2\sigma_x^2}\right)\mathrm{d}j \int_{-\infty}^{\infty} \exp\left(-\frac{k^2}{2\sigma_y^2}\right)\mathrm{d}k \right.$$

$$+ 2D^1_{nx}D^1_{ny} \int_{-\infty}^{\infty} \exp(j) \exp\left(-\frac{j^2}{2\sigma_x^2}\right) dj \int_{-\infty}^{\infty} \exp(k) \exp\left(-\frac{k^2}{2\sigma_y^2}\right) dk$$
$$(7.118)$$
$$+ D^2_{ny} \int_{-\infty}^{\infty} \exp\left(-\frac{j^2}{2\sigma_x^2}\right) dj \int_{-\infty}^{\infty} \exp(2k) \exp\left(-\frac{k^2}{2\sigma_y^2}\right) dk \Bigg\}.$$

Now the integral definition (7.83) is applied to (7.118) and the final expression for collection growth in terms of number concentration is

$$N_{\mathrm{T}x}AC_y = \frac{\pi}{4} E_{xy} |\bar{V}_{\mathrm{TN}x} - \bar{V}_{\mathrm{TN}y}| N_{\mathrm{T}x} N_{\mathrm{T}y} \times \Bigg[D^2_{nx} \exp(2\sigma_x^2) \qquad (7.119)$$
$$+ 2D_{nx}D_{ny} \exp\left(\frac{\sigma_x^2}{2}\right) \exp\left(\frac{\sigma_y^2}{2}\right) + D^2_{ny} \exp(2\sigma_y^2) \Bigg].$$

7.13 Approximations for terminal-velocity differences

7.13.1 Wisner approximation

The Wisner *et al.* (1972) approximation, mentioned briefly above, is a means to simplify the integration of the general collection equation. First, the mass- or number-weighted means of the terminal velocities for each of species x and y are computed. Then the absolute value of the difference of mass-weighted mean terminal velocities is taken and this is assumed to be independent of diameter. Then this quantity can be moved to outside the integral. The absolute value of the difference of the mass-weighted means of terminal velocities of species x and y is

$$\Delta \bar{V}_{\mathrm{T}Qxy} = \left| \bar{V}_{\mathrm{T}Qx} - \bar{V}_{\mathrm{T}Qy} \right|, \qquad (7.120)$$

and the absolute value of the difference of number-weighted means of the terminal velocities of species x and y is

$$\Delta \bar{V}_{\mathrm{TN_T}xy} = \left| \bar{V}_{\mathrm{TN_T}x} - \bar{V}_{\mathrm{TN_T}y} \right|. \qquad (7.121)$$

Many have found the above approximations to be flawed as pointed out by Flatau *et al.* (1989), Ferrier (1994), Curic and Janc (1997), etc. and probably most decisively by Verlinde *et al.* (1990). The approximation fails severely when mass- or number-weighted mean terminal velocities of each species x and y are similar. When the mass- or number-weighted mean terminal-velocity differences between the species are large, the Wisner approximation performs much better. These are well demonstrated in examples provided by Verlinde *et al.* (1990) for cases when terminal-velocity differences are relatively large (e.g. the two species are raindrops and cloud drops) and relatively small (e.g. the two species are high-density graupel and raindrops).

7.13.2 Murakami and Mizuno approximations

The modification following Murakami (1990) is given as an approximation to produce solutions close to the analytical solution to the collection equation, especially when V_{TQx} is close to V_{TQy}. (Note V_{TQx} is the terminal velocity of species x in terms of mixing ratio, and V_{TQy} is the terminal velocity of species y in terms of mixing ratio.) The Murakami (1990) approximation is given as the following for snow and rain,

$$\Delta\bar{V}_{TQxy} = \left|\bar{V}_{TQx} - \bar{V}_{TQy}\right| = \left(\left\{\bar{V}_{TQx} - \bar{V}_{TQy}\right\}^2 + 0.04\bar{V}_{TQx}\bar{V}_{TQy}\right)^{1/2}, \quad (7.122)$$

and it is suggested that in terms of number concentration the same form is used following Milbrandt and Yau (2005),

$$\Delta\bar{V}_{TN_Txy} = \left|\bar{V}_{TN_Tx} - \bar{V}_{TN_Ty}\right| = \left(\left\{\bar{V}_{TN_Tx} - \bar{V}_{TN_Ty}\right\}^2 + 0.04\bar{V}_{TN_Tx}\bar{V}_{TN_Ty}\right)^{1/2}. \quad (7.123)$$

The Mizuno (1990) approximation is very similar and is given in terms of mixing ratio as

$$\Delta\bar{V}_{TQxy} = \left|\bar{V}_{TQx} - \bar{V}_{TQy}\right| = \left(\left\{\alpha\bar{V}_{TQx} - \beta\bar{V}_{TQy}\right\}^2 + 0.08\bar{V}_{TQx}\bar{V}_{TQy}\right)^{1/2}, \quad (7.124)$$

where $\alpha = 1.2$ and $\beta = 0.95$.

Again it is suggested that the same form is used for number concentration,

$$\overline{\Delta V}_{TNxy} = \left|\bar{V}_{TNx} - \bar{V}_{TNy}\right| = \left(\left\{\alpha\bar{V}_{TNx} - \beta\bar{V}_{TNy}\right\}^2 + 0.08\bar{V}_{TNx}\bar{V}_{TNy}\right)^{1/2}. \quad (7.125)$$

For the Mizuno (1990), as is for the Murakami (1990), the approximation is for rain accreting snow and vice versa.

Using the Wisner *et al.* (1972) approximation for two particles that have about the same fallspeed distribution, there is a large region of the parameter space with large errors. A stripe of 40% errors cuts diagonally through the parameter space, whilst there is a large region of the parameter space with $<40\%$ errors for larger particles collecting smaller particles.

7.13.3 Weighted root-mean-square approximation

Next is an examination of the weighted root-mean-square terminal-velocity approach of Flatau *et al.* (1989). In this approach the approximation for the velocity difference is given by the square root of squared mean velocity. The form of the terminal-velocity equation is usually designed to be as general as possible through the use of power laws,

$$\overline{\Delta V}^2_{Txy} = \frac{\int_0^\infty \int_0^\infty \left[V_T(D_x) - V_T(D_y)\right]^2 w_{xy} dD_x dD_y}{\int_0^\infty \int_0^\infty w_{xy} dD_x dD_y}, \tag{7.126}$$

where w_{xy} is the weighting function given as

$$w_{xy} = m(D_x)(D_x + D_y)^2 n(D_x) n(D_y). \tag{7.127}$$

Integration and algebra lead to the following for the numerator,

$$\begin{aligned} &\int_0^\infty \int_0^\infty \left[V_T(D_x) - V_T(D_y)\right]^2 w_{xy} dD_x dD_y \\ &= N_x N_y m(D_{nx}) \left[C_1 V_T^2(D_{nx}) - 2C_2 V_T(D_{nx}) V_T(D_{ny}) + V_T^2 C_3(D_{ny})\right]. \end{aligned} \tag{7.128}$$

The constants are written in vector notation as $\mathbf{C} = \mathbf{FD}$ or,

$$\mathbf{F} = \begin{bmatrix} F_x(2d_x + b_x + 2)F_y(0) & F_x(2d_x + b_x + 1)F_y(1) & F_x(2d_x + b_x)F_y(2) \\ F_x(d_x + b_x + 2)F_y(d_y) & F_x(d_x + b_x + 1)F_y(d_y + 1) & F_x(d_x + b_x)F_y(d_y + 2) \\ F_x(b_x + 2)F_y(2d_y) & F_x(b_x + 1)F_y(2d_y + 1) & F_x(b_x)F_y(2d_y + 2) \end{bmatrix}, \tag{7.129}$$

where F_x and F_y are related to the type of distribution. In this case, it is the modified gamma distribution. For example, the liquid-water content, L, for a spherical particle can be written assuming the gamma distribution as

$$L = \frac{\pi}{6} \rho N D_n F(3) = \frac{\pi}{6} \rho N D_n \frac{\Gamma(\nu + p)}{\Gamma(\nu)} \tag{7.130}$$

where $p = 3$ (this is related to the third moment of the distribution from 0 to ∞).

To complete the solution, the value of the (3×1) column vector $\mathbf{D}$ needs to be defined,

$$\mathbf{D} = \left(D_{nx}^2, 2D_{nx} D_{ny}, D_{ny}^2\right)^T, \tag{7.131}$$

where T denotes the transpose of the matrix. The (1×3) row vector $\mathbf{V}_\mathrm{T}$ needs to be defined,

$$\mathbf{V}_\mathrm{T} = \left[V_{\mathrm{T}x}^2(D_{nx}), -2V_{\mathrm{T}x}(D_{nx})V_{\mathrm{T}y}\left(D_{ny}\right), V_{\mathrm{T}y}^2\left(D_{ny}\right)\right]. \quad (7.132)$$

Using (7.129)–(7.132), the value of $\overline{\Delta V}_{\mathrm{T}xy}$ from (7.128) can be written as

$$\overline{\Delta V}_{\mathrm{T}xy} = \left(\vec{V}_\mathrm{T} \bullet \mathbf{FD}\right)^{1/2} \left(\frac{\rho_0}{\rho}\right)^{1/2}, \quad (7.133)$$

where the typical density correction $(\rho_0/\rho)^{(1/2)}$ has been employed and $\overline{\Delta V}_{\mathrm{T}xy}$ replaces the terminal velocity term in (7.126).

7.14 Long's kernel for rain collection cloud

Next, Long's kernel is derived following Cohard and Pinty (2000) for collection of cloud water by rain for a modified gamma distribution. Long's kernel has been used previously in a parameterization numerical model by Ziegler (1985). First define the kernel, K,

$$K(D_1, D_2) = \begin{cases} k_2\left(D_1^6 + D_2^6\right) & \text{if } D_1 \leq 100 \text{ microns} \\ k_1\left(D_1^3 + D_2^3\right) & \text{if } D_2 > 100 \text{ microns} \end{cases}, \quad (7.134)$$

with $k_2 = 2.59 \times 10^{15}\,\mathrm{m}^{-3}\,\mathrm{s}^{-1}$ for $D < 100$ microns and $k_1 = 3.03 \times 10^3\,\mathrm{m}^{-3}\,\mathrm{s}^{-1}$ for $D > 100$ microns. At present these are considered only approximations of the kernels but seem to suffice for the problem (Pruppacher and Klett 1997). Cohard and Pinty write the following for part of the stochastic collection equation that is important and maintains mass conservation,

$$I = C_n \int_0^{D_m} D^n \frac{\partial N_\mathrm{cw}(D)}{\partial t} \mathrm{d}D. \quad (7.135)$$

Substituting for the rate term,

$$I \approx C_n \int_0^{D_m} D^n AC(N_\mathrm{cw}(D))\mathrm{d}D, \quad (7.136)$$

where I is the integral, AC is accretion, C_n is a constant in the collection equation. Substituting for the accretion term,

$$I \approx C_n \int_0^{\infty} D^n(N_\mathrm{cw}(D_1, t)) \int_0^{\infty} K(D_1, D_2) N_\mathrm{rw}(D_2, t)\mathrm{d}D_1 \mathrm{d}D_2. \quad (7.137)$$

Now based on (7.137), the tendency equation for any moment can be developed,

$$IAC = \int_0^\infty D_1^n(N_{cw}(D_1,t)) \int_0^\infty K_i\left(D_1^{3i},D_2^{3i}\right)N_{rw}(D_2,t)\mathrm{d}D_2\mathrm{d}D_1, \tag{7.138}$$

where the index $i = 0$ for the zeroth moment and $i = 3$ for the third moment; AC is the accretion rate. In (7.138) the D_2 integral, following Cohard and Pinty, is the sum of moments $M_{rw}(0)$ and $M_{rw}(3i)$,

$$IAC = \int_0^\infty D_1^n(N_{cw}(D_1,t))K_iN_{rw}\left[D_1^{3i}M_{rw}(0) + M_{rw}(3i)\right]N_{rw}\mathrm{d}D_1. \tag{7.139}$$

This leads to a generalized gamma function solution for N_{cw} that is an n-order and $n + 3i$ moment scheme.

$$IAC = K_iN_{rw}\left[M_{rw}(0)M_{cw}(n+3i) + M_{rw}(3i)M_{cw}(n)\right]. \tag{7.140}$$

For small raindrops with $D < 100$ microns, from the general accretion formula from Appendix B of Cohard and Pinty (2000),

$$IAC = \int_0^\infty D_1^n n_{cw}(D_1,t) \int_0^\infty K_2\left(D_1^6 + D_2^6\right)n_{rw}(D_2,t)\mathrm{d}D_2\mathrm{d}D_1. \tag{7.141}$$

Then the modified gamma function

$$n(D) = \frac{N_T}{\Gamma(\nu_{rw})}\frac{1}{D_n}\left(\frac{D}{D_n}\right)^{\nu_{rw}-1}\exp\left(-\frac{D}{D_n}\right) \tag{7.142}$$

is substituted into (7.141),

$$\begin{aligned} IAC = K_2\frac{N_{Tcw}}{\Gamma(\nu_{cw})}\frac{N_{Trw}}{\Gamma(\nu_{rw})}\int_0^\infty &\left\{\left(\frac{D}{D_{ncw}}\right)^{\nu_{cw}-1}\exp\left(-\frac{D_{cw}}{D_{ncw}}\right)\right. \\ \times \int_0^\infty D_{cw}^6 D_{nrw}^0 &\left(\frac{D_{rw}}{D_{nrw}}\right)^{\nu_{rw}-1+0} + D_{nrw}^6\left(\frac{D_{rw}}{D_{nrw}}\right)^{\nu_{rw}-1+6}\left.\exp\left(-\frac{D}{D_n}\right)\right\} \\ \times\, \mathrm{d}&\left(\frac{D_{rw}}{D_{nrw}}\right)\mathrm{d}\left(\frac{D_{cw}}{D_{ncw}}\right), \end{aligned} \tag{7.143}$$

where D_n is the characteristic diameter.

Using the definition of the gamma function gives

$$IAC = \frac{K_2 N_{\text{Trw}} N_{\text{Tcw}}}{\Gamma(\nu_{\text{rw}})\Gamma(\nu_{\text{cw}})} \int_0^{\infty} \left(\frac{D}{D_{n\text{cw}}}\right)^{\nu_{\text{cw}}-1}$$
$$\exp\left(-\frac{D}{D_{n\text{cw}}}\right)\left(D_{\text{cw}}^{6} D_{n\text{rw}}^{0} \Gamma(\nu_{\text{rw}}) + D_{nr}^{6} \Gamma(\nu_{\text{rw}} + 6)\right) \mathrm{d}\left(\frac{D_{\text{cw}}}{D_{n\text{cw}}}\right). \tag{7.144}$$

Expanding,

$$IAC = \frac{K_2 N_{\text{Trw}} N_{\text{Tcw}}}{\Gamma(\nu_{\text{rw}})\Gamma(\nu_{\text{cw}})} \int_0^{\infty} \left(D_{n\text{cw}}^{6} D_{n\text{rw}}^{0} \Gamma(\nu_{\text{rw}}) \left(\frac{D_{\text{cw}}}{D_{n\text{cw}}}\right)^{\nu_{\text{cw}}-1+6}\right.$$
$$\left. + D_{n\text{rw}}^{6} D_{n\text{cw}}^{0} \Gamma(\nu_{\text{rw}} + 6) \left(\frac{D_{\text{cw}}}{D_{n\text{cw}}}\right)^{\nu_{\text{cw}}-1+0}\right) \exp\left(-\frac{D}{D_{n\text{cw}}}\right) \mathrm{d}\left(\frac{D_{\text{cw}}}{D_{n\text{cw}}}\right). \tag{7.145}$$

Then integrating,

$$IAC = K_2 N_{\text{Trw}} N_{\text{Tcw}} \left[\frac{\Gamma(\nu_{\text{cw}} + 6) D_{n\text{cw}}^{6} \Gamma(\nu_{\text{rw}})}{\Gamma(\nu_{\text{rw}})\Gamma(\nu_{\text{cw}})} + \frac{\Gamma(\nu_{\text{rw}} + 6) D_{n\text{rw}}^{6} \Gamma(\nu_{\text{cw}})}{\Gamma(\nu_{\text{rw}})\Gamma(\nu_{\text{cw}})}\right]. \tag{7.146}$$

Thus the collection equation in terms of number concentration using the modified gamma distribution is

$$N_{\text{rw}} AC_{\text{cw}} = K_2 N_{\text{Trw}} N_{\text{Tcw}} \left[\frac{\Gamma(\nu_{\text{cw}} + 6) D_{n\text{cw}}^{6}}{\Gamma(\nu_{\text{cw}})} + \frac{\Gamma(\nu_{\text{rw}} + 6) D_{n\text{rw}}^{6}}{\Gamma(\nu_{\text{rw}})}\right]. \tag{7.147}$$

For the prediction of mixing ratio, Q (setting $n = 3$; third moment), following similar procedures, the following collection equation using the modified gamma distribution is obtained,

$$Q_{\text{rw}} AC_{\text{cw}} = \frac{\pi}{6} \frac{\rho_{\text{cw}}}{\rho} K_2 N_{\text{Trw}} N_{\text{Tcw}} D_{n\text{cw}}^{3}$$
$$\times \left[\frac{\Gamma(\nu_{\text{cw}} + 9) D_{n\text{cw}}^{6}}{\Gamma(\nu_{\text{cw}})} + \frac{\Gamma(\nu_{\text{cw}} + 3)\Gamma(\nu_{\text{rw}} + 6) D_{n\text{rw}}^{6}}{\Gamma(\nu_{\text{rw}})\Gamma(\nu_{\text{cw}})}\right]. \tag{7.148}$$

For larger raindrops with $D > 100$ microns, also from the general accretion formula from Appendix B of Cohard and Pinty, it can be written

$$N_{\text{rw}} AC_{\text{cw}} = K_1 N_{\text{Trw}} N_{\text{Tcw}} \left[\frac{\Gamma(\nu_{\text{cw}} + 3) D_{n\text{cw}}^{3}}{\Gamma(\nu_{\text{cw}})} + \frac{\Gamma(\nu_{\text{rw}} + 3) D_{n\text{rw}}^{3}}{\Gamma(\nu_{\text{rw}})}\right]. \tag{7.149}$$

For the prediction of mixing ratio (setting $n = 3$; third moment), following similar procedures as above, the following is obtained

$$Q_{\mathrm{rw}}AC_{\mathrm{cw}} = \frac{\pi}{6}\frac{\rho_{\mathrm{cw}}}{\rho}K_2 N_{\mathrm{Trw}} N_{\mathrm{Tcw}} D_{n\mathrm{cw}}^3 \times \left[\frac{\Gamma(\nu_{\mathrm{cw}}+6)D_{n\mathrm{cw}}^3}{\Gamma(\nu_{\mathrm{cw}})} + \frac{\Gamma(\nu_{\mathrm{cw}}+3)\Gamma(\nu_{\mathrm{rw}}+3)D_{n\mathrm{rw}}^3}{\Gamma(\nu_{\mathrm{rw}})\Gamma(\nu_{\mathrm{cw}})}\right]. \tag{7.150}$$

7.15 Analytical solution to the collection equation

Flatau *et al.* (1989) and Verlinde *et al.* (1990) present an analytical solution to the collection equation. Lookup tables as a function of D_{nx}, D_{ny}, $V_{\mathrm{T}x}$, and $V_{\mathrm{T}y}$ can be built and bilinear interpolation in D_{nx} and D_{ny} can be used to obtain a very accurate solution using the analytical solutions of Verlinde *et al.* (1990) or numerical integration as discussed later.

Let us first assume that the terminal velocities are represented by simple power laws,

$$\Delta\bar{V} = \left[V(D_x) - V(D_y)\right] \tag{7.151}$$

which change sign when

$$c_x D_x^{d_x} = c_y D_y^{d_y}, \tag{7.152}$$

where d_x and d_y are the ratios of terminal-velocity powers. With some algebra this can be rewritten as

$$D_{xy} = f_{xy} D_y^{d_{xy}}, \tag{7.153}$$

where,

$$f_{xy} = \left(\frac{c_x}{c_y}\right)^{(1/d_{xy})} \tag{7.154}$$

and

$$d_{xy} = \frac{d_x}{d_y}. \tag{7.155}$$

Following Flatau *et al.* (1989) and Verlinde *et al.* (1990), using piecewise integration with regard to D_y of the collection equation such that the velocity difference has the same sign over each part, the following equation in terms of the mixing ratio using the gamma distribution can be written,

$$Q_y AC_x = \frac{1}{\rho_0}\frac{\pi}{4}\bar{E}_{xy} J\left(\frac{\rho_0}{\rho}\right)^{1/2}. \tag{7.156}$$

Now defining the variable J as the integral is simple, as it is just

$$J = \int_0^\infty m(D_x)(J_1 - J_2)n(D_x)\mathrm{d}D_x. \tag{7.157}$$

The integrals represented by J_1 and J_2 are given as

$$J_1 = \int_0^{D_{xy}} (D_x + D_y)^2 \left[V_\mathrm{T}(D_x) - V_\mathrm{T}(D_y)\right] n(D_y)\mathrm{d}D_y \tag{7.158}$$

and

$$J_2 = \int_{D_{xy}}^\infty (D_x + D_y)^2 \left[V_\mathrm{T}(D_x) - V_\mathrm{T}(D_y)\right] n(D_y)\mathrm{d}D_y. \tag{7.159}$$

Again, following Flatau *et al.* (1989), integrating for J_1 gives,

$$\begin{aligned} J_1 = \frac{N_{\mathrm{T}y}}{\Gamma(v_y)} \Big\{ & V_\mathrm{T}(D_x)\left[D_x^2 G_1(0, D_{xy}) + 2D_x D_{ny} G_1(1, D_{xy}) G_1(0, D_{xy}) + D_{ny}^2 G_1(2, D_{xy})\right] \\ & - V_\mathrm{T}(D_{ny})\left[D_x^2 G_1(b_y, D_{xy}) + 2D_x D_{ny} G_1(b_y + 1, D_{xy}) G_1(0, D_{xy}) + D_{ny}^2 G_1(b_y + 2, D_{xy})\right]\Big\}, \end{aligned} \tag{7.160}$$

where

$$G_1(p, q) = \gamma(p + 1, q) - \Gamma(p + 1, q), \tag{7.161}$$

where γ and Γ are partial gamma functions. Flatau *et al.* (1989) explains that the integral for J_2 is similar to J_1 and G_2 is similar to G_1.

The remaining integral over D_x (the definition of J) is quite difficult to solve. The possibilities for finding a solution for the collection equation are (i) to follow Verlinde *et al.* (1990) for an analytical solution, or (ii) to integrate numerically following Flatau *et al.* (1989). It seems that because of some of the difficulties with solving this equation, scientists have opted for using a hybrid bin model for solutions and storing them in lookup tables at the start of a model simulation. This procedure has been called the hybrid parameterization – or the hybrid bin model approach by some. It is very efficient and easier to utilize.

7.16 Long's kernel self-collection for rain and cloud

Cohard and Pinty (2000) incorporate Long's kernel to find a solution that is straightforward to derive and apply from equations given earlier in this chapter. They make use of the modified gamma distribution with $v \neq 1$, $\mu = 1$, $\alpha = 1$. For cloud droplets where $D < 100$ microns,

$$C_{\mathrm{cw}} SC_{\mathrm{cw}} = k_2 N_{\mathrm{Tcw}}^2 \frac{\Gamma(v_{\mathrm{cw}} + 6) D_{n\mathrm{cw}}^6}{\Gamma(v_{\mathrm{cw}})}, \tag{7.162}$$

where SC is the self-collection rate. For drizzle and for raindrops where $D > 100$ microns,

$$C_{\mathrm{rw}} SC_{\mathrm{rw}} = k_1 N_{\mathrm{Trw}}^2 \frac{\Gamma(v_{\mathrm{rw}} + 3) D_{n\mathrm{rw}}^3}{\Gamma(v_{\mathrm{rw}})}. \tag{7.163}$$

7.17 Analytical self-collection solution for hydrometeors

Verlinde and Cotton's (1993) analytical solution is not extremely difficult, but requires the hyper-geometric function, ${}_2F_1$. The formulation of Verlinde and Cotton's (1993) self-collection equation for change in number concentration, which is a loss term for raindrops in this example, is given by

$$\begin{aligned} C_{\mathrm{rw}} SC_{\mathrm{rw}} = & -\frac{\pi}{8} c_{\mathrm{rw}} D_{n\mathrm{rw}}^{b+2} N_{\mathrm{Trw}}^2 E_{\mathrm{rwrw}} \\ & \times \sum_{n=0}^{2} \left[\frac{2}{v+n} \Gamma(\eta)\, {}_2F_1(v+n, \eta; v+n+1; -1) - \Gamma(v+n) \right. \\ & \qquad \left. \times \Gamma(d+v-n+2) \right] \\ & + \sum_{n=0}^{2} \left[\frac{2}{v+d+n} \Gamma(\eta)\, {}_2F_1(v+d+n, \eta; v+d+n+1; -1) \right. \\ & \qquad \left. - \Gamma(v+d+n)\Gamma(v-n+2) \right], \end{aligned} \tag{7.164}$$

where E_{rwrw} is the collection efficiency of rain water collecting rain water, c is the leading coefficient for the power law for terminal velocity, and d is the power. In addition, $\eta = d + 2v + 2$.

In their earlier paper, Verlinde *et al.* (1990) showed that H was derived from the special case of self-collection for the general case of two-body interactions. In that case, they showed,

$$\frac{\mathrm{d}Q}{\mathrm{d}t} = -\frac{\pi}{4\rho} E_{xx} J, \tag{7.165}$$

where

$$J = \frac{1}{2} m(D_n) V_{\mathrm{T}}(D_n) N_{\mathrm{T}}^2 D_n^2 H_{\mathrm{mass}}. \tag{7.166}$$

Also note $V_T = cD_n^d$. Substitution of (7.166) into (7.165) gives

$$\frac{dQ}{dt} = -\frac{\pi}{8\rho} E_{xx} c D_n^{d+2} a D_n^b N_T^2 H_{mass}, \tag{7.167}$$

where H_{mass} represents the summation for mass changes by self-collection. This equation is very similar to the equation in terms of number concentration,

$$\frac{dN_T}{dt} = -\frac{\pi}{8\rho} c D_n^{d+2} N_T^2 E_{xx} H_{number}, \tag{7.168}$$

such that

$$\frac{dm}{dt} \approx m(D_n)\frac{dN_T}{dt}, \tag{7.169}$$

where,

$$H_{mass} = \sum_{n=0}^{2}\left[\frac{2}{v+n}\Gamma(\eta)_2F_1(v+n,\eta;v+n+1;-1) - \Gamma(v+n)\Gamma(b+d+v-n+2)\right]$$
$$+\sum_{n=0}^{2}\left[\begin{array}{l}\frac{2}{v+d_v+n}\Gamma(\eta)_2F_1(v+d+n,\eta;v+d+n+1;-1)\\ -\Gamma(v+d_v+n)\Gamma(b+v-n+2)\end{array}\right], \tag{7.170}$$

and where,

$$\eta = b + d + 2v + 2. \tag{7.171}$$

Note that there are two different exponents from the power law in this case. One, *b*, is from the mass power law whereas, *d* is from the velocity power law. Verlinde *et al.* (1990) claim that this form is used for the mass computation (i.e. the amount of total mass involved in self-collection).

7.18 Reflectivity change for the gamma distribution owing to collection

The reflectivity owing to collection can be approximated following Milbrandt and Yau (2005b) by

$$Z_x AC_y = \frac{G(v_x)}{\frac{\pi}{6}\rho_x}\rho^2\left[2\frac{Q_x}{N_{Tx}}Q_x AC_y - \left(\frac{Q_x}{N_{Tx}}\right)^2 N_{Tx} AC_y\right], \tag{7.172}$$

where $G(v_x)$ is given in Chapter 2 as is the derivation of this equation. Alternative forms including α and μ for the gamma distribution can be derived as well as for the log-normal distribution.

7.19 Numerical solutions to the quasi-stochastic collection equation

As pointed out by Pruppacher and Klett (1981, 1997) there is a plethora of numerical approximation techniques to solve the approximate stochastic collection equation. Herein, we will cover two numerical interpolation approaches including one by Berry (1967) and Berry and Reinhardt (1974a–d), and the simple and fast Kovetz and Olund (1969) method and attempts to modify it. Then the "method of moments" techniques by Bleck (1970) and Danielsen *et al.* (1972), which are single-moment approximations, and that by Tzivion *et al.* (1987), which is a multiple-moment approximation, will be covered. Pruppacher and Klett (1981, 1997) write that these are some of the most widely used techniques in the literature; that is the reason they will be covered here. Finally, Bott's (1998) flux method for solving the stochastic collection equation will be examined.

It is interesting to see what type of solution can be parameterized from bin model results. Khairoutdinov and Kogan (2000) ran a bin model and computed the collection rates for rain as a function of Q_{cw} and Q_{rw} and found that the explicit bin model results for drizzling stratocumulus could be well represented by two parameterized functions, one slightly better than the other (Fig. 7.8). This should be tried for deeper convection and for ice clouds to see if bin model results could be used for collection of cloud, rain and crystals by other hydrometeors.

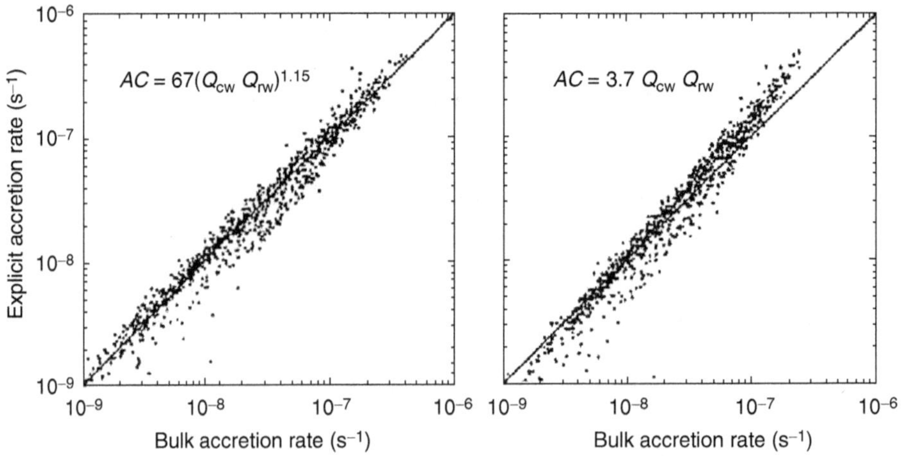

Fig. 7.8. Scatterplots of the bulk accretion rates given by the x–y axes versus the corresponding rates obtained from the explicit mode. Note that only every twentieth data point is shown. (From Khairoutdinov and Kogan 2000; courtesy of the American Meteorological Society.)

7.19.1 Methods of interpolation for the stochastic collection equation

7.19.1.1 The Kovetz and Olund method for the stochastic collection equation

Though hardly used any more, the Kovetz and Olund (1969) scheme is a simple and efficient interpolation method to study the collision–coalescence of particles. It has received criticism for being too diffusive a scheme. Moreover, it is claimed that the Kovetz and Olund method is not a true stochastic collection scheme. Though Scott and Levin (1975) argued that all schemes are approximations to true stochastic collection equations. The author of this book agrees with the opinion that the Kovetz and Olund scheme is too diffusive a scheme compared to several newer modern schemes; however, it is presented here for completeness. Moreover the Kovetz and Olund scheme is a mass-conserving scheme, which is a desirable quality.

The stochastic collection equation for the Kovetz and Olund scheme is written such that

$$N_{\mathrm{T}}(r_i, t+\Delta t) = N_{\mathrm{T}}(r_i, t) + \sum_{n=1}^{i-1} \sum_{m=n+1}^{i} B(n,m,i) P(n,m) N_{\mathrm{T}}(r_n, t) N_{\mathrm{T}}(r_m, t) - \sum_{n=1}^{M} P(i,n) N_{\mathrm{T}}(r_i, t) N_{\mathrm{T}}(r_n, t), \tag{7.173}$$

where $P(n, m)$ is the coalescence probability for particles with radii r_n and r_m. The term B is an exchange coefficient to move particles from one bin to another and is given by

$$B(n,m,i) = \begin{cases} \left(r_n^3 + r_m^3 - r_{i-1}^3\right) / \left(r_i^3 - r_{i-1}^3\right) & \text{for } r_{i-1}^3 \le r_n^3 + r_m^3 \le r_i^3 \\ \left(r_{i+1}^3 - r_n^3 - r_m^3\right) / \left(r_{i+1}^3 - r_i^3\right) & \text{for } r_i^3 \le r_n^3 + r_m^3 \le r_{i+1}^3 \\ 0 & \text{for } r_n^3 + r_m^3 \le r_{i-1}^3 \\ & \text{or } r_{i+1}^3 < r_n^3 + r_m^3 \end{cases} . \tag{7.174}$$

This scheme for $B(n, m, i)$ preserves the mass of water. A brief discussion comparing the Kovetz and Olund (1969) scheme to Golovin's analytical solution and the Berry and Reinhardt (1974a–d) scheme is given below (Fig. 7.9) from Scott and Levin (1975). They claim that the Kovetz and Olund (1969) solutions are not prohibitively erroneous and that neither the Kovetz and Olund nor the Berry and Reinhardt scheme are perfect at representing the true stochastic collection process. In the first example, Golovin's analytical solution is compared to the Kovetz and Olund scheme. The peaks in the Kovetz and Olund scheme are slightly lower than with Golovin's

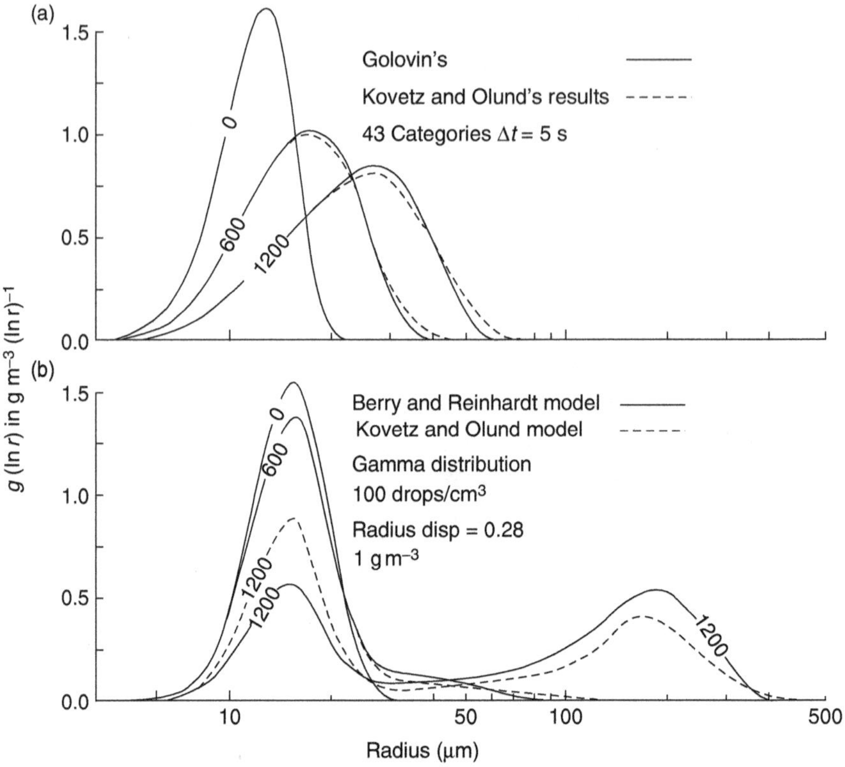

Fig. 7.9. Comparison of Kovetz and Olund's (1969) results with (a) Golovin's analytical solution and (b) Berry and Reinhardt's (1974a–d) gamma distribution. The graphs plot number concentration versus the drop radius. See text for details. (From Scott and Levin 1975; courtesy of the American Meteorological Society.)

solution, whereas the tails at large drop sizes are slightly longer indicating the known spreading by the Kovetz and Olund scheme. In the comparison with the Berry and Reinhardt (1974a–d) scheme, there is a larger difference between the solutions, with shallower peaks for the Kovetz and Olund scheme and more prominent undesirable spreading at 600 and 1200 seconds at the large drop tails.

7.19.1.2 Berry and Reinhardt method for the stochastic collection equation

The method of interpolation is quite useful if high-order interpolation polynomials are used for accuracy and enough bin categories cover the spectrum adequately. Unfortunately, given their accuracy [at least the Berry (1967), and Berry and Reinhardt (1974a–d) methods], interpolation methods do not preserve any mass moments including the zeroth (number concentration) and first (water content). A practical problem with any method is determining the number of bin categories, and spacing the bin categories to have enough resolution at both small and large sizes. Linearly spaced bins over four orders of magnitude in diameter would be prohibitively expensive if enough resolution for the small-sized end of the spectrum were covered and these small bin category sizes were used throughout the spectrum for the larger end of the spectrum. On the other hand, use of coarser bin category sizes would give poor solutions. At least somewhere around 40 or more bin category sizes are required for accurate solutions (Pruppacher and Klett 1997 and many others). In recent years all interpolation methods have been abandoned as they do not handle sharp discontinuities with the high-order polynomials needed for accurate smooth solutions.

One of the most common solutions to the problem of bin category size resolution was proposed by Berry (1967) by using an exponential subdivision method. For bin category J the sizes range as

$$r(J) = r_0 \exp\left(\frac{J-1}{J_R}\right), \tag{7.175}$$

where $J = 1, 2, 3 \ldots J_{max}$, r_0 is the smallest radius, and J_R is a distribution spacing parameter typically between 3 and 7 or so depending on r_0 and the size range to be covered. The mass coordinates given by this bin category division method are given as

$$m(J) = m_0 \exp\left(\frac{3[J-1]}{J_R}\right), \tag{7.176}$$

where m_0 is the smallest mass corresponding to the smallest radius by

$$m_0 = \frac{4}{3}\pi\rho_L r_0^3, \tag{7.177}$$

where ρ_L is the density of liquid water (though any density of any particle could be inserted). Note that the lower limit of integration is now no longer 0 but m_0 in the stochastic collection equation (and later in the stochastic breakup equation).

The following closely follows Pruppacher and Klett (1997) and Berry and Reinhardt (1974a–d). First the stochastic collection equation given by

$$\frac{dN_k}{dt} = \frac{1}{2}\sum_{i=1}^{k-1} A_{i,k-i} N_i N_{k-i} - N_k \sum_{i=1}^{\infty} A_{ik} N_i \tag{7.178}$$

is rewritten as

$$\left(\frac{\partial n(m,t)}{\partial t}\right)_{\mathrm{C}} = \int_0^{m/2} [K(m_c, m')n(m_c, t)n(m', t)]\mathrm{d}m' - n(m,t)\int_0^{\infty} [K(m,m')n(m',t)]\mathrm{d}m', \tag{7.179}$$

where m, m' and m_c are different masses, n is the number of droplets, t is the time, and C means collection. Symmetry of the collection kernel K allows the first integral to be written as a single integral rather than a double integral, and $m/2$ in the first integral is an integer J_{up} (upper limit), following Pruppacher and Klett's (1997) convention, where

$$m(J_{\mathrm{up}}) = m(J/2), \tag{7.180}$$

or

$$J_{\mathrm{up}} = J - [J_R \ln(2)]/3, \tag{7.181}$$

where J_R is a parameter to control the size of the exponential. Therefore, each drop mass is $2^{1/J_R}$ times the preceding mass category or

$$m(J) = m_0 2^{\left(\frac{[J-1]}{J_R}\right)}, \tag{7.182}$$

or each mass category is twice the mass of the previous two mass categories; i.e. mass doubles every two mass categories.

This results in a new distribution function for number concentration bin categories $n(J)$ so that

$$n(J,t)\mathrm{d}J = n(m,t)\mathrm{d}m. \tag{7.183}$$

Note also that, $m_c = m - m'$, so that we can write J_c, which is not an integer and corresponds to m_c,

$$J_c = J + \frac{J_R}{\ln(2)}\ln\left[1 - 2^{(J'-J)/J_R}\right]. \tag{7.184}$$

With these definitions, the above equations result in a time-dependent equation for $n(J)$,

$$\left(\frac{\partial n(J,t)}{\partial t}\right)_{\mathrm{C}} = m(J)\int_1^{J-J_R}\left[\frac{K(m_c,m')}{m(J_c)}n(J_c,t)n(J',t)\right]\mathrm{d}J' - n(J,t)\int_1^{J_{\max}}[K(J,J')n(J',t)]\mathrm{d}J'. \tag{7.185}$$

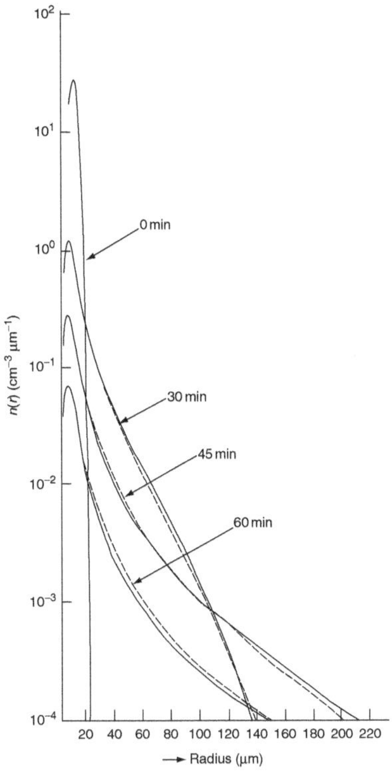

Fig. 7.10. Drop-size distribution as a function of the drop radius and time: solid lines, numerical solution; dashed lines, analytical solution. (From Ogura and Takahashi 1973; courtesy of the American Meteorological Society.)

Ogura and Takahashi (1973) solved this equation using a variety of different interpolating polynomials for finding $n(J_c)$ by interpolating on $n(J)$. By definition, remember that $J - 2 < J_c < J$. These interpolation polynomials included the following five examples, with (3) and (5) behaving equally well:

(1) three-point interpolation using $n(J-2)$, $n(J-1)$, and $n(J)$;
(2) three-point interpolation using $n(J-2)$, $n(J-1)$, and $n(J)$ for $J-2 < J_c < J-1$, $n(J-1)$, $n(J)$ and $n(J+1)$ for $J-1 < J_c < J$;
(3) four-point interpolation using $n(J-2)$, $n(J-1)$, $n(J)$, and $n(J+1)$;
(4) three-point interpolation using $n(J-2)$, $n(J-1)$, and $n(J)$ for $J-2 < J_c < J-1$, and four-point interpolation using $n(J-1)$, $n(J)$, $n(J+1)$, and $n(J+2)$ for $J-1 < J_c < J$;
(5) three-point interpolation using $n(J-2)$, $n(J-1)$, and $n(J)$ for $J-2 < J_c < J-1$, and four-point interpolation using $n(J-2)$, $n(J-1)$, $n(J)$, and $n(J+1)$ for $J-1 < J_c < J$.

The accuracy of using method (5) is shown for the Golovin kernel and initial condition in Fig. 7.10, where there is only 6% loss in mass after 60 min (Ogura and Takahashi 1973).

For reasons found in Berry and Reinhardt (1974a–d) and Pruppacher and Klett (1997) to use (7. 185) more effectively it is advantageous to work with a water content per unit ln(r) interval, which is related to $n(m, t)$ by

$$g(\ln r)\mathrm{d}(\ln r) = mn(m,t)\mathrm{d}m \tag{7.186}$$

or,

$$g(\ln r) = 3m^2 n(m,t). \tag{7.187}$$

(Note, though, that the author finds (7.185) even more accurate with higher-order interpolating polynomials.) Now defining $G(J)$,

$$G(J) \equiv g(\ln r), \tag{7.188}$$

the following is found,

$$G(J) \equiv J_R mn(J,t). \tag{7.189}$$

Now the equation for $G(J)$ follows directly from the above definitions as

$$\begin{aligned}\left(\frac{\partial G(J,t)}{\partial t}\right)_C = \left(\frac{m(J)}{J_R}\right)\Bigg[m(J) \int_0^{J-J_R} \frac{K(m_c, m')G(J_c,t)G(J',t)}{m^2(J_c)m(J')}\mathrm{d}J' \\ - \frac{G(J,t)}{m(J)} \int_1^{J_{\max}} \frac{K(J,J')G(J',t)}{m(J')}\mathrm{d}J'\Bigg].\end{aligned} \tag{7.190}$$

To employ this equation two types of numerical calculations need to be carried out. As J_c is not an integer, the values of $G(J_c)$ must be found from interpolation [these are described below using Lagrange polynomials from Berry and Reinhardt (1974a–d)]. In addition, integration must be carried out by numerical quadrature. Reinhardt (1972) devised adequate schemes for this purpose, and these are given in Berry and Reinhardt (1974a–d) as well.

We begin with the interpolation. As J_c is not an integer $G(J_c)$ is not known. A six-point Lagrange interpolation formula is employed in natural log space. The coefficients A_1 through A_6 are given by

$$\begin{aligned}
A_1 &= \left(-A^5 - 4A + 5A^2\right)/120 \\
A_2 &= \left(A^4 - 7A^3 + \left[A^5 - A^2 + 6A\right]\right)/24 \\
A_3 &= \left(-A^5 + 8A^2 - 12A - \left[2A^4 - 7A^3\right]\right)/12 \\
A_4 &= \left(3A^4 - 5A^3 - 12 + \left[A^5 - 15A^2 + 4A\right]\right)/12 \\
A_5 &= \left(-5A^5 + 16A^2 - \left[4A^4 - A^3 - 12A\right]\right)/24 \\
A_6 &= \left(5A^4 - 6A + \left[A^5 + 5A^3 - 5A^2\right]\right)/120,
\end{aligned} \tag{7.191}$$

where

$$A = J_c - J = \frac{2}{\ln 2} \ln\left[1 - 2^{(J'-J)/2}\right]. \quad (7.192)$$

The following example shows how the Lagrange interpolation formula is used. First, $G(J)$ in terms of natural logarithms is given by

$$G(J_c) = \exp\left[\begin{array}{c} A_1 \ln G(J-3) + A_2 \ln G(J-2) + A_3 \ln G(J-1) \\ +A_4 \ln G(J-0) + A_5 \ln G(J+1) + A_6 \ln G(J+2) \end{array}\right]. \quad (7.193)$$

These coefficients are valid when $(J - J') \geq 4$. When $(J - J') = 3$, the following set of coefficients must be used,

$$\begin{aligned} A_1 &= -B/[120(A+4)] \\ A_2 &= +B/[24(A+3)] \\ A_3 &= -B/[12(A+2)] \\ A_4 &= +B/[12(A+1)] \\ A_5 &= -B/[24(A)] \\ A_6 &= -B/[120(A-1)], \end{aligned} \quad (7.194)$$

where

$$B = (A-1)(A)(A+1)(A+2)(A+3)(A+4) \quad (7.195)$$

and

$$A = J_c - J. \quad (7.196)$$

The interpolation formula in natural log space becomes,

$$G(J_c) = \exp\left[\begin{array}{c} A_1 \ln G(J-4) + A_2 \ln G(J-3) + A_3 \ln G(J-2) \\ +A_4 \ln G(J-1) + A_5 \ln G(J+0) + A_6 \ln G(J+1) \end{array}\right]. \quad (7.197)$$

There are two sets of interpolating formula because of the following two ranges for J_c:

(1) when $(J - J') \geq 4$, $(J-1) < J_c < J$;
(2) when $(J - J') = 3$, $(J-2) < J_c < (J-1)$.

Also, note that when $(J - J') = 2$, $J_c = (J-2) = J'$, and when $(J - J') = 3$, the polynomial format need not be used, i.e.

$$A = J_c - J = \frac{2}{\ln 2} \ln\left[1 - 2^{(J'-J)/2}\right] = -1.258793747. \quad (7.198)$$

The results of integrating the stochastic collection equation with the Berry and Reinhardt scheme are very accurate and compare very favorably with analytical solutions, though this method does not conserve any integral properties. Moreover, the method does not permit use of empirical collisional breakup functions specified by Low and List (1982a, b).

Now for integration, three Lagrange integration coefficients are used. However, there are intervals where these will not work because the integrand becomes zero at certain values of J. So special methods are put in place near the zeros. The zeros occur when like-sized drops collide ($J = J'$), as the collection kernel then includes the fall-velocity difference. Consider the following special cases, where "aint" is a function that creates an integer from a real number.

(1) Even number of points: gain integral, $J_d = 8$

$$\begin{aligned}\int_1^8 \mathrm{aint}(J')\mathrm{d}J' &= (1/3)\mathrm{aint}(1) + (4/3)\mathrm{aint}(2) + (2/3)\mathrm{aint}(3) \\ &+ (4/3)\mathrm{aint}(4) + (1/3)\mathrm{aint}(5) + (3/8)\mathrm{aint}(5) \\ &+ (9/8)\mathrm{aint}(6) + (9/8)\mathrm{aint}(7) + (3/8)\mathrm{aint}(8).\end{aligned} \tag{7.199}$$

(2) Odd number of points: gain integral, $J_d = 9$

$$\begin{aligned}\int_1^9 \mathrm{aint}(J')\mathrm{d}J' &= (1/3)\mathrm{aint}(1) + (4/3)\mathrm{aint}(2) + (2/3)\mathrm{aint}(3) \\ &+ (4/3)\mathrm{aint}(4) + (1/3)\mathrm{aint}(5) + (14/45)\mathrm{aint}(5) \\ &+ (64/15)\mathrm{aint}(6) + (24/45)\mathrm{aint}(7) \\ &+ (64/45)\mathrm{aint}(8) + (14/45)\mathrm{aint}(9).\end{aligned} \tag{7.200}$$

(3) Even number of points through zero integrand: loss integral, $J = 8$, $J_m = 15$

$$\begin{aligned}\int_1^{15} \mathrm{aint}(J')\mathrm{d}J' &= (1/3)\mathrm{aint}(1) + (4/3)\mathrm{aint}(2) + (2/3)\mathrm{aint}(3) \\ &+ (4/3)\mathrm{aint}(4) + (1/3)\mathrm{aint}(5) + (3/8)\mathrm{aint}(5) \\ &+ (9/8)\ \mathrm{aint}(6) + (9/8)\ \mathrm{aint}(7) + (3/8)\mathrm{aint}(8) \\ &+ (3/8)\mathrm{aint}(8) + (9/8)\mathrm{aint}(9) + (9/8)\mathrm{aint}(10) \\ &+ (3/8)\mathrm{aint}(11) + (1/3)\mathrm{aint}(11) + (4/3)\mathrm{aint}(12) \\ &+ (2/3)\mathrm{aint}(13) + (4/3)\mathrm{aint}(14) + (1/3)\mathrm{aint}(15).\end{aligned} \tag{7.201}$$

(4) Odd number of points through zero integrand: loss integral, $J = 9$, $J_m = 16$

$$\int_1^{16} \mathrm{aint}(J')\mathrm{d}J' = (1/3)\mathrm{aint}(1) + (4/3)\mathrm{aint}(2) + (2/3)\mathrm{aint}(3)$$
$$\begin{aligned}&+ (4/3)\mathrm{aint}(4) + (1/3)\mathrm{aint}(5) + (14/45)\mathrm{aint}(5)\\ &+ (64/45)\mathrm{aint}(6) + (24/45)\mathrm{aint}(7) + (64/45)\mathrm{aint}(8)\\ &+ (14/45)\mathrm{aint}(9) + (3/8)\mathrm{aint}(9) + (9/8)\mathrm{aint}(10)\\ &+ (9/8)\mathrm{aint}(11) + (3/8)\mathrm{aint}(12) + (1/3)\mathrm{aint}(12)\\ &+ (4/3)\mathrm{aint}(13) + (2/3)\mathrm{aint}(14) + (4/3)\mathrm{aint}(15)\\ &+ (1/3)\ \mathrm{aint}(16).\end{aligned} \tag{7.202}$$

For further details the reader should see Berry and Reinhardt (1974a).

7.19.2 Method of moments for the stochastic collection equation

7.19.2.1 A one-moment method for the stochastic collection equation

The one-moment method of Bleck (1970) and Danielsen *et al.* (1972) has received considerable use in meteorology, as it is relatively simple and inexpensive to utilize. It does have its limitations as noted toward the end of this presentation of the method.

First, consider the equation written in (7.203) regarding the consideration of the completeness of the stochastic collection equation,

$$\frac{\mathrm{d}N_{\mathrm{T}k}}{\mathrm{d}t} = \frac{1}{2}\sum_{i=1}^{k-1} K_{i,k-i}N_{\mathrm{T}i}N_{\mathrm{T}k-i} - N_{\mathrm{T}k}\sum_{i=1}^{\infty} K_{ik}N_{\mathrm{T}i}. \tag{7.203}$$

Bleck (1970) and Danielsen *et al.* (1972) considered a drop spectrum described by Berry and Reinhardt (1974a–d), and later in discussion of multi-moment methods, so did Tzivion *et al.* (1987),

$$m_{k+1} = p_k m_k, \tag{7.204}$$

where k is the bin category index, m_k and m_{k+1} are the lower and upper bounds of the category, and p_k is a parameter describing the category width. This usually is given in terms of 2 to some power, such as

$$p_k = 2^{1/J}. \tag{7.205}$$

Alternatively, it was seen for the Berry and Reinhardt (1974a–d) method that

$$m(J) = m_0 \exp\left(\frac{3[J-1]}{J_R}\right), \tag{7.206}$$

with m_0 given as the smallest mass,

$$m_0 = \frac{4}{3}\pi\rho_L r_0^3, \tag{7.207}$$

and where r_0 is the smallest radius.

Now to obtain the solution more easily, (7.203) can be written in the following form,

$$\left(\frac{\partial n(m,t)}{\partial t}\right)_C = \int_0^{m/2} K(m_c, m')n(m_c, t)n(m', t)\mathrm{d}m' - n(m,t) \times \int_0^{\infty} K(m, m')n(m', t)\mathrm{d}m', \tag{7.208}$$

where C denotes collection, and $m/2$ in the first integral is an integer J_{up}, following Pruppacher and Klett's (1997) convention, where $m(J_{\mathrm{up}}) = m(J/2)$.

Note also that

$$m_c = m - m'. \tag{7.209}$$

As before, symmetry of the collection kernel K allows the first integral to be written as a single integral rather than a double integral.

To solve the stochastic collection equation with the one-moment method, Bleck and Danielson *et al.* both used subcategories to describe the spectrum defined by a mass-weighted mean value for the number density of the hydrometeor species in each mass category, given by

$$\bar{n}_k(t) = \int_{m_k}^{m_{k+1}} n(m,t)m\mathrm{d}m\left[\int_{m_k}^{m_{k+1}} m\mathrm{d}m\right]^{-1} = \frac{2}{m_{k+1}^2 - m_k^2}\int_{m_k}^{m_{k+1}} n(m,t)m\mathrm{d}m. \tag{7.210}$$

To find an equation for $\mathrm{d}\bar{n}_k/\mathrm{d}t$, both sides of (7.208) are multiplied by $m\mathrm{d}m$ and the resultant between m_k and m_{k+1}, is integrated; using (7.210) gives

$$\left(\frac{\partial \bar{n}_k(t)}{\partial t}\right)_C = \left[\frac{2}{m_{k+1}^2 - m_k^2}\right]\left[\int_{m_k}^{m_{k+1}} m\mathrm{d}m\int_0^{m/2} K(m-m', m')n(m-m', t)n(m', t)\mathrm{d}m'\right] - \left[\frac{2}{m_{k+1}^2 - m_k^2}\right]\left[\int_{m_k}^{m_{k+1}} m\mathrm{d}m\int_0^{\infty} K(m, m')n(m', t)n(m,t)\mathrm{d}m'\right]. \tag{7.211}$$

Given the definition

$$J_c \equiv \left(\frac{\partial n(m,t)}{\partial t}\right)_{\mathrm{C}}, \tag{7.212}$$

there is also the definition that

$$\left(\frac{\partial \bar{n}_k(t)}{\partial t}\right)_{\mathrm{C}} = \left[\frac{2}{m_{k+1}^2 - m_k^2}\right] \int_{m_k}^{m_{k+1}} J_c(m,t) m \mathrm{d}m. \tag{7.213}$$

Now Bleck made the approximation

$$n(m,t) = \bar{n}_k(t), \tag{7.214}$$

where k is such that,

$$m_k < m < m_{k+1}. \tag{7.215}$$

Now the continuous size distribution is replaced by a piecewise constant function with discontinuities at $m_k (k = 0, 1, 2, 3 \ldots)$,

$$\left(\frac{\partial n(m)}{\partial t}\right)_{\mathrm{C}} = \frac{2}{m_{k+1}^2 - m_k^2} \left[\int_{m_k}^{m_{k+1}} J_c(m,t) m \mathrm{d}m\right]. \tag{7.216}$$

Graphically, Bleck demonstrated that the term in square brackets in (7.216) could be given by

$$\int_{m_k}^{m_{k+1}} J_c(m,t) m \mathrm{d}m \approx \sum_{j=k-1}^{k} \sum_{i=1}^{k-1} a_{ijk} \bar{n}_j \bar{n}_i - \bar{n}_k \sum_{i=1}^{I} b_{ik} \bar{n}_i, \tag{7.217}$$

so that

$$\left(\frac{\partial n(m)}{\partial t}\right)_{\mathrm{C}} \approx \left[\frac{2}{m_{k+1}^2 - m_k^2}\right] \sum_{j=k-1}^{k} \sum_{i=1}^{k-1} a_{ijk} \bar{n}_j \bar{n}_i - \bar{n}_k \sum_{i=1}^{I} b_{ik} \bar{n}_i, \tag{7.218}$$

where I is the total number of bins or categories, and definitions for a_{ijk} and b_{ik} are given by Danielsen *et al.* (1972) and Brown (1983, 1985).

As this equation is normalized by the mass density distribution function, it does not conserve any other moments but the first one, mixing ratio or mass,

$$M_k^1 = \int_{m_k}^{m_{k+1}} m n(m,t) \mathrm{d}m. \tag{7.219}$$

The method has received perhaps unfair criticism, though it does accelerate larger particle drop growth (Tzivion *et al.* 1987). It does have an advantage over the Berry scheme in that various breakup parameterizations can easily be incorporated (e.g. Low and List 1982a, b; and Brown 1988).

7.19.2.2 *A multi-moment method for the stochastic collection equation*

Next the discussion turns to Tzivion *et al.*'s (1987) multi-moment or ν-moment approximation method where ν is the νth moment. The categories are defined as with the one-moment method,

$$m_{k+1} = p_k m_k \tag{7.220}$$

and

$$p_k = 2^{1/J}. \tag{7.221}$$

As noted above, ν is the νth moment of the distribution function $n(m, t)$ in category k,

$$M_k^\nu = \int_{m_k}^{m_{k+1}} m^\nu n(m,t)\mathrm{d}m. \tag{7.222}$$

Application of

$$\int_{m_k}^{m_{K+1}} m^\nu \mathrm{d}m \tag{7.223}$$

to

$$\left(\frac{\partial n(m)}{\partial t}\right)_{\mathrm{C}} = \int_0^{m/2} K(m_c, m')n(m_c)n(m')\mathrm{d}m' - n(m)\int_0^{\infty} K(m, m')n(m')\mathrm{d}m' \tag{7.224}$$

gives the following equations with respect to the moments in each category. The result is a system of equations, given by

$$\begin{aligned}\left(\frac{\partial M_k^\nu}{\partial t}\right) =& \frac{1}{2}\int_{m_k}^{m_{k+1}} m^\nu \mathrm{d}m \int_{m_0}^{m} K(m-m', m')n(m-m', t)n(m', t)\mathrm{d}m' \\ &- \sum_{i=1}^{I}\int_{m_k}^{m_{k+1}} m^\nu n_k(m,t)\mathrm{d}m \int_{m_i}^{m_{i+1}} K(m,m')n_i(m',t)\mathrm{d}m'\end{aligned} \tag{7.225}$$

and

$$\frac{\partial M_k^\nu(t)}{\partial t} = \sum_{i=1}^{k-1} \int_{m_i}^{m_{i+1}} n_i(m',t)\mathrm{d}m' \int_{m_k}^{m_{k+1}-m'} (m+m')^\nu K_{k,i}(m,m') n_k(m,t)\mathrm{d}m$$

$$+ \sum_{i=1}^{k-2} \int_{m_k}^{m_{k+1}} n_i(m',t)\mathrm{d}m' \int_{m_{k-1}-m'}^{m_{i+1}} (m+m')^\nu K_{k-1,i}(m,m') n_{k-1}(m,t)\mathrm{d}m$$

$$+ \frac{1}{2} \int_{m_{k-1}}^{m_k} n_{k-1}(m,t)\mathrm{d}m' \int_{m_{k-1}}^{m_k} (m+m')^\nu n_{k-1}(m,t) K_{k-1,k-1}(m,m')\mathrm{d}m \tag{7.226}$$

$$- \sum_{i=1}^{I} \int_{m_k}^{m_{k+1}} m^\nu n_k(m,t)\mathrm{d}m \int_{m_{k-1}-m'}^{m_{i+1}} n_k(m,t) K_{k,i}(m,m')\mathrm{d}m',$$

where K is the collection kernel. Now we let,

$$\xi_p = \frac{\int_{m_k}^{m_{k+1}} m^{\nu+1} n_k(m,t)\mathrm{d}m' \int_{m_i}^{m_{i+1}} m^{\nu-1} n_k(m,t)\mathrm{d}m}{\left[\int_{m_k}^{m_{k+1}} m^\nu n_k(m,t)\mathrm{d}m\right]^2}, \tag{7.227}$$

where

$$1 \le \xi_p \le \frac{(p_k+1)^2}{4p_k}, \tag{7.228}$$

and where p_k is a parameter describing the category width. Now using the mean value of ξ_p $(\bar{\xi}_p)$ the connection "between three neighboring moments" can be expressed as

$$M_k^{\nu+1} = \bar{\xi}_p \bar{m}_k^\nu M_k^\nu. \tag{7.229}$$

Now the zeroth moment, or number concentration, can be presented as

$$\frac{\partial N_k(t)}{\partial t} = \begin{bmatrix} +\frac{1}{2} \int_{m_{k-1}}^{m_k} n_{k-1}(m',t)\mathrm{d}m' \int_{m_{k-1}}^{m_k} K_{k-1,k-1}(m,m') n_{k-1}(m,t)\mathrm{d}m \\ + \sum_{i=1}^{k-2} \int_{m_i}^{m_{i+1}} n_i(m',t)\mathrm{d}m' \int_{m_k-m'}^{m_{i+1}} K_{k-1,i}(m,m') n_{k-1}(m,t)\mathrm{d}m \end{bmatrix}$$

$$
-\left[\begin{array}{l}
+\frac{1}{2}\int_{m_k}^{m_{k+1}} n_k(m',t)\mathrm{d}m' \int_{m_k}^{m_{k+1}} K_{k,k}(m,m')n_k(m,t)\mathrm{d}m \\
+\sum_{i=1}^{k-1} n_i(m',t)\mathrm{d}m' \int_{m_{k+1}-m'}^{m_{i+1}} K_{k,i}(m,m')n_k(m,t)\mathrm{d}m
\end{array}\right]
$$

$$
-\left[\begin{array}{l}
\frac{1}{2}\int_{m_k}^{m_{k+1}} n_k(m',t)\mathrm{d}m' \int_{m_k}^{m_k+1} K_{k,k}(m,m')n_k(m,t)\mathrm{d}m \\
+\sum_{i=k+1}^{I} \int_{m_k}^{m_{k+1}} n_i(m',t)\mathrm{d}m \int_{m_k}^{m_{k+1}} K_{k,i}(m,m')n_k(m,t)\mathrm{d}m'
\end{array}\right]
\qquad (7.230)
$$

and the first moment, or water content, as

$$
\frac{\partial M_k(t)}{\partial t} = \left[\begin{array}{l}
\frac{1}{2}\int_{m_{k-1}}^{m_k} n_{k-1}(m',t)\mathrm{d}m' \int_{m_{k-1}}^{m_k} (m+m')K_{k-1,k-1}(m,m')n_{k-1}(m,t)\mathrm{d}m \\
+\sum_{i=1}^{k-2}\int_{m_i}^{m_{i+1}} n_i(m',t)\mathrm{d}m' \int_{m_k-m'}^{m_{i+1}} (m+m')K_{i,k-1}(m,m')n_{k-1}(m,t)\mathrm{d}m
\end{array}\right]
$$

$$
-\left[\begin{array}{l}
\frac{1}{2}\int_{m_k}^{m_{k+1}} n_k(m',t)\mathrm{d}m' \int_{m_k}^{m_{k+1}} (m+m')K_{i,k}(m,m')n_k(m,t)\mathrm{d}m \\
+\sum_{i=1}^{k-1}\int_{m_i}^{m_{i+1}} n_i(m',t)\mathrm{d}m' \int_{m_{k+1}-m'}^{m_{i+1}} (m+m')K_{i,k}(m,m')n_k(m,t)\mathrm{d}m
\end{array}\right]
\qquad (7.231)
$$

$$
-\left[\begin{array}{l}
\sum\int_{m_k}^{m_{k+1}} m'n_k(m',t)\mathrm{d}m' \int_{m_{k+1}-m'}^{m_k} (m+m')K_{i,k}(m,m')n_k(m,t)\mathrm{d}m \\
-\sum_{i=k+1}^{I}\int_{m_i}^{m_{i+1}} n_i(m',t)\mathrm{d}m \int_{m_k}^{m_{k+1}} mK_{i,k}(m,m')n_i(m,t)\mathrm{d}m'
\end{array}\right].
$$

These two equations (7.230) and (7.231) were derived by Tzivion *et al.* (1987) to be interpreted as easily as possible in a physical sense. The autoconversion of the number of particles to category k as the result of coalescence between the number of particles in category $k-1$ with one another (term 1) and with the number of particles in the categories less than categories $k-1$ (term two) are the first two terms. The third and fourth terms represent the autoconversion of the number of particles to category k as the result of coalescence between the number of particles in category k with one another (term 3) and with the number of particles in categories less than k (term 4). The last two terms represent the loss in particles in category k during collisions with one another (term 5) and with the particles in categories larger than k (term 6).

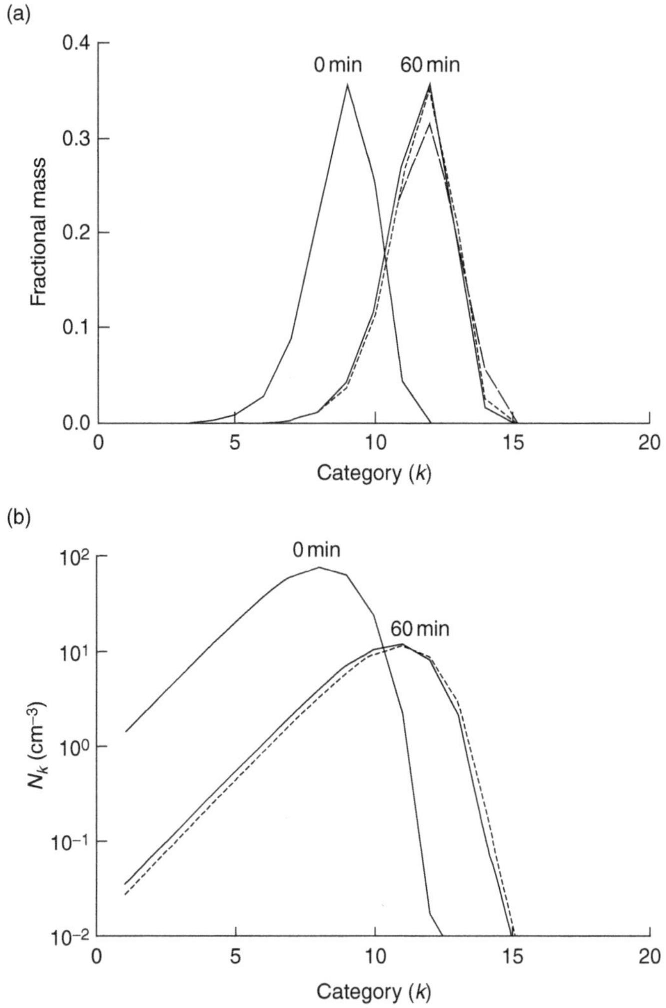

Fig. 7.11. A comparison of stochastic collection computations for the constant collection kernel [$K(x, y) = 1.1 \times 10^{-4}$ cm^{-3} s^{-1}] after 60 min of collection for (a) a fractional mass (M_k/LWC), where LWC is liquid-water content and (b) category number concentrations (N_k). The analytical solutions are represented by solid lines, Bleck's method by long dashed lines, and the proposed method by short dashed lines. The initial exponential distribution is indicated at the left by a solid line. (Tzivion *et al.* 1987; courtesy of the American Meteorological Society.)

In the mass equation (7.231) it is mass that is transferred rather than number concentration.

Tests with a constant kernel show general agreement between the one-moment Bleck scheme and two-moment Tzivion *et al.* scheme (Fig. 7.11) starting with an inverse exponential profile for the size distribution. There is

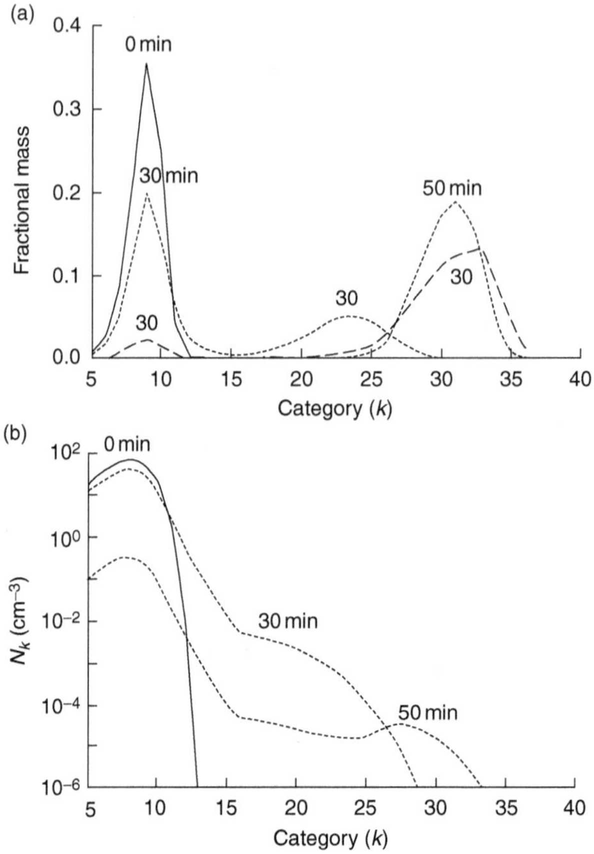

Fig. 7.12. A comparison of stochastic collection computations for the constant collection kernel [$K(x, y) = 1500(x+y)$ cm^{-3} s^{-1}] for (a) a fractional mass (M_k/LWC) and (b) category number concentrations (N_k). The analytical solutions are represented by solid lines, Bleck's method by long dashed lines, and the proposed method by short dashed lines. The initial Golovin's distribution is indicated at left by a solid line. Solutions are shown at 30 and 50 min collection time. Note the excellent fit obtained for the proposed method for both mass and number concentration and the tendency for the Bleck solution to accelerate the collection process. (From Tzivion *et al.* 1987; courtesy of the American Meteorological Society.)

some error with Bleck's method with the fractional mass. Also shown is Fig. 7.12 using the initial inverse exponential distribution that shows the failure of the one-moment scheme that occurs with rapid acceleration of larger drops (large dashed lines) as compared to Tzivion *et al.*'s method. Finally, the most enlightening result is found using a realistic kernel. In this case, the Tzivion *et al.* method considerably out-performs the Bleck method (Fig. 7.13). (Note number concentrations are not shown for Bleck's one-moment scheme, as it

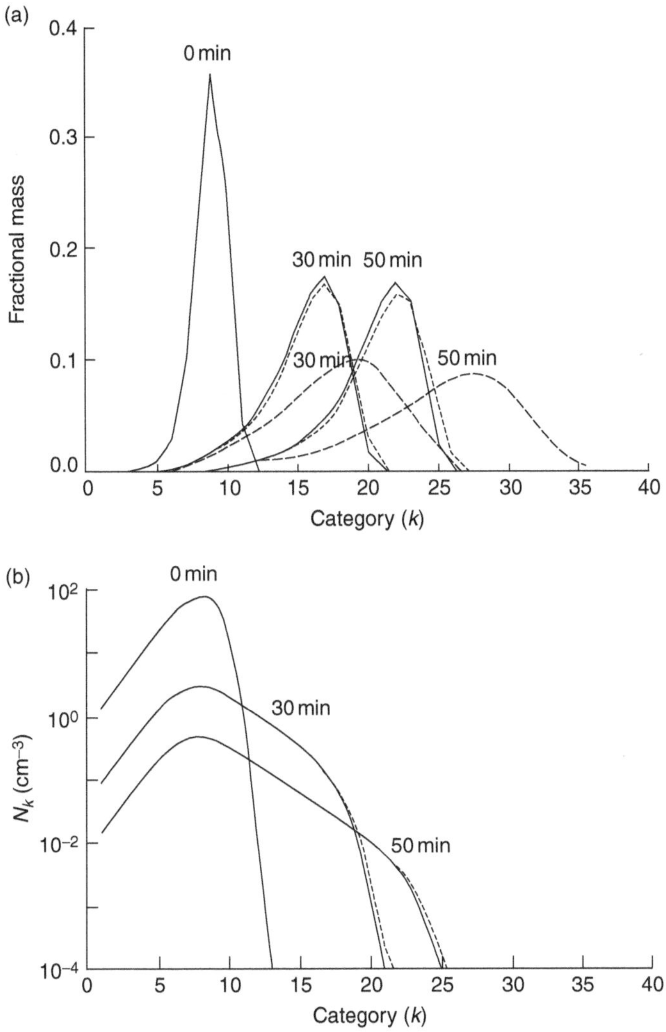

Fig. 7.13. A comparison of stochastic collection computations for a real collection kernel (see Long 1974) for (a) a fractional mass (M_k/LWC) and (b) category number concentrations (N_k). The analytical solutions are represented by solid lines, Bleck's method by long dashed lines, and the proposed method by short dashed lines. Evidently, the Bleck method enhances collection by a factor of about two compared with the proposed method. (From Tzivion *et al.* 1987; courtesy of the American Meteorological Society.)

does not solve a solution for the zeroth moment, which is the number concentration. It only solves for mass.)

To solve the equations, Long's kernel with integer-order polynomials is used to solve the integrals such as

$$\int_{m_{k+1}-m'}^{m_{k+1}} m^{\nu} n_k(m,t) \mathrm{d}m. \tag{7.232}$$

Using a linear distribution to approximate the integrand,

$$m^{\nu} n_k(m,t) = m_k^{\nu} f_k \left(\frac{m_{k+1} - m}{m_k} \right) + m_{k+1}^{\nu} \psi_k \left(\frac{m - m_k}{m_k} \right), \tag{7.233}$$

the integral can be solved as the following,

$$\int_{m_{k+1}-m'}^{m_{k+1}} m^{\nu} n_k(m,t) \mathrm{d}m = m_{k+1}^{\nu} \psi_k m' - \frac{m_k^{\nu}}{2m_k} (2^{\nu} \psi_k - f_k) m'. \tag{7.234}$$

The functions for f_k and ψ_k, used to describe the transfer of particles from one bin to another, have to be given in terms of the moments M_k^{ν}.

Substitution of (7.233) into the definition for the moments allows the following to be written,

$$M_k^{\nu} = \int_{m_k}^{m_{k+1}} m^{\nu} n(m,t) \mathrm{d}m, \tag{7.235}$$

where

$$f_k = \frac{2N_k}{m_k} \left(2 - \frac{\overline{m}_k}{m_k} \right), \tag{7.236}$$

and

$$\psi_k = \frac{2N_k}{m_k} \left(\frac{\overline{m}_k}{m_k} - 1 \right). \tag{7.237}$$

Tzivion *et al.* note that the proposed approximate distribution function is positive definite on (m_k, m_{k+1}) as long as the following is true,

$$m_k \leq \overline{m}_k \leq m_{k+1}. \tag{7.238}$$

If this does not hold because of truncation error, it is required that

$$f_k = 0; \ \psi(k) = 2\frac{N_k}{m_k} \text{ if } \overline{m}_k > m_{k+1} \tag{7.239}$$

and

$$f_k = 2\frac{N_k}{m_k}; \ \psi(k) = 0 \text{ if } \overline{m}_k < m_k. \tag{7.240}$$

Higher-order polynomials can be used, but are not really deemed necessary for typical high-resolution modeling with small timesteps. Nevertheless, though not presented here, a cubic approximation that performs well can be found in Tzivion *et al.* (1987).

7.19.2.3 The flux method for the stochastic collection equation

Bott (1998) introduced a new method for solving the stochastic collection equation. It conserves mass exactly, and is computationally very efficient compared to other methods. Bott's method is a two-step procedure. In step one, drops with mass m' that have been just formed by collision, are entirely added to grid box k of the numerical grid mesh with $m_k \leq m' \leq m_{k+1}$. In step two, a certain fraction of the water mass in box k is transported to box k+1. This transport is carried out as an advection procedure, which is unique to the method.

First Bott (1998) starts off with Pruppacher and Klett's (1997) definition of the stochastic collection equation written slightly differently,

$$\left(\frac{\partial n(m)}{\partial t}\right)_C = \int_0^{m/2} K(m_c, m')n(m_c)n(m')\mathrm{d}m' - \int_0^{\infty} K(m, m')n(m)n(m')\mathrm{d}m'. \quad (7.241)$$

Here, remember that n is the number concentration, K is the collection kernel describing the rate that a drop of mass

$$m_c = m - m' \quad (7.242)$$

is collected by a particle of mass m' forming a drop of mass m. Now, m_0 is the mass of the smallest particle involved in the collection process and m_1 is $= m/2$.

Following Berry (1967) a mass distribution function $g(y, t)$ is employed with

$$g(y, t)\mathrm{d}y = mn(m, t)\mathrm{d}m \quad (7.243)$$

and

$$n(m, t) = \frac{1}{3m^2} g(y, t). \quad (7.244)$$

The following definitions are made including $y = \ln r$ where r is particle radius with mass m. Substituting these into the modified Pruppacher and Klett (1997) description of the stochastic collection equation gives a somewhat familiar equation from the earlier discussion of the Berry and Reinhardt method,

$$\left(\frac{\partial g(y,t)}{\partial t}\right)_{\mathrm{C}} = \left[\int_{y_0}^{y_1} m^2 \frac{K(y_c, y')g(y_c, t)g(y', t)}{m_c^2 m'} \mathrm{d}y' - \int_{y_0}^{\infty} \frac{K(y, y')g(y, t)g(y', t)}{m'} \mathrm{d}y'\right]. \quad (7.245)$$

As usual, the first integral is the gain of a particle by collection between two particles, and the second integral is loss of particles with mass m owing to collection with particles of other sizes.

For a numerical solution, Bott uses a logarithmically equidistant mass grid mesh, where

$$m_{k+1} = \alpha m_k \quad k = 1, 2, 3 \ldots l, \quad (7.246)$$

where l is the total number of grid points. This gives a grid mesh in y that is equally spaced such that

$$\Delta y_k = \Delta y = \ln(\alpha/3), \quad (7.247)$$

where $\alpha = 2^{(1/2)}$. This represents a doubling of the particle mass with every two grid cells. Different values of α can be used. Discretizing the collision of particles of mass m_i with drops of mass m_j gives a change in the mass distributions g_i and g_j, such that

$$g_i(i,j) = g_i - g_i g_j \frac{\bar{K}(i,j)}{m_j} \Delta y \Delta t \quad (7.248)$$

and

$$g_j(j,i) = g_j - g_i g_j \frac{\bar{K}(j,i)}{m_i} \Delta y \Delta t. \quad (7.249)$$

The variables g_i and g_j are described as the mass distributions before collisions at grid points i and j, whilst $g_i(i,j)$ and $g_j(j,i)$ are the new mass distributions after collisions. In addition, $\bar{K}(i,j)$ is an average value of the collection kernel found by bilinear interpolation such that

$$\bar{K}(i,j) = \frac{1}{8}[K(i-1,j) + K(i,j-1) + 4K(i,j) + K(i+1,j) + K(i,j+1)]. \quad (7.250)$$

As occasionally is the case, the collection kernel is symmetric in i and j so that

$$\bar{K}(j,i) = \bar{K}(i,j) \quad (7.251)$$

and

$$g(j,i) = g(i,j). \quad (7.252)$$

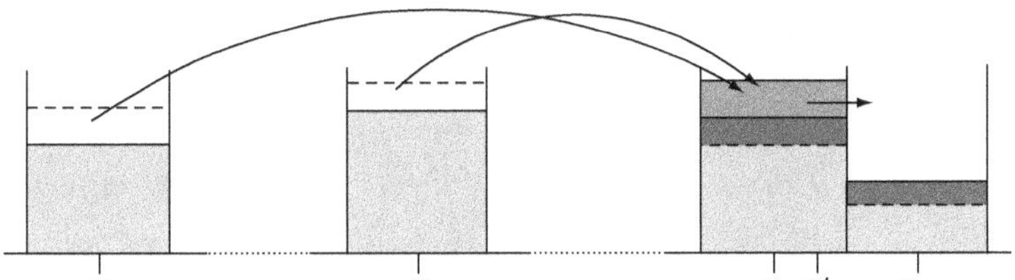

Fig. 7.14. Schematic of the flux method. Details in the text. (From Bott 1998; courtesy of the American Meteorological Society.)

Owing to collisions in box j new particles with mass $m'(i,j) = m_i + m_j$ are formed, so that the following is written,

$$g_i(i,j) = g_i g_j \bar{K}(i,j) \frac{m'(i,j)}{m_i m_j} \Delta y \Delta t. \tag{7.253}$$

As with most schemes the mass $x'(i,j)$ differs from the grid mesh x_k, as

$$x_k \le x'(i,j) \le x_{k+1}. \tag{7.254}$$

As a result, the mass density needs to be split up into grid cells k and $k+1$. This is where Bott (1998) uses a two-step procedure. In the first step, $g'(i,j)$ is added in its entirety to grid box k,

$$g'_k(i,j) = g_k + g'(i,j). \tag{7.255}$$

Now comes the tricky part. In step two, a fraction of the new mass, $g'_k(i,j)$ is transported into grid box $k+1$ by advection through the boundary $k + (1/2)$ between boxes k and $k+1$. Bott's schematic of this is shown in Fig. 7.14.

More formally, the advection step is given as

$$\begin{cases} g_k(i,j) = g'_k(i,j) - f_{k+1/2}(i,j) \\ g_{k+1}(i,j) = g'_k(i,j) + f_{k+1/2}(i,j), \end{cases} \tag{7.256}$$

where the mass flux through the boundary $k + (1/2)$ is

$$f_{k+1/2}(i,j) \frac{\Delta y}{\Delta t}. \tag{7.257}$$

Now the upstream approach can be used to find $f_{k+1/2}(i,j)$,

$$f_{k+1/2}(i,j) = c_k g'_k(i,j) w(i,j), \tag{7.258}$$

where c_k is like the Courant number and is calculated by

$$c_k = \frac{x'_k(i,j) - x_k}{x_{k+1} - x_k}, \tag{7.259}$$

and the weighting function is

$$w(i,j) = \frac{g'(i,j)}{g'_k(i,j)}. \tag{7.260}$$

Substituting (7.260) into (7.258) shows that the value of $f_{k+1/2}(i,j)$ is just,

$$f_{k+1/2}(i,j) = c_k g'(i,j), \tag{7.261}$$

resulting in the same partitioning of $g'(i,j)$ as in Kovetz and Olund (1969); but where they solved the stochastic collection equation for number distribution $n(m, t)$ instead of the mass distribution. This method has been called the upstream flux method by Bott (1998). Unfortunately the upstream flux method produces broad distributions, which, however, can be remedied by using higher-order advection schemes that are also positive definite like the upstream flux method.

To begin, consider grid box k, where the constant value of $g'_k(i,j)$ is replaced by a higher-order polynomial of order L,

$$g_{k,L}(z) = \sum_{s=0}^{L} a_{k,s} z^s, \tag{7.262}$$

where

$$z = (y - y_k)/\Delta y, \tag{7.263}$$

and $-1/2 \leq z \leq 1/2$.

Next $c_k g'_k(i,j)$ is replaced by the integral relation

$$\int_{1/2-c_k}^{1/2} g_{z,L}(z)\mathrm{d}z = \sum_{s=0}^{L} \frac{a_{k,s}}{(s+1)2^{s+1}}\left[1-(1-2c_k)^{s+1}\right]. \tag{7.264}$$

This gives the mass flux through $k + (1/2)$ as

$$f_{k+1/2}(i,j) = w(i,j)\sum_{s=0}^{L} \frac{a_{k,s}}{(s+1)2^{s+1}}\left[1-(1-2c_k)^{s+1}\right]. \tag{7.265}$$

For the upstream flux method, with $L = 0$,

$$a_{k,0} = g'_k(i,j). \tag{7.266}$$

For the linear flux method, with $L = 1$,

$$a_{k,1} = g_{k+1} - g'_k(i,j). \tag{7.267}$$

Finally for the parabolic method, with $L = 2$,

$$\begin{cases} a_{k,0} = -1/24\left[g_{k+1} - 26g'_k(i,j) + g_{k-1}\right] \\ a_{k,1} = +1/2[g_{k+1} - g_{k-1}] \\ a_{k,2} = -1/2\left[g_{k+1} - 2g'_k(i,j) + g_{k-1}\right]. \end{cases} \quad (7.268)$$

For positive definiteness to be guaranteed the following simple flux limiter is applied,

$$0 \leq f_{k+2}(i,j) \leq g'(i,j). \quad (7.269)$$

Now all the possible collisions need to be treated in timestep Δt. To do so, an iterative procedure is used. To start, the smallest particle involved in collisions is defined by index $i = i_0$, and the largest particle by $i = i_1$. In step one, following Bott (1998) closely, collisions between the smallest particle with particles of grid box $j = i_0 + 1$ are found using new values from calculating $g_{i0}\,(i_0,i_0 + 1)$, $g_{i0\,+\,1}(i_0,i_0 + 1)$, and $g_k(i_0,i_0 + 1)$.

In the following steps the collision of the remaining particles in $i = i_0$ having new mass distribution $g_{i0}\,(i_0,i_0 + 1)$ with particles in grid box $j = i_0 + 2$ are calculated. To complete the iteration this is continued until all collisions of particles in grid box $i = i_0$ with particles in $j = i_0 + 1, i_0 + 2, i_0 + 3, \ldots, i_1$ are completed.

The next step starts with particles in $i = i_0 + 1$ colliding with particles in $i_0 + 2, i_0 + 3, \ldots, i_1$ until all are completed.

Owing to the iterative approach after each collision process, the drop distribution is updated before the next collision process is calculated (Bott 1998). According to Bott, at least analytically, this is done by replacing (7.253)–(7.256) after the first collision process, g_i, g_j, g_k, and g_{k+1} with $g_i(i,j - 1)$, $g_j(i,j - 1)$, $g_k(i,j - 1)$, and $g_{k+1}(i,j - 1)$.

The timestep must be limited for positive definiteness by

$$\Delta t \leq \frac{x_j}{g_j(i,j-1)K(i,j)\Delta y}, \quad (7.270)$$

and for $j \neq k$ by

$$\Delta t \leq \frac{x_i}{g_i(i,j-1)K(i,j)\Delta y}. \quad (7.271)$$

The results of integrating show a definite improvement in solutions using the Bott (1998) method when using higher-and-higher-order polynomials for the advection with the parabolic flux method performing the best of the schemes tried, as shown below for Golovin's kernel and distribution. Other adjustments can be made to the linear flux method to improve its performance as described by Bott by using a different flux limiter resulting in the modified flux method. Moreover, Bott shows solutions from several different initial conditions that could be used for benchmarks for other methods (Fig. 7.15) in Bott (2000). Also, a flux method for the numerical solution of the stochastic collection equation extended for two-dimensional distribution is developed.

7.20 Collection, collision, and coalescence efficiencies

7.20.1 Rain collecting cloud water

Collection efficiencies, which are the product of collision efficiencies and coalescence efficiencies, have been studied for decades by many investigators. Cloud-drop and raindrop collection efficiencies seem to be those most commonly studied. This may be because wind tunnels for cloud drops and raindrops are easier to construct than those for other particles, which in turbulent free air do not tumble or follow irregular trajectories.

Possibilities when raindrops collide with cloud drops or other raindrops are coalescence, rebounding, and tearing away after coalescence, and finally drop breakup for raindrop–raindrop collisions between certain sizes of raindrops. Typically, for bulk parameterization, the collection efficiencies for rain collecting cloud range from the order of 0.55 in some models to as high as 1.0 in other models.

Some, such as Proctor (1987) have computed polynomials for raindrops (subscript rw) collecting cloud drops (subscript cw). Proctor's polynomial is given by

$$E_{\mathrm{rwcw}} = \min[(a_{\mathrm{cw}} + r_{\mathrm{cw}}(b_{\mathrm{cw}} + r_{\mathrm{cw}}(c_{\mathrm{cw}} + r_{\mathrm{cw}}(d_{\mathrm{cw}})))), 1.0], \tag{7.272}$$

where r_{cw} is the radius of the cloud drop and the coefficients for the collision efficiency are given as

$$\begin{aligned} a_{\mathrm{cw}} &= -0.27544 \\ b_{\mathrm{cw}} &= 0.26249 \times 10^{6} \\ c_{\mathrm{cw}} &= -1.8896 \times 10^{10} \\ d_{\mathrm{cw}} &= 4.4626 \times 10^{14}. \end{aligned}$$

This polynomial (7.272) works remarkably well when compared to the efficiencies given in Rogers and Yau (1989; Fig. 7.16), especially for medium-sized cloud drops. Figure 7.17 shows coalescence efficiencies from Low and List (1982a). A comprehensive figure of self-collection, breakup, and accretion probabilities is shown in Fig. 7.18.

For bin models, Cooper *et al.* (1997) came up with collection efficiencies for all sizes of drops up to about 5 mm in diameter, collecting all sizes of drops up to about 5 mm in diameter (Fig. 7.19). The collection efficiencies are

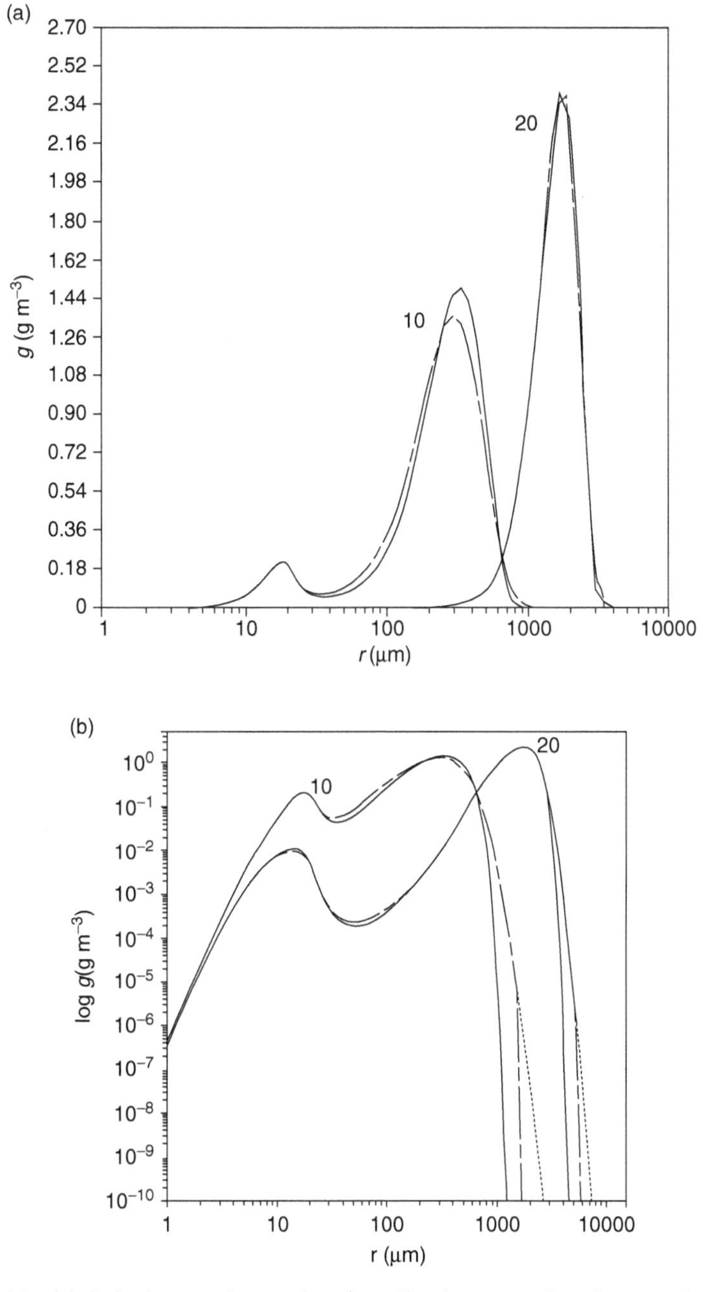

Fig. 7.15. (a) Solution to the stochastic collection equation for $L = 2\ \mathrm{g\ m^{-3}}$ and $\bar{r} = 15\ \mu\mathrm{m}$, with different versions of the flux method in comparison to the BRM (full curve). Short dashed curve: LFM; long dashed curve: MFM. Curves shown after 10 and 20 min. BRM is the Berry and Reinhardt method. LFM is the linear flux method; MFM is the modified linear flux method. (From Bott 1998; courtesy of the American Meteorological Society.) (b) Same as (a) except with a logarithmic scale of the ordinate. (From Bott 1998; courtesy of the American Meteorological Society.)

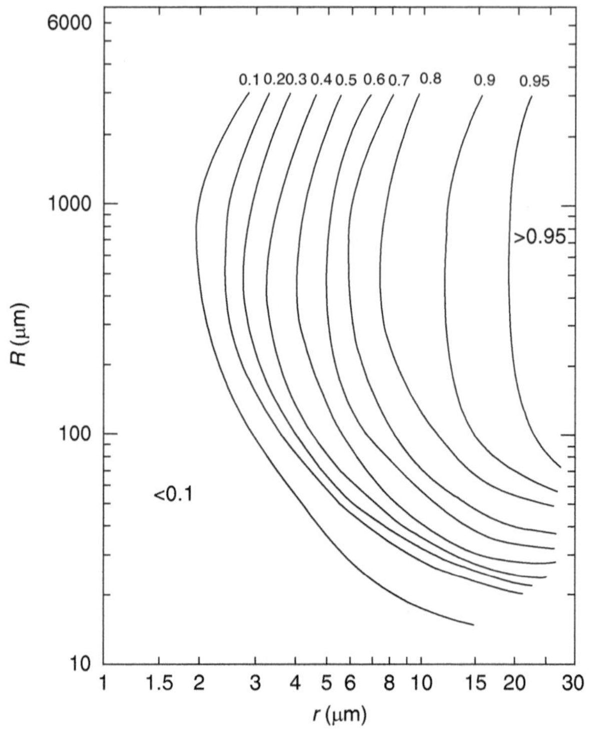

Fig. 7.16. Collision efficiencies for larger particle collecting smaller particles; R is the radius of the drop doing the collecting, and r is the radius of the drop collected. (From Rogers and Yau 1989; courtesy of Elsevier.)

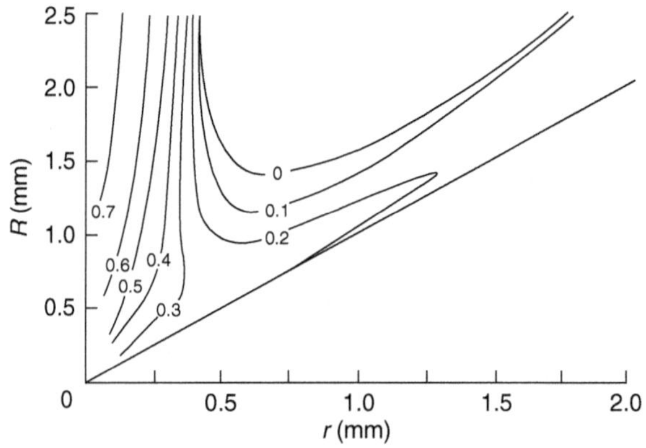

Fig. 7.17. Empirical coalescence efficiencies. (From Low and List 1982a; courtesy of the American Meterological Society.)

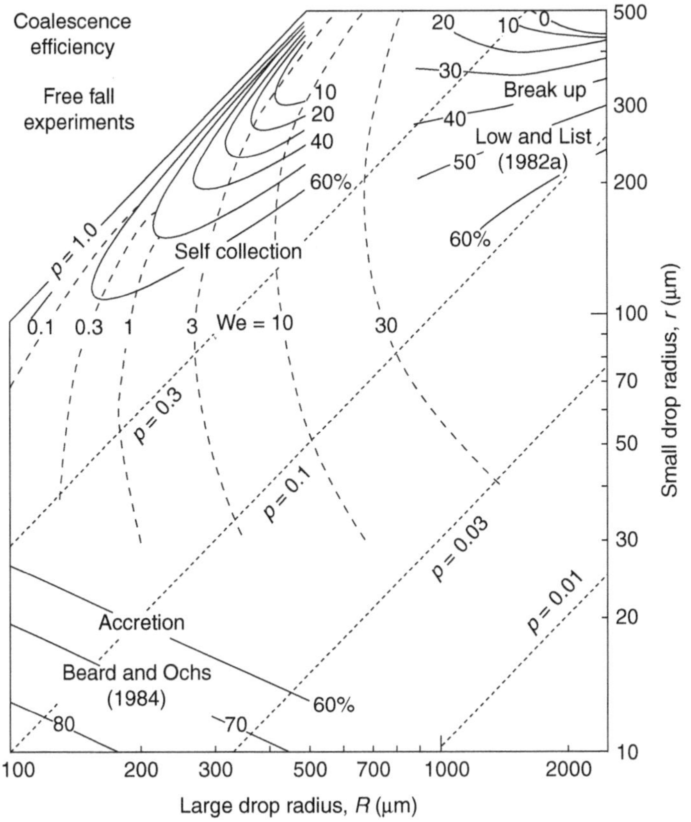

Fig. 7.18. Accretion-efficiency, self-collection-efficiency, and breakup percentages; We is the Weber number, the *P*'s indicate probability. (Cooper *et al.* 1997; courtesy of the American Meteorological Society.)

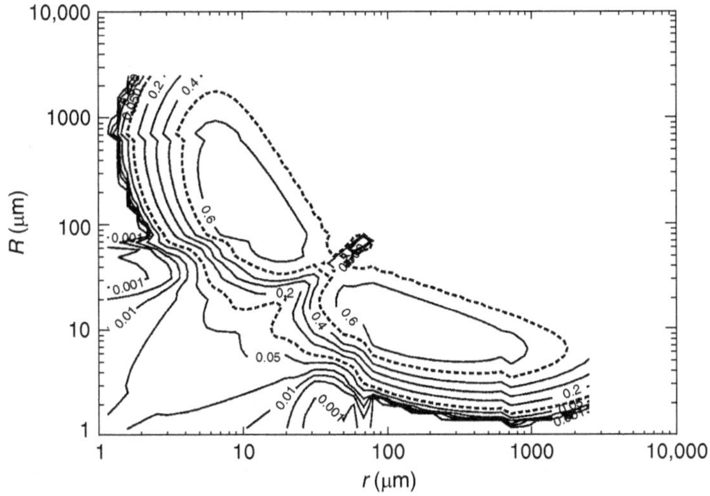

Fig. 7.19. Collection efficiencies between drops and droplets. (Cooper *et al.* 1997; courtesy of the American Meteorological Society.)

shown as two lobes of values higher than about 0.6 and then much smaller values for like-sized particles and small particles collecting small particles. In addition, very small particles are difficult for any sized particle to collect.

7.20.2 Larger ice hydrometeors collecting cloud water

Cloud water collection efficiencies for larger ice hydrometeors are difficult to measure, so researchers often resort to numerical techniques. For graupel and hail, values are either set to one, or to a function developed by Milbrandt and Yau (2005b) using data from Macklin and Bailey (1962),

$$E_{x\mathrm{cw}} = \exp\left[-8.68 \times 10^{-7} D_{\mathrm{cwmv}}^{-1.6} D_{x\mathrm{mv}}\right], \tag{7.273}$$

where D_{cwmv} and $D_{x\mathrm{mv}}$ are the mean-mass diameters of cloud water and graupel or hail, respectively, in meters. For example if D_{cwmv} and D_{hwmv} are 2.5×10^{-5} m and 9×10^{-3} m, respectively, then the collection efficiency is 0.835. For snow collecting cloud water collection efficiencies are usually taken to be about one.

7.20.3 Ice crystals collecting cloud water

Pruppacher and Klett (1997) show the onset size that various ice crystals must achieve before they can collect cloud drops, as seen by observing the size of the rimed ice crystal. Plates start readily collecting cloud droplets at about 200 mm, sectors at about 300–400 mm, and dendrites at about 800 mm. The size of cloud droplets collected by plates between about 1000 and 1600 mm ranges from $10 < D < 42$ mm, where D is diameter, with a peak number between $20 < D < 30$ mm. The size of cloud droplets collected by dendrites with sizes of between 2900 mm and 4650 mm ranges from $10 < D < 78$ mm with a peak number for the smaller of these dendrites near $20 < D < 40$ mm; for the larger of these dendrites, the peak numbers occur for sizes of cloud drops that are around $D \approx 50$ mm. A summary is shown in Fig. 7.20. Pruppacher and Klett (1997) also present some graphs from Wang (2002) of different sizes of different types of ice crystals and the efficiencies with which they collect cloud drops. These are rather preliminary results, though they could be made into lookup tables for bin or bulk microphysical parameterizations. Saleeby and Cotton (2008) have done this, as have Straka *et al.* (2009b). A comparison between an infinite column and a finite column (Fig. 7.21) is unique in that most early calculations employed infinite columns. The results in Fig. 7.21 show that collection of $D = 10$ to 15 micron cloud droplets is best for all sizes

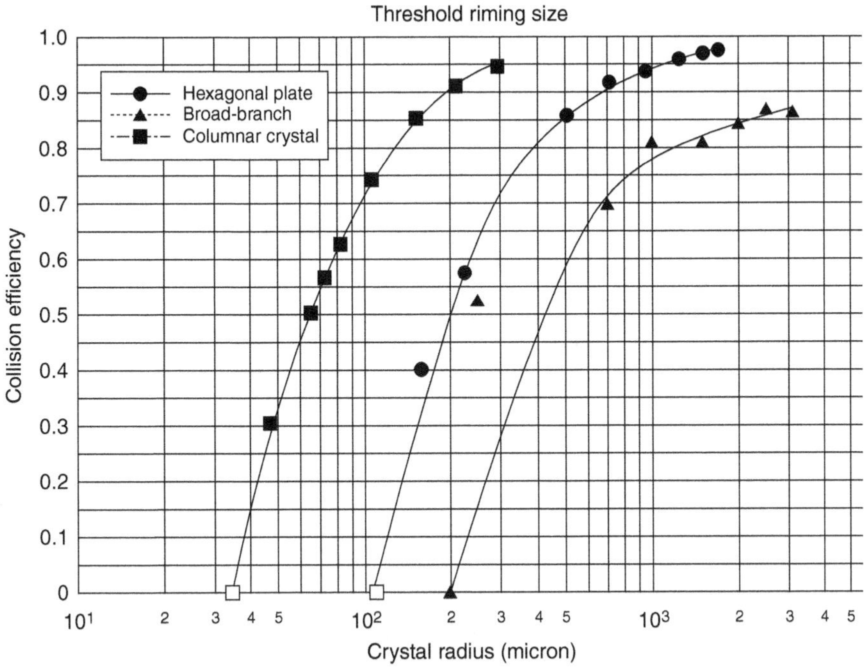

Fig. 7.20. Cut-off riming ice crystal sizes, extrapolated. For broad-branched crystals, the data point for crystal radius 2.5 μm was ignored when performing the best fit. (From Wang 2002; courtesy of Elsevier.)

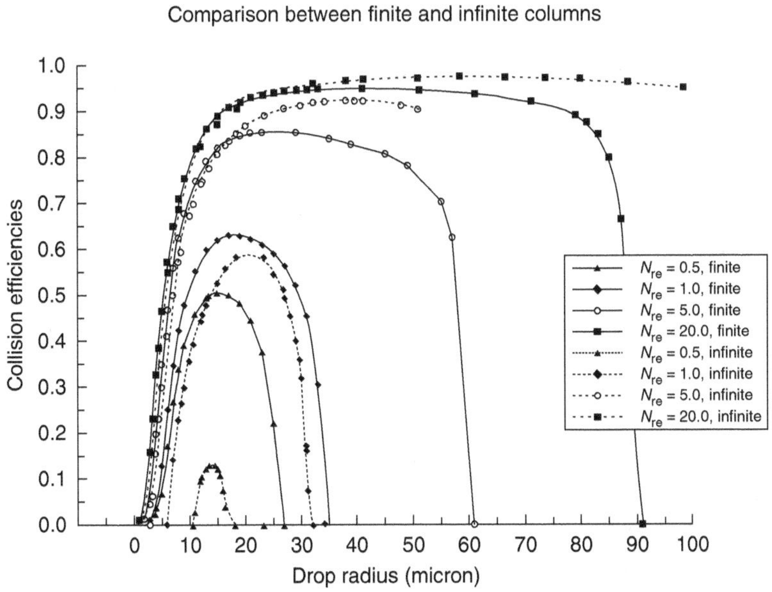

Fig. 7.21. Comparison between the collision efficiencies for finite cylinders (present results) and infinite cylinders (Schlamp *et al.* 1975) for N_{re} 0.5, 1.0, 5, and 20. (From Wang 2002; courtesy of Elsevier.)

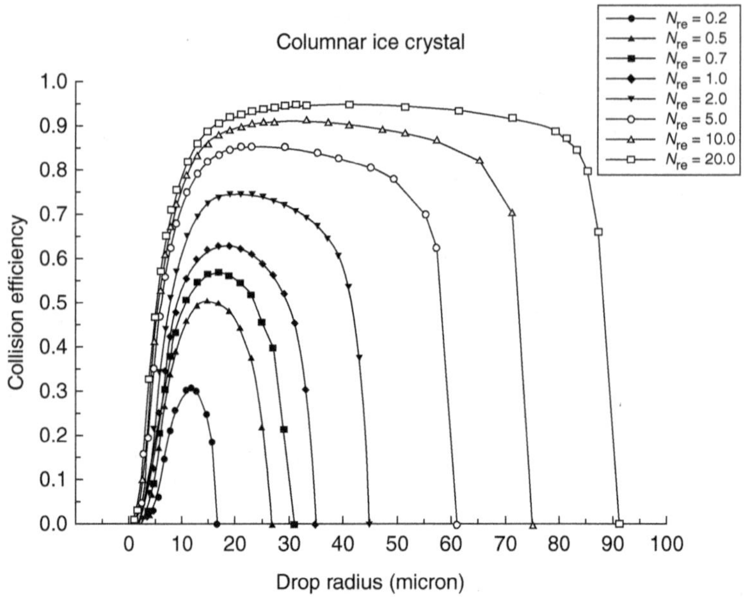

Fig. 7.22. Collision efficiencies of columnar crystals colliding with supercooled water drops. The last data points (at the large drop-size end) for N_{re} 0 to 20 are extrapolated. (From Wang 2002; courtesy of Elsevier.)

of columns, and infinite and finite at the columns' smaller sizes. The collection efficiencies for finite columns collecting cloud droplets are summarized in Fig. 7.22, broad-branched plate crystals collecting cloud droplets in Fig. 7.23, and plates collecting cloud droplets are shown in Fig. 7.24. These are only the collision efficiencies, but what Wang and Ji showed is really a first step forward after many years of attempts to describe collision efficiencies of ice crystals collecting cloud droplets.

7.20.4 Ice particles collecting ice particles

Ice particles collecting ice particles are probably the most difficult to measure in any manner and the functions researchers use are only loose approximations. One measurement that stands out is that dendrites collect other dendrites with collection efficiencies of about 140 percent. Cotton *et al.* (1986) also present a function from Hallgren and Hosler's (1960) results for other types of crystals collecting crystals as well as other ice hydrometeors; it is given by

$$E_{ix} = \min\left[10^{(0.035\{T-273.15\}-0.7)}, 0.2\right], \qquad (7.274)$$

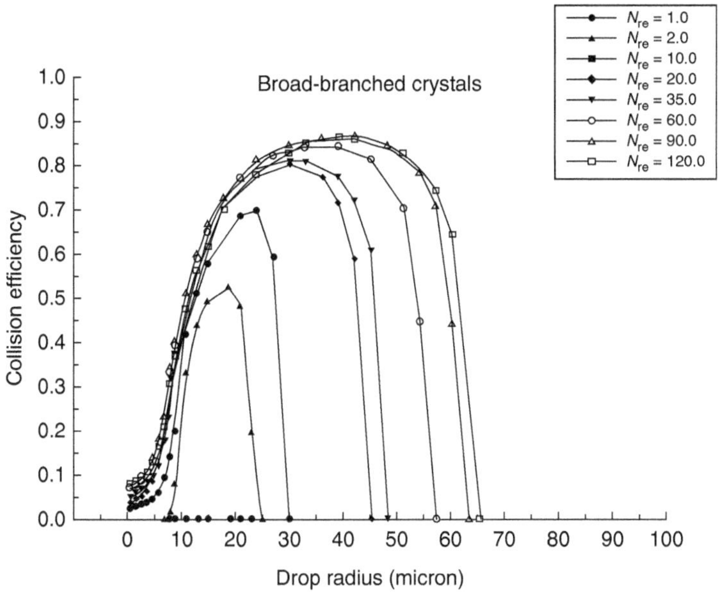

Fig. 7.23. Collision efficiencies of broad-branched crystals colliding with supercooled water drops. The last data points (at the large drop-size end) for $N_{re} = 1$ to 120 are extrapolated. (From Wang 2002; courtesy of Elsevier.)

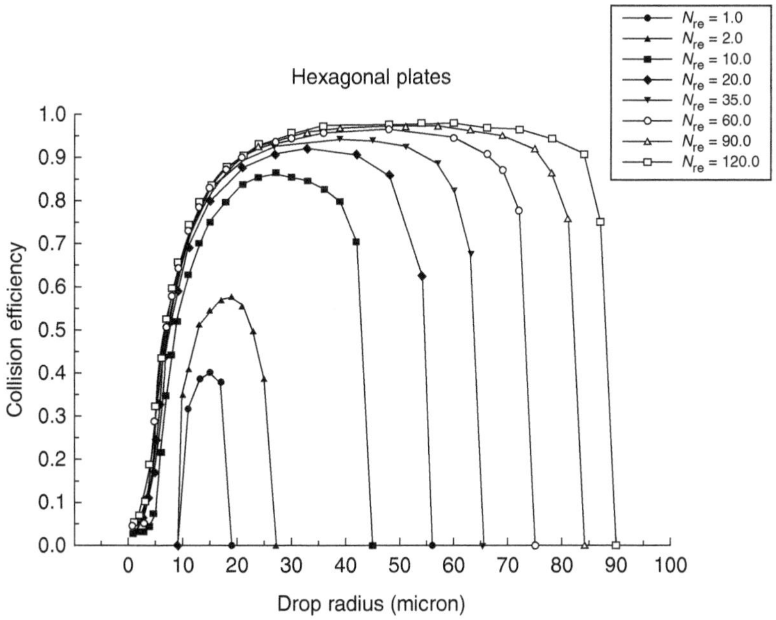

Fig. 7.24. Collision efficiencies of hexagonal ice plates colliding with supercooled water drops. The last data points (at the large drop-size end) for N_{re} 0 to 120 are extrapolated. (From Wang 2002; courtesy of Elsevier.)

where x is ice, snow, graupel, or hail, and the temperature used is that of the warmer particle. Besides this function, Milbrandt and Yau (2005b) and others use the following functions for dry collection as given by,

$$E_{\mathrm{cisw}} = \min[0.05\exp(T - 273.15\mathrm{K}), 1.0], \tag{7.275}$$

where subscripts ci and sw denote ice crystals and snow aggregates, respectively.

For larger ice particles collecting ice crystals, the following is used,

$$E_{\mathrm{cigw}} = E_{\mathrm{sgw}} = E_{\mathrm{cihw}} = E_{\mathrm{shw}} = \min[0.01(T - 273.15\mathrm{K}), 1.0]. \tag{7.276}$$

Lin *et al.* (1983) used slightly different numbers for these collection efficiencies, including

$$E_{\mathrm{cisw}} = \min[0.025\exp(T - 273.15\mathrm{K}), 1.0] \tag{7.277}$$

and

$$E_{\mathrm{cihw}} = E_{\mathrm{swhw}} = \min[0.09(T - 273.15\mathrm{K}), 1.0]. \tag{7.278}$$

During wet growth of hail, the collection of ice crystals and snow aggregates is set to one. Many of these originated with Lin *et al.* (1983) who used these functional forms.

8

Drop breakup

8.1 Introduction

The characteristics of hydrodynamic instability are called bag breakup. A very large raindrop (D = 8 to 9 mm) accelerates, and a concave region forms on the underside of the drop, which amplifies with time. At some critical diameter the concave region grows explosively. An annular ring of water at its base supports the bag of liquid water. As the bag grows, it thins and bursts and many small droplet fragments are formed. A few larger droplets or drops may form from breakup of the annular ring. The maximum size a drop can achieve before bag breakup occurs under conditions where drag forces exceed surface-tension forces. The drag force or stress is

$$F_{\mathrm{d}} = c_{\mathrm{d}} \rho u_{\infty}^2, \tag{8.1}$$

where c_{d} is the drag coefficient, ρ is the density of air, and u_{∞} is the fallspeed at terminal velocity; whereas the surface-tension stress is

$$F_{\mathrm{t}} = \frac{4\sigma}{D}. \tag{8.2}$$

The scales of the variables are, $c_{\mathrm{d}} = 0.85$, $\sigma = 7.6 \times 10^{-2}\ \mathrm{J\,m^{-2}}$, $\rho = 1\ \mathrm{kg\,m^{-3}}$, $u_{\infty} = 9\ \mathrm{m\,s^{-1}}$. With these $D_{\max}$ can be defined as

$$D_{\max} = \frac{8\sigma}{c_{\mathrm{d}} \rho u_{\infty}^2} = 8.8 \times 10^{-3}\ \mathrm{m}, \tag{8.3}$$

or about 8 to 9 mm depending on the values used for C_{d}, ρ, and values used for u_{∞}.

Liquid raindrops generally are not observed to be this size. In addition, raindrops are rarely larger than 3 to 5 mm in diameter. The reason for these conditions is that collisional breakup occurs much more frequently than bag breakup.

8.2 Collision breakup of drops

There are three primary forms of collisional breakup. These include neck or filament breakup (27%), sheet breakup (55%), and disk breakup (18%). Schematics of each type of breakup are shown in Fig. 8.1.

8.2.1 Neck or filament breakup

Neck or filament breakup occurs by glancing collisions between a smaller and a larger drop. As the smaller drop makes contact with the larger drop and the large drop falls away a neck or filament of water keeps the two drops momentarily attached. Eventually, the filament breaks and the two drops retain much of their original mass. However, two to ten fragments usually form, with five the most common number (including the original drops). The number of drops formed by filament, sheet, and disk breakup increases with the increasing collisional kinetic E_{CKE} energy or larger differences in sizes of drops. These types of breakup occur over a wide range of drop pair sizes. Schematics of the various forms of breakup are shown in Fig. 8.2 and discussed below.

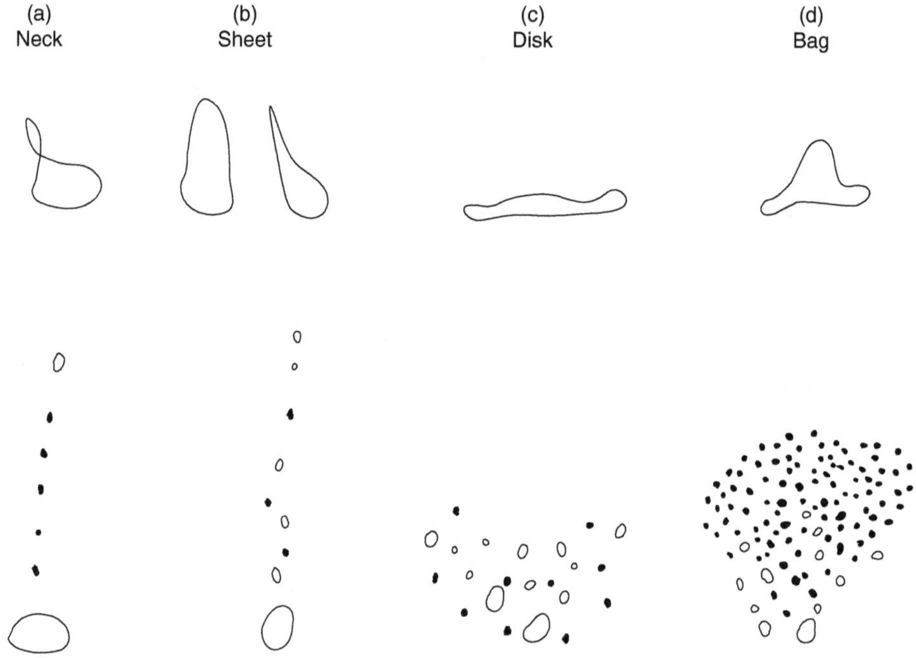

Fig. 8.1. Schematic of the common types of breakup. (From McTaggert-Cowan and List 1975; courtesy of the American Meteorological Society.)

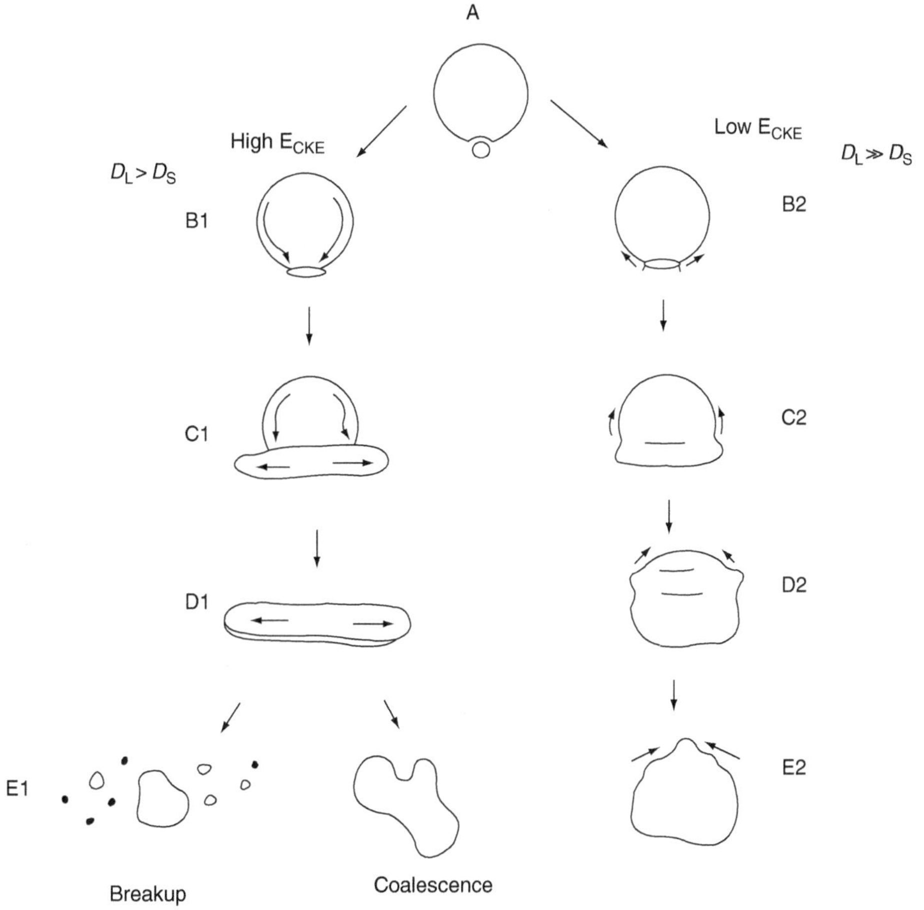

Fig. 8.2. "Collision with D_S (small drop) hitting D_L (large drop) in the center. The left branch in the schematic shows disk formation with liquid flowing radially outward; the right branch shows collapse of the cavity to form a surface wave traveling to the top of the D_L drop." (From Low and List 1982b; courtesy of the American Meteorological Society.)

8.2.2 Sheet breakup

Sheet breakup occurs when a smaller drop is impinged upon by a much faster falling large drop such that the larger drop is broken up into primarily two pieces or a sheet of water is ripped off. The onset of sheet breakup occurs at larger small-drop sizes of the drop-size pairs than with filament breakup. After breakup, the small drop is usually indistinguishable from the original small drop that caused sheet breakup. The large drop is severely distorted from its original size. With sheet breakup two to ten fragments form, with eight the most common number, including the original drops. The number of drops formed by sheet breakup increases with the increasing collisional energy or larger differences in sizes of drops.

8.2.3 Disk breakup

Disk breakup occurs when the smaller drop strikes the larger drop along its center line (Fig. 8.2). After collision, a disk forms and extends the original drop to two to three times its original diameter. The aerodynamic forces then act to form a bowl-shaped particle, which sheds droplets. Few fragments are formed if the collisional energy is low. If the collisional energy is high, then up to 50 fragments may form.

8.3 Parameterization of drop breakup

The breakup of raindrops plays an important role in describing the hydrometeor distribution in the real atmosphere and can lead to the so-called Marshall–Palmer distribution or negative-exponential distribution, in the mean over the number of collision events during the experiment. However, two to three modes in the distribution can develop in as little as five to ten minutes owing to breakup by particles 1 to 2 mm in diameter colliding with larger particles 3 to 5 mm in diameter. One of the most common types of breakup is sheet breakup, followed by filament breakup, and lastly, disk breakup (Low and List 1982a). A starting place for parameterization of one of these is presented by Brown (1997), though it is too complicated to consider for bulk microphysical parameterizations. Hydrodynamic breakup (Pruppacher and Klett 1997) is rare (Rogers and Yau 1989; Pruppacher and Klett 1997) as few, if any, drops ever get large enough for this mechanism to operate. However, in models without provisions for other forms of breakup, there is nothing to limit the size of the particles and they are able to reach sizes at which hydrodynamic breakup becomes important. Again this parameterization is most appropriate for bin model parameterizations, as it is too difficult to parameterize for bulk model parameterizations.

Rather than include the net effect of all the drop-breakup mechanisms in the stochastic collection equation, they are interpreted as a perturbation in the self-collection equation (Cohard and Pinty 2000). Therefore, we follow the simple formulation by Verlinde and Cotton (1993) to describe the breakup, which is designed to represent the most common breakup mechanisms. This parameterization limits the mean drop diameter of $D_{\rm m} = \nu D_n$ to 900 microns, and does so based on adjusting the self-collection efficiency for distributions when collection occurs. In other words, the application of the parameterization is an adjustment to the collection efficiency in the self-collection calculation,

$$E_{xx} = \begin{cases} 1 & D_{\rm m} < 600 \text{ microns} \\ 2 - \exp(2300[D_{\rm m} - D_{\rm cut}]) & D_{\rm m} > 600 \text{ microns.} \end{cases} \tag{8.4}$$

For example, with the cut-off diameter $D_{\rm cut} = 6 \times 10^{-4}$ m, $E_{xx} = 1.0$ for $D_{\rm m} < 6 \times 10^{-4}$ m, then as the mean drop size increases, E_{xx} decreases to 0.0 at $D_{\rm m} = 9 \times 10^{-4}$ m. At particle sizes larger than 9×10^{-4} m, the efficiency exponentially becomes more negative, which implies quick breakup. For example: with $D_{\rm m} = 1 \times 10^{-3}$ m, $E_{xx} = -0.51$; with $D_{\rm m} = 1.1 \times 10^{-3}$ m, $E_{xx} = -1.16$; with $D_{\rm m} = 1.2 \times 10^{-3}$ m, $E_{xx} = -1.97$. These numbers show the quick breakup of large drops and produce a number concentration source for rain. Breakup of particles other than rain is not permitted. Melting aggregates that are to a large extent liquid (>50%) might break up too, but these particles become redefined as melt rain in the model and then breakup can occur if they are large enough. Note that the parameterization of E_{xx} also can force the self-collection equation to act as a number concentration sink causing smaller drops to coalesce and number concentration to decrease as described above.

There are other similar formulations for drop breakup that are worth mentioning; they follow the same philosophy as above. The first of these is by Ziegler (1985) whose parameterization is similar to Verlinde and Cotton's (1993), but has a zero E_{xx} at diameters greater than 1000 microns.

Another drop breakup formulation is by Cohard and Pinty (2000), where the self-collection efficiency is adjusted when $D_{\rm m} > 600$ microns,

$$\begin{cases} 1 & \text{if } D_{\rm m} < 600 \text{ microns} \\ \exp\{-2.5 \times 10^3 (D_{\rm m} - 6 \times 10^{-4})\} & \text{if } 600 \le D_{\rm m} \le 2000 \text{ microns} \\ 0 & \text{if } D_{\rm m} > 2000 \text{ microns.} \end{cases} \tag{8.5}$$

This scheme does not have the feature of relaxing drops back to the equilibrium or largest permitted mean diameter when drops get very large, as in the Verlinde and Cotton formulation.

8.3.1 Stochastic breakup equations

8.3.1.1 Hydrodynamic instability

If stochastic collection equations are integrated indefinitely they generate drops which are far too large compared to what is generally observed. A mean steady rain spectrum such as the Marshall and Palmer (1948) type is not possible. In general, it is possible to understand only the physics of the early development of the first raindrops. Beyond the initial period of significant growth, the stochastic collection equation alone begins to fail. In order to accommodate more reality in understanding precipitation development, stochastic breakup equations have been developed.

As raindrops grow they usually break up by collisions. However, if there are no collisions (Rauber *et al.* 1991), then when aerodynamic forces exceed surface-tension forces typically around drop diameters of 8 to 9 mm, they break up spontaneously. This has been studied in the laboratory and the results have been repeated numerous times by Blanchard (1950) and Komabayasi *et al.* (1964), among others. Komabayasi *et al.* (1964) proposed the following parameterizations for bin models based on empirical evidence and came up with one parameterization where the probability that a drop of mass m, has the probability of breaking up, $P_B(m)$, which is described by the following,

$$P_B(m) = 2.94 \times 10^{-7} \exp(34 r_m), \tag{8.6}$$

where the units of $P_B(m)$ are s^{-1}, r_m is radius in cm of the drop of mass m. A second term given by $Q_B(m', m)$ is defned so that $Q_B(m', m)$ is the number of drops of mass m to $m + dm$ formed by the breakup of one large drop of mass m'. Komabayasi *et al.* (1964) defined $Q_B(m', m)$ as

$$Q_B(m', m) = 10^{-1} r_m'^3 \exp(-15.6 r_m) \tag{8.7}$$

where r_m' is in cm and is the size of the drop of m' and r_m again is the size in cm of the drop of mass m.

Equations (8.6) and (8.7) can be used to write the breakup equation which represents the case where drag forces exceed surface-tension forces as

$$\left.\frac{\partial N(m,t)}{\partial t}\right|_B = -N(m,t)P_B(m) + \int_m^\infty N(m',t)Q_B(m',m)P_B(m')dm. \tag{8.8}$$

Unfortunately Srivastava (1971) found that this parameterization for $Q_B(m',m)$ fails to conserve liquid water, which renders it not useful for numerical models and theoretical work. Therefore, he took the data given by Komabayasi *et al.* (1964) and reanalyzed them to develop a new parameterization given by

$$Q_{\mathrm{B}}(m', m) = \frac{ab}{3m}\left(\frac{r_m}{r_{m'}}\right)\exp\left(-b\frac{r_m}{r_{m'}}\right), \tag{8.9}$$

where $b = 7$ fits the data reasonably well and some algebra gives $a = 62.3$. Srivastava (1971) then made adequate simulations of drop breakup using initially the Marshall–Palmer distribution with various initial rain rates. Kogan (1991) also successfully used this parameterization.

Spontaneous breakup is useful, but not the complete answer. First, rarely, if ever, do drops reach spontaneous or hydrodynamic breakup sizes. In addition, the drop-size spectrum that results from the stochastic collection equation and the stochastic breakup equation for spontaneous/hydrodynamic breakup is unrealistically flat with a bias toward large drops, found first by Srivastava (1971) as described by Pruppacher and Klett (1997).

8.3.2 Parameterization of collisional breakup by Low and List

The following is a description of the parameterization of the collisional breakup of water drops including sheet, bag, filament or neck, and disk breakup described by Low and List (1982b) using the data of Low and List (1982a). This parameterization is for bin models and a similar formulation is impossible in all likelihood for bulk parameterizations. Brown (1986) noted that the parameterization is very difficult to implement and requires very careful programming. The scheme was implemented by Hu and Srivastava (1995) who found that it might be deficient in several ways in that it produces multiple peaks in the size-distribution spectrum. The scheme was followed up in part by an attempt at mass conservation by Brown (1997) using the histograms of larger-drop breakups. Brown's (1997) parameterization also is appropriate for bin models. McFarguhar (2004) also developed an updated version of the parameterization.

The variable $P_i(D_i)$ is used to give the average number of fragments of diameter D_i on the interval $D_i \pm \Delta D_i/2$ for a collision between a small drop of diameter, D_{S}, and one large drop of diameter, D_{L}, as averaged over at least 100 collisions according to Low and List (1982b). In other words, this can be stated with the equation,

$$P_i(D_i) = \frac{\text{Total number of fragments of size } D_i(=N_i)}{\text{Total number of collisions}} \times \frac{1}{\Delta D_i}. \tag{8.10}$$

The continuous fragment number $P_i(D_i)$ is a number density function. The fragment number distribution is given by

$$P(D_i) = \int_{-\infty}^{D_i} P_i \mathrm{d}D. \tag{8.11}$$

Now, $P_{\mathrm{f}i}(D_i)$, $P_{\mathrm{s}i}(D_i)$, and $P_{\mathrm{d}i}(D_i)$ represent the fragment number density functions for filament, sheet, and disk breakup, respectively, whilst the variable C_{b} equals the total number of breakups by collisions; it follows that, C_{f}, C_{s}, and C_{d} are each the total number of breakups for filaments, sheets, and disks, which when summed is C_{b}. Next $N_{\mathrm{f}i\Delta}$ is the number of fragments per size interval of filament breakups, $N_{\mathrm{s}i\Delta}$ is the number of fragments per size interval of sheet breakups, and $N_{\mathrm{d}i\Delta}$ is the number of fragments per size interval of disk breakup; the sum of breakups can be subdivided as

$$P_{\mathrm{f}i}(D_i) = N_{\mathrm{f}i\Delta} C_{\mathrm{f}}^{-1}, \tag{8.12}$$

$$P_{\mathrm{s}i}(D_i) = N_{\mathrm{s}i\Delta} C_{\mathrm{s}}^{-1}, \tag{8.13}$$

and

$$P_{\mathrm{d}i}(D_i) = N_{\mathrm{d}i\Delta} C_{\mathrm{d}}^{-1}. \tag{8.14}$$

Now an expression of the contribution of each density function to the total breakup density function is desired. Thus the fraction R_j (j = f, s, d) of the total of each type of breakup can be written as

$$R_{\mathrm{f}} = \frac{\text{Total number of filament breakups}}{\text{Total number of breakup collisions}} = \frac{C_{\mathrm{f}}}{C_{\mathrm{b}}}, \tag{8.15}$$

as well as

$$R_{\mathrm{s}} = \frac{C_{\mathrm{s}}}{C_{\mathrm{b}}} \tag{8.16}$$

and

$$R_{\mathrm{d}} = \frac{C_{\mathrm{d}}}{C_{\mathrm{b}}}. \tag{8.17}$$

The sum of the breakup number distribution functions is given as follows,

$$P_{\mathrm{b}i} = R_{\mathrm{f}} P_{\mathrm{f}i} + R_{\mathrm{s}} P_{\mathrm{s}i} + R_{\mathrm{d}} P_{\mathrm{d}i}. \tag{8.18}$$

Then empirical fits must be found defined for each type of breakup, such as

$$\bar{F} = \frac{1}{C} \sum_i N_i = \sum_i P_i(D_i) \Delta D_i, \tag{8.19}$$

or

$$\bar{F} = \sum_j R_j F_j = R_f F_f + R_s F_s + R_d F_d. \tag{8.20}$$

The average number of breakups $\bar{F}_j$ is now expressed in functional form from data collected given by Low and List (1982a). In addition, the values of P_j are given.

8.3.2.1 Filament breakup

The empirical fit for filament breakups includes the large drop and small drop, with the small drop having limits described after the empirical fit. The two empirical fits are given for filament breakups as follows, for large small drops and smaller small drops. For larger small drops, the following fit is considered,

$$\bar{F}_{f1} = \left[-2.25 \times 10^4 (D_L - 0.403)^2 - 37.9\right] D_S^{2.5} + 9.67(D_L - 0.170)^2 + 4.95, \tag{8.21}$$

where diameters are in centimeters. This equation is valid for $D_S > D_{S0}$, assuming that each breakup produces a minimum of two fragments. For $D_S < D_{S0}$, the following is given,

$$\bar{F}_{f2} = a'' D_S^{b''} + 2, \tag{8.22}$$

where $a'' = 1.02 \times 10^4$ and $b'' = 2.83$.

The limit of D_S (i.e. D_{S0}) is computed as the intersection of $\bar{F}_{f1}$ and $\bar{F}_{f2}$, or

$$D_{S0} = \left(\bar{F}_{f1}/b''\right)^{1/a''}. \tag{8.23}$$

Now the density functions for filament breakup need to be developed. They are given in Low and List (1982a). These have to be computed for all modes to give a complete tally of the number occurring in any mode, be it filament, sheet, or disk breakup.

The density function P_{f1} is given as

$$P_{f1}(D_i) = H_{f1} \exp\left[-0.5\left(\frac{D_i - \mu}{\sigma_{f1}}\right)^2\right], \tag{8.24}$$

where

$$H_{f1} = 50.8 D_L^{-0.718}. \tag{8.25}$$

The value of μ is equal to the mode of distribution given by D_L. The value of σ_{f1} is the standard deviation, which is a dependent variable found by iteration described by Low and List (1982b).

A Gaussian curve similar to that used for the large drop is used for the small drop so that the density function component for the small drop, P_{f2}, is given as

$$P_{f2}(D_i) = H_{f2} \exp\left[-0.5\left(\frac{D_i - \mu}{\sigma_{f2}}\right)^2\right], \tag{8.26}$$

where H_{f2} is given as the following with D_S the mode,

$$H_{f2} = 4.18 D_S^{-1.17}, \tag{8.27}$$

and

$$\sigma_{f2} = \left(\sqrt{2\pi} H_{f2}\right)^{-1}. \tag{8.28}$$

The third component to the total filament number density function, $P_{f3}(D_i)$, results from disintegration of a bridge of water that connects the two main fragments of the collision pair. Given a log-normal density function, $P_{f3}(D_i)$ is the following,

$$P_{f3}(D_i) = \frac{H_{f3}}{D_i} \exp\left[-0.5\left(\frac{\ln D_i - \mu_{f3}}{\sigma_{f3}}\right)^2\right], \tag{8.29}$$

where H_{f3} is a constant, μ_{f3} is the natural log of the mode; σ_{f3} is related to the mode by

$$\mu_{f3} = \ln(D_{ff3}) + \sigma_{f3}^2, \tag{8.30}$$

where σ_{f3} is found by iteration (see Low and List 1982b). The value of D_{ff3} is related to the small-drop diameter D_S as given by

$$D_{ff3} = 0.241 D_S + 0.0129, \tag{8.31}$$

and is also the modal diameter.

Further details on the three local maximum values in the curves are discussed in Low and List (1982b) and are included briefly here. The Gaussian function (8.29) above represents the maximum value $P_{f3,0}$ of the density curve which depends on both D_L and D_S. The variable $P_{f3,0}$ is composed of three parts. The first, $P_{f3,01}$, is given as

$$P_{f3,01} = 1.68 \times 10^5 D_S^{2.33} \tag{8.32}$$

for $D_S < D_{S0}$.

The second, $P_{f3,02}$, is given as

$$P_{f3,02} = \left[43.4(D_L + 1.81)^2 - 159\right] D_S^{-1} - \left[3870(D_L - 0.285)^2\right] - 58.1 \tag{8.33}$$

for $D_S > 1.2\, D_{S0}$.

Finally, the third, $P_{f3,03}$, is given as

$$P_{f3,03} = \alpha P_{f3,01} + (1 - \alpha) P_{f3,02}, \tag{8.34}$$

where

$$\alpha = \left(D_S - D_{S,0}\right) \times \left(0.2 D_{S0}\right)^{-1}. \tag{8.35}$$

The drops that remain after filament breakup are shown schematically in Fig. 8.1.

8.3.2.2 Sheet breakup

Low and List (1982b) note that to determine the average fragment number, the area under the curve given by P_{si} needs to be computed from the experimental data (Fig. 8.4). The area is said to be equal to $\bar{F}_s$ or the average number of fragments. The value $\bar{F}_s$ is found to be a function of the total surface energy S_T, with $\sigma_{L/A} = 7.28 \times 10^{-2}$ N m^{-1} as the surface tension of liquid in air (which is really a function of temperature, Pruppacher and Klett 1997) as follows,

$$S_T = \pi \sigma_{L/A}\left(D_L^2 + D_S^2\right). \tag{8.36}$$

The limiting value of fragments on the small end is two, and the value of $\bar{F}_s$ is

$$\bar{F}_s = 5\,\mathrm{erf}\left(\frac{S_T - 2.53 \times 10^{-6}}{1.85 \times 10^{-6}}\right) + 6. \tag{8.37}$$

The density function for sheet breakup is given in two parts. The part $P_{s1}(D_i)$ gives the distribution of the large fragment around D_L, and one other function $P_{s2}(D_i)$ represents the rest of the fragments as the initial small drop is no longer recognizable. A Gaussian represents $P_{s1}(D_i)$ as

$$P_{s1} = H_{s1} \exp\left[-0.5 \frac{(D_i - D_L)}{\sigma_{S1}}\right], \tag{8.38}$$

which is centered at D_L, with height of H_{s1} and a spread of σ_{s1}. The value of H_{s1} is given by

$$H_{s1} = 100.0 \exp(-3.25 D_S). \tag{8.39}$$

The fragment number of the cloud-droplet part of the breakup is given by a log-normal function with a peak at

$$D_{ss2} = 0.254 D_S^{0.413} \exp\left[3.53 D_S^{-2.51}(D_L - D_S)\right], \tag{8.40}$$

with

$$P_{s2} = 0.23 D_S^{-3.93} D_L^{b^*}, \tag{8.41}$$

and b^* given by,

$$b^* = 14.2 \exp(-17.2 D_S). \tag{8.42}$$

The drops that remain after sheet breakup are shown schematically in Fig. 8.1.

8.3.2.3 Disk breakup

The number of fragments from disk breakups is most numerous compared to other means of breakup. Again the minimum number of fragments is two. The number of fragments for disk breakups is found to be closely related to the collision kinetic energy E_{CKE} (in joules) and given by

$$\bar{F}_d = 297.5 + 23.7 \ln E_{CKE}. \tag{8.43}$$

The density function for disk breakup is given in two parts and is similar to sheet breakup with many small fragments. First, the large drop may break into several drops with the large drop often no longer recognizable. The fragment density around D_L is given as $P_{d1}(D_i)$, after definitions in (8.44)–(8.47). The mode for the Gaussian is given as

$$D_{dd1} = D_L\{1 - \exp[-3.70(3.10 - W_1)]\}, \tag{8.44}$$

where, in joules, the Weber number of energies is used,

$$W_1 = \frac{E_{CKE}}{S_c}, \tag{8.45}$$

and S_c is given as the surface energy of the coalesced drops,

$$S_c = \pi \sigma_{L/A} \left(D_L^3 + D_S^3\right)^{2/3}. \tag{8.46}$$

Finally the height of the distribution is

$$H_{d1} = 1.58 \times 10^{-5} E_{CKE}^{-1.22}. \tag{8.47}$$

The number density function is given as,

$$P_{d1} = H_{d1} \exp\left[-0.5 \frac{(D_i - D_{dd1})^2}{\sigma_{d1}}\right], \tag{8.48}$$

where σ_{d1} is found by iteration as described in Low and List (1982b).

The rest of the disk breakup fragments can be described by a log-normal distribution, as with sheet breakup. The mode D_{dd2} of $P_{d2}(D_i)$ is dependent on drop sizes that collide and is to the left of D_S as given by

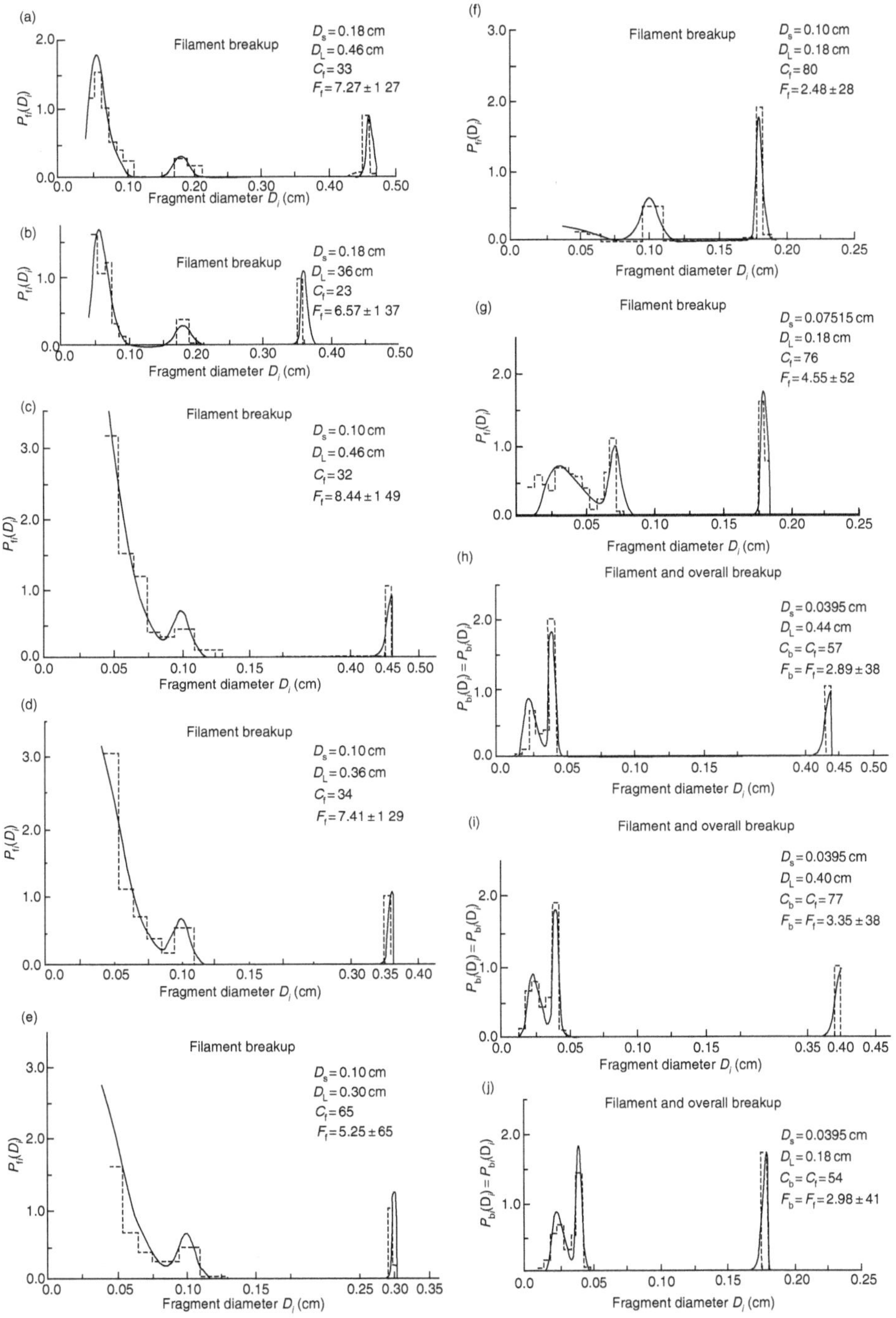

Fig. 8.3. "Histograms with average probability fragment size distributions $P_{fi}(D_i)$ in numbers per fragment size interval (0.01 cm), for filament

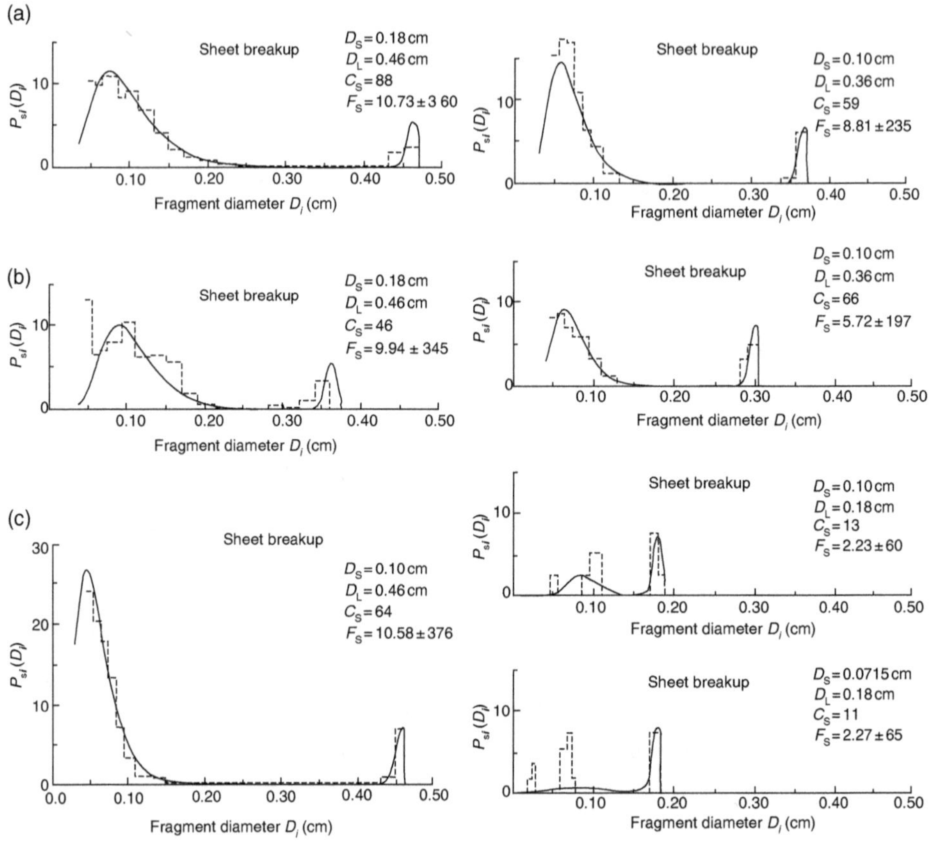

Fig. 8.4. Histograms with average fragment size distributions $P_{si}(D_i)$ in numbers per fragment size interval (0.01 cm), for sheet collision/breakup configurations for seven drop sizes used in the experiments of Low and List (1982a). The solid lines represent the parameterized approximation of the whole data set; C_j represents the number of collisions of each type. (From Low and List 1982b; courtesy of the American Meteorological Society.)

Caption for Fig. 8.3. (*cont.*)
collision/breakup configurations for all 10 drop sizes used in the experiments of Low and List (1982a). The solid lines represent the parameterized approximation of the whole data set; C_j represents the number of collisions of each type." (From Low and List 1982b; courtesy of the American Meteorological Society.)

$$D_{dd2} = \exp\left[-17.4D_S - 0.671(D_L - D_S)\right]D_S. \tag{8.49}$$

The height therefore is as before found from

$$P_{d2}(D_{dd2}) = \frac{H_{d2}}{D_{dd2}}\exp\left[-0.5\sigma_{d1}^2\right] = 0.0884D_S^{-2.52}(D_L - D_S)^{b^*}, \tag{8.50}$$

where b^* is given by

$$b^* = 0.007D_S^{-2.54}. \tag{8.51}$$

A schematic of low collisional kinetic energy and high collisional kinetic energy breakups for the disk breakup is shown in Fig. 8.2. The types of distributions found in disk breakup are given in Fig. 8.5.

8.3.2.4 Overall breakup

The fraction of breakups that occur as filament breakups (in all of the breakups) is given as follows,

$$R_f = \begin{cases} 1.11 \times 10^{-4}E_{CKE}^{-0.654} & \text{for } E_{CKE} \geq E_{CKE_0} = 0.893\mu J \\ 1.0 & \text{for } E_{CKE} < E_{CKE_0} \end{cases}. \tag{8.52}$$

There is a specific number of breakups given by overall sheet breakups. The fraction is a function of the ratio W_2 of the E_{CKE} to the sum of surface energies of two original drops, given as

$$W_2 = \frac{E_{CKE}}{S_T}. \tag{8.53}$$

The fraction of breakups that occur as sheets in all of the breakups is given as follows,

$$R_s = \begin{matrix} 0.685\left\{1 - \exp[-1.63(W_2 - W_0)]\right\} & \text{for } W \geq W_0 = 0.86 \\ 1.0 & \text{for } W < W_0. \end{matrix} \tag{8.54}$$

With the values for R_f and R_s defined then that for R_d can be defined simply as

$$R_d = \begin{cases} 1.0 - (R_f + R_s) & \text{for } R_f + R_s \leq 1 \\ 0.0 & \text{for } R_f + R_s > 1. \end{cases} \tag{8.55}$$

To get the overall fragment number distribution for an average collision the values of $P_b(D_i)$ must be adjusted by the fraction of the sum of collisions that make two or more fragments. Low and List (1982b) give the breakup efficiency by

$$\left[1 - E_{coal}\right] \tag{8.56}$$

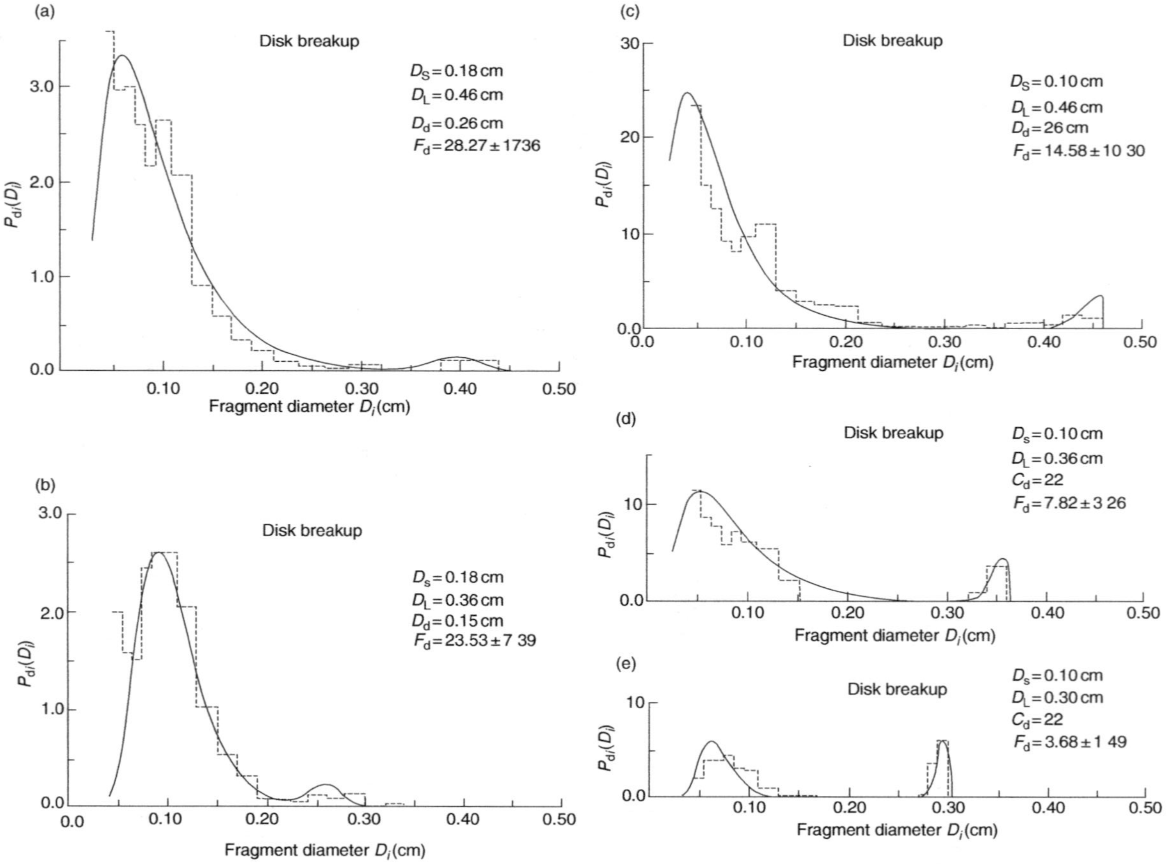

Fig. 8.5. Histograms with average fragment size distributions $P_{di}(D_i)$ in numbers per fragment size interval (0.01 cm), for disk collision/breakup configurations for five drop sizes used in the experiments of Low and List (1982a). The solid lines represent the parameterized approximation of the whole data set; C_j represents the number of collisions of each type. (From Low and List 1982b; courtesy of the American Meteorological Society.)

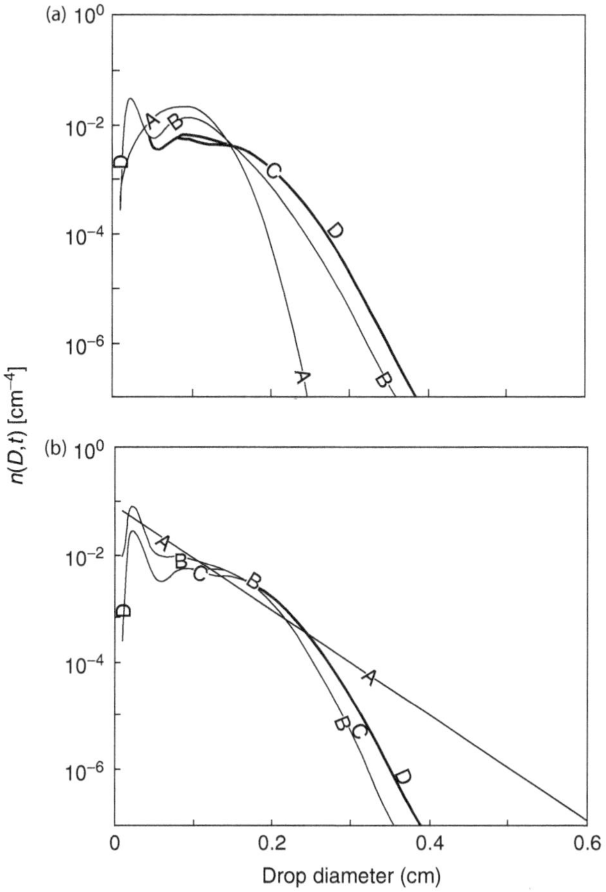

Fig. 8.6. Evolution of drop-size distribution by coalescence and breakup for (a) initial exponential in mass distribution, and (b) initial Marshall–Palmer distribution. The initial distributions (A) and the distributions for 10 (B) 40 (C), and 60 (D) min are shown. (From Hu and Srivastava 1995; courtesy of the American Meteorological Society.)

where E_{coal} is the coalescence efficiency given in parameterized form in List and Low (1982a) so that an overall equation is parameterized as

$$P_i(D_i) = [R_{\mathrm{f}}P_{\mathrm{f}}(D_i) + R_{\mathrm{s}}P_{\mathrm{s}}(D_i) + R_{\mathrm{d}}P_{\mathrm{d}}(D_i)][1 - E_{\mathrm{coal}}] + \delta(D_{\mathrm{coal}})E_{\mathrm{coal}}, \quad (8.57)$$

and

$$\begin{aligned} \delta(D_{\mathrm{coal}}) &= 1 \quad \text{for } D_i = D_{\mathrm{coal}} \\ &= 0 \qquad \text{otherwise.} \end{aligned} \quad (8.58)$$

Hu and Srivastava (1995) modeled these breakup mechanisms and found the resultant distributions for an initial gamma distribution shown in Fig. 8.6a, and an initial negative-exponential distribution (Fig. 8.6b). For the gamma distribution disk breakup (curve D) seems to dominate over the other breakup mechanisms. In tests with single distributions acting, disk and filament seem to be most important (see Fig. 8.7a), where mixed pairs of breakup occur, disk and filament together show the most dramatic changes in the initial distribution (Fig. 8.7b).

8.3.3 Mass conservation with the collisional breakup parameterization

8.3.3.1 One-moment method of collisional breakup

From the initial stochastic collection experiments realistic distributions did not develop. It was decided by investigators that breakup into drop fragments by collisions was missing from the stochastic breakup equation. The experimental results of McTaggert-Cowan and List (1975) were first used for more realistic breakup frequencies. Later others such as Brown (1983, 1985, 1986, 1987, 1988, 1991, 1997, 1999), Tzivion *et al.* (1987, 1989), Feingold *et al.* (1988), and Hu and Srivastava (1995) used more up-to-date experimental results of List and Low (1982b) to describe various forms of drop breakup as described earlier, as well as different formulations to represent breakup. For collision breakup using the stochastic breakup equation, the form given below is used,

$$\left(\frac{\partial n(m,t)}{\partial t}\right)_{\mathrm{B}} = \frac{1}{2}\int_0^\infty n(m',t)\mathrm{d}m' \int_0^\infty K(m',m'')[1 - E_{\mathrm{c}}(m',m'')]Q(m;m',m'')n(m'')\mathrm{d}m'' \quad (0 < m < m'+m'')$$
$$- n(m,t)\int_0^\infty \frac{n(m'',t)K(m,m'')[1 - E_{\mathrm{c}}(m,m'')]}{m+m''}\int_0^{m+m''} m'Q(m';m,m'')\mathrm{d}m' \quad (0 < m < \infty). \tag{8.59}$$

In this equation $K(m',m'')$ is the collision kernel of an m'-drop with an m''-drop, and $E_{\mathrm{c}}(m',m'')$ is the coalescence efficiency for the kernel. Furthermore $[1 - E_{\mathrm{c}}(m,m'')] = p(m,m'')$ is the breakup probability for an m-drop that collides with an m''-drop. In this equation $Q(m;m',m'')$ is the mean number of fragments of m-drops produced by a collision between an m'-drop and an m''-drop. As described by Pruppacher and Klett (1997) the first term on the right-hand side of (8.59) is the gain of m-drops generated by collision of all masses m' and m''. The 1/2 factor keeps from double counting the same (m,m'') pair twice. The second term on the right-hand side of (8.59) is loss of m-drops resulting from collision and then breakup of drops of mass m and m''. For mass conservation (Pruppacher and Klett 1997) it is required that

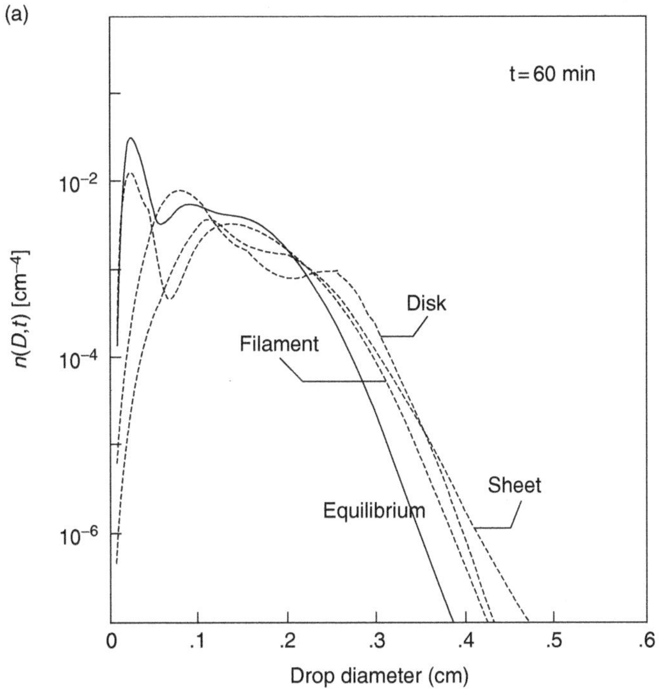

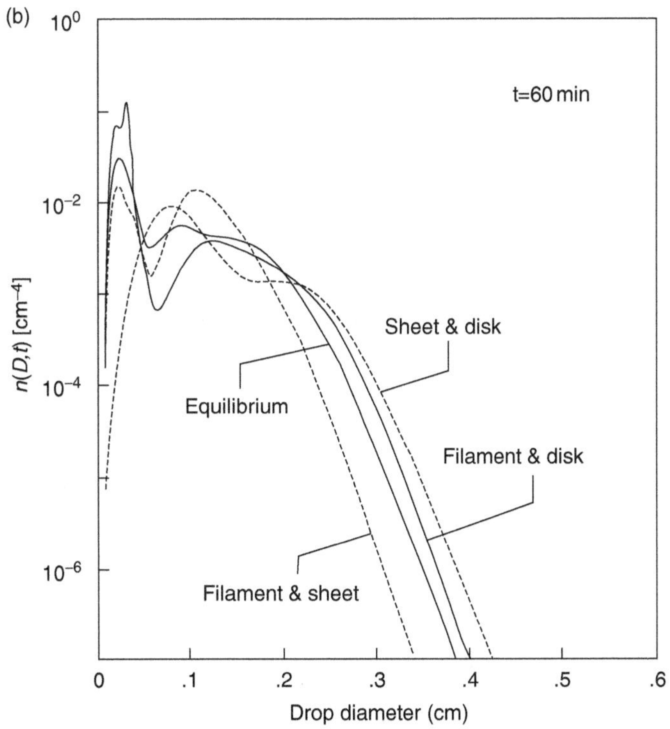

Fig. 8.7. (a) Equilibrium distribution with only one type of breakup and coalescence operating, and (b) equilibrium distributions with only two types

$$\int_0^{m+m''} mQ(m; m', m'')\mathrm{d}m = m' + m''. \tag{8.60}$$

Bleck's (1970) one-moment method can be applied to the stochastic breakup equation following the procedure used for the one-moment method for the stochastic collection equation (List and Gillespie 1976). First, the mass coordinate is separated into bins as with the stochastic collection equation (in the same form as the stochastic collection equation if used with the stochastic collection equation). The subcategories are defined as before and applied

$$\int_{m_k}^{m_{k+1}} m\mathrm{d}m, \tag{8.61}$$

to both sides of (8.59). The number density mean in each mass category is defined, according to Bleck and Danielson *et al.* to get a stochastic collection equation with drop breakup,

$$\bar{n}_k(t) = \int_{m_k}^{m_{k+1}} n(m,t)m\mathrm{d}m \left[\int_{m_k}^{m_{k+1}} m\mathrm{d}m \right]^{-1} \tag{8.62}$$

and

$$\bar{n}_k(t) = \frac{2}{m_{k+1}^2 - m_k^2} \int_{m_k}^{m_{k+1}} n(m,t)m\mathrm{d}m. \tag{8.63}$$

The result then is given quite simply by

$$\left(\frac{\partial \bar{n}_k(t)}{\partial t} \right)_{\mathrm{B}} = \left[\frac{2}{m_{k+1}^2 - m_k^2} \right] \int_{m_k}^{m_{k+1}} B(m,t)m\mathrm{d}m, \tag{8.64}$$

where B is the breakup probability for a particle at mass m and time t. Following List and Gillespie (1976) the integral is just

$$\int_{m_k}^{m_{k+1}} B(m,t)m\mathrm{d}m \approx \sum_{j=1}^{I} \sum_{i=1}^{j} p_{ijk}\bar{n}_j\bar{n}_i - \bar{n}_k \sum_{i=1}^{I} q_{ik}\bar{n}_i. \tag{8.65}$$

Caption for Fig. 8.7. (*cont.*)
of breakup and coalescence operating. The equilibrium distribution with all three types of breakup and coalescence operating is marked "equilibrium". The water content is 1 g m^{-3}. (From Hu and Srivastava 1995; courtesy of the American Meteorological Society.)

Then, following the procedure used again,

$$\left(\frac{\partial n(m)}{\partial t}\right)_{\mathrm{B}} \approx \left[\frac{2}{m_{k+1}^2 - m_k^2}\right] \sum_{j=1}^{I} \sum_{i=1}^{j} p_{ijk}\bar{n}_j\bar{n}_i - \bar{n}_k \sum_{i=1}^{I} q_{ik}\bar{n}_i. \tag{8.66}$$

The coefficients for p_{ijk} and q_{ik} are given by List and Gillespie (1976) and Brown (1983).

8.3.3.2 Multi-moment method of collisional breakup

The multi-moment method of breakup was generalized by Feingold *et al.* (1988) and Tzivion *et al.* (1989). This was done by taking multiple moments of the stochastic breakup equation. First, applying the following,

$$\int_{m_k}^{m_{k+1}} m^{v} \mathrm{d}m \tag{8.67}$$

to both sides of (8.59), gives

$$\begin{aligned}
\left.\frac{\partial M_k^v(t)}{\partial t}\right|_{\mathrm{B}} = {} & \frac{1}{2}\int_{m_k}^{m_{k+1}} m^v \mathrm{d}m \int_0^{\infty} n(m',t)\mathrm{d}m' \int_0^{\infty} K(m',m'')[1 - E_{\mathrm{c}}(m',m'')]Q(m;m',m'')n(m'')\mathrm{d}m'' \\
& (0 < m < m' + m'') \\
& - \int_{m_k}^{m_{k+1}} n(m,t)m^v \mathrm{d}m \int_0^{\infty} \frac{n(m'',t)K(m,m'')[1 - E_{\mathrm{c}}(m,m'')]}{m + m''} \int_0^{m+m''} m'Q(m';m,m'')\mathrm{d}m', \\
& (0 < m < \infty)
\end{aligned} \tag{8.68}$$

where M_k^v is the mass of moment v for bin k.

Following Feingold *et al.* (1988), Tzivion *et al.* (1989), Pruppacher and Klett (1997), and the steps made by Bleck, a transformation results so that

$$\begin{aligned}
\left.\frac{\partial M_k^v(t)}{\partial t}\right|_{\mathrm{B}} = {} & \sum_{i=1}^{I}\sum_{j=1}^{i-1} \int_{m'_i}^{m'_{i+1}} n_i(m',t)\mathrm{d}m' \int_{m'_j}^{m'_{j+1}} n_j(m'',t)K_{i,j}(m',m'')[1 - E_{\mathrm{c}}(m',m'')]\mathrm{d}m'' \\
& \times \int_{m'_k}^{m'_{k+1}} m^v Q_{k,i,j}(m;m',m'')\mathrm{d}m \\
& + \frac{1}{2}\sum_{i=1}^{I} \int_{m'_i}^{m'_{i+1}} n_i(m',t)\mathrm{d}m' \int_{m'_i}^{m'_{i+1}} n_i(m'',t)K_{i,j}(m',m'')[1 - E_{\mathrm{c}}(m',m'')]\mathrm{d}m''
\end{aligned}$$

$$
\times \int_{m'_k}^{m'_{k+1}} m^{v} Q_{k,i,i}(m; m', m'')\mathrm{d}m
$$

$$
-\sum_{i=1}^{I}\sum_{j=1}^{k} \int_{m'_k}^{m'_{k+1}} m^{v} n_k(m,t)\mathrm{d}m \int_{m'_j}^{m'_{j+1}} \frac{n_j(m'',t)K_{i,j}(m',m'')}{m+m''}[1-E_{\mathrm{c}}(m,m'')]\mathrm{d}m''
$$

$$
\times \int_{m'_i}^{m'_{i+1}} m' Q_{i,k,j}(m'; m, m'')\mathrm{d}m'
$$

$$
-\sum_{i=1}^{I}\sum_{j=k+1}^{I} \int_{m'_k}^{m'_{k+1}} m^{v} n_k(m,t)\mathrm{d}m \int_{m'_j}^{m'_{j+1}} \frac{n_j(m'',t)K_{j,k}(m'',m)}{m+m''}[1-E_{\mathrm{c}}(m'',m)]\mathrm{d}m''
$$

$$
\times \int_{m'_i}^{m'_{i+1}} m' Q_{i,k,j}(m'; m'', m)\mathrm{d}m' \tag{8.69}
$$

for which $Q(m;m',m'')=0$ for $m>m'+m''$, and $Q(m';m,m'')=0$ for $m'>m+m''$. This equation is different from the stochastic collection equation in its application, as the integrals are complete over the categories so no approximations to n_k need to be made. Kernels are approximated using polynomials to close the system. This equation can be written for the two moments simply by making $v=0$ for the $\mathrm{d}N_k(t)/\mathrm{d}t$ equation and making $v=1$ for the $\mathrm{d}M_k(t)/\mathrm{d}t$. Making these approximations are left to the reader. A surprising finding using this equation is that the one-moment stochastic breakup equation and the two-moment stochastic breakup equation provide essentially the same solutions for breakup only. The formula (Feingold *et al.* 1988; Tzivion *et al.* 1989) used for $Q(m;m',m'')=0$ is given by $Q(m';m,m'')=g^2(m'+m'')\exp(-gm)$. This permits an analytical solution when $g=n_{N_0}/M_0$, where n is the integer number of fragments. The values of N_0 and M_0 are the initial drop number concentration and liquid-water content. The reason an analytical solution is possible with g defined as above is that the stochastic breakup equation is a forward-progressing model.

9

Autoconversions and conversions

9.1 Introduction

An autoconversion or conversion scheme represents hydrometeors changing from one species/habit to another. The change could be a phase change such as homogeneous freezing of water drops. Or it could be a change within a phase, but a change of species dependent on diameter, such as cloud droplets to drizzle or raindrops. It also could be a graupel of one density becoming more or less dense and subsequently reclassified as a different density owing to the riming it experienced.

One reason that autoconversions and conversions are so difficult to parameterize is that autoconversions and conversions are not well-observed processes, though they can be simulated approximately using a hybrid bin model (see Feingold *et al.* 1998). Furthermore, conversions of ice crystals or snow aggregates to graupels of particular densities are terribly difficult to parameterize, as there are few accurate measurements in nature or from the laboratory on this topic.

Multi-dimensional, Eulerian models incorporate autoconversion and conversion schemes of varying complexity to try to capture the physics changes on the sub-grid scale in terms of grid-scale quantities, much the way turbulence is parameterized (Stull 1988). This has been done for cloud, mesoscale, synoptic, and global models with complexity usually decreasing with increasing scale (Wisner *et al.* 1972; Koenig and Murray 1976; Cotton *et al.* 1982; Cotton *et al.* 1986; Cotton *et al.* 2001; Lin *et al.* 1983; Farley *et al.* 1989; Ferrier 1994, Straka and Mansell 2005 among many others) (this is the nature of microphysical parameterizations).

An unsolved problem in applying these types of parameterizations is that autoconversion or conversion often does not commence, at least in nature, until a sufficient "aging" period has passed such that the actual physics occurs based on its trajectory and time along that trajectory. This means that for autoconversions and conversions in bulk models to occur some local natured physical process needs to be active for more than a model timestep. Unfortunately the nature of microphysics is generally anything but local at a grid point. With vertical and horizontal drafts in model flow fields this means that to truly capture an accurate estimate of autoconversion or conversion along a trajectory requires keeping track of the process. This could involve releasing trajectories every timestep at each grid point in a model. Embarking on the use of Lagrangian trajectories in Eulerian-based microphysical models is, quite frankly, out of the question for long-term, or even short-term, simulations for significant use. The result is that many microphysical parameterizations in Eulerian models, especially autoconversion and conversion processes, fail to capture explicitly the Lagrangian history of the growth of cloud and precipitation particles.

An example of a microphysical process that depends on an aging period is the conversion of cloud into drizzle or cloud into rain (Cotton 1972; Cotton and Anthes 1989). The need to account for the relevant physics including diffusional growth and collection growth for a cloud droplet to become a drizzle or a rain particle is never truly captured in Eulerian-model parameterizations. Cotton (1972a) and Cotton and Anthes (1989) note that all the conversion and autoconversion schemes developed thus far have neglected the Lagrangian "aging" period. As a result, models produce drizzle or raindrops far too quickly and at too low an altitude in simulated clouds (Simpson and Wiggert 1969; Cotton 1972a; Cotton and Anthes 1989). In a seemingly successful attempt to account for the Lagrangian aging of a particle, Cotton (1972a) used a one-dimensional, Lagrangian model to develop an autoconversion parameterization scheme that includes the age of a parcel with cloud drops in it to determine when autoconversion should occur. However, this parameterization by Cotton has not been implemented in two- or three-dimensional models that are Eulerian in nature (Cotton and Tripoli 1978; Tripoli and Cotton 1980; Cotton *et al.* 1982; among many, many others). Tripoli and Cotton noted that failure to account for this aging might make it difficult to simulate certain storm phenomena such as a thunderstorm's weak echo region.

Then Straka and Rasmussen (1997) came up with a methodology to predict the age of a process starting at some grid point, as well as where the parcel started and where it presently is; finally, they determined how to find time weight mean exposures to the variables allowed in the process. For example, the mean cloud content a parcel has experienced at any time or location may be useful for autoconversion. The breakthrough by Straka and Rasmussen (1997)

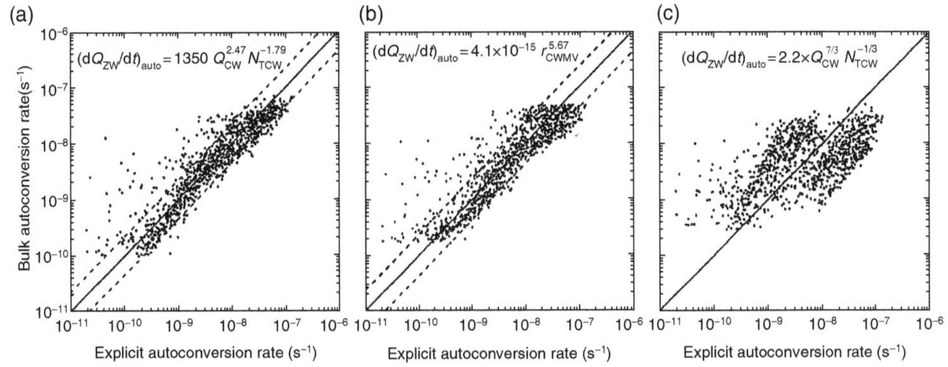

Fig. 9.1. Scatterplots of the bulk autoconversion rates or conversions from cloud drops, which grow by collection to become drizzle drops, given by the *x–y* (or *z*) axes versus the corresponding rates obtained from the explicit mode. The dashed line represents a factor of two deviations from the perfect match. Note that only every twentieth data point is shown. (From Khairoutdinov and Kogan 1994; courtesy of the American Meteorological Society.)

might someday lead to better autoconversion and conversion parameterizations. The methodology of this proposed scheme is presented in Chapter 2 (sections 2.7.5.1 and 2.7.5.2). Results with Lagrangian-determined positions of a parcel, and mean mixing ratios of cloud water, are shown to be exceptionally well simulated by the Eulerian model with the Straka and Rasmussen method accounting for the Lagrangian nature of autoconversion physics (Straka and Rasmussen 1997). The parcels exiting the cloud are where the solutions diverge between the true Lagrangian- and Eulerian-based-Lagrangian predicted values.

9.2 Autoconversion schemes for cloud droplets to drizzle and raindrops

Autoconversion schemes for cloud droplets to rain droplets ideally encompass the growth of cloud drops by diffusion until they are large enough to grow predominately by collection into embryonic precipitation particles. From theoretical modeling, the size necessary for collection growth was found to be about 81 microns (Berry 1967; Berry 1968b; Berry and Reinhardt 1974a–d; and others). It can be argued that the most troubling aspects of essentially all autoconversion schemes are that they develop unrealistically mature rain spectra immediately after the autoconversion schemes activate; and that numerical models do not represent embryonic precipitation particles well, if at all. After all, an embryonic precipitation particle is more likely to be more akin to a drizzle drop than a raindrop. First, the history of the development of cloud to rain autoconversion schemes of varying complexity will be examined. In some of the more recent models, drizzle is predicted (Straka *et al.* 2009a) along with rain, so that the conversion from cloud drops to drizzle is possible, not necessarily cloud drops to raindrops. For raindrops to grow, drizzle drops need to grow to 500 microns in radius.

Described below are some of the more popular autoconversion and "so-called autoconversion" schemes developed in the late 1960s, 1970s, and 1980s. (Some of these are "so-called autoconversion" schemes, as they were not true autoconversion schemes. Based on timescale analyses, they contained accretion and self-collection processes.) Later in the 2000s there was another wave of attempts at autoconversion-scheme development.

The Berry (1967, 1968b) scheme was one of the first autoconversion schemes introduced and is based on the time for the sixth moment of the diameter to reach a size of 80 microns, which is about the size of an embryonic precipitation particle, that is, a drizzle particle. The equation that Berry presented is

$$Q_{\text{rw}}CN_{\text{cw}} = \frac{Q_{\text{cw}}^2}{60}\left(2 + \frac{0.266\,N_{\text{Tcw}}}{\gamma_{\text{cw}}\;Q_{\text{cw}}}\right)^{-1}, \tag{9.1}$$

where $Q_{\text{rw}}CN_{\text{cw}}$ is the conversion of mixing ratio of bulk cloud water converted to rain water. The *CN* is used to denote conversion. The *Q* represents mixing ratio and subscript cw represents cloud water. The number concentration of cloud droplets is N_{Tcw}, and, γ_{cw}, is the dispersion of the cloud droplet size distribution, which can range from 0.1 to 0.2 for continental clouds to 0.3 to 0.4 for maritime clouds. As pointed out by Cotton and Anthes (1989) this scheme has a cubic dependence of cloud-water mixing ratio making it fairly non-linear. The scheme does embody some desirable features of the warm-rain process that includes the use of a specified number concentration to represent the number of drops nucleated for a given environment. Notice that when cloud water forms it is rapidly converted to rain. A lack of a so-called "aging" (Cotton 1972a; and Straka and Rasmussen 1997) of cloud droplets causes this scheme to very rapidly and very unrealistically convert cloud drops to embryonic raindrops. Moreover, some have argued that this scheme, based on timescale arguments, contains accretion and self-collection processes as well as autoconversion processes.

One of the most widely used autoconversion schemes still in use is the simple Kessler (1969) formulation,

$$Q_{\text{rw}}CN_{\text{cw}} = K_1 H(Q_{\text{cw}} - Q_{\text{cw0}}). \tag{9.2}$$

In this scheme, K_1 is a rate constant and H is the Heaviside step function. However this scheme suffers from the inability to distinguish types of initial aerosol or cloud condensation nuclei concentrations, but does account for the broadening of cloud-drop spectra to raindrop spectra. Without fine-tuning this scheme, it is impossible to distinguish maritime from continental regimes. Initially, values of the constants were assigned as $K_1 = 1 \times 10^{-3}\ \mathrm{s}^{-1}$, and $Q_{\mathrm{cw}0} = 5 \times 10^{-4}$ to $1 \times 10^{-3}\ \mathrm{kg\ kg}^{-1}$. Note that Cotton and Anthes (1989) found that this scheme has some possible non-linear behavior. It is unfortunate that so many choose this scheme for autoconversion as it produces embryonic precipitation particles at far too low an altitude in continental clouds for the given constant values; it may be difficult for embryonic precipitation particles to form drizzle as very often cloud-droplet mixing ratios do not exceed $5 \times 10^{-4}\ \mathrm{kg\ kg}^{-1}$ in maritime boundary layers. It should be noted that Kessler developed this scheme apparently with deep convective clouds in tropical regions in mind. Thus, without modification it is probably not appropriate for continental clouds. Nevertheless, it has been employed in numerous models, from that of Klemp and Wilhelmson (1978) to Reisner *et al.* (1998), for example.

Gilmore and Straka (2008) examined the Berry and Reinhardt (1974b) scheme very carefully and found some mistakes and misconceptions about the scheme. They corrected typographical errors in terms of equations and units of terms to reconcile the scheme's behavior in numerous models.

The Berry and Reinhardt (1974b) scheme has been widely adopted as a parameterization of rain production as derived from a bin model (described in Berry and Reinhardt 1974a). The Berry and Reinhardt (1974b) scheme was not originally intended to be a parameterization of smaller cloud drops becoming raindrops. The investigation by Gilmore and Straka (2008) demonstrates how differences in the Berry and Reinhardt (1974b) formulation influence initial rain development. These authors show differences between versions that result from typesetting errors (some from the original Berry and Reinhardt (1974b) paper), derivation errors, and methodology. The differences are important to point out as they influence the initiation of rain.

The Berry and Reinhardt (1974b) scheme is a more complex scheme when compared to others because it embeds details about the cloud-droplet distribution that can affect collision/coalescence, which leads cloud droplets to form raindrops. These distribution attributes are mean cloud-droplet size and dispersion. The Berry and Reinhardt (1974b) scheme also has an equation to approximate rain number-concentration rates, whilst some new schemes (e.g. Liu and Daum 2004) do not. It is common knowledge that the scheme incorporates autoconversion – rain production via collision–coalescence of cloud droplets (e.g. Berry and Reinhardt 1974a). Although Pruppacher and Klett (1981) and most subsequent authors loosely referred to this entire parameterization as "autoconversion", the scheme also includes the further accretion of cloud water by those growing small raindrops and the collision–coalescence (or self-collection) of the growing small raindrops (Berry and Reinhardt 1974b).

Whilst the Berry and Reinhardt (1974b) parameterization has been around for over 30 years, it is still the mainstay of many models owing to its relative completeness, yet minimal cost. This scheme is utilized in the Weather Research and Forecast (WRF) Model (Thompson *et al.* 2004) and the Regional Atmospheric Modeling System (RAMS; Version 3b). Versions of the Berry and Reinhardt (1974b) parameterization have been presented and used by Nickerson *et al.* (1986); Proctor (1987); Verlinde and Cotton (1993); Walko *et al.* (1995); Meyers *et al.* (1997); Carrió and Nicolini (1999); Cohard and Pinty (2000); Thompson *et al.* (2004); and Milbrandt and Yau (2005b).

Berry and Reinhardt (1974b) presented curve fits that related mean mass and mass-relative variance of an initial S1 distribution to the mean mass and number concentration of an S2 distribution that resulted from all accretion and self-collection processes during a characteristic timescale. The characteristic timescale T_2 is defined as the time when the radius of the predominant rain mass, r_{gr}, of the developing S2 distribution first reaches 50 microns in the bin model (Berry and Reinhardt 1974b; their Table 1). This definition was made so that Berry and Reinhardt (1974b) could establish mass and number concentrations for the S1 and S2 distributions.

The following are other important properties valid at time T_2 (from Berry and Reinhardt 1974b):

- L'_2 and N'_2 are S2's total mass and number concentration, respectively;
- S2 first attains a Golovin shape;
- S2's relative mass variance is 1 (increasing from prior smaller values);
- S2 obtains a mean-mass radius of ~41 mm; and
- $r(g_m)$ is the threshold radius corresponding to the minimum in the mass between the two modes of the total liquid spectrum.

The determination of T_2 is important to modelers because T_2 was the only time for which Berry and Reinhardt (1974b) tabulated both the rain mass and rain number concentration from a bin model. The other timescales defined by Berry and Reinhardt (1974b) are: T_H ($\sim 1.1T_2$), which denotes the time at which the developing S2 mass distribution curve forms a hump; and T ($\sim 1.25T_2$), which denotes the time at which the radius of the predominant mass of the joint S1 + S2 (bimodal) distribution first reaches a radius of 50 mm.

The focus first will be on the earliest timescale presented in Berry and Reinhardt (1974b), T_2, as both mass and number concentration are defined. After converting to SI units, making some corrections to typesetting errors in Berry and Reinhardt, and converting radius to diameter, the resulting timescale is given as

$$\hat{T}_2\{\mathrm{s}\} = 3.72\{\mathrm{s\ kg\ m^{-3}\ \mu m}\} \times \left[0.5 \times 10^6 \left\{\frac{\mu\mathrm{m}}{\mathrm{m}}\right\} D_\mathrm{b}^0\{\mathrm{m}\} - 7.5\{\mu\mathrm{m}\} L^0\{\mathrm{kg\ m^{-3}}\}\right]^{-1}. \tag{9.3}$$

The second equation involved in the Berry and Reinhardt (1974b) scheme is shown in their Eq. 18, Fig. 9, as the following for total mass,

$$\hat{L}_2\{\mathrm{gm^{-3}}\} = \left(10^{-4}\left[10^4\right]^4 \left\{\frac{\mu\mathrm{m}}{\mathrm{cm}}\right\}^{-1} \left(r_\mathrm{b}^0\right)^3 \{\mathrm{cm}\}^3 \left(r_\mathrm{f}^0\right)\{\mathrm{cm}\} - 0.4\{\mu\mathrm{m}\}^4\right) \times \left(2.7 \times 10^{-2}\right)\{\mu\mathrm{m}\}^{-4} L^0\{\mathrm{g\ m^{-3}}\}. \tag{9.4}$$

Pruppacher and Klett (1981) were perhaps first to suggest that (9.3) and (9.4) could be combined to obtain an average rate of change in rain mixing ratio during $\hat{T}_2$ for a bulk microphysics model:

$$Q_\mathrm{zw} CN_\mathrm{cw} = \frac{\mathrm{d}Q_\mathrm{zw}}{\mathrm{d}t}\{\mathrm{kg\ kg^{-1}\ s^{-1}}\} = \frac{1}{\rho}\frac{\max\left(\hat{L}'_2, 0\right)}{\max\left(\hat{T}'_2, 0\right)}. \tag{9.5}$$

Berry and Reinhardt (1974b) do not propose this average mixing ratio rate; this is probably because they only consider those curve fits to data as an intermediate step (Berry and Reinhardt 1974b, p. 1825) to their parameterization, and because these authors (Berry and Reinhardt 1974c,d) present a way to evaluate precise rates at any arbitrary time (rather than average rates via a characteristic timescale). Nevertheless, the simple form is what all subsequent bulk microphysics modelers have used and what is herein designated as the "Berry and Reinhardt (1974b) parameterization" or "Berry and Reinhardt (1974b) scheme".

There are some limitations to the Berry and Reinhardt (1974b) scheme. First, it unfortunately does not give the remaining cloud-water number concentration N_Tcw at T_2, and therefore an average $-\mathrm{d}N_\mathrm{Tcw}/\mathrm{d}t$ (owing to S1 self-collection and S1 accretion by S2) cannot be derived. Next, adequate mass and number concentration rates are difficult to define since the S1 and S2 distributions overlap. Also, Cohard and Pinty (2000) have noted that cloud accretion by rain and rain self-collection both appear twice for some size ranges: once implicitly within the Berry and Reinhardt (1974b) scheme, and again when explicitly parameterized. This results in double counting of particles. Finally, Berry and Reinhardt (1974b) write that the parameterization is based upon initial values of cloud droplets only in the range from

$$20 \leq D_f^0 \leq 36\,\mu\text{m}, \tag{9.6}$$

where $f = L/Z$ (L is the liquid-water content, Z is the reflectivity); and only with the relative variance *var M* given as lying between

$$0.25 \leq var\ M \leq 1, \tag{9.7}$$

which corresponds to a Golovin-distribution shape parameter with limits $0 < \nu_{\text{cw}} < 3$.

Cohard and Pinty's (2000) formulation is given as

$$\hat{T}_{\text{zw,inv}}\{\text{s}^{-1}\} = \left(10^6 \frac{1}{2} D_{\text{cw}} (var\ M)^{1/6} - 7.5\right) Q_{\text{cw}}\rho/3.72, \tag{9.8}$$

$$\hat{L}_{\text{zw}}\{\text{kg}\,\text{m}^{-3}\} = \left(\frac{10^{20}}{16} D_{\text{cw}}^4 (var\,M)^{1/2} - 0.4\right) \times 2.7 \times 10^{-2} Q_{\text{cw}}\rho, \tag{9.9}$$

and

$$\frac{\mathrm{d}q_{\text{zw}}}{\mathrm{d}t}\{\text{kg}\,\text{kg}^{-1}\,\text{s}^{-1}\} = \max\left(\hat{L}_{\text{zw}}, 0\right) \times \max\left(\hat{T}_{\text{zw,inv}}, 0\right)/\rho, \tag{9.10}$$

where ρ is the density of air; and where

$$D_{\text{cw}}\{\text{m}\} = \left(6\rho Q_{\text{cw}}/\pi\rho_{\text{cw}} N_{\text{Tcw}}\right)^{1/3}, \tag{9.11}$$

is the mean volume diameter of the cloud water, and

$$var\,M\{\text{no dim}\} = \frac{\Gamma(\nu_{\text{cw}})\Gamma(\nu_{\text{cw}} + 6/x_{\text{cw}})}{\Gamma(\nu_{\text{cw}} + 3/x_{\text{cw}})^2} - 1, \tag{9.12}$$

is the mass-relative variance of the generalized gamma distribution of cloud water.

The distribution parameters ν_{cw} and x_{cw} are typically chosen such that the value of *var M* is kept within the 0.25 to 1 range that Berry and Reinhardt (1974b) empirically tested. The reduction of the cloud mixing ratio is equal to the drizzle gain,

$$\frac{\mathrm{d}Q_{\text{cw}}}{\mathrm{d}t}\{\text{kg}\,\text{kg}^{-1}\,\text{s}^{-1}\} = -\frac{\mathrm{d}Q_{\text{zw}}}{\mathrm{d}t}, \tag{9.13}$$

where the subscript zw denotes drizzle. Note that this Berry and Reinhardt (1974b) autoconversion scheme implicitly includes autoconversion, accretion of cloud by drizzle, and drizzle self-collection, but such double counting is typically ignored by modelers.

For number concentration, Cohard and Pinty (2000) as well as Milbrandt and Yau (2005b) used the diameter $\hat{D}_{\mathrm{H}}$ (corresponding to the hump in the S2 mass distribution, which was found at the later time, T_{H}) instead of the time used by Berry and Reinhardt (1974b),

$$\hat{D}_{\mathrm{H}} = \frac{1.26 \times 10^{-3}\{\mathrm{m\,\mu m}\}}{0.5 \times 10^{6}\left\{\frac{\mu\mathrm{m}}{\mathrm{m}}\right\} D_{b}^{0}\{\mathrm{m}\} - 3.5\{\mu\mathrm{m}\}}. \tag{9.14}$$

This equation has been converted to SI units. In addition, a factor number of 10 is included for a correction to the Berry and Reinhardt (1974b) equation. To get the number concentration, the following equation is used,

$$D_{x} = \max\left[\max\left(82 \times 10^{-6}, \hat{D}_{\mathrm{H}}\right), D_{\mathrm{zw}}\right], \tag{9.15}$$

where D_{zw} is drizzle mean volume diameter. Then mass M_{zw} is defined

$$M_{\mathrm{zw}} = \frac{\pi}{6}\rho_{\mathrm{L}} D_{\mathrm{zw}}^{3}, \tag{9.16}$$

where ρ_{L} is the density of liquid, from which the rate of drizzle or raindrop production can be found using $Q_{\mathrm{zw}}CN_{\mathrm{cw}}$,

$$C_{\mathrm{zw}}CN_{\mathrm{cw}} = \frac{\rho Q_{\mathrm{zw}}CN_{\mathrm{cw}}}{M_{\mathrm{zw}}}. \tag{9.17}$$

The Berry and Reinhardt (1974a,b) scheme is an excellent one to get started with but ultimately there are better ways to approach autoconversion, with the best perhaps the hybrid-bin approach.

In an attempt to represent different continental as well as maritime regimes, Manton and Cotton (1977) and Tripoli and Cotton (1980) developed a scheme given as

$$Q_{\mathrm{zw}}CN_{\mathrm{cw}} = f_{\mathrm{cw}}\bar{Q}_{\mathrm{cw}}H(Q_{\mathrm{cw}} - Q_{\mathrm{cw0}}). \tag{9.18}$$

The factor, f_{cw}, is related to the collection frequency among cloud drops, which become raindrops. As collection is largely responsible for the autoconversion of cloud drops to embryonic raindrops, the development of this scheme was an attempt to bring some of this information into the formulation. The threshold value of Q_{cw0} is given as

$$Q_{\mathrm{cw0}} = \frac{\pi}{6} D_{\mathrm{cw0}} \frac{\rho_{\mathrm{L}} N_{\mathrm{Tcw}}}{\rho}, \tag{9.19}$$

where D_{cw0} is approximately 40 microns. The mean frequency of collisions among cloud drops is approximated by

$$f_{cw} = \frac{\pi}{4} D_{cw}^2 E_{cw} V_{Tcw} N_{Tcw}, \tag{9.20}$$

where the collection efficiency is $E_{cw} = 0.55$, V_{Tcw} can be found from Stokes' law. The diameter D_{cw} is given as

$$D_{cw} = \left(\frac{\pi Q_{cw} \rho}{4 \rho_L N_{Tcw}} \right)^{1/3}. \tag{9.21}$$

As noted in Cotton and Anthes (1989) care must be taken in choosing the appropriate mean Q_{cw}.

In an apparent attempt to correct the problem of producing too much rain too fast, Lin *et al.* (1983) represented the Berry (1967) scheme with changes to delay the production of raindrops. However, their tests have shown that rain still unrealistically forms too fast for continental clouds. Nevertheless, as the Berry scheme is often used in this modified form it is presented (e.g. Ferrier 1994 and Straka and mansell 2005) in CGS units as

$$Q_{rw} CN_{cw} = \frac{\rho (Q_{cw} - Q_{cw0})^2}{\left[\frac{1.2 \times 10^{-4} + 1.569 \times 10^{-12} N_{Tcw}}{\gamma_{cw} (Q_{cw} - Q_{cw0})} \right]}, \tag{9.22}$$

where Q_{cw0} is the cloud-water mixing ratio present before rain can develop. In addition to Lin *et al.* (1983), Orville and Kopp (1977) also made modifications to Berry's original scheme (Berry 1968b) in an attempt to capture better the development of first echoes in simulations, though they note that even this modification does not suppress raindrop development when it should. For example, Lin *et al.* (1983) note that Dye *et al.* (1974) stated from observations that cloud-droplet collision–coalescence rarely leads to the formation of rain on the high plains of the United States. For this reason Lin *et al.* (1983) actually turned off the autoconversion model process for cold-based, continental-type, high-plains storm simulations. Typical values of Q_{cw0}, N_{Tcw}, and γ_{cw} used by modelers of continental clouds are 2×10^{-3} g g^{-1}, 1000 cm^{-3}, and 0.15, respectively. Other values can be used to simulate other types of clouds in different climatic regimes. Some of these values are presented in Proctor's (1987) discussion of the various uses of the Berry and Reinhardt (1974b) scheme.

Using large eddy simulations of stratocumulus clouds, Khairoutdinov and Kogan (2000) developed a two-moment autoconversion scheme for cloud water into drizzle particles. This is one of the few schemes where conversion is explicitly stated to go from cloud water to drizzle. The form of the equation is determined from a least-squares fit for parameters *a*, *b*, and *c* from bin model information,

$$\frac{dQ_{zw}}{dt} = c\phi^a\psi^b. \tag{9.23}$$

The best-fit autoconversion scheme was found to be

$$\frac{dQ_{zw}}{dt} = 1350Q_{cw}^{2.47}N_{Tcw}^{-1.79}. \tag{9.24}$$

A one-parameter form was also presented in terms of the drop mean volume radius r_{cwmv}

$$\frac{dQ_{zw}}{dt} = 4.1 \times 10^{-15} r_{cwmv}^{5.67}, \tag{9.25}$$

where r_{cwmv} is in terms of microns. With this scheme, autoconversion occurs when the mean volume radius is between 7 and 19 microns.

When Khairoutdinov and Kogan went further and did a regression analysis of their model information they found another representation for autoconversion,

$$\frac{dQ_{zw}}{dt} = 2.2Q_{cw}^{7/3}N_{Tcw}^{-1/3}. \tag{9.26a}$$

With this expression they conclude that the average collision efficiency is about 0.04. Two other forms include,

$$\frac{dQ_{zw}}{dt} = 1350Q_{cw}^{2.47}N_{Tcw}^{-1.79} \tag{9.26b}$$

$$\frac{dQ_{zw}}{dt} = 4.1 \times 10^{-15} r_{cwmv}^{5.67} \tag{9.26c}$$

were r_{cwmv} is the radius of the mean volume cloud water droplet in microns at a grid point (Fig. 9.1 shows the bulk autoconversion rates plotted against the explicit autoconversion rates from the Kogan bin microphysics parameterization model).

In order to use these schemes, an equation for number concentration tendency is required. By assuming all newly formed drizzle particles have a radius $r_{cw0} = 25$ microns the number concentration rate is given as

$$\left.\frac{dN_{Tzw}}{dt}\right|_{auto} = \frac{\left.\frac{dQ_{zw}}{dt}\right|_{auto}}{\left(\frac{4\pi\rho_L r_{cw0}^3}{3\rho_0}\right)}. \tag{9.27}$$

9.3 Self-collection of drizzle drops and conversion of drizzle into raindrops

Drizzle enters the liquid-drop spectrum from autoconversion of cloud droplets at a diameter of about 82 microns via the Berry and Reinhardt (1974b) autoconversion parameterization. It then experiences self-collection via Long's kernel (see Cohard and Pinty 2000) or by the analytical solution by Verlinde and Cotton (1993) and Flatau *et al.* (1989) as well as by accretion of cloud water (Mizuno 1990 approach).

The conversion of drizzle to rain water occurs after sufficient broadening of the distribution via the warm-rain process (diffusion and coalescence). The amount of mass transferred to the warm-rain distribution from the drizzle distribution is computed using the Farley *et al.* (1989) approach. The mass and number concentration of drizzle particles with diameters greater than 500 microns are transferred using (9.28) and (9.29). The equations for mixing ratio and number concentration (assuming limits from $D_{\min} = 500$ microns (5×10^{-4} m) to infinity) are given as

$$Q_{\mathrm{rw}}CN_{\mathrm{zw}} = D^3_{\mathrm{nzw}}\left\{\frac{\pi\rho_z N_{\mathrm{Tzw}}}{6\Delta t\rho}\frac{\gamma([b_z + \nu_z\mu_z]/\mu_z; D_{\min}/D_{n\mathrm{zw}})}{\Gamma(\nu_z)}\right\} \tag{9.28}$$

and

$$N_{\mathrm{rw}}CN_{\mathrm{zw}} = \left\{\frac{N_{\mathrm{Tzw}}}{\Delta t}\frac{\Gamma(\nu_z; D_{\min}/D_{n\mathrm{zw}})}{\Gamma(\nu_z)}\right\}. \tag{9.29}$$

Alternatively, with a hybrid-bin approach, a similar procedure could be carried out as for the conversion of cloud droplets to drizzle as mentioned earlier.

9.4 Conversion of ice crystals into snow crystals and snow aggregates

Autoconversion and conversion of ice crystals and snow aggregates is one of the more severe cruxes of cloud modeling. The conditions under which autoconversion and conversion occur can make the difference between a realistic simulation or not (e.g. Cotton *et al.* 1986; DeMott *et al.* 1994).

A simple autoconversion parameterization for cloud ice crystals to snow aggregates is given by Lin *et al.* (1983) and Rutledge and Hobbs (1983). In their formulations, cloud ice crystals are converted to snow aggregates when the snow mixing ratios exceed a certain threshold much like the Kessler autoconversion scheme. The equation given by Lin *et al.* (1983) appears as

$$Q_{\mathrm{sw}}CN_{\mathrm{ci}} = 0.001E_{\mathrm{ii}}\max(Q_{\mathrm{ci}} - Q_{\mathrm{ci0}}, 0), \tag{9.30}$$

where Q_{ci0} is $= 0.001$ kg kg^{-1} and $E_{ii} = \exp[0.025(T-273.15)]$. This equation does not show well the dependence of aggregation efficiency on crystal structure, which is dependent on temperature (Lin *et al.* 1983).

The Cotton *et al.* (1986) scheme was presented a few years later and took into account the self-collection occurring between ice crystals to produce snow aggregates. The equation for number tendency takes the form,

$$\left.\frac{dN_{Tci}}{dt}\right|_{auto} = -K_i N_{Tci}^2, \tag{9.31}$$

where K_i is the collection cross-section,

$$K_i = \frac{\pi D_i}{6} V_{Tci} E_{ii} \chi, \tag{9.32}$$

where V_{Tci} is the terminal velocity for cloud ice, and $\chi = 0.25$ is proportional to the variance in particle fallspeed. Or, as Cotton *et al.* put it, it represents the dispersion of the fallspeed spectrum.

The conversion rate of ice-crystal mixing ratio to snow aggregates is given simply as

$$Q_{ci}CN_{sw} = -\frac{m_{ci}}{\rho}\left.\frac{dN_{Tci}}{dt}\right|_{auto} = +K_i N_{Tci} Q_{ci}, \tag{9.33}$$

where

$$m_{ci} = \frac{Q_{ci}\rho}{N_{Tci}}, \tag{9.34}$$

and E_{ii} can be chosen to be that used by Lin *et al.* or an approximation to Hallgren and Hosler (1960) by Cotton *et al.*, given as

$$E_{ii} = \min\left[10^{0.035(T-273.15)-0.7}, 0.2\right]. \tag{9.35}$$

This representation does not allow the coalescence efficiency to exceed 0.2, whereas the Lin *et al.* scheme approaches 1.0 at temperatures of about 273.15 K. Cotton *et al.* note that Pruppacher and Klett (1981) only show Rogers (1973) efficiencies at values as high as 0.6. It should be noted that Passarelli and Srivastava (1979) inferred from aircraft measurements collection efficiencies as high as 140% in temperature ranges of −12 to −15 °C. Presumably this has to do with the predominance of the production of dendrites in this temperature range. To summarize quickly, very little is known about collection efficiencies among ice-crystal particles.

An alternate form for aggregation among ice crystals to form snow aggregates is given by Murakami (1990) as the following

$$N_{\text{Tci}}AC_{\text{sw}} = \frac{Q_{\text{ci}}}{\Delta\tau_1}, \tag{9.36}$$

where $\Delta\tau_1$ is given as

$$\Delta\tau_1 = \frac{2}{C_1}\log\left(\frac{r_{\text{ci}}}{r_{\text{sw0}}}\right)^3, \tag{9.37}$$

with r_{ci} equal to radius of ice given by

$$r_{\text{ci}} = \left(\frac{3\rho Q_{\text{ci}}}{4\pi\rho_{\text{ci}}N_{\text{Tci}}}\right)^{1/3}. \tag{9.38}$$

Also, C_1 is given as

$$C_1 = \rho Q_{\text{ci}}a_{\text{ci}}E_{\text{ii}}\chi, \tag{9.39}$$

with $a_{\text{ci}} = 700$, $E_{\text{ii}} = 0.1$, and $\chi = 0.25$.

One of the more complex conversion schemes of ice crystals converted into snow aggregates is based on the length of time, $\Delta\tau$, it takes for an ice crystal to grow from a mass m_{ci} to a mass m_{sw0} through riming and vapor deposition growth (Reisner *et al.* 1998), which is given by the following,

$$\Delta\tau = \frac{N_{\text{Tci}}(m_{\text{sw0}} - m_{\text{ci}})}{(Q_{\text{ci}}DP_{\text{sw}} + Q_{\text{ii}}AC_{\text{cw}})}, \tag{9.40}$$

where m_{sw0} is the mass of the smallest snow particle given by

$$m_{\text{sw0}} = \frac{\pi}{6}\rho_{\text{sw}}D_{\text{sw0}}^3, \tag{9.41}$$

with $D_{\text{sw0}} = 1.5 \times 10^{-4}$ m. In this equation $Q_{\text{ci}}AC_{\text{cw}}$ is given by

$$Q_{\text{ci}}AC_{\text{cw}} = \min(Q_{\text{ci}}AC_{\text{cw}}, Q_{\text{ci}}DP_{\text{cw}}). \tag{9.42}$$

Now the amount of cloud ice converted to snow aggregates is defined when $m_{\text{ci}} < 0.5m_{\text{sw0}}$, and the equation used is

$$Q_{\text{sw}}CN_{\text{ci}} = \frac{1}{\Delta\tau}\rho Q_{\text{ci}} = \frac{m_{\text{ci}}}{m_{\text{sw0}} - m_{\text{ci}}}(Q_{\text{ci}}DP_{\text{v}} + Q_{\text{ci}}AC_{\text{ci}}), \tag{9.43}$$

and when $m_{\text{ci}} > 0.5m_{\text{sw0}}$ the equation used is

$$Q_{\text{sw}}CN_{\text{ci}} = \frac{Q_{\text{ci}}}{2\Delta t} = \left(1 - \frac{0.5m_{\text{sw0}}}{m_{\text{ci}}}\right)(Q_{\text{ci}}DP_{\text{v}} + Q_{\text{ci}}AC_{\text{ci}}). \tag{9.44}$$

9.5 Conversion of ice crystals and snow aggregates into graupel by riming

Perhaps the simplest conversion rate of snow aggregates into graupel is that used by Lin *et al.* (1983) and Rutledge and Hobbs (1983) and again is similar to Kessler (1969). Again, in their formulations, snow is converted to graupel when the snow content exceeds a certain mixing-ratio threshold. The equation given by Lin *et al.* appears as

$$Q_{gw}CN_{sw} = 0.001\alpha \max(Q_{sw} - Q_{sw0}, 0), \tag{9.45}$$

where Q_{sw0} is $= 0.0006$ kg kg^{-1} and $\alpha = \exp[0.09(T - 273.15)]$.

Another method for conversion of ice crystals or snow aggregates into graupel at temperatures < 273.15 K during riming is to assume that the production rate is related to the vapor deposition rate and riming rate as follows. First, the vapor deposition rate and riming rate are computed for snow aggregates. The conversion rate is equal to the difference of the riming rate and the vapor deposition rate when the riming rate is the larger of the two,

$$Q_{gw}CN_{sw} = \max(Q_{sw}AC_{cw} - Q_{sw}DP_{v}, 0). \tag{9.46}$$

The same approach can be used with ice crystals,

$$Q_{gw}CN_{ci} = \max(Q_{ci}AC_{cw} - Q_{ci}DP_{v}, 0). \tag{9.47}$$

More complicated forms of the above equations were proposed by Murakami (1990) for the production of graupel from riming ice and snow aggregates, respectively.

Murakami's (1990) formulation for the production of graupel from riming ice crystals is

$$Q_{gw}CN_{ci} = \frac{\rho Q_{ci}}{\Delta m_{gi}} \max\left(\frac{Q_{gw}AC_{ci} - Q_{ci}DP_{v}, 0.0}{N_{Tci}}\right), \tag{9.48}$$

where

$$\Delta m_{gi} = m_{g0} - m_{ci}, \tag{9.49}$$

and $m_{g0} = 1.6 \times 10^{-10}$ kg is the mass of the smallest graupel particle. The number-concentration change of ice crystals converted to graupel is given by

$$N_{gw}CN_{ci} = \rho \max\left(\frac{Q_{gw}CN_{ci} + Q_{gi}AC_{cw}}{m_{g0}}\right). \tag{9.50}$$

The number conversion rate is given by Milbrandt and Yau (2005b) as

$$N_{gw}CN_{ci} = \left(\frac{\rho}{M_{gw0}}\right) Q_{gw}CN_{ci}, \tag{9.51}$$

where

$$M_{\mathrm{gw0}} = 1.6 \times 10^{-10}\,\mathrm{kg}. \tag{9.52}$$

Murakami (1990) proposed an equation for conversion to graupel from riming snow aggregates,

$$Q_{\mathrm{gw}}CN_{\mathrm{sw}} = \max\left(\frac{\rho_{\mathrm{sw}}}{\rho_{\mathrm{gw}} - \rho_{\mathrm{sw}}} Q_{\mathrm{gwsw}}AC_{\mathrm{cw}}, 0.0\right). \tag{9.53}$$

The number conversion rate given by Milbrandt and Yau (2005b) is

$$N_{\mathrm{gw}}CN_{\mathrm{sw}} = \left(\frac{\rho}{M_{\mathrm{gw0}}}\right) Q_{\mathrm{gw}}CN_{\mathrm{sw}}, \tag{9.54}$$

where again (9.52) holds.

It is also noted that the riming for the growth of snow aggregates during this three-body process is

$$Q_{\mathrm{swsw}}AC_{\mathrm{cw}} = Q_{\mathrm{sw}}AC_{\mathrm{cw}} - Q_{\mathrm{gwsw}}AC_{\mathrm{cw}}, \tag{9.55}$$

where $Q_{\mathrm{sw}}AC_{\mathrm{cw}}$ can be from any of the forms of the collection equation presented in Chapter 7.

Cotton *et al.* (1986) devised a parameterization for the conversion of snow aggregates to graupel that was activated when the mixing ratio of rimed aggregates is the same as the mixing ratio of a population of graupel particles. Then the tendency difference between the aggregate riming tendency and the growth tendency of the graupel (where the former is greater than the latter), when the temperature is colder than 273.15 K, can be written as

$$Q_{\mathrm{gw}}CN_{\mathrm{sw}} = \max\left[Q_{\mathrm{sw}}AC_{\mathrm{cw}} - Q_{\mathrm{gw}}AC_{\mathrm{cw}}\left(Q_{\mathrm{gw}} = Q_{\mathrm{sw}}\right), 0\right]. \tag{9.56}$$

Another conversion scheme of snow aggregates to graupel follows that of Farley *et al.* (1989). In this scheme a three-body procedure is developed where some of the rimed snow aggregates remain as snow aggregates and some are converted to graupel depending on the amount of riming. The amount of cloud water that is rimed by snow aggregates and converted to graupel is given by

$$Q_{\mathrm{gwsw}}AC_{\mathrm{cw}} = \frac{1}{4}\pi E_{\mathrm{swcw}} Q_{\mathrm{cw}} N_{\mathrm{Tsw}} D_{n\mathrm{sw}}^{2+b_{\mathrm{sw}}} a_{\mathrm{sw}} \left(\frac{\rho_0}{\rho}\right)^{1/2} \gamma\left(2 + b_{\mathrm{sw}} + \nu; \frac{D_0}{D_{n\mathrm{sw}}}\right), \tag{9.57}$$

with γ the partial gamma function, and

$$Q_{\mathrm{swsw}}AC_{\mathrm{cw}} = Q_{\mathrm{sw}}AC_{\mathrm{cw}} - Q_{\mathrm{gwsw}}AC_{\mathrm{cw}}. \tag{9.58}$$

The amount of snow aggregates converted to the new particle gw can be written from the definition of mixing ratio of sw as

$$Q_{\mathrm{gwcw}}AC_{\mathrm{sw}} = \frac{\pi \rho_{\mathrm{sw}} N_{\mathrm{Tsw}} D_{n\mathrm{sw}}^3 \Gamma\left(\beta_{\mathrm{sw}} + v_{\mathrm{sw}}, \frac{D_{\mathrm{swmin}}}{D_{n\mathrm{sw}}}\right)}{6\rho \Delta t \Gamma(v_{\mathrm{sw}})}, \tag{9.59}$$

where E_{swcw} is the collection efficiency of cloud water by snow and ρ is the reference density of air; where subscript cw represents the sum of all the liquid that collects snow sw. Here Γ is the partial gamma function. The corresponding equation for N_{T} is given as,

$$N_{\mathrm{gwcw}}AC_{\mathrm{sw}} = \frac{N_{\mathrm{Tsw}} \Gamma(v_{\mathrm{sw}}, D_{\mathrm{swmin}}/D_{n\mathrm{sw}})}{\Delta t \Gamma(v_{\mathrm{sw}})}. \tag{9.60}$$

Seifert and Beheng (2005) developed parameterizations for conversion of cloud ice to graupel when ice crystals and snow aggregates rime sufficiently. The conversion of cloud ice to graupel occurs when plate-like crystals are $>$ 500 microns in diameter, column-like crystals are $>$ 50 microns, and snow aggregates $>$ 250 microns (along the a-axis for ice crystals). The critical amount of rime can be written as

$$X_{\mathrm{critical\_rime}} = spacefill\ \rho_{\mathrm{L}}\ \max\left(\frac{\pi}{6} D_{\mathrm{ni}}^3 - \frac{X_{\mathrm{i}}}{\rho_{\mathrm{i}}}\right), \tag{9.61}$$

where X_{i} is given by

$$X_{\mathrm{i}} = \rho \frac{q_{\mathrm{i}}}{N_{\mathrm{Ti}}}. \tag{9.62}$$

The parameter *spacefill* is from Beheng (1981) and Seifert and Beheng (2005) is equal to 0.68 for ice crystals and 0.01 for snow for rapid conversion of snow to graupel when riming occurs. The value of tau, τ, for conversion is

$$X_{\mathrm{tau\_conv}} = \frac{X_{\mathrm{critical\_rime}} N_{\mathrm{Ti}}}{\rho Q_{\mathrm{ci}} AC_{\mathrm{cw}}}, \tag{9.63}$$

which gives mixing-ratio and number-concentration rates of

$$Q_{\mathrm{gi}}AC_{\mathrm{cw}} = \frac{Q_{\mathrm{ci}}}{X_{\mathrm{tau\_conv}}} \tag{9.64}$$

and

$$N_{\mathrm{gi}}AC_{\mathrm{cw}} = \frac{\rho}{X_i} Q_{\mathrm{gi}}AC_{\mathrm{cw}}. \tag{9.65}$$

This process is a three-body interaction, so not all ice crystals or snow aggregates are converted to graupel, and some of the ice crystals and snow aggregates are left behind with some riming. The equations for the increase of mass of ice crystals and snow aggregates then become

$$Q_{\text{ci}}AC_{\text{cw}} = Q_{\text{ci}}AC_{\text{cw}} - Q_{\text{gi}}AC_{\text{cw}} \tag{9.66}$$

and

$$N_{\text{ci}}AC_{\text{cw}} = N_{\text{ci}}AC_{\text{cw}} - N_{\text{gi}}AC_{\text{cw}}. \tag{9.67}$$

9.6 Conversion of graupel and frozen drops into small hail

The conversion of graupel and frozen drops into small hail also can be cast as a three-body interaction following Farley *et al.* (1989).

There is no corresponding number change. The amount of ice y converted to the new particle z can be written from the definition of mixing ratio of y as,

$$Q_{z\text{L}}AC_y = \frac{\pi\rho_y N_{\text{T}y} D_{ny}^3 \Gamma\left(\beta_y + \nu_y, \frac{D_{y\min}}{D_{ny}}\right)}{6\rho_0 \Delta t \Gamma(\nu_y)}, \tag{9.68}$$

where subscript L represents the sum of all the liquid that collects ice y. The corresponding equation for N_{T} is given as,

$$N_{z\text{L}}AC_y = \frac{N_{\text{T}y}\Gamma(\nu_y, D_{y\min}/D_{ny})}{\Delta t \Gamma(\nu_y)}. \tag{9.69}$$

In some models (Straka *et al.* 2009b) a particle is initiated at the mean volume diameter and the continuous growth equation is integrated to see if the particle grows by the time-weighted mean water content estimate and the Lagrangian time estimate (see Chapter 2). If a particle reaches a minimum diameter by continuous collection growth with the procedure above the conversion occurs.

Ziegler (1985) used the model of Nelson (1983) to derive a variable D_{w} to indicate the onset of wet growth. He showed that when $D < D_{\text{w}}$, then dry growth continued. However when $D > D_{\text{w}}$, wet growth began.

$$D_{\text{w}} = \exp\left[\frac{T(^\circ\text{C})}{1.1 \times 10^7 \rho Q_{\text{cw}} - 1.3 \times 10^6 \rho Q_{\text{ci}} + 1}\right] - 1. \tag{9.70}$$

Ziegler's wet- and dry-growth equations involved using incomplete gamma functions, with hail designated by particle size with $D > D_{\text{w}}$ and graupel particles defined as size $D < D_{\text{w}}$, both represented by the same size distribution. The same scheme was implemented by Milbrandt and Yau (2005b), though they used a similar equation [in SI units and included collection of rain, (9.71)] to determine when a particle would become hail or remain as graupel. Unlike Ziegler, they carried prognostic equations for both graupel and hail particles.

$$D_{\mathrm{w}} = 0.01\left\langle \exp\left[\frac{T(^{\circ}\mathrm{C})}{1.1\times10^{4}(Q_{\mathrm{cw}}+Q_{\mathrm{rw}}) - 1.3\times10^{3}Q_{\mathrm{ci}} + 1\times10^{-3}}\right] - 1\right\rangle. \quad (9.71)$$

The conversion rate that Milbrandt and Yau use, incorporating (9.71) to delineate graupel from hail, is given as

$$Q_{\mathrm{hw}}CN_{\mathrm{gw}} = \frac{D_{\mathrm{mvgw}}}{D_{\mathrm{hw0}}}\left(Q_{\mathrm{gw}}AC_{\mathrm{cw}} + Q_{\mathrm{gw}}AC_{\mathrm{rw}} + Q_{\mathrm{gw}}AC_{\mathrm{ci}}\right). \quad (9.72)$$

Milbrandt and Yau note that $D_{\mathrm{hw0}} = D_{\mathrm{w}}$ can sometimes be smaller than the mean volume diameter of graupel D_{mvgw}. This can happen with high liquid water contents at relatively high temperatures. To prevent the total mass converted to hail from becoming larger than the total graupel mass plus the mass of liquid and ice that graupel collects, the following is incorporated, though it is somewhat artificial,

$$Q_{\mathrm{hw}}CN_{\mathrm{gw}} = \min\left[Q_{\mathrm{hw}}CN_{\mathrm{gw}}, Q_{\mathrm{gw}} + \left(Q_{\mathrm{gw}}AC_{\mathrm{cw}} + Q_{\mathrm{gw}}AC_{\mathrm{rw}} + Q_{\mathrm{gw}}AC_{\mathrm{ci}}\right)\right]. \quad (9.73)$$

Other limits are needed when wet growth is not expected, and graupel does not convert to hail. This occurs at relatively cold temperatures and low liquid-water contents. Thus, a lower limit is placed upon the ratio $D_{\mathrm{gwmv}}/D_{\mathrm{w}}$ such that it does not go below 0.1. When it does go below 0.1, the conversion rate is set to zero. The number conversion rate is then given by

$$N_{\mathrm{hw}}CN_{\mathrm{gw}} = \left(\frac{\rho}{M_{\mathrm{hw0}}}\right)Q_{\mathrm{hw}}CN_{\mathrm{gw}}, \quad (9.74)$$

where

$$M_{\mathrm{hw0}} = c_x D_{\mathrm{hw0}}^{d_x}. \quad (9.75)$$

9.7 Conversion of three graupel species and frozen drops amongst each other owing to changes in density by collection of liquid particles

9.7.1 Graupel and frozen drops collecting cloud water

For snow aggregates, graupel, and frozen drops collecting cloud water, a simplified approach somewhat like Ferrier (1994) can be used. First, the amount of riming that takes place is computed as Q_xAC_{cw}. Next the rime density of cloud water collected is computed following data archived by Pflaum [1980; (9.76a,b)] and Heymsfield and Pflaum [1985; (9.76c)] as

$$\rho_{x,\mathrm{rime}} = 261\left(-\frac{0.5D_{\mathrm{cw}}V_{\mathrm{T,impact},x}}{T(^\circ\mathrm{C})}\right)^{0.38}, \tag{9.76a}$$

$$\rho_{x,\mathrm{rime}} = 110\left(-\frac{0.5D_{\mathrm{cw}}V_{\mathrm{T,impact},x}}{T(^\circ\mathrm{C})}\right)^{0.76}, \tag{9.76b}$$

or

$$\rho_{x,\mathrm{rime}} = 300\left(-\frac{0.5D_{\mathrm{cw}}V_{\mathrm{T,impact},x}}{T(^\circ\mathrm{C})}\right)^{0.44}, \tag{9.76c}$$

where the value of $V_{\mathrm{T,impact},x}$ is the impact velocity of cloud drops on the ice and is given as approximately 0.6 $V_{\mathrm{T}x}$ after Pflaum and Pruppacher (1979). The new density of the ice particle undergoing collection of cloud water after a 90 s (60 to 120 s) period Δt_{rime} is given by

$$\rho_{x,\mathrm{new}} = \left(\frac{Q_x\rho_x + \Delta t_{\mathrm{rime}}Q_xAC_{\mathrm{cw}}}{Q_x + \Delta t_{\mathrm{rime}}Q_xAC_{\mathrm{cw}}}\right). \tag{9.77}$$

Consider medium-density graupel collecting cloud water. Low-density rime can be added to become low-density graupel if $\rho_{\mathrm{gm,rime}} < 0.5(\rho_{\mathrm{gl}} + \rho_{\mathrm{gm}})$; remain added to medium-density graupel if $0.5(\rho_{\mathrm{gl}} + \rho_{\mathrm{gm}}) < \rho_{\mathrm{gm,rime}} < 0.5(\rho_{\mathrm{gm}} + \rho_{\mathrm{gh}})$; can be added to high-density graupel if $0.5(\rho_{\mathrm{gm}} + \rho_{\mathrm{gh}}) < \rho_{\mathrm{gm,rime}} < 0.5(\rho_{\mathrm{gh}} + \rho_{\mathrm{fw}})$; or added to frozen drops if $0.5(\rho_{\mathrm{gh}} + \rho_{\mathrm{fw}}) < \rho_{\mathrm{gm,rime}}$. This can be done for all three graupel species and frozen drops. Any species can be converted to one of the other by either low- or high-density riming.

9.7.2 Graupel and frozen drops collecting drizzle or rain water

Consider low-density graupel collecting drizzle or rain water. The particle source can collect drizzle or rain and remain as low-density graupel if $\rho_{\mathrm{gl,rime}} < 0.5(\rho_{\mathrm{gl}} + \rho_{\mathrm{gm}})$; become medium-density graupel if $0.5(\rho_{\mathrm{gl}} + \rho_{\mathrm{gm}}) < \rho_{\mathrm{gl,rime}} < 0.5(\rho_{\mathrm{gm}} + \rho_{\mathrm{gh}})$; high-density graupel if $0.5(\rho_{\mathrm{gm}} + \rho_{\mathrm{gh}}) < \rho_{\mathrm{gl,rime}} < 0.5(\rho_{\mathrm{gh}} + \rho_{\mathrm{fw}})$; and frozen drops if $0.5(\rho_{\mathrm{gh}} + \rho_{\mathrm{fw}}) < \rho_{\mathrm{gl,rime}}$. This can be done for all three graupel species and frozen drops. Any particle can be converted to the other by high-density riming owing to the collection of drizzle and rain. In general particles do not have their densities reduced by collection of drizzle and rain.

9.8 Heat budgets used to determine conversions

From Walko *et al.*'s (2000) extensive work, an implicit system of each equation of each species to close the system of diffusive fluxes together with a temperature equation for hydrometeor surfaces are derived. The diffusive flux of heat and vapor between hydrometeor species and the air depends on differences in the vapor and temperature over the surfaces of hydrometeor species and the air (Walko *et al.* 2000). It is noted that all the hydrometeors must be treated interactively as all the hydrometeor categories compete for excess moisture. Walko *et al.*'s (2000) method is similar to Hall's (1980) implicit, iterative approach; but with some algebra, the implicit approach can be transformed so that it can be solved directly.

The values of Q_m and T_m represent the mixing ratio and temperature of any hydrometeor species indexed by m. Each step of this approach is started with updated values of temperature for air T_a, vapor in air Q_v owing to advection and diffusion etc. Hydrometeor temperature and mixing ratios are all updated by all other processes from advection and diffusion to accretion and freezing or melting.

The Walko *et al.* (2000) approach makes use of the ice–liquid potential temperature θ_{il} system that the CSU RAMS (Colorado State University Regional Area Modeling System) model uses (Cotton *et al.* 2003). The θ_{il} system was developed by Tripoli and Cotton (1981) and has been re-examined by Bryan and Fritsch (2004). An advantage of the θ_{il} system is that θ_{il} is conserved following adiabatic motion with or without internal phase changes of water. With this system the air temperature is given,

$$T_a = T_{il}\left[1 + \frac{q_{lat}}{c_p \max(T_a, 253)}\right], \tag{9.78}$$

where T_{il} is given by

$$T_{il} = \theta_{il}\left(\frac{p_0}{p}\right)^{\frac{R_d}{c_p}}, \tag{9.79}$$

following Tripoli and Cotton (1981) and Walko *et al.* (2000). The value of q_{lat} is the enthalpy released in the conversion from vapor to all ice and liquid in a parcel, and L_v and L_s are enthalpies of vaporization and sublimation,

$$q_{lat} = \left[\sum_{l} Q_l + \sum_{i}(1 - f_{r,i})Q_l\right]L_v + \left[\sum_{i} f_{r,i}Q_i\right]L_s, \tag{9.80}$$

where $f_{r,i}$ is the fraction of ice. The equation for T_a above can be rewritten with the variable A_l by linearizing as follows,

$$T_a^{t+\delta T}(°C) - T_a^*(°C) = A_l\left(Q_v^* - Q_v^{t+\delta T}\right), \tag{9.81}$$

where A_l is given by,

$$A_l = \begin{cases} \frac{T_{il}\bar{L}}{c_p 253} & T_a^*(^\circ\text{C}) < -20^\circ\text{C} \\ \frac{T_{il}\bar{L}}{c_p\left(2T_a^* - T_{il}\right)} & T_a^*(^\circ\text{C}) > -20^\circ\text{C} \end{cases}. \tag{9.82}$$

According to Walko *et al.* (2000) the total loss of water vapor by diffusion is the total gain of water mass over all species,

$$Q_v^* - Q_v^{t+\delta t} = -\sum_m \left(Q_m^* - Q_m^{t+\delta t}\right). \tag{9.83}$$

The change in mixing ratio of any species m over a model step is

$$Q_m^{t+\delta t} - Q_m^* = U_m\left(Q_v^{t+\delta t} - Q_{sm}^{t+\delta t}\right) - V_m Q_m^*, \tag{9.84}$$

where Q_{sm}^t is the saturation mixing ratio over a particle at its temperature. The U_m term describes the vapor diffusion growth or evaporation of some hydrometeor species. To keep from evaporating more water vapor than is present in the system, it switches to using a U_m and a V_m to prevent over-depletion; U_m and V_m are given below as the following,

$$U_m = \begin{cases} 4\pi\Delta t\psi N_m (N_{re})_m & \text{if species } m \text{ is not depleted} \\ 0 & \text{if species } m \text{ is depleted,} \end{cases} \tag{9.85}$$

and

$$V_m = \begin{cases} 0 & \text{if species } m \text{ is not depleted} \\ 1 & \text{if species } m \text{ is depleted.} \end{cases} \tag{9.86}$$

Next the term Q_{sm} is defined. When m is hydrometeor species vapor, its elimination needs to be accounted for using a linearized form of the Clausius–Clapeyron equation,

$$Q_{sm}^{t+\delta t} = Q_{sRm}(T_{sm}) + \frac{dQ_{sRm}}{dT_{Rm}}\left(T_m^{t+\delta t} - T_{Rm}\right), \tag{9.87}$$

where T_{Rm} is a reference temperature close to the final temperature T of species m. The reference temperature is approximated by Walko *et al.* (2000) using the following,

$$T_{Rm} = T_a(^\circ\text{C}) - \min(25,\ 700[Q_{sm} - Q_v]). \tag{9.88}$$

Note that T_{Rm} is limited to a maximum of 0 °C for ice species hydrometeors. Now the following can be written with simple substitution of (9.87) and (9.88) into (9.84) as

$$Q_m^{t+\delta t} - Q_m^* = U_m\left(Q_v^{t+\delta t} - Q_{sRm} - \frac{dQ_{sRm}}{dT_{Rm}}\left[T_{sm}^{t+\delta t} - T_{Rm}\right]\right) - V_m Q_m^*. \tag{9.89}$$

Making the substitution of (9.89) into (9.83) gives

$$Q_{\mathrm{v}}^{t+\delta t} = \frac{Q_{\mathrm{v}}^{*} + \sum_{m} U_m \left(Q_{\mathrm{sR}m} + \frac{\mathrm{d}Q_{s\mathrm{R}m}}{\mathrm{d}T_{\mathrm{R}m}} \left[T_{\mathrm{s}m}^{t+\delta t} - T_{\mathrm{R}m} \right] \right) - \sum_{m} V_m Q_m^{*}}{1 + \sum_{m} U_m}. \tag{9.90}$$

Now if the future temperature is to be obtained, a "hydrometeor energy equation" is needed for each species. Walko *et al.* (2000) give the internal energy of hydrometeor species m as,

$$I_m = f_{r,m} c_{\mathrm{i}} T_m + \left(1 - f_{r,m}\right)\left(c_{\mathrm{L}} T_m + L_{\mathrm{f}}\right), \tag{9.91}$$

where I_m is the internal energy of hydrometeor species m and c_{i} and c_{L} are the specific heat with respect to ice and liquid, respectively. In addition, the terms $f_{r,m}$ are fractions of ice. At 0 °C for pure ice, the internal energy is specified to be zero. To get the internal energy of each of the hydrometeor species m the following is written, $I_m Q_m$, which is internal energy per kilogram of air.

The heat budget for each hydrometeor species m is written in terms of I_m as

$$I_m^{t+\delta t} Q_m^{t+\delta t} - I_m^{*} Q_m^{*} = 4\pi \Delta t N_{\mathrm{T}m} (N_{\mathrm{re}})_m K \left(T_{\mathrm{a}}^{t+\delta t} - T_m^{t+\delta t}\right) + left(Q_m^{t+\delta t} - Q_m^{*})\left(L_j - I_m^{t+\delta t}\right), \tag{9.92}$$

where $N_{\mathrm{T}m}$ is the number concentration, K is thermal conductivity of air, $(N_{\mathrm{re}})_m$ is the product of the ventilation coefficient, shape factor, and hydrometeor diameter integrated over the hydrometeor species m size spectrum, and L_j is the enthalpy of phase change j. Rearranging gives

$$\left(I_m^{t+\delta t} - I_m^{*}\right) Q_m^{*} = 4\pi \Delta t N_{\mathrm{T}m} (N_{\mathrm{re}})_m K \left(T_{\mathrm{a}}^{t+\delta t} - T_m^{t+\delta t}\right) + \left(Q_m^{t+\delta t} - Q_m^{*}\right) L_j. \tag{9.93}$$

Walko *et al.* (2000) note that as hydrometeors like rain, graupel, frozen drops, and hail can come out of equilibrium with the environment, the value of the predicted internal energy is stored for the next timestep. Cloud droplets and ice crystals come into equilibrium temperature-wise with the environment nearly immediately and have essentially no heat storage; so it is up to the modeler whether to store the predicted internal energy for them. This implies that there is a balance between sensible and latent-heat diffusion for these smaller hydrometeors at temperatures below 0 °C. This means that the terms of (9.93) on the right-hand side are equal to zero for the small negligible heat-storage particles. Melting ice hydrometeor species m have an internal energy at the start of the process that is zero and the internal energy at the end of the step for these hydrometeor species can be evaluated with (9.93).

Using the right case for each hydrometeor species, and substituting (9.81) into (9.93) to eliminate the new air temperature in °C, and substituting (9.83) into (9.93) to eliminate the new r_j results in

$$I_m^{t+\delta t} D_m + T_m^{t+\delta t} E_m = Q_{\mathrm{v}}^{t+\delta t} F_m + G_m. \tag{9.94}$$

Mixed-phase particles have a new temperature equal to 0 °C. All-liquid and all-ice particles have a temperature predicted from

$$I_m^{t+\delta t} = c_m T_m^{t+\delta t} + L_{ilm}, \tag{9.95}$$

where here L_{ilm} is the enthalpy of fusion for ice–liquid mixtures, and for all-liquid particles $L_{il} \neq 0$ and for ice while $c_m = c_{\mathrm{l}}$ and c_{L} are specific heats for liquid and ice particles, respectively.

The goal is to find a temperature equation for each of the hydrometeor species. The short number of steps for this is found in Walko *et al.* (2000). After some algebra, a temperature equation is found that is explicit,

$$T_m^{t+\delta t} = \left(S_m Q_{\mathrm{v}}^{t+\delta t} + W_m\right) M_m, \tag{9.96}$$

where the Heaviside step function $M_m = 0$ for mixed-phase particles, or $M_m = 1$ for all-liquid or all-ice particles. To choose the right M_m, start by setting it to 1 and assume the old and new Q_{v} are equal. If the new $T_m > 0$, then $M_m = 0$, and $T_m = 0$. For low-heat-storage hydrometeor species such as ice or snow, then the test is done with $H_m = 0$. If the value of $T_m = 0$, then $H_m = 1$. Further details on choosing M_m can be found in Walko *et al.* (2000). The following are variables in the temperature budget,

$$S_m = \frac{F_m}{(C_m D_m + E_m)} \tag{9.97}$$

and

$$W_m = \frac{(G_m - K_m D_m)}{(C_m D_m + E_m)}, \tag{9.98}$$

where

$$C_m = c_L, c_{\mathrm{l}}; \tag{9.99}$$

where $K_m = L_{il}$ for all liquid species, and 0 for all ice species, and the variable to be used is the one that is appropriate for each of the ice and liquid phases. Next,

$$D_m = H_j Q_m^*, \tag{9.100}$$

where H is a factor of 1 or 0 for hydrometeors such as ice and aggregates at $T < 0\,^{\circ}\mathrm{C}$ and for cloud water, 1 for ice, aggregates at $T = 0\,^{\circ}\mathrm{C}$, and rain, graupel, frozen drops, and hail. Now we define

$$F_m = U_m L_m - 4\pi \Delta t N_{\mathrm{T}m} (\mathrm{f_{re}})_m K A_1, \tag{9.101}$$

$$E_m = U_m L_m \frac{\mathrm{d}Q_{\mathrm{sR}m}}{\mathrm{d}T_{\mathrm{R}m}} - B_m K, \tag{9.102}$$

where $B_m = 4\pi \delta t N_{\mathrm{T}m}$ and

$$G_m = \left(U_m L_m \frac{\mathrm{d}Q_{\mathrm{sR}m}}{\mathrm{d}T_{\mathrm{R}m}} T_{\mathrm{R}m} - Q_{\mathrm{sR}m} \right) + B_m K \left(T_{\mathrm{a}}^{*}(^{\circ}\mathrm{C}) + A_1 Q_{\mathrm{v}}^{*} \right) + J_m I_m^{*} Q_m^{*} - V_m L_m Q_m^{*}, \tag{9.103}$$

where $J_m = 0$ for cloud water, ice, or aggregates, and 1 for rain, graupel, or hail.

Finally a closed solution for Q_{v} is found from

$$Q_{\mathrm{v}}^{t+\delta t} = \frac{Q_{\mathrm{v}}^{*} + \sum_m U_m Y_m + \sum_m V_m Q_m^{*}}{1 + \sum_m U_m Z_m}, \tag{9.104}$$

where

$$Y_m = \frac{\mathrm{d}Q_{\mathrm{sR}m}}{\mathrm{d}T_{\mathrm{R}m}} \left[\frac{\left(G_m - K_m H_m Q_m^{*} \right)}{\left(C_m H_m Q_m^{*} + U_m L_m \frac{\mathrm{d}Q_{\mathrm{sR}m}}{\mathrm{d}T_{\mathrm{R}m}} + K B_m \right)} \right], \tag{9.105}$$

and

$$Z_m = 1 - \frac{\mathrm{d}Q_{\mathrm{sR}m}}{\mathrm{d}T_{\mathrm{R}m}} S_m M_m. \tag{9.106}$$

Once the new vapor mixing ratio is updated, then the new hydrometeor species m temperatures are calculated. In addition, each new mixing ratio of hydrometeor species m is then solved.

In computing (9.104) it is assumed that no hydrometeor evaporates or sublimes completely and therefore that $U_m \neq 0$, and $V_m = 0$. Now if a hydrometeor species m completely evaporates/sublimes, then U_m and V_m are changed to their alternate values and (9.104) is evaluated and Q_m is set to zero. Walko *et al.* (2000) do the above in the order: first cloud droplets; then ice crystals; snow aggregates; rain; graupel; and hail. Then the temperature equation (9.94) is evaluated. This requires I_m to be evaluated from (9.93) if $M_m = 1$ or (9.91) if $M_m = 0$. Walko *et al.* (2000) state that it is I_m that is used to determine how much mass and number concentration is transferred between species in collisions, melting, and shedding of rain by hail. The air temperature is then evaluated by (9.96). Usually the trial M_m is used if the temperature equation is correct. Walko *et al.* (2000) explain that when an ice species hydrometeor has an I_m value that is very near to zero, a slight error can be found, and for $M_m = 1$ when $M_m = 0$ the temperature from (9.96) will be a bit above 0 °C. In this case the value of I_m for ice will be too low, so I_m is bound by the value zero.

9.9 Probabilistic (immersion) freezing

The probabilistic freezing of liquid-water drops forms high-density ice-water particles (Bigg 1953; Wisner *et al.* 1972; Lin *et al.* 1983; Ferrier 1994; and others) and is a heterogeneous freezing process owing to the presence of freezing nuclei in the liquid-water drops. Frozen cloud drops are assumed to begin immediately developing into the crystal habit representative of the supersaturation and temperature regime where they form. Frozen drizzle and raindrops are a source for lower- and higher-density graupel, through rapid low- to high-density riming, respectively, or into frozen drizzle and frozen rain if high-density riming occurs. As pointed out by Wisner *et al.*, laboratory experiments suggest that drop freezing is likely a stochastic process, and a function of the volume of the liquid-water particle and the number of ice nuclei that can activate in droplets or drops at a given temperature. Following Bigg (1953) and Wisner *et al.* (1972), an equation can be derived for the probability, ρ, of freezing of a drop with volume V and temperature T,

$$\ln(1 - P) = B'Vt\{\exp[A'(T_0 - T) - 1]\}, \tag{9.107}$$

where T_0 is the freezing temperature (273.15 K), t is the time, and A' and B' are coefficients, $A' = 0.60$ and $B' = 0.01$.

9.9.1 Parameterization for Bigg freezing of raindrops

Bigg first published his design of the basics of the data fit for the freezing of raindrops in 1953. Subsequently Wisner *et al.* (1972) considered an equation for the number of drops of diameter D that are frozen considering only differentials with time t and number of drops N varying. The following equation then can be given,

$$-\frac{\mathrm{d}[n(D)\mathrm{d}D]}{\mathrm{d}t} = \frac{\pi B' N_{\mathrm{T}} D^3}{6}\{\exp[A'(T_0 - T) - 1]\}. \tag{9.108}$$

Parameterizing the equation over the complete distribution can be done to give

$$N_{\mathrm{Tfw}} FZ_x = \int_0^\infty -\frac{\mathrm{d}[n(D)\mathrm{d}D]}{\mathrm{d}t}, \tag{9.109}$$

where FZ_x is the freezing of a species x, where $x =$ cw, or rw.

The tendency equation for mixing ratio is simply given as

$$Q_{\mathrm{fw}}FZ_x = \int_0^\infty -\frac{\mathrm{d}[n(D)\mathrm{d}D]}{\mathrm{d}t}\frac{\pi}{6}D^3\rho_x. \tag{9.110}$$

9.9.1.1 Gamma distribution parameterization for Bigg freezing

To obtain the generalized gamma distribution parameterization, the minus sign in (9.109) is removed, and integration gives,

$$\begin{aligned} N_{\mathrm{Tfw}}FZ_x = &\int_0^\infty \frac{\pi B' N_{\mathrm{T}x} D_x^3}{6}\{\exp[A'(T_0 - T) - 1]\}\frac{1}{\Gamma(\nu_x)}\left(\frac{D_x}{D_{nx}}\right)^{\nu_x - 1} \\ &\times \exp\left(-\left[\frac{D_x}{D_n}\right]\right)\mathrm{d}\left(\frac{D_x}{D_{nx}}\right). \end{aligned} \tag{9.111}$$

The complete gamma distribution equation is

$$\begin{aligned} N_{\mathrm{Tfw}}FZ_x = &\int_0^\infty \left\{\frac{\pi\mu_x\alpha_x^{\nu_x} B' N_{\mathrm{T}x} D_x^3}{6}\{\exp[A'(T_0 - T) - 1]\}\right. \\ &\left.\times \frac{1}{\Gamma(\nu_x)}\left(\frac{D_x}{D_{nx}}\right)^{\nu_x\mu_x - 1}\exp\left(-\alpha_x\left[\frac{D}{D_n}\right]^{\mu_x}\right)\mathrm{d}\left(\frac{D_x}{D_{nx}}\right)\right\}, \end{aligned} \tag{9.112}$$

which upon integration gives

$$N_{\mathrm{Tfw}}FZ_x = \frac{\Gamma(3+\nu_x)\pi B' N_{\mathrm{T}x} D_{nx}^{2+\nu_x}}{6\Gamma(\nu_x)}\{\exp(A'\max[T_0 - T], 0.0) - 1\} \tag{9.113}$$

or

$$N_{\mathrm{Tfw}}FZ_x = \frac{\dfrac{\Gamma\left(\frac{3+\nu_x\mu_x}{\mu_x}\right)}{\alpha_x^{\left(\frac{3+\nu_x\mu_x}{\mu_x}\right)}}\alpha_x^{\nu_x}\pi B' N_{\mathrm{T}x} D_{nx}^{\frac{2+\nu_x\mu_x}{\mu_x}}}{6\Gamma(\nu_x)}\{\exp(A'\max[T_0 - T], 0.0) - 1\}. \tag{9.114}$$

Similarly the mixing ratio equation can be parameterized as the following by removing the minus sign in (9.110) and integrating,

$$\begin{aligned} Q_{\mathrm{fw}}FZ_x = &\int_0^\infty \frac{\pi^2 B' N_{\mathrm{T}x} D_x^6}{36}\frac{\rho_{\mathrm{L}}}{\rho_0}\{\exp[A'(T_0 - T) - 1]\} \\ &\times \frac{1}{\Gamma(\nu_x)}\left(\frac{D_x}{D_{nx}}\right)^{\nu_x - 1}\exp\left(-\left[\frac{D_x}{D_{nx}}\right]\right)\mathrm{d}\left(\frac{D_x}{D_{nx}}\right). \end{aligned} \tag{9.115}$$

For the complete gamma distribution,

$$Q_{\text{fw}}FZ_x = \int_0^\infty \frac{\pi^2 \mu_x \alpha_x^{\nu_x} B' N_{\text{T}x} D_x^6}{36} \frac{\rho_{\text{L}}}{\rho} \{\exp[A'(T_0 - T) - 1]\} \times \frac{1}{\Gamma(\nu_x)} \frac{D_x^{\nu_x \mu_x - 1}}{D_{nx}} \exp\left(-\alpha_x \left[\frac{D_x}{D_{nx}}\right]^{\mu_x}\right) \mathrm{d}\left(\frac{D_x}{D_{nx}}\right), \tag{9.116}$$

which upon integration gives

$$Q_{\text{fw}}FZ_x = \frac{\Gamma\left(\frac{6+\nu_x\mu_x}{\mu_x}\right) \alpha_x^{\nu_x} \pi^2 B' N_{\text{T}x} D_{nx}^{\frac{5+\nu_x\mu_x}{\mu_x}}}{36 \alpha_x^{\left(\frac{6+\nu_x\mu_x}{\mu_x}\right)} \Gamma(\nu_x)} \left(\frac{\rho_{\text{L}}}{\rho}\right) \{\exp(A' \max[T_0 - T], 0.0) - 1\}. \tag{9.117}$$

The modified gamma distribution is

$$Q_{\text{fw}}FZ_x = \frac{\Gamma\left(\frac{6+\nu_x\mu_x}{\mu_x}\right) \pi^2 B' N_{\text{T}x} D_{nx}^{\frac{5+\nu_x\mu_x}{\mu_x}}}{36 \Gamma(\nu_x)} \left(\frac{\rho_{\text{L}}}{\rho}\right) \{\exp(A' \max[T_0 - T], 0.0) - 1\}. \tag{9.118}$$

Finally, the gamma distribution form is

$$Q_{\text{fw}}FZ_x = \frac{\Gamma(6 + \nu_x) \pi^2 B' N_{\text{T}x} D_{nx}^{5+\nu_x}}{36 \Gamma(\nu_x)} \left(\frac{\rho_{\text{L}}}{\rho}\right) \{\exp(A' \max[T_0 - T], 0.0) - 1\}. \tag{9.119}$$

In these equations, the subscript L denotes liquid cloud droplets, drizzle, or any of the possible raindrop categories that might be used.

9.9.1.2 Log-normal distribution for Bigg freezing

For the log-normal distribution, we first start out with the definition of the freezing equation for number concentration, as was carried out for the gamma function. The prognostic equation for N_{T} for Bigg freezing can be written as

$$N_{\text{T}x}FZ_y = \int_0^\infty m(D) B' \exp[A'(T_0 - T) - 1] n(D) \mathrm{d}D, \tag{9.120}$$

where A' and B' are constants and $m(D)$ can be expressed as,

$$m(D_x) = a_x D_x^{b_x}. \tag{9.121}$$

The log-normal distribution can be written,

$$n(D_x) = \frac{N_{\text{T}x}}{\sqrt{2\pi}\sigma_x D_x} \exp\left[-\frac{\ln(D_x/D_{nx})}{\sqrt{2}\sigma_x}\right]^2, \tag{9.122}$$

where σ controls the nature of the log-normal curve and is typically set to 0.5. Substituting (9.121) and (9.122) into (9.120) gives

$$N_{Tx}FZ_y = \frac{N_{Tx}a_xB'\exp[A'(T_0 - T) - 1]}{\sqrt{2\pi}\sigma_x}\int_0^\infty D^{b_x-1}\exp\left(-\frac{[\ln(D_x/D_{nx})]^2}{2\sigma_x^2}\right)dD_x. \quad (9.123)$$

Dividing all D_x terms by D_{nx} gives

$$N_{Tx}FZ_y = \frac{N_{Tx}a_xB'\exp[A'(T_0 - T) - 1]D_{nx}^{b_x}}{\sqrt{2\pi}\sigma_x} \times \int_0^\infty \left(\frac{D_x}{D_{nx}}\right)^{b_x-1}\exp\left(-\frac{[\ln(D_x/D_{nx})]^2}{2\sigma_x^2}\right)d\left(\frac{D_x}{D_{nx}}\right). \quad (9.124)$$

Now letting $u = D_x/D_{nx}$,

$$N_{Tx}FZ_y = \frac{N_{Tx}a_xB'\exp[A'(T_0 - T) - 1]D_{nx}^{b_x}}{\sqrt{2\pi}\sigma_x}\int_0^\infty u^{b_x-1}\exp\left(-\frac{[\ln u]^2}{2\sigma_x^2}\right)du. \quad (9.125)$$

By letting $y = \ln(u)$, $u = \exp(y)$, $du/u = dy$, so

$$\frac{dN_{Tx}}{dt} = \frac{N_{Tx}a_xB'\exp[A'(T_0 - T) - 1]D_{nx}^{b_x}}{\sqrt{2\pi}\sigma_x}\int_{-\infty}^\infty \exp(b_xy)\exp\left(-\frac{y^2}{2\sigma_x^2}\right)dy, \quad (9.126)$$

where the limits of the integral are that as u approaches zero from positive values, $\ln(u)$ approaches negative infinity, and as u approaches positive infinity, $\ln(u)$ approaches positive infinity.

Now the following integral definition is applied:

$$\int_{-\infty}^\infty \exp(2b'x)\exp(-a'x^2)dx = \sqrt{\frac{\pi}{a'}}\exp\left(\frac{b'^2}{a'}\right), \quad (9.127)$$

by allowing $y = x$, $a' = 1/(2\sigma_x^2)$, $b' = b_x/2$; therefore (9.126) becomes the prognostic equation for N_{Tx} for the freezing process,

$$N_{Tx}FZ_y = N_{Tx}a_xB'\exp[A'(T_0 - T) - 1]D_{nx}^{b_x}\exp\left(\frac{b_x^2\sigma_x^2}{2}\right). \quad (9.128)$$

The prognostic equation for mixing ratio for Bigg freezing can be written as

$$Q_xFZ_y = \frac{1}{\rho_0}\int_0^\infty B'\exp[A'(T_0 - T) - 1][m(D)]^2 n(D)dD \quad (9.129)$$

where A' and B' are constants and $m(D)$ is as in (9.121). The log-normal distribution is given in (9.122). Substituting (9.121) and (9.122) into (9.129) results in

$$Q_xFZ_y = \frac{N_{Tx}B'\exp[A'(T_0 - T) - 1]a_x^2}{\rho_0\sqrt{2\pi}\sigma_x}\int_0^\infty D^{2b_x-1}\exp\left(-\frac{[\ln(D_x/D_{nx})]^2}{2\sigma_x^2}\right)dD_x. \quad (9.130)$$

Dividing all D_x terms by D_{nx},

$$Q_xFZ_y = \frac{N_{Tx}B'\exp[A'(T_0 - T) - 1]a_x^2 D_{nx}^{2b_x}}{\rho_0\sqrt{2\pi}\sigma_x} \times \int_0^\infty \left(\frac{D_x}{D_{nx}}\right)^{2b_x-1}\exp\left(-\frac{[\ln(D_x/D_{nx})]^2}{2\sigma_x^2}\right)d\left(\frac{D_x}{D_{nx}}\right). \quad (9.131)$$

Now letting $u = D_x/D_{nx}$,

$$Q_xFZ_y = \frac{N_{Tx}B'\exp[A'(T_0 - T) - 1]a_x^2 D_{nx}^{2b_x}}{\rho_0\sqrt{2\pi}\sigma_x}\int_0^\infty u^{2b_x-1}\exp\left(-\frac{[\ln u]^2}{2\sigma_x^2}\right)du. \quad (9.132)$$

By letting $y = \ln(u)$, $u = \exp(y)$, $du/u = dy$, so

$$Q_xFZ_y = \frac{N_{Tx}B'\exp[A'(T_0 - T) - 1]a_x^2 D_{nx}^{2b_x}}{\rho_0\sqrt{2\pi}\sigma_x}\int_{-\infty}^\infty \exp(2b_x y)\exp\left(-\frac{y^2}{2\sigma_x^2}\right)dy, \quad (9.133)$$

where the limits of the integral change as u approaches zero from positive values, $\ln(u)$ approaches negative infinity. Likewise, the upper limit, as u approaches positive infinity, $\ln(u)$ approaches positive infinity.

Now the following integral definition (9.127) is applied, where $y = x$, $a' = 1/(2\sigma_x^2)$, $b' = b$; therefore (9.133) becomes the prognostic equation for Q for the Bigg freezing process,

$$Q_xFZ_y = \frac{1}{\rho_0}N_{Tx}B'\exp[A'(T_0 - T) - 1]a_x^2 D_{nx}^{2b_x}\exp(2b_x^2\sigma_x^2). \quad (9.134)$$

With Bigg freezing, Khain *et al.* (2000) showed that at –20 °C, 2 s, 2×10^3 s, and 2×10^6 s were required using the Bigg data to freeze drops with radii of 1000 μm, 100 μm, and 10 μm, respectively thus leaving many small unfrozen drops at –20 °C. At –30 °C, the times reduce to 2×10^{-3} s, 2 s, 2×10^3 s for drops 1000 μm, 100 μm, and 10 μm.

9.10 Immersion freezing

Another form of immersion freezing is that proposed by Vali (1975) that also is temperature dependent, and similar to Bigg freezing,

$$N_{\rm im} = N_{\rm im0}(0.1[T - 273.15])^{\gamma},$$

where $N_{\rm im0} = 1 \times 10^7$ m^{-3}, and $\gamma = 4.4$ for cumulus clouds. More recently, Vali (1994) came up with a time-dependent form of an equation for freezing rates. However it is severely limited by lack of actual knowledge of the content of the freezing nuclei. The Straka and Rasmussen (1997) method can provide the Lagrangian age or time with this scheme.

9.11 Two- and three-body conversions

A two-body interaction is, for example, as simple as hail collection of cloud water. A three-body interaction is a bit more complex. It involves (i) two bodies interacting and producing one hydrometeor or another type of hydrometeor altogether; or (ii) as described later, the production of more of the same collector body, plus another body. An example of the former type of three-body interaction is described as one hydrometeor, such as rain, collecting snow. If the raindrops are very small (the raindrop amount is very small in some models and does not exceed a threshold), then only rimed snow may be produced. However, if the rain is large (the raindrop amount exceeds some threshold) then perhaps graupel or a frozen drop might be formed.

Several methods have been developed for models to decide how much mass should transfer from one ice category to another ice category during riming or collection of rain. First Ferrier's (1994) model is examined. The larger ice habits that are allowed are snow, graupel, and frozen drops. Ferrier assumes that there is a transfer of mixing ratio and number concentration of a given hydrometeor based upon a riming age, nominally chosen as 120 s. To put that into perspective, in a thunderstorm with an average updraft of 30 to 40 m s^{-1}, the particle might rise 3600 to 4800 m! This certainly complicates the problem as a particle's temperature environment might change on the order of 20 °C or more, which is a huge change when considering rime density changes. Ferrier's model can permit the following changes of snow to snow (i.e. no change), snow to graupel, and snow to hail, all in one timestep, owing to riming. Ferrier (1994) argues that precipitation particles should be categorized by their new densities upon riming cloud drops for a period of time or collecting raindrops over a timestep. For example, snow aggregates collecting small raindrops might remain as snow aggregates, though they might become graupel or frozen drops if they collect more massive raindrops. Similarly he argues that graupel collecting small drops should be characterized as graupel, or classified as frozen drops if it collides with a similar-sized or larger raindrop, as follows

$$\frac{\pi}{6}\left(\rho_x D_x^3 + \rho_{\mathrm{rw}} D_{\mathrm{rw}}^3\right) = \frac{\pi}{6}\rho_y D_x^3. \tag{9.135}$$

In Ferrier's (1994) parameterization, liquid collected is evenly distributed throughout an ice particle before freezing to ice. First, equating masses of the ice particle is done where ρ_x is the density of the original ice, ρ_{L} is the density of liquid and ρ_y is the new density of the mixture. This is categorized as snow if $\rho_y < 0.5(\rho_{\mathrm{sw}} + \rho_{\mathrm{gw}})$, graupel if $0.5(\rho_{\mathrm{sw}} + \rho_{\mathrm{gw}}) < \rho_y < 0.5(\rho_{\mathrm{gw}} + \rho_{\mathrm{fw}})$, or frozen drops if $\rho_y > 0.5(\rho_{\mathrm{gw}} + \rho_{\mathrm{fw}})$. Now these values are substituted into (9.135) to give a range from $D_1 < D_{\mathrm{rw}} < D_2$ as a function of the colliding particle D_x.

Ferrier (1994) has perhaps one of the more elaborate schemes to handle riming of snow, graupel, and frozen drops for particles to either lose or retain their species. For example, the following are possible: for snow to remain as snow or become graupel or frozen rain; graupel to remain as graupel or become frozen rain; and frozen rain to become graupel or remain as frozen rain. A sufficient amount of riming must occur at a different density than the density of the riming particle for a species conversion to occur. There are two diameters involved: D_{1xy} is the minimum size of the converted species; and D_{2xy} is the size at which the particle mass doubles in the time interval $\Delta t_{\mathrm{rime}} = 60$ to 180 s (120 s in Ferrier) such that only smaller particles have their densities sufficiently altered. For rime collected on particles smaller than D_{1xy} and larger than D_{2xy} the particles do not change into other species. However, for $D_{1xy} < D < D_{2xy}$, riming is a means by which the species in question becomes a different species. Partial gamma functions need to be used to define $Q_{y\mathrm{cw}}\,AC_x$ and $N_{y\mathrm{cw}}\,AC_x$ along with limits defined as the diameters D_{1xy} and D_{2xy}. Now the key is to obtain the two diameters D_{1xy} and D_{2xy}. The diameter of D_{1xy} needs to be solved by numerical iteration, and lookup tables are made for efficiency. The minimum diameter threshold is 0.0005 m, which is a function of mean cloud droplet diameter, cloud temperature, and height. The particle diameter D_{2xy} is solved in a different manner than that for D_{1xy}. (Remembering that this diameter is the size that the mass doubles in time Δt_{rime} and that is by using a simplified continuous collection equation). Rime densities come from Heymsfield and Pflaum (1985) and impact velocities from Rasmussen and Heymsfield (1985). The continuous collection equation is fairly straightforward, starting with,

$$\frac{\mathrm{d}m_x}{\mathrm{d}t} = 0.25\pi\rho_0 Q_{\mathrm{cw}} a_x \left(\frac{\rho_0}{\rho}\right)^{1/2} D_x^{2+b_x}. \tag{9.136}$$

Now making the assumption that the density of the rime that has been accreted is that of the converted particle with density of ρ_x,

$$\frac{dm_x}{dt} = \frac{d\left(\pi/6\rho_y D_x^3\right)}{dt}, \tag{9.137}$$

where ρ_y is the mixture of density of rime ice and the density of the particle experiencing riming.

Combining (9.136) and (9.137), integrating over the rime time interval Δt_{rime}, and specifying the final size of the particle by D_f, gives

$$D_{\text{f}xy}^{1-b_x} = D_{2xy}^{1-b_x} = \frac{(1-b_x)E_{x\text{cw}}\rho Q_{\text{cw}} a_x \left(\frac{\rho_0}{\rho}\right)^{1/2} \Delta t_{\text{rime}}}{\rho_y}, \tag{9.138}$$

where $E_{x\text{cw}}$ is the collection efficiency of x accreting cloud water.

If the mass has doubled in time Δt_{rime}, then

$$\frac{\pi}{6}\rho_y\left(D_{\text{f}xy}^3 - D_{2xy}^3\right) = \frac{\pi}{6}\rho_x D_{2xy}^3. \tag{9.139}$$

The equation for D_{2xy} can be solved by combining the equations above, which gives an equation as a function of water content and height (the density factor multiplied by the terminal velocity),

$$D_{2xy} = \left(\tau_{xy}\Delta t_{\text{rime}}\rho Q_{\text{cw}}\left(\frac{\rho_0}{\rho}\right)^{1/2}\right)^{1/(1-b_x)}, \tag{9.140}$$

where the variable τ_{xy} is given by the constant

$$\tau_{xy} = \frac{(1-b_x)E_{x\text{cw}}a_x}{2\rho_y}\left[\left(1+\rho_x/\rho_y\right)^{(1-b_x)/3} - 1\right]^{-1}. \tag{9.141}$$

By this method, riming conversion only occurs when $D_{1xy} < D_{2xy}$ and $D_{2xy} >$ 0.0002 m.

Milbrandt and Yau (2005b) simplified this parameterization somewhat for a faster, more simple model. They decide that raindrops freeze when they come into contact with ice particles. But first they assume, like Ferrier (1994), that during particle contact the liquid is uniformly dispersed throughout the particle and increases in its mass, but does not change its bulk volume. Equation (9.142) is used and the destinations are as given above, except that mass-weighted mean diameters are used,

$$\frac{\pi}{6}\left(\rho_x D_{mx}^3 + \rho_{\text{rw}} D_{mrw}^3\right) = \frac{\pi}{6}\rho_z D_{mz}^3, \tag{9.142}$$

and $D_{mz} = \max(D_{mx}, D_{mr})$. The destination category for number concentration of the new particle is a mass-weighted sum given by

$$N_{zx}CL_{rw} = \frac{\rho\delta_{zxrw}(Q_x AC_{rw} + Q_{rw}AC_x)}{\left(\frac{\pi}{6}\right)\rho_z \max(D_{mrw}, D_{mx})^3}, \tag{9.143}$$

where ρ_z is the density of the actual density of the species z, not the density computed from (9.142) above.

This methodology is used in our model for transfers among snow, low-density, medium-density and high-density graupel, and frozen drops, which all can collect rain water of various types, and drizzle. For larger ice particles collecting cloud water a riming time Δt_{rime} is chosen (60–120 s) and a rime amount and rime density are computed using (2.229) and (9.76) or prognosed from (2.231).

The conversion from one hydrometeor species to two different hydrometeor species is also called a three-body interaction or three-component interaction, and was re-examined by Farley *et al.* (1989). Their approach takes into account the amount lost of the particle collected Q_y that is gained by the two other different bodies Q_x and Q_z. Farley *et al.* (1989) consider the transfer from snow to graupel/hail by riming of snow, which is a bit unrealistic when it is considered that the conversion size threshold is 7 mm. This is larger than almost all graupel particles and is the size of embryonic hail particles (Pruppacher and Klett 1997). It is not clear that snow actually rimes to become hail without becoming graupel first. In the example herein, let us consider rain collected by graupel, and conversion of the graupel to hail, with a conversion size threshold of 5 mm.

The collection equation for Q and N of the collector (graupel) collecting the collectee (rain) is given by a partial gamma distribution (with $c = 1$) for the graupel part of the double integral,

$$Q_{zx}AC_y = \frac{0.25\pi E_{xy} N_{Tx} Q_y \Delta\bar{V}_{TxyQ}\left(\frac{\rho_0}{\rho}\right)^{1/2}}{\Gamma(\nu_x)\Gamma(B_y + \nu_y)} \times \begin{bmatrix} \Gamma\left(2 + \nu_x, \frac{Hdia_{\text{th}}}{D_{nx}}\right)\Gamma(3 + \nu_y)D_{nx}^2 \\ +2\Gamma\left(1 + \nu_x, \frac{Hdia_{\text{th}}}{D_{nx}}\right)\Gamma(4 + \nu_y)D_{nx}D_{ny} \\ +\ \Gamma\left(0 + \nu_x, \frac{Hdia_{\text{th}}}{D_{nx}}\right)\Gamma(5 + \nu_y)D_{ny}^2 \end{bmatrix}, \tag{9.144}$$

and

$$N_{zx}AC_y = \frac{0.25\pi E_{xy} N_{\mathrm{T}x} N_{\mathrm{T}y} \Delta\bar{V}_{\mathrm{T}xyN} \left(\frac{\rho_0}{\rho}\right)^{1/2}}{\Gamma(\nu_x)\Gamma(\nu_y)} \times \begin{bmatrix} \Gamma\left(2+\nu_x, \frac{Hdia_{\mathrm{th}}}{D_{nx}}\right)\Gamma(0+\nu_y)D_{nx}^2 \\ +2\Gamma\left(1+\nu_x, \frac{Hdia_{\mathrm{th}}}{D_{nx}}\right)\Gamma(1+\nu_y)D_{nx}D_{ny} \\ +\ \Gamma\left(0+\nu_x, \frac{Hdia_{\mathrm{th}}}{D_{nx}}\right)\Gamma(2+\nu_y)D_{ny}^2 \end{bmatrix}, \tag{9.145}$$

where $Hdia_{\mathrm{th}}$ is the cut-off threshold, z is hail, x is graupel and y is rain. The ratio of $Hdia_{\mathrm{th}}$ to D_{nx} can be stored in a lookup table for the partial gamma function, or it can be computed, which is expensive. Alternatively, a numerical approach using bins and integrating using Simpson's one-third rule as suggested by Farley *et al.* (1989) could be incorporated. For example, using the gamma distribution with μ and $\alpha = 1$,

$$Q_{zx}AC_y = \frac{0.25\pi E_{xy} N_{\mathrm{T}x} Q_y \Delta\bar{V}_{\mathrm{T}xyQ} \left(\frac{\rho_0}{\rho}\right)^{1/2}}{\Gamma(\nu_x)\Gamma(B_y+\nu_y)} \times \begin{bmatrix} \left\{\sum_{D_{\mathrm{hail},0}}^{D_\infty} \left(D_x^{2+\nu x}\exp[-D_x/D_{nx}]\Delta D_x\right)\right\}\Gamma(3+\nu_y) \\ +2\left\{\sum_{D_{\mathrm{hail},0}}^{D_\infty} \left(D_x^{1+\nu x}\exp[-D_x/D_{nx}]\Delta D_x\right)\right\}\Gamma(4+\nu_y)D_{ny} \\ +\left\{\sum_{D_{\mathrm{hail},0}}^{D_\infty} \left(D_x^{\nu x}\exp[-D_x/D_{nx}]\Delta D_x\right)\right\}\Gamma(5+\nu_y)D_{ny}^2 \end{bmatrix}. \tag{9.146}$$

These equations are not activated unless graupel and frozen drops grow for two minutes by riming, starting from the mass-weighted mean diameter to a diameter of 5 mm via the continuous growth collection equation for a gamma distribution.

An identical set of procedures is used to grow hail from graupel and hail from frozen drops into large hail ($D > 20$ mm). The equations above (9.145) and (9.146) are not activated unless hail from graupel or hail from frozen drops grows for 2 min by riming, starting from the mass-weighted mean diameter to a diameter of 20 mm via the continuous growth collection equation for a gamma distribution,

$$N_{zx}AC_y = \frac{0.25\pi E_{xy} N_{Tx} N_{Ty} \Delta \bar{V}_{TxyN} \left(\frac{\rho_0}{\rho}\right)^{1/2}}{\Gamma(\nu_x)\Gamma(\nu_y)} \times \left[\begin{array}{l} \left\{ \sum_{D_{\text{hail},0}}^{D_\infty} \left(D_x^{2+\nu x} \exp\left[-D_x/D_{nx}\right] \Delta D_x \right) \right\} \Gamma(0+\nu_y) \\ +2 \left\{ \sum_{D_{\text{hail},0}}^{D_\infty} \left(D_x^{1+\nu x} \exp\left[-D_x/D_{nx}\right] \Delta D_x \right) \right\} \Gamma(1+\nu_y) D_{ny} \\ + \left\{ \sum_{D_{\text{hail},0}}^{D_\infty} \left(D_x^{\nu x} \exp\left[-D_x/D_{nx}\right] \Delta D_x \right) \right\} \Gamma(2+\nu_y) D_{ny}^2 \end{array} \right]. \quad (9.147)$$

The amount of the rain that stays with graupel instead of being transferred to hail is

$$Q_{\text{gwgw}}AC_{\text{rw}} = Q_{\text{gw}}AC_{\text{rw}} - Q_{\text{hwgw}}AC_{\text{rw}}. \quad (9.148)$$

Next the question is how much of the graupel stays with graupel and how much graupel is transferred to hail in this three-body interaction. First, take the definition of the mixing ratio of graupel, using integration by parts to find the solution as in Farley *et al.* (1989), and divide the timestep to get a rate equation,

$$Q_{\text{rw}}AC_{\text{gw}} = \frac{Q_{\text{gw}}}{\Delta t} = \int_{D_{\text{hail},0}}^{D_\infty} \frac{N_{\text{Tgw}}}{\Delta t \Gamma(\nu_x)} \frac{\pi}{6} \frac{\rho_{\text{sw}}}{\rho} \left(\frac{D}{D_{nx}}\right)^{3+\nu_x} \exp\left(-\frac{D}{D_{nx}}\right) d\frac{D}{D_{nx}} \quad (9.149)$$

or simply employ the use of a partial gamma function for the integration solution,

$$Q_{\text{rw}}AC_{\text{gw}} = \frac{Q_{\text{gw}}}{\Delta t} = \frac{N_{\text{Tgw}}}{\Delta t \Gamma(\nu_x)} \frac{\pi}{6} \frac{\rho_{\text{sw}}}{\rho} D_{ng}^{3+\nu_x} \Gamma\left(3 + \nu_x, \frac{Hdia_{\text{th}}}{D_{n\text{gw}}}\right). \quad (9.150)$$

The same procedure is done for number concentration,

$$N_{\text{rw}}AC_{\text{gw}} = \frac{N_{\text{Tgw}}}{\Delta t} = \frac{N_{\text{Tgw}}}{\Delta t \Gamma(\nu_x)} \frac{\pi}{6} \frac{\rho_{\text{sw}}}{\rho} \Gamma\left(\nu_x, \frac{Hdia_{\text{th}}}{D_{n\text{gw}}}\right). \quad (9.151)$$

Farley *et al.* (1989) require the collector mixing ratio (snow) to be $> 1 \times 10^{-4}$ kg kg^{-1} and the collectee mixing ratio threshold (cloud water) to be $> 1 \times 10^{-3}$ kg kg^{-1}. For a three-body interaction between graupel or frozen drops and raindrops, perhaps similar thresholds might be used. Alternatively, rain thresholds reduced to a range between 1×10^{-4} kg kg^{-1} and 5×10^{-4} kg kg^{-1} might be more appropriate as there is very little rain above the $-10\,^{\circ}$C to $-15\,^{\circ}$C levels.

9.12 Graupel density parameterizations and density prediction

Very little is actually known about graupel growth as it is not readily amenable to studies by remote-sensing polarimetric radar, and as aircraft can only make limited observations. Although these observations are useful, wind tunnels in laboratories remain the best observational tool (Heymsfield 1978). For graupel, and probably equally valid for frozen drops, Pflaum and Pruppacher (1979) suggest the following empirical equation for graupel-rime density, which is somewhat larger than expected when compared to Macklin and Bailey's (1962) equation,

$$\rho_{\mathrm{rime}} = 261\left(\frac{-rV_{\mathrm{T,impact}}}{T_{\mathrm{s}}}\right)^{0.38}. \tag{9.152}$$

In this equation r is the mean volume cloud droplet radius in microns, $V_{\mathrm{T,impact}}$ is the impact velocity in m s^{-1}, and T_{s} is surface temperature of the hailstone. At this time, the best approximation for impact velocity for bulk microphysical models is represented by 0.6 times the terminal velocity.

A set of prognostic equations can be developed for variable density growth for graupel and frozen drops. The general form of the equations is the same as that in Chapter 13 for hail-rime density and variable density growth of hail and the procedure is to use appropriate densities and mass weight sources/sinks with mixing ratio tendencies times time,

$$\rho_{\mathrm{gw}}^{n+1} = \frac{\rho_{\mathrm{gw}}^{n} Q_{\mathrm{gw}}^{n} + \Delta t \rho_{\mathrm{gw,conv}}^{n} \left.\frac{\mathrm{d}Q_{\mathrm{gw}}}{\mathrm{d}t}\right|_{\mathrm{conv}}^{n} \Delta t \rho_{\mathrm{gw,rime}}^{n} \left.\frac{\mathrm{d}Q_{\mathrm{gw}}}{\mathrm{d}t}\right|_{\mathrm{rime}}^{n} + \Delta t \rho_{900} \left.\frac{\mathrm{d}Q_{\mathrm{gw}}}{\mathrm{d}t}\right|_{\mathrm{rain}}^{n} + \Delta t \rho_{\mathrm{gw}} \left.\frac{\mathrm{d}Q_{\mathrm{gw}}}{\mathrm{d}t}\right|_{\mathrm{sub/dep}}^{n}}{Q_{\mathrm{gw}}^{n} + \Delta t \left.\frac{\mathrm{d}Q_{\mathrm{gw}}}{\mathrm{d}t}\right|_{\mathrm{conv}}^{n} + \Delta t \left.\frac{\mathrm{d}Q_{\mathrm{gw}}}{\mathrm{d}t}\right|_{\mathrm{rime}}^{n} \Delta t \left.\frac{\mathrm{d}Q_{\mathrm{gw}}}{\mathrm{d}t}\right|_{\mathrm{rain}}^{n} + \Delta t \left.\frac{\mathrm{d}Q_{\mathrm{gw}}}{\mathrm{d}t}\right|_{\mathrm{sub/dep}}^{n}}, \tag{9.153}$$

where n and $n + 1$ are time level t and $t + \mathrm{d}t$, and the subscript gw refers to graupel of any density. This equation can be expanded to accommodate the source and sink terms in the mass or mixing ratio budget equations. The new density can be used to determine if the particle should be transferred from a low-density particle to a high-density particle in a bulk microphysical model, or just tracked in a bin model to compute terminal velocity, diameters, etc., for a given mass representing a bin. Notice in (9.153) above that the density of the budget related to accreting rain is that of frozen ice, i.e. approximately 900 kg m^{-3}, and that for sublimation and deposition the particle density does not change.

Some models have exchanges across lines of density for different hydrometeors. For example, Schoenberg-Ferrier's model uses very low-density snow, medium-density graupel, and high-density frozen drops. Similarly, Milbrandt and Yau (2005b) use very low-density snow, medium-density graupel, and high-density hail. In another case, Straka and Mansell (2005) and Straka *et al.* (2009b) employ very low-density snow, low-density graupel, medium-density graupel, high-density graupel, and frozen drops. In addition, Straka *et al.* (2007) include lower-density hail from graupel origins and higher-density hail from frozen-drop origins, slushy hail, and large hail.

9.13 Density changes in graupel and frozen drops collecting cloud water

For snow aggregates, graupel, and frozen drops collecting cloud water a simplified approach somewhat like Schoenberg-Ferrier (1994) is used. First the amount of riming that takes place is computed as Q_xAC_{cw}. Next, the rime density of cloud water collected is computed following Heymsfield and Pflaum (1985) as

$$\rho_{x,\mathrm{rime}} = 300\left(-\frac{0.5D_{\mathrm{cw}}V_{\mathrm{T,impact}}}{T(^{\circ}\mathrm{C})}\right)^{0.44}, \tag{9.154}$$

where the value of $V_{\mathrm{T,impact}}$ is the impact velocity of cloud drops on the ice and is given as approximately 0.6 $V_{\mathrm{T},x}$ in Pflaum and Pruppacher (1980). Here D_{cw} is in microns. The new density of the ice particle undergoing collection of cloud water after a 90 s period (Δt_{rime}) is given by

$$\rho_{x,\mathrm{new}} = \left(\frac{Q_x\rho_x + \Delta t_{\mathrm{rime}}Q_xAC_{\mathrm{cw}}}{Q_x + \Delta t_{\mathrm{rime}}Q_xAC_{\mathrm{cw}}}\right). \tag{9.155}$$

Now consider medium-density graupel collecting cloud water. Its source can collect low-density rime to become low-density graupel if $\rho_{\mathrm{gm,rime}} < 0.5(\rho_{\mathrm{gl}} + \rho_{\mathrm{gm}})$; remain added to medium-density graupel if $0.5(\rho_{\mathrm{gl}} + \rho_{\mathrm{gm}}) < \rho_{\mathrm{gm,rime}} < 0.5(\rho_{\mathrm{gm}} + \rho_{\mathrm{gh}})$; can be added to high-density graupel if $0.5(\rho_{\mathrm{gm}} + \rho_{\mathrm{gh}}) < \rho_{\mathrm{gm,rime}} < 0.5(\rho_{\mathrm{gh}} + \rho_{\mathrm{fw}})$; or added to frozen drops if $0.5(\rho_{\mathrm{gh}} + \rho_{\mathrm{fw}}) < \rho_{\mathrm{gm,rime}}$. This can be done for all three graupel species and frozen drops. Any species can be converted to one of the other by either low- or high-density riming.

9.14 Density changes in graupel and frozen drops collecting drizzle or rain water

Now consider medium-density graupel collecting drizzle or rain water. Its new density can be determined following Milbrandt and Yau (2005b) as,

$$\rho_{x,\text{new}} = \left(\frac{Q_x \rho_x + \Delta t Q_x A C_z}{Q_x + \Delta t Q_x A C_z} \right). \tag{9.156}$$

As an example a particle source can collect rain and stay as medium-density graupel if $\rho_{\text{gm,rime}} < 0.5(\rho_{\text{gl}} + \rho_{\text{gm}})$; remain as medium-density graupel if $0.5(\rho_{\text{gl}} + \rho_{\text{gm}}) < \rho_{\text{gm,rime}} < 0.5(\rho_{\text{gm}} + \rho_{\text{gh}})$; remain as high-density graupel if $0.5(\rho_{\text{gm}} + \rho_{\text{gh}}) < \rho_{\text{gm,rime}} < 0.5(\rho_{\text{gh}} + \rho_{\text{fw}})$; and frozen drops if $0.5(\rho_{\text{gh}} + \rho_{\text{fw}}) < \rho_{\text{gm,rime}}$. This can be done for all three graupel species and frozen drops. Almost any can be converted to the other by either sufficient low- or high-density riming. However, in collecting drizzle and rain, the collected water probably freezes solid and most changes are to high densities.

9.15 More recent approaches to conversion of ice

One of the newest approaches to conversion of ice from one species to another is one developed by Morrison and Grabowski (in press), where the amount of rime collected, and the amount of vapor deposited are predicted. The benefit of this new method is that it does not need any thresholds per se, like many older methods previously discussed. Rather it allows particles to be exchanged with a rather smooth transition over a wide range of rimed fractions collected. It can be applied to bin or bulk parameterization models. For a bulk model the equations predicted are for N_{T}, Q_{rime}, and Q_{dep},

$$\frac{\partial N_{\text{T}m}}{\partial t} = -\frac{\partial u_i N_{\text{T}m}}{\partial x_i} + N_{\text{T},m}\frac{\partial u_i}{\partial x_i} + \frac{\partial}{\partial x_i}\left(K_{\text{h}}\frac{\partial N_{\text{T}m}}{\partial x_i}\right) + \frac{\partial(\bar{V}_{\text{T}_N} N_{\text{T}m})}{\partial x_3} + SN_m, \tag{9.157}$$

where SN_m is given by

$$\begin{aligned} SN_m = {} & \left(\frac{\partial N_{\text{T}m}}{\partial t}\right)_{\text{nuc}} + \left(\frac{\partial N_{\text{T}m}}{\partial t}\right)_{\text{sub}} + \left(\frac{\partial N_{\text{T}m}}{\partial t}\right)_{\text{frz}} + \left(\frac{\partial N_{\text{T}m}}{\partial t}\right)_{\text{col}} \\ & + \left(\frac{\partial N_{\text{T}m}}{\partial t}\right)_{\text{mlt}} + \left(\frac{\partial N_{\text{T}m}}{\partial t}\right)_{\text{mltc}} + \left(\frac{\partial N_{\text{T}m}}{\partial t}\right)_{\text{mult}} \end{aligned} \tag{9.158}$$

and

$$\begin{aligned} \frac{\partial Q_{\text{rime},m}}{\partial t} = {} & -\frac{1}{\rho}\frac{\partial \rho u_i Q_{\text{rime},m}}{\partial x_i} + \frac{Q_{\text{rime},m}}{\rho}\frac{\partial \rho u_i}{\partial x_i} + \frac{\partial}{\partial x_i}\left(\rho K_{\text{h}}\frac{\partial Q_{\text{rime},m}}{\partial x_i}\right), \\ & + \frac{1}{\rho}\frac{\partial(\rho \bar{V}_{\text{T}Q} Q_{\text{rime},m})}{\partial x_3} + SQ_{\text{rime},m}, \end{aligned} \tag{9.159}$$

where $SQ_{\mathrm{rime},m}$ is given by

$$SQ_{\mathrm{rime},m} = \left(\frac{\partial Q_{\mathrm{rime},m}}{\partial t}\right)_{\mathrm{frz}} + \left(\frac{\partial Q_{\mathrm{rime},m}}{\partial t}\right)_{\mathrm{mlt}} + \left(\frac{\partial Q_{\mathrm{rime},m}}{\partial t}\right)_{\mathrm{sub}} + \left(\frac{\partial Q_{\mathrm{rime},m}}{\partial t}\right)_{\mathrm{colc}} + \left(\frac{\partial Q_{\mathrm{rime},m}}{\partial t}\right)_{\mathrm{colr}} + \left(\frac{\partial Q_{\mathrm{rime},m}}{\partial t}\right)_{\mathrm{mltc}} \tag{9.160}$$

and

$$\frac{\partial Q_{\mathrm{dep},m}}{\partial t} = -\frac{1}{\rho}\frac{\partial \rho u_i Q_{\mathrm{dep},m}}{\partial x_i} + \frac{Q_{\mathrm{dep},m}}{\rho}\frac{\partial \rho u_i}{\partial x_i} + \frac{\partial}{\partial x_i}\left(\rho K_{\mathrm{h}}\frac{\partial Q_{\mathrm{dep},m}}{\partial x_i}\right) + \frac{1}{\rho}\frac{\partial\left(\rho \bar{V}_{\mathrm{T}Q} Q_{\mathrm{dep},m}\right)}{\partial x_3} + SQ_{\mathrm{dep},m}, \tag{9.161}$$

where $SQ_{\mathrm{dep},m}$ is given by

$$SQ_{\mathrm{dep},m} = \left(\frac{\partial Q_{\mathrm{dep},m}}{\partial t}\right)_{\mathrm{nuc}} + \left(\frac{\partial Q_{\mathrm{dep},m}}{\partial t}\right)_{\mathrm{dep}} + \left(\frac{\partial Q_{\mathrm{dep},m}}{\partial t}\right)_{\mathrm{sub}} + \left(\frac{\partial Q_{\mathrm{dep},m}}{\partial t}\right)_{\mathrm{frz}} + \left(\frac{\partial Q_{\mathrm{dep},m}}{\partial t}\right)_{\mathrm{mlt}} + \left(\frac{\partial Q_{\mathrm{dep},m}}{\partial t}\right)_{\mathrm{mltc}}. \tag{9.162}$$

The subscripts are defined as follows: nuc is nucleation, dep is deposition, sub is sublimation, frz is drop freezing, mlt is meltwater, mltc is meltwater that is recaptured, mult is ice multiplication, and col is accretion or collection of cloud (colc) or rain (colr).

In addition to these approaches, prediction of rime density and particle density discussed earlier can be used to make transformations smooth. These are expensive to use for numerical weather prediction but soon the cost may become minimized by petaflop machines in the near future.

Straka *et al.* (2009b) took a slightly different approach and predicted just the rime collected along with the rime's density, but also the total mixing ratio of the particle, number concentration, and, if desired, reflectivity. This too allowed for smoother transitions from one graupel type to another of snow to graupel. Ice crystals were not assumed to grow into graupel in their model unless particles had a mean volume diameter greater than 500 mm for planar-type crystals and 50 mm for columnar-type crystals. These conditions are not often met except perhaps in the upper reaches of intense convection.

10

Hail growth

10.1 Introduction

The formation of hailstones is an intriguing aspect of precipitation development studies owing to the unique cloud systems in which hailstones form. As equal amounts are added of hailstones, for a given density the shells of equal mass will be of different thickness (Fig. 10.1). In particular, the formation of very large hailstones, some of which are greater than 51 mm in diameter, is of great interest. Models of hail growth can be very simple or imply very detailed processes (Takahashi 1976 and Fig. 10.2).

Hailstones are typically defined as solid or nearly solid ice particles that are greater than 5 mm in diameter. The National Weather Service in the United States classifies hailstones as constituting severe hail if they are larger than 19 mm (3/4″) in diameter, and constituting very severe hail if the hailstones are larger than 51 mm (2″). These larger hailstones develop most frequently from rapid riming of higher-density graupel particles and/or large frozen drops. Studies suggest that high-plains storms produce most of their hail from graupel, and Southern Plains storms produce most from frozen drops (Fig. 10.3). It is not known exactly why this is so, but some have speculated that high-plains storms do not produce large water drops above the freezing level because they have cold cloud bases cooler than 5 °C (278.15 K) (Fig. 10.3) and the 500 mb temperature in the updraft core is perhaps 260.15 to 263.15 K. Thus, many of the hailstone embryos are perhaps formed from graupel. On the other hand, Southern Plains storms produce many large drops owing to an active collision and coalescence regime in the storm in the warm, cloud-water laden updraft, where cloud bases are typically warmer than 15 to 20 °C (288.15 to 293.15 K) and temperatures at 500 mb in the updraft core may be close to about 270.15 to 273.15 K.

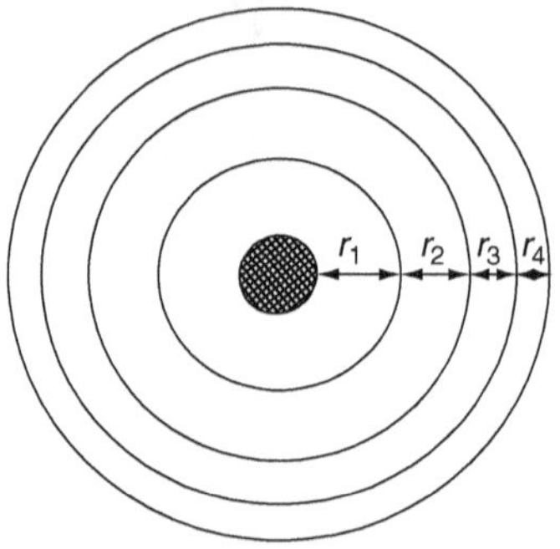

Fig. 10.1. Concentric layers of equal mass (m) illustrating thinning of layer thickness (r is radius). (Pflaum 1980; courtesy of the American Meteorological Society.)

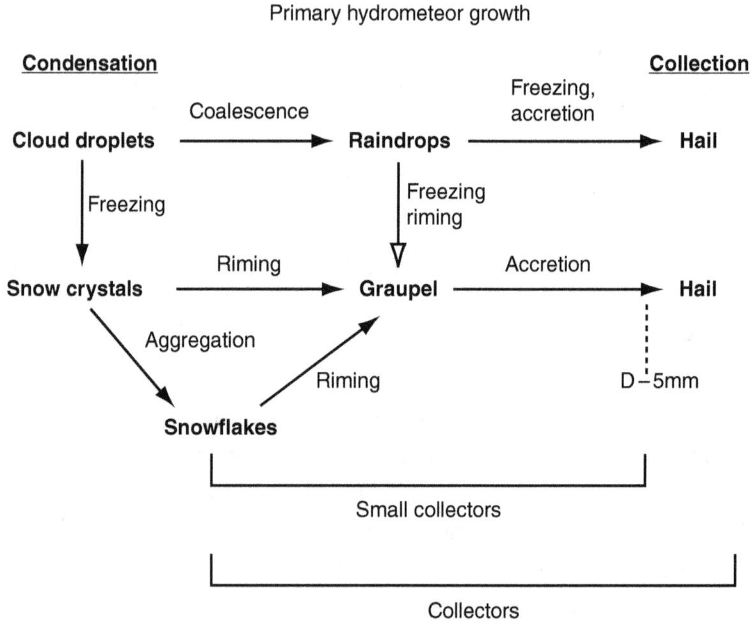

Fig. 10.2. Flow-chart showing hailstones growing from freezing of raindrops and from production graupel. (Knight and Knight 2001; courtesy of the American Meteorological Society.)

The growth of a hailstone can be thought of in very simple terms with a simple continuous equation to gauge the hailstone growth by riming. Hail growth models also can be very detailed considering low-density growth (Fig. 10.4) leading to smaller terminal velocity for the same mass. In addition, drag coefficient values can play a role in hail growth (Fig. 10.5).

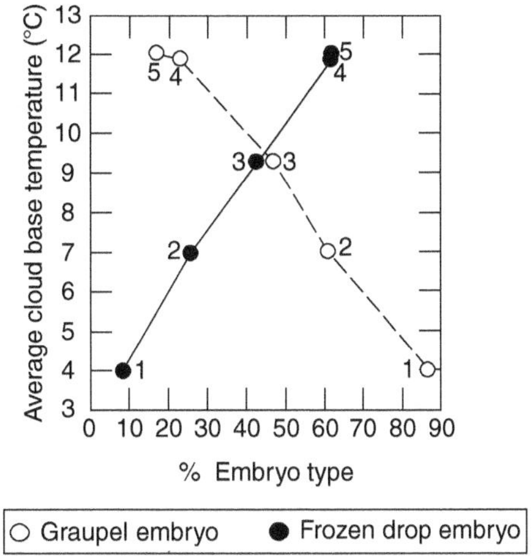

Fig. 10.3. Percentage of graupel and frozen drop embryos as a function of average cloud base temperature, as measured in the following locations: 1, the National Hail Research Experiment, from representative surroundings; 2, Alberta; 3, Switzerland; 4, South Africa; 5, Oklahoma. (From Knight and Knight 1981; courtesy of the American Meteorological Society.)

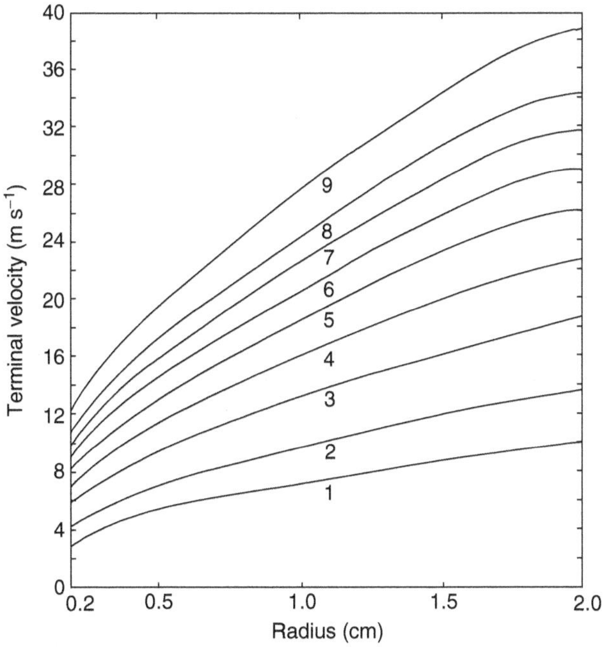

Fig. 10.4. Terminal velocities of various sizes of spherically smooth graupel and hailstones as a function of their densities. An ambient temperature of 253.15 K and an ambient pressure of 500 mb have been used in the calculations. Particle densities (units of kg m^{-3}): (1) 50; (2) 100; (3) 200; (4) 300; (5) 400; (6) 500; (7) 600; (8) 700; (9) 900. (From Pflaum 1980; courtesy of the American Meteorological Society.)

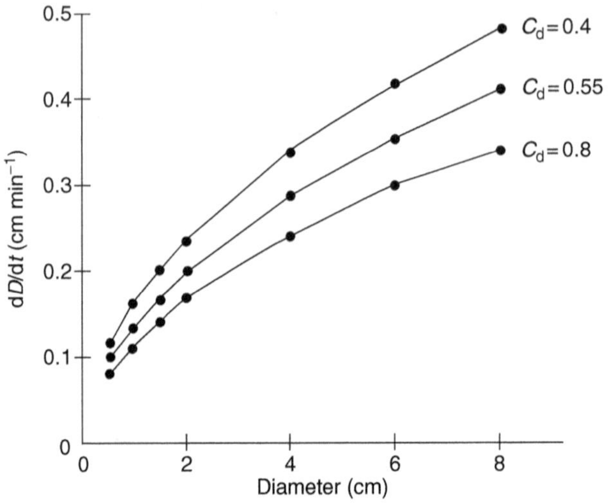

Fig. 10.5. Growth of hailstones from 5 mm to 8 cm assuming different terminal velocity drag coefficients (C_d). (From Knight and Knight 1981; courtesy of the American Meteorological Society.)

The hailstone definition may be related to the smallest size an ice particle can grow by wet growth, which means not all of the collected liquid freezes immediately upon contact. This results in hailstones almost always being warmer than the environmental temperature where the hail is found at temperatures below freezing. To find the growth rate of a hailstone, along with its temperature and fraction of liquid that freezes, a heat budget and temperature equation must be solved. When all collected liquid water freezes instantaneously upon contact the growth is said to be dry growth. These and other examples are covered in detail below. A special example of wet growth is in a mixed-phase cloud, where cloud water and snow crystals coexist. In this case, snow particles that would normally bounce off a hailstone because they are dry would stick upon collision during wet growth owing to the liquid-water presence on the hailstone's surface.

As mentioned above another interesting aspect about hail growth that is covered in this chapter is low-density riming. If a hailstone's rime is small compared to solid ice water, then its size will increase with a smaller increase in terminal velocity, which would permit weaker updraft requirements to keep a hailstone lofted. This idea has yet to have been proven well enough to be accepted, though it is logically and pragmatically based.

Nelson (1987) wrote that hailstorms that produce many hailstones differ kinematically and microphysically from hailstorms that produce fewer very large hailstones. The microphysics and kinematic aspects of these different

Table 10.1. *Microphysics and kinematic aspects of different hailstorms Nelson (1987)*

Large hail		**Large amounts of hail**	
Microphysics	Kinematics	Microphysics	Kinematics
Large values of supercooled liquid water	Light horizontal flow across the updraft	High embryo concentration	Large contiguous updraft area with $w = 20$ to 40 m s^{-1} (a, c, d, f, h)
Wet growth in mixed phase region (b, f)	Large contiguous updraft area with $w = 20$ to 40 m s^{-1} (c, d, f, h)	Ample supercooled liquid water	Flow field that injects embryos across a broad updraft front (f)
Low-density growth (c, g)	Optimal trajectories (including embryo trajectories; a, c, d, e, f, h)		
Large embryos (c, d, f)	Favorable updraft gradients (f)		

References:
(a) Danielson *et al.* (1972); (b) Dennis and Musil (1973); (c) English (1973); (d) Foote (1984); (e) Heymsfield *et al.* (1980); (f) Nelson (1983); (g) Pflaum (1980); (h) Ziegler *et al.* (1983).

systems are summarized in Table 10.1 with references for each of the findings where appropriate. Notice that both storms that produce large hail and large amounts of hail require strong updrafts of 20 to 40 m s^{-1} and optimal trajectories (though different types of trajectories from each other). It may be safe to state that at this time we do not know if microphysical variability plays a significant role in modeling microphysics of hail growth. However, there seems to be a growing consensus that just the right trajectories need to be followed for very large hail to form. Bulk microphysical models can even make distinctions between non-severe, severe, and giant hailstones with appropriate number concentrations in simulations (e.g. Straka *et al.* 2009b).

10.2 Wet and spongy hail growth

Wet growth of a hailstone occurs when the accreted liquid cannot all freeze immediately, that is, part of it remains unfrozen. There are two modes of wet growth. Following Rasmussen and Heymsfield (1987a), there is high-density wet growth, as well as spongy wet growth, both of which are defined below. The modes are relevant to modeling hail for detailed models of hail growth and radar calculations as well as some of the newer multi-class bulk parameterizations.

With high-density wet growth, all frozen water has a density of 910 kg m^{-3}. The unfrozen portion, or liquid portion, may remain on the surface much as it does during melting of solid ice. At diameters of less than 20 mm the same mixed-phase morphology and shedding physics occurs as with melting ice.

Spongy wet growth occurs when the liquid water collected in part freezes and in part remains liquid. The factors that influence spongy wet growth and inherent shedding include icing conditions, rotation rates, and nutation precession rate for a sphere (Pruppacher and Klett 1997). The liquid-water part freezes in a dendritic structure with a density of 450 to 750 kg m^{-3}. The remaining liquid part fills the spaces between this dendritic structure like a sponge at the freezing temperature (Pruppacher and Klett 1997). If more liquid is accreted than can be absorbed by the spongy-like dendritic structure then it may remain on the surface much like a solid ice particle. Rasmussen and Heymsfield (1987a) parameterized the minimum amount of liquid water that could be shed by a hailstone formed by spongy wet growth. To do so, the ice fraction of liquid frozen was reduced for a rotation rate of 0.5 Hz by 0.2. This allowed the spongy hailstones to behave like a particle with a rotation rate of 5 to 7 Hz as observed. At this rotation rate, the ice fraction and shedding both reached a minimum. This is because this rotation rate allowed more liquid to be in the spongy dendritic ice and less to accumulate on the bulge of water near the equatorial region of the hailstone. The details of wet-growth parameters for spongy wet-growth particles, such as the density and the ice fraction of the particle, are found in Rasmussen and Heymsfield (1987a), and are reproduced below,

$$\rho_{\text{spongy}} = (1 - 0.08\,F_{\text{ice}})F_{\text{ice}}, \tag{10.1}$$

where ρ_{spongy} is the density of spongy-growth particles (in CGS units) and F_{ice} is the fraction of ice. The fraction of ice for $Q_{\text{cw}} \geq 2$ g m^{-3} is given by,

$$F_{\text{ice}} = 0.25 + \left[(1 - 0.25)/\left(1 + 0.1789\,\text{g m}^{-3}\{Q_{\text{cw}} - 2.0\,\text{g m}^{-3}\}\right)\right], \tag{10.2}$$

otherwise,

$$F_{\text{ice}} = 1. \tag{10.3}$$

10.3 Heat-budget equation

The heat budget for an ice particle, used to determine melting or wet growth, is based on a balance of heat budget. Here, a heat budget for ice crystals represents primarily graupel, frozen drops, or hail. The heat budget accounts for heating by conduction, vapor deposition, sensible heat, and enthalpy of freezing of rain or riming of cloud drops, and sensible heat with collection of ice crystals (Dennis and Musil 1973). The budget equation is given as,

$$\frac{dq_{\text{cond}}}{dt}+\frac{dq_{\text{diff}}}{dt}+\frac{dq_{\text{rime}}}{dt}+\frac{dq_{\text{ice}}}{dt}=0. \tag{10.4}$$

The first term is the heating by conduction and is given by

$$\frac{dq_{\text{cond}}}{dt}=2\pi DK(T_{\text{ice}}-T)f_{\text{h}}, \tag{10.5}$$

where T_{ice} is the surface temperature of hail, κ is the thermal conductivity and f_{h} is the heat ventilation coefficient. This term and the next may have ventilation coefficients greater than reported below as heat transfer is enhanced on the frontal laminar flow side and the rearward turbulent wake. Expressions for the heat and vapor ventilation coefficients are

$$f_{\text{h}}=0.78+0.308N_{\text{pr}}^{1/3}N_{\text{re}}^{1/2} \tag{10.6}$$

and

$$f_{\text{v}}=0.78+0.308N_{\text{sc}}^{1/3}N_{\text{re}}^{1/2}, \tag{10.7}$$

respectively, where N_{sc} is the Schmidt number and N_{re} the Reynolds number.

The second term is the heating by deposition/sublimation and is given as

$$\frac{dq_{\text{diff}}}{dt}=2\pi DL_{\text{s}}\psi\rho(Q_{\text{v}}-Q_{\text{s0}})f_{\text{v}}, \tag{10.8}$$

where Q_{v} is the vapor mixing ratio of air, Q_{s0} is the saturation mixing ratio at 0 °C, L_{s} is the enthalpy of sublimation, ρ is the density of air, D is the diameter, and ψ is water-vapor diffusivity. Next the third term is the heating with freezing of supercooled water and sensible heat transfer between the water and ice particle. The third term is

$$\frac{dq_{\text{rime}}}{dt}=\frac{dM_{\text{liquid}}}{dt}(F_{\text{f}}L_{\text{f}}+c_{\text{L}}[T-T_{\text{ice}}]), \tag{10.9}$$

where c_{L} is the specific heat of liquid, L_{f} is the enthalpy of freezing, and F_{f} is the fraction of liquid frozen. This term may need modification on more than one occasion. Such an occasion occurs when some liquid is retained in the interior of the hailstone or as its shell; then an appropriate F_{f} must be chosen. If some of the liquid accreted by the hailstone is shed by a fraction E_{shed}, then the liquid may extract some heat when leaving the hailstone and the equation above becomes

$$\frac{dq'_{\text{rime}}}{dt}=\frac{dM_{\text{liquid}}}{dt}(F_{\text{f}}L_{\text{f}}+c_{\text{L}}[T-T_{\text{ice}}]+E_{\text{shed}}c_{\text{L}}[T_{\text{L}}-T_{0}]), \tag{10.10}$$

where T_L is the temperature of the water shed, T_0 is $T = 0$ °C, and E_{shed} is $1 - E_{retain}$, where E_{shed} is the fraction of collected drop mass shed in the geometric sweep-out volume per second.

Finally the last term is given as the heat change owing to collisions with ice of different temperatures and sticking,

$$\frac{dq_{ice}}{dt} = \frac{dM_{ice}}{dt}(c_i[T - T_{ice}]). \tag{10.11}$$

This term accounts for the energy exchange associated with hail collecting ice.

$$\frac{dq_{ice}}{dt}\left(1 - \frac{c_i[T - T_{ice}]}{L_f + c_L[T - T_{ice}]}\right). \tag{10.12}$$

Setting the sum of (10.8), (10.10), (10.11), and (10.12) to zero and solving for dQ_{wet}/dt, the wet-growth mixing ratio rate can be found. In practice, the ice particle is assumed to be at temperature of 273.15 K, and the vapor pressure over the ice particle corresponds to that of a wet surface at 273.15 K, which requires using L_v instead of L_s in dq_{diff}/dt.

After some algebra, an equation for T_{ice} can be derived from the above heat terms summed to zero following Nelson's (1980) procedure. This equation gives the temperature of a hailstone, assuming no heat storage.

Making these assumptions results in the following for a sphere, which is assumed to represent the shape of graupel, frozen drops, and hailstone particles.

For the complete gamma distribution,

$$\begin{aligned} Q_{xwet} &= \frac{2\pi N_{Tx}\alpha_x^{\nu_x}(\rho L_v\psi[Q_{v,ice} - Q_{s0}] - K[T - T_{ice}])}{\rho(L_f + c_L[T - T_{ice}])} \\ &\times \left[0.78 D_{nx}\frac{\Gamma\left(\frac{1+\nu_x\mu_x}{\mu_x}\right)}{\alpha_x^{\left(\frac{1+\nu_x\mu_x}{\mu_x}\right)}} + 0.308 N_{sc}^{1/3}\nu^{-1/2}D_{nx}^{\frac{3+d_x}{2}}c_x^{1/2}\frac{\Gamma\left(\frac{3+d_x}{2\mu_x} + \frac{\nu_x\mu_x}{\mu_x}\right)}{\alpha_x^{\left(\frac{3+d_x}{2\mu_x} + \frac{\nu_x\mu_x}{\mu_x}\right)}}\left(\frac{\rho_0}{\rho}\right)^{1/4}\right] \\ &+ \left(1 - \frac{c_i[T - T_{ice}]}{L_f + c_L[T - T_{ice}]}\right)\left.\frac{dQ}{dt}\right|_{ice,snow}. \end{aligned} \tag{10.13}$$

The modified gamma distribution form is the following,

$$\begin{aligned} Q_{xwet} &= \frac{2\pi N_{Tx}(\rho L_v\psi[Q_{v,ice} - Q_{s0}] - K[T - T_{ice}])}{\rho(L_f + c_L[T - T_{ice}])} \\ &\times \left[0.78 D_{nx}\Gamma\left(\frac{1 + \nu_x\mu_x}{\mu_x}\right) + 0.308 N_{sc}^{1/3}\nu^{-1/2}D_{nx}^{\frac{3+d_x}{2}}c_x^{1/2}\Gamma\left(\frac{3 + d_x}{2\mu_x} + \frac{\nu_x\mu_x}{\mu_x}\right)\left(\frac{\rho_0}{\rho}\right)^{1/4}\right] \\ &+ \left(1 - \frac{c_i[T - T_{ice}]}{L_f + c_L[T - T_{ice}]}\right)\left.\frac{dQ}{dt}\right|_{ice,snow}. \end{aligned} \tag{10.14}$$

Finally, the gamma distribution gives

$$Q_{x\text{wet}} = \frac{2\pi N_{\text{T}x}\left(\rho L_{\text{v}}\psi\left[Q_{\text{v,ice}} - Q_{\text{s0}}\right] - K[T - T_{\text{ice}}]\right)}{\rho(L_{\text{f}} + c_{\text{L}}[T - T_{\text{ice}}])}$$
$$\times \left[0.78 D_{nx}\Gamma(1 + \nu_x) + 0.308 N_{\text{sc}}^{1/3}\nu^{-1/2} D_{nx}^{\frac{3+d_x}{2}} c_x^{1/2}\Gamma\left(\frac{3 + d_x}{2} + \nu_x\right)\left(\frac{\rho_0}{\rho}\right)^{1/4}\right] \qquad (10.15)$$
$$+ \left(1 - \frac{c_{\text{i}}[T - T_{\text{ice}}]}{L_{\text{f}} + c_{\text{L}}[T - T_{\text{ice}}]}\right)\left.\frac{\text{d}Q}{\text{d}t}\right|_{\text{ice,snow}}.$$

The inverse-exponential version of (10.15) has $\nu_x = 0$.

10.4 Temperature equations for hailstones

As stated above, an equation for T_{ice} can be derived following Nelson (1980). This equation gives the temperature of a hailstone assuming no heat storage,

$$T_{\text{ice}} = \frac{2\pi D f_{\text{h}} K T - 2\pi D f_{\text{v}} L_{\text{v}}\rho(Q_{\text{vh}x} - Q_{\text{v}}) + c_{\text{i}} T \frac{\text{d}M_{\text{ice}}}{\text{d}t} + (L_{\text{f}} + c_{\text{L}} T)\frac{\text{d}M_{\text{liquid}}}{\text{d}t}}{2\pi D f_{\text{h}} K + c_{\text{i}} T \frac{\text{d}M_{\text{ice}}}{\text{d}t} + c_{\text{L}}\frac{\text{d}M_{\text{liquid}}}{\text{d}t}}, \qquad (10.16)$$

where $Q_{\text{vh}x}$ is the vapor mixing ratio over hail of various types. As the equation is implicit in the term T_{ice}, it can be solved easily by iteration using the Newton–Raphson technique with convergence in three to five iterations. This procedure is given below for completeness.

First some variable definitions are set,

$$H_0 = 2\pi D f_{\text{h}} K, \qquad (10.17)$$

$$H_1 = 2\pi D T, \qquad (10.18)$$

$$H_2 = \left(c_{\text{i}} T \frac{\text{d}M_{\text{ice}}}{\text{d}t} + \frac{\text{d}M_{\text{liquid}}}{\text{d}t}\left[L_{\text{f}} + c_{\text{L}}(T - T_0) + c_{\text{i}} T_0\right]\right), \qquad (10.19)$$

$$H_3 = 2\pi D f_{\text{h}} \text{K} + c_{\text{i}}\left(\frac{\text{d}M_{\text{liquid}}}{\text{d}t} + \frac{\text{d}M_{\text{ice}}}{\text{d}t}\right), \qquad (10.20)$$

$$H_4 = -2\pi D f_{\text{v}}\psi L_{\text{v}}\rho, \qquad (10.21)$$

and

$$H_5 = \frac{-17.27 H_4}{H_3}. \qquad (10.22)$$

Now the following iteration loop is executed,

$$T_{\mathrm{ice,old}} = T_{\mathrm{ice}}, \tag{10.23}$$

$$T_1 = T_{\mathrm{ice,old}} - 273.15, \tag{10.24}$$

$$T_2 = T_{\mathrm{ice,old}} - 35.86, \tag{10.25}$$

$$Q_{\mathrm{vh}x} = \frac{380}{p} \exp\left(\frac{17.27[T - 273.15]}{[T - 35.86]}\right), \tag{10.26}$$

where p is pressure (Pa), and

$$H_6 = H_4(Q_{\mathrm{vh}x} - Q_{\mathrm{v}}). \tag{10.27}$$

For the Newton–Raphson iteration loop the parameters f and f' are needed,

$$f = \frac{T_{\mathrm{ice,old}} - (H_1 + H_2 + H_6)}{H_3}, \tag{10.28}$$

$$f' = 1 + H_5\left([T_2 - T_1]/T_2^2\right)Q_{\mathrm{vh}x}, \tag{10.29}$$

and

$$T_{\mathrm{ice}} = T_{\mathrm{ice,old}} - \frac{f}{f'}. \tag{10.30}$$

If T_{ice} is less than 273.15 K then the hail growth is computed using $\mathrm{d}M_{\mathrm{ice}}/\mathrm{d}t$ and $\mathrm{d}M_{\mathrm{liquid}}/\mathrm{d}t$. If not, then the fraction of liquid frozen F_{f} needs to be computed; this can be done following Nelson (1980) as well,

$$F_{\mathrm{f}} = \frac{2\pi D f_{\mathrm{h}} K(T_{\mathrm{ice}} - T_a) + 2\pi D f_{\mathrm{v}} L_{\mathrm{v}} \rho(Q_{\mathrm{vh}x} - Q_{\mathrm{v}}) + c_{\mathrm{i}}(T_{\mathrm{ice}} - T)\frac{\mathrm{d}M_{\mathrm{ice}}}{\mathrm{d}t} + [c_{\mathrm{L}}(T_{\mathrm{ice}} - T_a)]\frac{\mathrm{d}M_{\mathrm{liquid}}}{\mathrm{d}t}}{L_{\mathrm{f}}\dfrac{\mathrm{d}M_{\mathrm{liquid}}}{\mathrm{d}t}}. \tag{10.31}$$

Once F_{f} is calculated the growth rate is computed from $F_{\mathrm{f}}\,\mathrm{d}M_{\mathrm{l}}/\mathrm{d}t$ and $\mathrm{d}M_{\mathrm{i}}/\mathrm{d}t$.

10.5 Temperature equation for hailstones with heat storage

Of interest in modeling hailstone growth is the prediction of their temperature to determine if they are growing by dry, spongy, or wet growth. This section establishes equations associated with the temperature change of individual dry and wet hailstones. Following Dennis and Musil (1973), the heat content of a dry hailstone is given by

$$MT_{\mathrm{s}}c_{\mathrm{i}}, \tag{10.32}$$

where c_{i} is the specific heat of dry ice and T_{s} the surface temperature of the hailstone, which is always negative (°C). The mass of the hailstone is given by M. This can be written as a heat-budget equation,

$$\frac{\mathrm{d}}{\mathrm{d}t}(MT_{\mathrm{s}}c_{\mathrm{i}}) = \frac{\mathrm{d}q_{\mathrm{T}}}{\mathrm{d}t}. \tag{10.33}$$

From (10.32) above the following can be written,

$$\frac{\mathrm{d}T_{\mathrm{s}}}{\mathrm{d}t} = -\frac{T_{\mathrm{s}}}{M}\frac{\mathrm{d}M}{\mathrm{d}t} + \frac{1}{Mc_{\mathrm{i}}}\frac{\mathrm{d}q_{\mathrm{T}}}{\mathrm{d}t}. \tag{10.34}$$

The total heat budget for wet and dry hailstones was given previously as

$$\frac{\mathrm{d}q_{\mathrm{T}}}{\mathrm{d}t} = \frac{\mathrm{d}q_{\mathrm{cond}}}{\mathrm{d}t} + \frac{\mathrm{d}q_{\mathrm{diff}}}{\mathrm{d}t} + \frac{\mathrm{d}q_{\mathrm{rime}}}{\mathrm{d}t} + \frac{\mathrm{d}q_{\mathrm{ice}}}{\mathrm{d}t}, \tag{10.35}$$

where the heat with conduction is q_{cond}, heat with sublimation/deposition or evaporation/condensation is q_{diff}, heat with accretion of liquid q_{rime} and ice q_{ice}.

Let us first consider dry hailstones. By substituting in the heat-transfer terms, the following temperature equation is found

$$\begin{aligned}\frac{\mathrm{d}T_{\mathrm{s}}}{\mathrm{d}t} &= -\frac{T_{\mathrm{s}}}{M}\frac{\mathrm{d}M}{\mathrm{d}t} + \frac{1}{Mc_{\mathrm{i}}}(2\pi Df[K\{T - T_{\mathrm{s}}\} - L_{\mathrm{v}}\psi\rho(Q_{\mathrm{v}} - Q_{\mathrm{s0}})]) \\ &+ \frac{\mathrm{d}M_{\mathrm{liquid}}}{\mathrm{d}t}(L_{\mathrm{f}} + c_{\mathrm{L}}T) + \frac{\mathrm{d}M_{\mathrm{i}}}{\mathrm{d}t}c_{\mathrm{i}}T = \left(\frac{T - T_{\mathrm{s}}}{M}\right)\frac{\mathrm{d}M}{\mathrm{d}t} \\ &+ \frac{L_{\mathrm{f}} + (c_{\mathrm{L}} - c_{\mathrm{i}})T}{Mc_{\mathrm{i}}}\frac{\mathrm{d}M_{\mathrm{liquid}}}{\mathrm{d}t} + \frac{2\pi Df}{Mc_{\mathrm{i}}}[K\{T - T_{\mathrm{s}}\} - L_{\mathrm{v}}\psi\rho(Q_{\mathrm{v}} - Q_{\mathrm{s0}})].\end{aligned} \tag{10.36}$$

For wet hailstones the temperature change is $\mathrm{d}T_{\mathrm{s}}/\mathrm{d}t = 0$ as liquid- and ice-water mixtures have a temperature of 0 °C. But the fraction of liquid water is often desired. When a hailstone is at the freezing temperature, all the liquid water accreted cannot freeze and some of the mass remains as liquid.

The heat content for a wet hailstone, is written similarly to that for the dry hailstone, with the exception that the fraction of liquid water by mass F_{L} and the enthalpy of freezing L_{f} are included and c_{i} and T are omitted,

$$ML_{\mathrm{f}}F_{\mathrm{L}}. \tag{10.37}$$

Differentiation of this equation results in

$$\frac{\mathrm{d}}{\mathrm{d}t}(ML_{\mathrm{f}}F_{\mathrm{L}}) = \frac{\mathrm{d}q_{\mathrm{T}}}{\mathrm{d}t}, \tag{10.38}$$

or upon applying the chain rule, the fraction of the mass of the liquid can be obtained from

$$\frac{\mathrm{d}F_{\mathrm{L}}}{\mathrm{d}t} = -\frac{F_{\mathrm{L}}}{M}\frac{\mathrm{d}M}{\mathrm{d}t} + \frac{1}{ML_{\mathrm{f}}}\frac{\mathrm{d}q_{\mathrm{T}}}{\mathrm{d}t}. \tag{10.39}$$

Now substituting the terms on the right-hand side of the heat-budget equation, and with $T_s = 0\,°\mathrm{C}$, the following equation for the liquid fraction of the mass of a hailstone is found for wet growth,

$$\frac{\mathrm{d}F_\mathrm{L}}{\mathrm{d}t} = -\frac{F_\mathrm{L}}{M}\frac{\mathrm{d}M}{\mathrm{d}t} + \frac{1}{ML_\mathrm{f}}(2\pi Df[K\{T - T_\mathrm{s}\} - L_\mathrm{v}\psi\rho(Q_\mathrm{v} - Q_\mathrm{s0})]) \\ + \frac{\mathrm{d}M_\mathrm{liquid}}{\mathrm{d}t}(L_\mathrm{f} + c_\mathrm{L}T) + \frac{\mathrm{d}M_\mathrm{ice}}{\mathrm{d}t}c_\mathrm{i}T. \tag{10.40}$$

Simulations with a prognostic hail temperature equation show a very fast relaxation time to the 0 °C spongy or wet growth regime; it was found to be on the order of 10 s (Pellett and Dennis 1974) with conditions rapidly approaching equilibrium temperature conditions for hail temperature (using an iterative technique for the solution). Differences in final sizes of an initial 1 cm hailstone at 500 mb, −20 °C, and 3 g m^{-3} of liquid are only 1.1% different as found from the prognostic and iterative methods, with the hail-temperature prediction case producing only slightly larger hail of about 1.55 cm after a nominal 150 s. Whilst the prognosis of hail temperature is cheaper to compute than the iteration procedure of the equilibrium temperature method, it requires extra memory storage for a temperature variable that the equilibrium temperature method does not need to retain. It should be noted that Hitchfield and Stauder (1967) found that a 1.1 cm hailstone may be up to 12 °C colder than the freezing temperature. Wet growth may be delayed by falling as much as 2 km vertically in a cloud using a prognostic equation for temperature of a hailstone. On the other hand, large hailstones were found to produce very similar solutions for either the equilibrium temperature method or the temperature prognosis method. Therefore, it seems that either method is acceptable to use to determine spongy wet growth conditions, with timing differences on actual temperatures of hailstones of order of 10 s or so for large cloud contents.

Little information was given in either study for conditions of small water contents < 1 g m^{-3}. Interestingly, in Johnson and Rasmussen's (1992) modeling study of wet and dry growth of hail, it was found that there is a hysteresis associated with the onset of wet growth that was possibly related to the drag and heat transfer. The effect of this was to make the onset of wet growth slightly more difficult to achieve and the cessation of wet growth easier to reach.

10.6 Schumann–Ludlam limit for wet growth

The Schumann (1938) and Ludlam (1958) limit is the demarcation between dry growth and wet growth and is based on the temperature and liquid-water

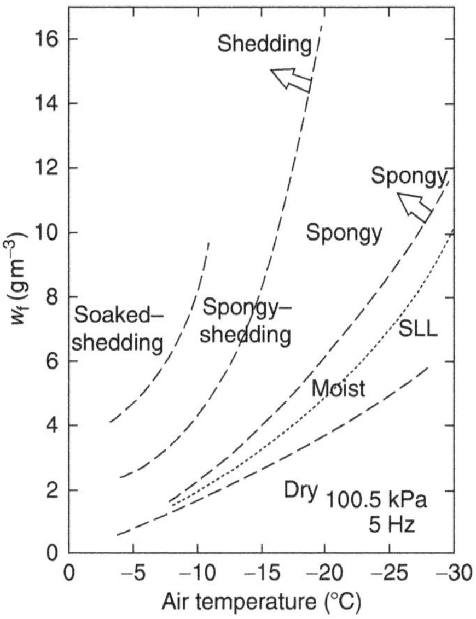

Fig. 10.6. The five observed hailstone growth regimes as a function of liquid-water content (LWC) and air temperature; regimes: dry, moist, spongy, spongy–shedding, and soaked–shedding, based on 28 experiments at laboratory pressure (100.5 kPa) and nutation/precession and spin frequencies of 5 Hz. The dotted line is the theoretically derived Schumann–Ludlam limit (SLL). (From Lesins and List 1986; courtesy of the American Meteorological Society.)

content available for growth of a hailstone of a given size. The smallest particle that can grow by wet growth is typically less than 5 mm in diameter, and thus 5 mm diameter is the threshold size between graupel and hail.

Lesins and List (1986) studied nutation/precession and spinning frequency of 5 Hz and > 20 Hz of hailstones. They found that for 5 Hz five growth modes for hail existed (Fig. 10.6).

The first of the five modes is the *dry regime* where the water collected freezes on contact, and the ice fraction is unity. The surface is completely dry and deposit temperatures are less than 0 °C. No shedding occurs, and the net collection efficiency is essentially unity. The deposit is opaque indicating that there is air encapsulated below the outer ice surface. This is because cloud drops freeze nearly instantaneously upon contact, which leaves air spaces. Thus, there is no equator-to-pole water movement on the surface, which may have millimeter-sized blobs and lobes where water makes contact at an oblique angle.

The *moist regime* is somewhat similar to the dry regime as the net collection efficiency and ice fraction are essentially unity. No shedding is observed. The difference between the moist and dry regime is that a band of transparent ice exists around the equatorial region of the ice. There are no surface-roughness elements in this band. The ice temperature of the liquid that is accreted is approximately 0 °C. Moreover, in the equatorial zone, the liquid collected is freezing at the slowest rate of any part of the hailstone. The moist regime is the narrow transition region between solid ice deposits and spongy ice deposits. It formally is most consistent with the Schumann–Ludlam limit. At this limit, the ice fraction is still unity and the temperature of the ice deposit is 0 °C.

Next is the *spongy–no-shedding regime*. The ice fraction is finally less than unity and this regime is split into three parts, one without shedding, which is discussed here, and two with shedding, which are discussed below. For spongy ice, a no-shedding regime exists toward temperatures closer to 0 °C or at warmer temperatures, and a collection efficiency of one is found after passing the Schumann–Ludlam limit. Heat transfer is not adequate to freeze all of the liquid collected, and instead of liquid water being shed, it is trapped into air spaces of ice deposits where it produces spongy ice. The original oblate spheroid of the hailstone changes now with ridges from pole to pole. The shape change results from mobile liquid water on the surface and prevents liquid from building up near the equator where shedding would occur if it accumulated there, but it does not in this mode. This is because of the net transfer of ice toward the pole regions. In this regime the ice fraction is 0.8 to 1.0 and most of the sponginess is found in pole-to-pole ridges on the surfaces.

The second spongy regime is the *spongy–shedding regime*. In this regime, the ice fraction is 0.5 to 0.7, and collection efficiencies are less than unity. The spongy deposit is unable to incorporate all of the liquid collected and excess water is shed as 1-mm-sized drops. These shed drops originate from the back half of the hailstone with respect to the flow and pass through the wake zone of the flow past the hailstone. Another zone of shedding is the torus near the equatorial region. This type of hailstone appears similar to that of the spongy regime.

The third spongy regime is the *soaked–shedding regime* where ice fractions are less than 0.5, a minimum value. If the temperature is increased all the collected water is shed.

Finally there is a sixth regime only for very high rotation rates (> 20 Hz), the *dry–shedding regime* (Fig. 10.7). Centrifugal forces cause all unfrozen liquid to be shed as 1-mm liquid drops. The ice fraction is unity and the net collection efficiency is less than unity.

10.7 Collection efficiency of water drops for hail

Very few studies of collection efficiencies have been carried out between hail and cloud drops. Probably the data that most are familiar with, and perhaps from the only study, are those by Macklin and Bailey (1966).

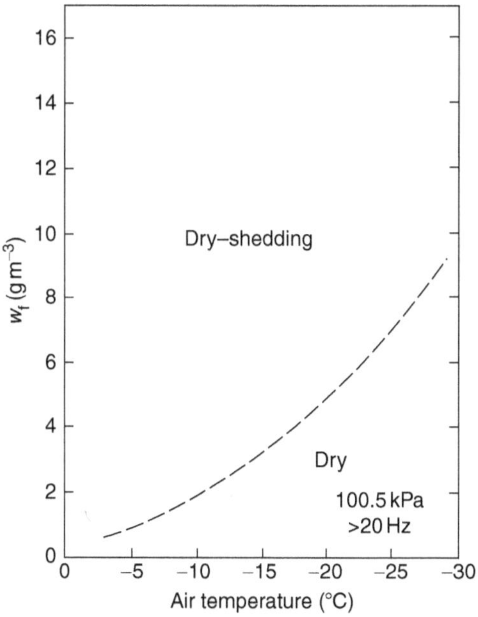

Fig. 10.7. Observed hailstone growth regimes as a function of liquid-water content W_f and air temperature, for high rotation rates (> 20 Hz); experiments at laboratory pressure (100.5 kPa) and spin frequency equal to nutation/precession frequencies. All ice fractions are unity and the deposit temperatures were probably < 0 °C. (From Lesins and List 1986; courtesy of the American Meteorological Society.)

A simple parameterization for these data was developed by Milbrandt and Yau (2005b), and can be used for graupel and hail,

$$E_{\mathrm{hwcw}} = \exp\left(-8.68 \times 10^{-7} D_{m\mathrm{cw}} D_{m\mathrm{gw,hw}}\right), \tag{10.41}$$

where E_{hwcw} is the collection efficiency of cloud water by hail water.

Certainly future experiments to measure collection efficiencies more effectively for hailstones collecting cloud water and raindrops would be very useful.

10.8 Hail microphysical recycling and low-density riming

Researchers have been examining hail growth for many years, trying to find ways in which hail often can get very large (> 50 mm in diameter), and what causes repeating layers of apparently low- and high-density ice. One attempt to explain the growth of large hail is particularly interesting and is the focus of a method put forth by Pflaum and Pruppacher (1979). In this method, hailstones grow by switching back and forth between low-density riming and wet growth; together this is called microphysical recycling.

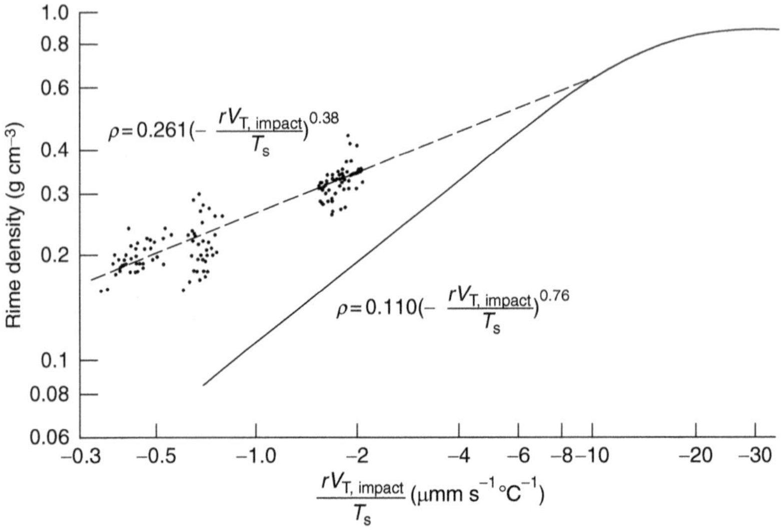

Fig. 10.8. Rime density versus predicting parameter $r\,V_{\mathrm{T,impact}}\,/\,T_{\mathrm{s}}$ where the dashed line represents a least-square fit to experimental data and the solid line represents Macklin and Bailey's (1962) empirical relationship. All data for ice particle surface temperatures colder than −5 °C. (From Pflaum and Pruppacher 1979; courtesy of the American Meteorological Society.)

The mode of operation of microphysical recycling is that first nascent hailstones on the order of 5 mm in diameter grow via low-density dry growth. This often results from some or all of the following including small cloud-droplet size, cold hailstone temperatures, and small cloud-droplet impact velocities, which correspond to lower terminal velocities. The density of accreted cloud water, which then freezes to ice water or rather hail rime, has been empirically fitted to data by Macklin and Bailey (1962) and Pflaum and Pruppacher (Fig. 10.8) to be

$$\rho_{\mathrm{rime}} = 110\left(\frac{-rV_{\mathrm{T,impact}}}{T_{\mathrm{s}}}\right)^{0.76}, \tag{10.42}$$

$$\rho_{\mathrm{rime}} = 261\left(\frac{-rV_{\mathrm{T,impact}}}{T_{\mathrm{s}}}\right)^{0.76} \tag{10.43}$$

or empirically fitted by Heymsfield and Pflaum (1985) to be

$$\rho_{\mathrm{rime}} = 300\left(\frac{-rV_{\mathrm{impact}}}{T_{\mathrm{s}}}\right)^{0.44}. \tag{10.44}$$

Only rarely does ρ_{rime} fall below 170 kg m^{-3}, and has an upper limit of 900 kg m^{-3}. Pflaum and Pruppacher (1979) note that smaller rime densities

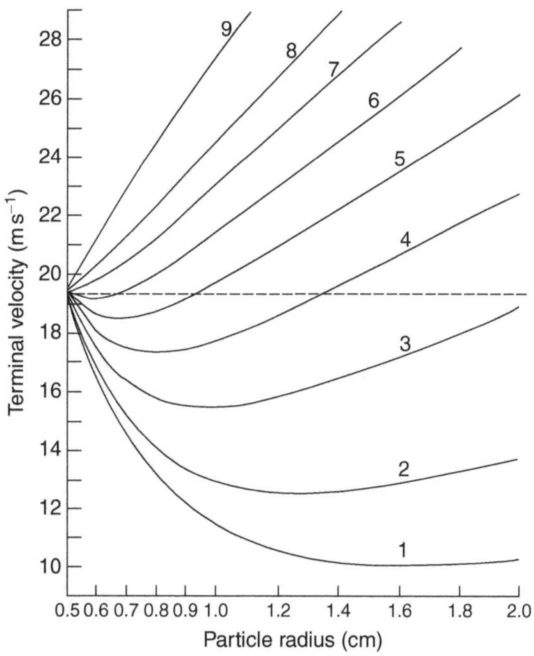

Fig. 10.9. Theoretical effect on terminal velocities of adding a progressively thicker coat of various constant-density smooth rime to an initial ice sphere (r = 900 kg m^{-3}) of 0.5 cm radius. An ambient temperature of 253.15 K and an ambient pressure of 500 mb have been used in the calculations. Rime density acquired (units of kg m^{-3}): (1) 50; (2) 100; (3) 200; (4) 300; (5) 400; (6) 500; (7) 600; (8) 700; (9) 900. The dashed line provides a reference for constant terminal velocity. (From Pflaum 1980; courtesy of the American Meteorological Society.)

might be possible if electrical effects are responsible. In these equations r is the mean volume cloud-droplet radius in microns, V_{t0} is the impact velocity in m s^{-1}, and T_s is surface temperature of the hailstone. The impact velocity is the velocity component of the hail relative to the cloud drop when they make contact. A head-on collision between hailstone and cloud-water droplet would make a large impact velocity close to the terminal velocity of a hailstone. On the other hand, a cloud droplet making an off-center or glancing strike with a hailstone produces a smaller impact velocity with a hailstone. Impact velocities can vary substantially from the typical factor of 0.6 times the terminal velocity. In a bulk microphysical parameterization it is most prudent to use an impact velocity of 0.6 times terminal velocity.

Low-density hailstones can grow rapidly given sufficient liquid water, as they become larger more quickly with reduced requirements for updraft strength; therefore, they can have longer growth times compared to otherwise

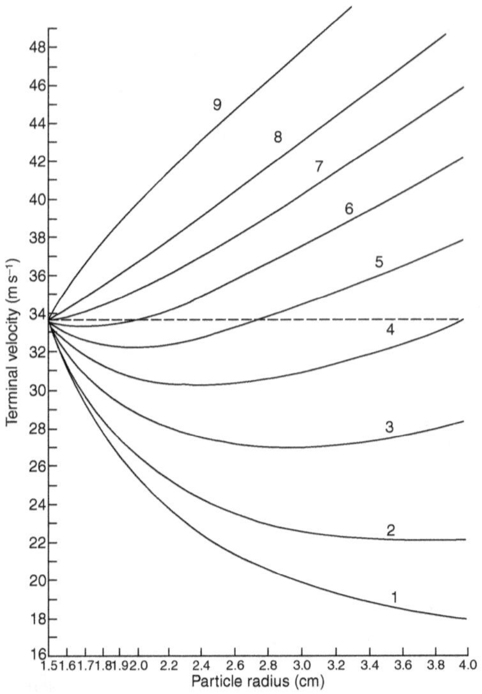

Fig. 10.10. Theoretical effect on terminal velocities of adding a progressively thicker coat of various constant-density smooth rime to an initial ice sphere ($r = 900$ kg m^{-3}) of 1.5 cm radius. An ambient temperature of 253.15 K and an ambient pressure of 500 mb have been used in the calculations. Rime density acquired (units of kg m^{-3}): (1) 50; (2) 100; (3) 200; (4) 300; (5) 400; (6) 500; (7) 600; (8) 700; (9) 900. (From Pflaum 1980; courtesy of the American Meteorological Society.)

high-density hail growth modes that often are used to explain hail growth. Even though low-density hailstones have a slower fallspeed for a given mass compared to their high-density counterparts, the diameter for a given low-density hailstone mass has a wider sweep-out cross-section (D^2). This makes up for the smaller fall velocity ($cD^{1/2}$) through the lower particle density contained in the constant c. After a period, the low-density hailstone is advected or falls back into a zone of high liquid-water contents and grows by high-density wet growth. The high water contents result in unfrozen liquid that soaks the porous ice particle and its surface. This increases the density of the porous pockets and produces a layer of frozen liquid water on the surface of the hailstone, as it is re-advected to colder temperatures; then the low-density growth, high-density/re-densification wet-growth process is repeated. Examination of this problem by Pflaum (1980) gave some spectacular results and they are shown in Figs. 10.9 and 10.10 below. In response to these conclusions by Pflaum (1980), Farley (1987) included 20 variable-density bins representing the spectrum of ice particles from snow, graupel, and hail. The results of the research by Pflaum and Pruppacher (1979) and Pflaum (1980) suggest that models should include variable-density growth for graupel and hailstones, as has been carried out by Farley (1987), Straka and Mansell (2005), and Straka *et al.* (2009b).

To take into account variable-density hail growth, a set of prognostic equations can be developed. Usage comes down to applying appropriate densities and mass weight sources/sinks with mixing ratio tendencies times the time,

$$\rho_{\mathrm{hx}}^{n+1} = \frac{\rho_{\mathrm{hx}}^{n} Q_{\mathrm{hx}}^{n} + \Delta t \rho_{\mathrm{hx,conv}}^{n} \left.\frac{\mathrm{d}Q_{\mathrm{hx}}}{\mathrm{d}t}\right|_{\mathrm{conv}}^{n} + \Delta t \rho_{\mathrm{hx,rime}}^{n} \left.\frac{\mathrm{d}Q_{\mathrm{hx}}}{\mathrm{d}t}\right|_{\mathrm{rime}}^{n} + \Delta t \rho_{900}^{n} \left.\frac{\mathrm{d}Q_{\mathrm{hx}}}{\mathrm{d}t}\right|_{\mathrm{rain}}^{n} + \Delta t \rho_{\mathrm{hx}}^{n} \left.\frac{\mathrm{d}Q_{\mathrm{hx}}}{\mathrm{d}t}\right|_{\mathrm{sub/dep}}^{n}}{Q_{\mathrm{hx}}^{n} + \Delta t \left.\frac{\mathrm{d}Q_{\mathrm{hx}}}{\mathrm{d}t}\right|_{\mathrm{conv}}^{n} + \Delta t \left.\frac{\mathrm{d}Q_{\mathrm{hx}}}{\mathrm{d}t}\right|_{\mathrm{rime}}^{n} + \Delta t \left.\frac{\mathrm{d}Q_{\mathrm{hx}}}{\mathrm{d}t}\right|_{\mathrm{rain}}^{n} + \Delta t \left.\frac{\mathrm{d}Q_{\mathrm{hx}}}{\mathrm{d}t}\right|_{\mathrm{sub/dep}}^{n}} \tag{10.45}$$

where n and $n + 1$ are time levels t and $t + \mathrm{d}t$. This equation can be expanded to accommodate the source and sink terms in the mass or mixing-ratio budget equations. The new density can be used to determine if the particle should be transferred from a low-density particle to a high-density particle in a bulk microphysical model, or just tracked in a bin model to compute terminal velocity, diameters, etc., for a given mass representing a bin. Notice in (10.45) that the density related to accreting rain is that of frozen ice, approximately 900 kg m^{-3}, and that the particle density related to sublimation and deposition does not change.

11
Melting of ice

11.1 Introduction

A heat budget is used to derive the melting equation; this accounts for heating owing to conduction, vapor diffusion, and sensible heat from the collection of rain, drizzle, and cloud drops that might be warmer than the collector ice-water particle. The collection process is complicated in most models if the temperature of any of the hydrometeor types is not predicted. In this case, the temperature of liquid hydrometeors is assumed to relax to the environmental temperature instantaneously. In the case where hydrometeor temperatures are predicted, the rate of condensation or evaporation on a melting ice-water particle can be more accurately computed (Walko *et al.* 2000). Diagnosing the instantaneous hydrometeor-species' temperature can lessen the accuracy of melting computations as compared to predicting hydrometeor temperatures. The influence of energy storage and the relaxation times of the hydrometeor-species' temperature to the temperature of the environment will be investigated later in the chapter.

Melting of ice-water hydrometeors can be made very simplistic; or for the case of particle trajectory models, hybrid-bin models and bin models, can be quite sophisticated. For parameterizations, difficulties arise when the Reynolds number of a particle is taken into account. The equations from Rasmussen and Heymsfield (1987a) generally cannot be used for bulk parameterization models whereas they can be used for bin-type models where each bin has its own characteristics. For bulk parameterization models, melting can be treated with one set of equations, say for frozen drizzle, which has a small Reynolds number N_{re} compared to larger particles. For larger particles, another equation for particle species such as snow aggregates, graupel, frozen drops, or hail, which have larger Reynolds numbers, a different variant of the melting equation should be used. For Lagrangian trajectory-type models, and various sorts of bin-type models, the more sophisticated equations for melting should be incorporated if at all possible, as they take into account the internal circulations of liquid water in smaller ice-water–liquid-water mixtures, such as melting graupel and frozen drops. It should be noted that particle transfer from bin to bin must explicitly be taken into account with any particle bin-type model.

Finally, the issue of porous graupel, frozen drops, hailstones soaking liquid water, and shedding liquid-water drops (for at least some sizes of hailstone) needs to be taken into account. In the past, collected cloud water and rain water were just shed from graupel, frozen drops, and hailstones. More recently attempts to incorporate all these effects have been tried in various forms, from the very simplistic to the very complex for use in bulk parameterization models. However, the procedure is fairly standard for particle bin trajectory and various types of bin models.

11.2 Snowflakes and snow aggregates

Mitra *et al.* (1990) identified a common pattern in the melting of snowflakes. First, small droplets form on the tips of the crystal. Second, these drops move by capillary forces and by surface-tension effects to the central part of the flake or to linkages. The branches tend to remain basically liquid-water free. Third, the central region of the crystal begins to melt and cause structural distortions in the branches. The many small openings in the flake lose definition and only a few larger openings remain. Fourth, the ice frame collapses and becomes a water drop (Knight 1979; Fig. 11.1). Interestingly, breakup was not observed (Mitra *et al.* 1990).

Snow aggregates of 5 to 11 mm in diameter have been observed to melt into raindrops of 1.1 to 2.6 mm in diameter and fall at 4.3 to 7.6 m s^{-1} (Locetalli and Hobbs 1974; Stewart *et al.* 1984). Melting generally occurs at temperatures between 0 and 5 °C. The formation of liquid-water drops from melting of snow aggregates can produce various rain size distribution changes. Melting snow aggregates do not shed drops if there are no collisions. Rather, they first soak liquid water in their ice lattices and then become liquid-water-coated until their ice cores melt substantially and completely collapse (Knight 1979).

Low relative humidities can substantially slow the melting of ice. This is shown well in Fig. 11.2.

11.3 Graupels and hailstones

For larger ice-water particles, such as the various forms of hail, melt water can both soak into the ice lattice and exist at the surface, just as with graupels

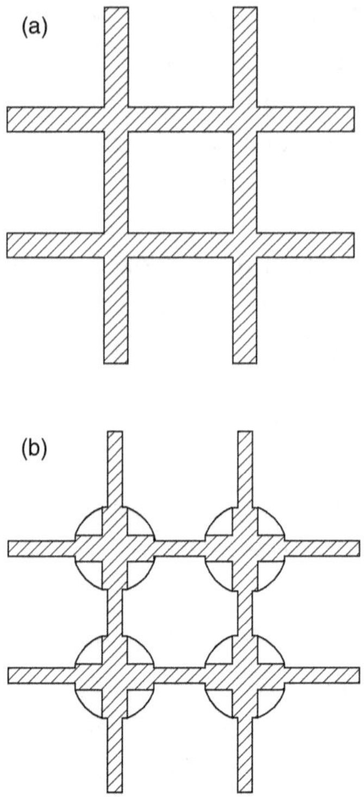

Fig. 11.1. Schematic representation of a stage of melting, which is preliminary to breakup of an aggregate or dendrite: (a) before melting and (b) during melting. (From Knight 1979; courtesy of the American Meteorological Society.)

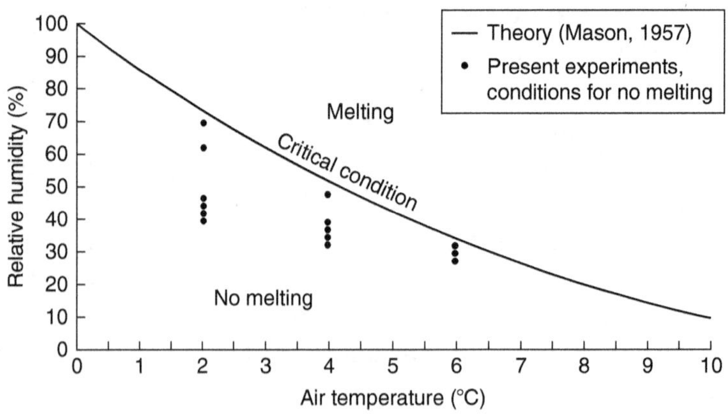

Fig. 11.2. Conditions (relative humidity of air and air temperature) at which an ice sphere does/does not melt due to evaporation cooling of the ice-sphere surface. Comparison with theory of Mason (1957). (From Rasmussen and Pruppacher 1982; courtesy of the American Meteorological Society.)

and aggregates. For diameters between about 9 and 19 mm, hailstones shed drops of 0.5 to 2 mm with a mode of 1 mm. This also has been inferred by polarimetric radar (Hubbert *et al.* 1998; Straka *et al.* 2000; Loney *et al.* 2002). The melting of smaller hailstones can result in very large raindrops as often inferred at the leading edges of thunderstorms using polarimetric radar [van den Broeke *et al.* (2008) and Kumjian and Ryzhkov (2008)]. Therefore a separate category is recommended for liquid-water drops shed from hailstones where the mode is constrained by values of sizes observed in the laboratory (Rasmussen *et al.* 1984; Rasmussen and Heymsfield 1987a). Melting graupels do not shed drops if there are few or no collisions between each other or with melting hail. As a result these graupels also can result in large raindrops if the graupels have large diameters (3 to 5 mm).

11.4 Melting of graupel and hail

The heat budget used to determine melting for ice particles is similar to that for wet growth. Here ice particles represent primarily ice crystals, snow aggregates, graupel, frozen drizzle or drops, or hail (Fig. 11.3). The heat budget accounts for heating by conduction, vapor deposition, and sensible heat from the collection of rain, drizzle, and cloud drops. Sensible heat with collection of ice crystals is ignored. The heat-budget equation terms are written as follows,

$$\frac{dq_{\text{cond}}}{dt} + \frac{dq_{\text{diff}}}{dt} + \frac{dq_{\text{liquid}}}{dt} = 0. \tag{11.1}$$

The first term is the heating by conduction and is given by

$$\frac{dq_{\text{cond}}}{dt} = 2\pi DK(T - T_{\text{ice}})f_{\text{h}}, \tag{11.2}$$

where T_{ice} is the surface temperature of ice, K is the thermal conductivity of air, and f_{h} is the heat ventilation for conduction.

The second term is the heating by condensation/evaporation and is given by

$$\frac{dq_{\text{diff}}}{dt} = 2\pi DL_{\text{s}}\psi\rho(Q_{\text{v}} - Q_{\text{s0}}(T = 273.15))f_{\text{v}} \tag{11.3}$$

where Q_{v} is the mixing ratio, L_{s} is the enthalpy of sublimation, ψ is water vapor diffusivity. The Q_{s0} is the saturation mixing ratio at 273.15 K.

Next, the third term, is the sensible heating associated with the collection of cloud, drizzle, and rain, which are assumed to be at the ambient temperature (unless temperature is predicted, then that temperature is used).

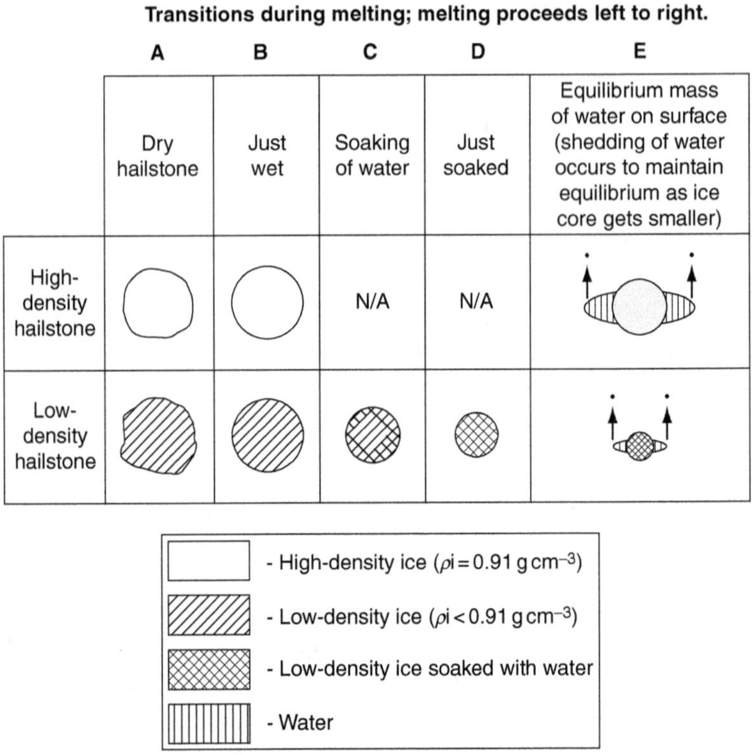

Fig. 11.3. Schematic diagram showing the stages of melting experienced by high- and low-density particles. The left-most panel shows a dry particle; panels progressively to the right show stages encountered with increasing melting. The density of the particle refers to the initial ice density. Columns D and E represent hailstones with a density between ice and liquid. (From Rasmussen and Heymsfield 1987a; courtesy of the American Meteorological Society.)

$$\frac{dq_{\text{liquid}}}{dt} = \frac{dM_{\text{liq}}}{dt}(c_L[T - T_{\text{ice}}]), \tag{11.4}$$

where c_L is the specific heat of liquid.

Setting the sum of these equations (11.2)–(11.4) equal to $dq/dt = 0$, and solving for dQ_{melt}/dt, the melting mixing ratio rate can be found. What is done in practice is that the ice particle is assumed to be at temperature of 273.15 K, and the vapor pressure over the ice particle corresponds to that for a wet surface at 273.15 K, which requires using the enthalpy of vaporization L_v instead of L_s in dq_{diff}/dt, (11.3). The vapor ventilation coefficient is given in terms of the Schmidt number N_{sc} for both conduction and diffusion terms,

$$f_v = 0.78 + 0.308 N_{sc}^{1/3} N_{re}^{1/2}. \tag{11.5}$$

Thus, melting of ice water involves three terms in general, including thermal conduction, vapor diffusion, and sensible heat transfer. These are all incorporated in the following equation, which is from the heat-budget equation for an ice-water particle that is melting. It should be noted that this is very similar to the heat budget used in wet growth, as we will see later. Also, for extremely dry conditions, freezing can be predicted owing to evaporation at temperatures above 273.15 K if the evaporation overcomes the conduction and sensible heat terms. At present this is not permitted in any models to the author's knowledge and in this case the melting term is set to zero. With the above heat-budget equation, the melting equation can be written as

$$\begin{aligned}\frac{\partial M(D_x)}{\partial t} = & -\frac{2\pi D_x[K(T - T_0) + \rho\psi L_v(Q_v - Q_{s0})]f_v}{L_f} \\ & - \frac{c_L(T - T_0)}{L_f}\left(\frac{dM(D_x)}{dt}\right)\bigg|_{AC_L},\end{aligned} \tag{11.6}$$

where L_f is the enthalpy of freezing.

This equation for melting is a generic one that generally works well and is based on work by Mason (1957). This equation was based on studies of graupel and larger hail. Moreover, the particles considered were not falling at terminal velocity in the studies [Rasmussen and Heymsfield (1987a)]. Still (11.6) represents the most widely used equation in bulk parameterization models. More detailed work on the melting of large ice particles ($0.3 < D < 2.5$ cm) that incorporates the set of heat-transfer equations set forth by Rasmussen and Heymsfield (1987a) is summarized below for use in many models types, although they are difficult to incorporate in bulk parameterization models.

To begin, Rasmussen and Heymsfield consider particles with a Reynolds number range of $N_{re} < 250$,

$$\frac{dq}{dt} = [-2\pi D\kappa(T_\infty - T_0)f_h - 2\pi D\psi L_v\rho(Q_{v,\infty} - Q_{s0})f_v]2 - c_L(T_\infty - T_0)\frac{dm}{dt}_{AC_L}, \tag{11.7}$$

where T_∞ is the temperature at an infinite distance from the particle, $T_0 = 273.15\,\text{K}$, $Q_{v,\infty}$ is the mixing ratio of vapor at an infinite distance from the particle; for Reynolds number range of $250 < N_{re} < 3000$,

$$\frac{dq}{dt} = -2\pi D\kappa(T_\infty - T_0)f_h - 2\pi D\psi L_v\rho(Q_{v,\infty} - Q_{s0})f_v - c_L(T_\infty - T_0)\frac{dm}{dt}_{AC_L}, \tag{11.8}$$

and for Reynolds number range of $3000 < N_{re} < 6000$, the equation needs to be solved iteratively,

$$\frac{dq}{dt} = -\frac{2\pi D_d D_i \kappa_w (T_0 - T_a[D_i]) f_h}{(D_d - D_i)}$$

$$= -2\pi D \kappa (T_\infty - T_a[D_i]) f_h - 2\pi D \psi L_v \rho \times (Q_{v,\infty} - Q_{s0}) f_v - c_L (T_\infty - T_r) \frac{dm}{dt}_{AC_L}, \tag{11.9}$$

where T_a is the temperature at radius $r = a$, D_d is the overall diameter of the ice-liquid mixture, D_i is the diameter of ice, and κ_w is the thermal conductivity of water.

Now for the Reynolds number range of $6000 < N_{re} < 20\,000$,

$$\frac{dq}{dt} = 0.76\left[-2\pi D_i \kappa (T_\infty - T_0) f_h - 2\pi D_i \psi L_v \rho (Q_{v,\infty} - Q_{s0}) f_v\right] - c_L (T_\infty - T_0) \frac{dm}{dt}_{AC_L} \tag{11.10}$$

and for the Reynolds number range of $N_{re} > 20\,000$,

$$\frac{dq}{dt} = \chi\left[-2\pi D_i \kappa (T_\infty - T_0) f_h - 2\pi D_i \psi L_v \rho (Q_{v,\infty} - Q_{s0}) f_v\right] - c_L (T_\infty - T_0) \frac{dm}{dt}_{AC_L}, \tag{11.11}$$

where χ is $0.57 + 9 \times 10^{-6} N_{re}$.

11.4.1 Melting of small spherical ice particles with diameter greater than one thousand microns

The melting of a small spherical ice-water particle ($D < 1000$ mm) requires solving Fick's second generalized law for temperature. The equation starts with an equation proposed by Mason (1957) and used by Rasmussen and Pruppacher (1982). The temperature equation will depend upon the liquid water that is shed (if any, at ice sizes < 9 mm), the geometric arrangement of melt liquid water around the melting ice core, and internal circulations in the melt water surrounding the ice core. Bulk density of the ice water and its structural makeup will be important too. Important variables are T, T_0, $T_T(a_i)$, and a_T. The starting equation given by Mason (1957) is

$$\frac{\partial T}{\partial t} + \vec{u} \bullet \nabla T = \kappa \nabla^2 T, \tag{11.12}$$

where κ is the conductivity of heat through the liquid water, and the advective term is to account for internal circulations. By first assuming steady state, and no internal circulations, then the remaining term is the diffusion term written using spherical symmetry as

$$\nabla^2 T = \frac{d^2 T}{dr^2} + \frac{2}{r}\frac{dT}{dr} = \frac{d^2(rT)}{dr^2} = 0. \tag{11.13}$$

Integration of the diffusion term gives

$$T = \frac{c_1}{r} + c_2, \tag{11.14}$$

where the constants of integration are found using $T = T_0$ at $r = a_i$ (radius of ice core) and $T = T_T$ at $r = a_T$, which is the radius of the ice-water–liquid-water mixture. Note that $T_0 = 273.15$ K $= 0\,°$C. By making substitutions, the following can be found for the constants c_1 and c_2,

$$c_1 = \frac{(T_0 - T_T)a_i a_T}{(a_T - a_i)}, \tag{11.15}$$

and

$$c_2 = \frac{a_i T_0 - a_T T_T}{(a_i - a_T)}. \tag{11.16}$$

Now from (11.14)–(11.16), T can be obtained with relative ease,

$$T = \frac{(T_0 - T_T)a_i a_T}{r(a_T - a_i)} + \frac{a_i T_0 - a_T T_T}{(a_i - a_T)}. \tag{11.17}$$

Now $\partial T/\partial r|_{r=a_i}$ can be solved,

$$\left(\frac{\partial T}{\partial r}\right)_{r=a_i} = \frac{(T_0 - T_T)a_T a_i}{a_i^2(a_T - a_i)}. \tag{11.18}$$

The next goal is to find da_i/dt. The primary assumptions are that (i) the melting proceeds in steady-state conditions, (ii) the overall radius a_T remains constant, (iii) the ice-water particle that is melting is approximately spherical, (iv) the ice core and liquid coating remain spherical for all the time, and that internal circulations can be neglected and (v) and that the heat transfer occurs by molecular conduction. With these assumptions, it can be written, with some algebra that

$$\frac{da_i}{dt} = \frac{\kappa_w[T_0 - T_T(a_i)]a_T}{\rho_i L_f[a_T - a_i]a_i}, \tag{11.19}$$

where κ_w is the thermal conductivity of water. From these relations Drake and Mason (1966) and Rasmussen and Pruppacher (1982) give the total melt time as

$$t_m = \int_0^{t_m} dt = -\frac{\rho_i L_f}{\kappa_w a_T} \int_{a_i=0}^{a_i=a_T} \frac{[a_T - a_i]a_i}{[T_0 - T_T(a_i)]} da_i. \tag{11.20}$$

To solve for t_m, it is necessary to find $T_T(a_i)$, which can be found from an equation given by Pruppacher and Klett (1981) as

$$\begin{aligned} \frac{4\pi\kappa_w[T_0 - T_T(a_i)]a_T a_i}{[a_T - a_i]} &= -4\pi a_T \kappa_w [T_\infty - T_T(a_i)] f_h \\ &\quad -4\pi a_T L_v \psi \left[\rho_\infty - \rho_{v_{a_T}}\right] f_v, \end{aligned} \tag{11.21}$$

where ψ is the vapor diffusivity, ρ_{va_T} is the vapor density at the surface of the ice-liquid water mix, and ρ_∞ is the vapor density at some infinite distance (large distance) from the surface – in effect the environmental vapor density. Note that the typical time taken for a 700- to 900-mm-diameter-sized particle to melt is approximately 50 to 70 s.

Laboratory experiments suggest considerable disagreement with the times predicted by Mason (1957); the particles in the laboratory melted more quickly than he predicted. This is perhaps primarily a result of the asymmetry of the melting and internal circulations in the melt water, which were not considered by Mason (1957). Figure 11.4 shows an idealization of what was observed in the wind-tunnel experiment. Internal water circulations owing to drag of the surface water as the particle falls demonstrate the importance of advective/convective influences on small-ice-particle melting. Rasmussen and Heymsfield (1987a) include a correction factor to make predicted melting times in better agreement with observations.

11.4.2 Melting equation for snowflakes and snow aggregates

Based on observations of aggregates of crystals including dendrites, Mitra *et al.* (1990) found that aggregates are typically more planar or oblate than spherical in shape, and thus applied the electrostatic analog in developing a melting equation. The authors started with the standard heat-budget equation, stated as

$$\frac{dq}{dt} = -L_f \frac{dM_{ice}}{dt} - L_v \frac{dM_{liq}}{dt}. \tag{11.22}$$

The collection of water was not considered as a heat source. Then Mitra *et al.* (1990) considered a melting snowflake or snow aggregate to have a temperature of 0 °C. The resulting equation, assuming the electrostatic analog for an oblate spheroid, is

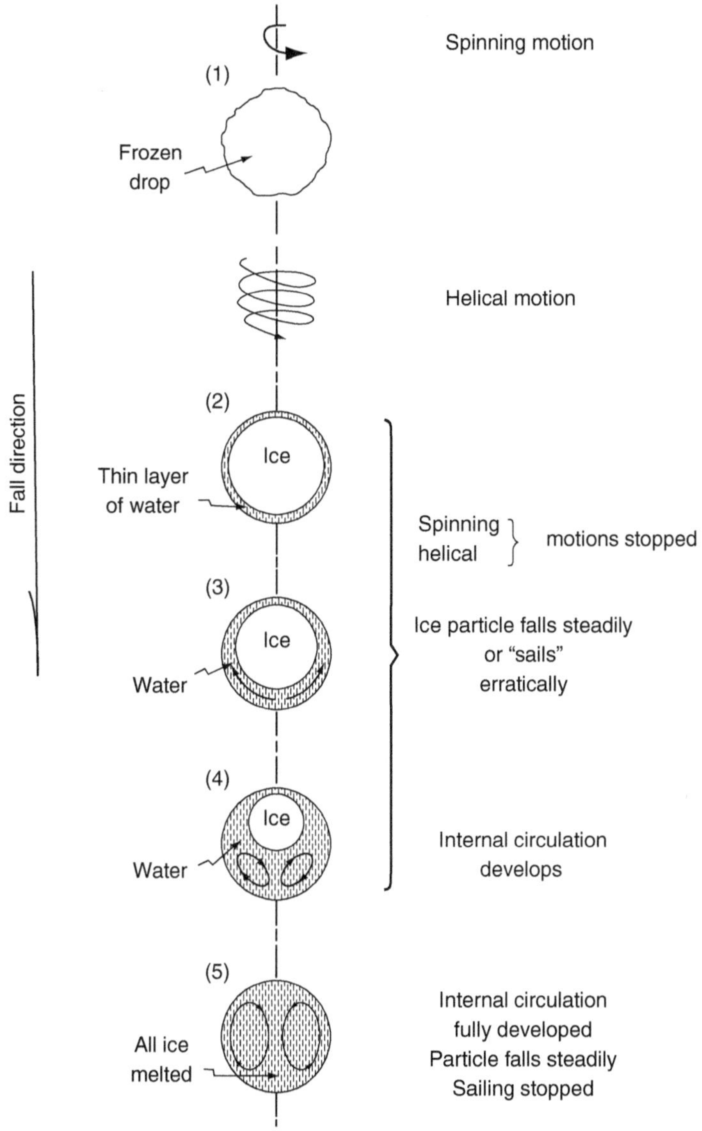

Fig. 11.4. Schematic of the melting of a spherical ice particle, as revealed by motion pictures. (From Rasmussen and Pruppacher 1982; courtesy of the American Meteorological Society.)

$$\frac{\mathrm{d}M_{\mathrm{ice}}}{\mathrm{d}t} = -\frac{4\pi fC}{L_{\mathrm{f}}}\left(K[T_{\infty}(t) - T_0] + \frac{\psi L_{\mathrm{v}}}{R_{\mathrm{v}}}\left[\frac{(RH/100)e_{\mathrm{SL}}(T_{\infty}(t))}{T_{\infty}(t)} - \frac{e_{\mathrm{SL}}(T_0)}{T_0}\right]\right). \quad (11.23)$$

In this equation, C is the capacitance, f is the ventilation coefficient, e_{SL} is the saturation vapor pressure over liquid, RH is the relative humidity, $T_{\infty}(t)$ is the temperature of air which varies with time and $T_0 = 273.15$ K is the temperature of the snow.

The ventilation coefficient is based on a length scale given by surface area, Ω, divided by the perimeter P,

$$L = \frac{\Omega}{P}, \tag{11.24}$$

where L is the length scale of the snow. The perimeter is given by

$$P = 2\pi a, \tag{11.25}$$

where a is the major axis radius of the particle. The surface area for an oblate spheroid is given as

$$\Omega = \pi a^2 \left[2 + \pi \frac{b}{a\epsilon} \ln \left(\frac{1+\epsilon}{1-\epsilon} \right) \right], \tag{11.26}$$

with b given as the minor axis radius and ϵ given by

$$\epsilon = \left[1 - \left(\frac{b}{a} \right)^2 \right]^{1/2}. \tag{11.27}$$

The ventilation coefficient is given for the assumed length scale L by

$$f_L = \begin{cases} 1 + 0.14\chi^2 & \chi < 1 \\ 0.86 + 0.28\chi & \chi = 1 \end{cases}, \tag{11.28}$$

where,

$$\chi = N_{\mathrm{sc}}^{1/3} N_{\mathrm{re},L}^{1/2}. \tag{11.29}$$

The capacitance is taken from Pruppacher and Klett (1997) as

$$C_0 = \frac{a\varepsilon}{\sin^{-1}\epsilon}, \tag{11.30}$$

using the value of a given by Mitra *et al.* (1990) as

$$a = \left[\frac{3M_{\mathrm{ice}}}{4\pi\rho_{\mathrm{ice}} b/a} \right]^{1/3}. \tag{11.31}$$

Mitra *et al.* (1990) assumed that the mass of the ice varied linearly with $M_{\mathrm{liq}}/M_{\mathrm{sw}}$ where M_{sw} is the mass of the initial snowflake or snow aggregate and M_{liq} is the mass of the liquid. The density of a dry snowflake or snow aggregate was assumed to be 20 kg m^{-3}, whilst that for a melted particle was assumed to be 1000 kg m^{-3}. The axis ratio b/a was assumed to vary linearly from 0.3 for a dry snowflake or snow aggregate to 1.0 for a completely melted particle or raindrop. Furthermore, the capacitance was assumed to vary linearly with $M_{\text{liq}}/M_{\text{sw}}$ from an initial value of $C/C_0 = 0.8$ to $C/C_0 = 1.0$ for $M_{\text{liq}}/M_{\text{sw}} = 1.0$.

11.4.3 Melting equation parameterizations

11.4.3.1 Gamma distribution parameterization

The melting equation is parameterized as follows for the complete gamma distribution

$$\frac{1}{\rho_0}\int_0^\infty \frac{\mathrm{d}M(D_x)n(D_x)\mathrm{d}D_x}{\mathrm{d}t}$$
$$= -\frac{1}{\rho_0}\int_0^\infty \frac{2\pi D_x N_{\mathrm{T}x}\mu_x \alpha_x^{\nu_x}}{\Gamma(\nu_x)}\frac{[K(T-T_0)+\rho\psi L_{\mathrm{v}}(Q_{\mathrm{v}}-Q_{\mathrm{s}0})]\left[0.78+0.308N_{\mathrm{sc}}^{1/3}\left(\frac{D_x V_{\mathrm{T}x}}{\nu}\right)^{1/2}\right]}{L_{\mathrm{f}}} \tag{11.32}$$
$$\times\left(\frac{D_x}{D_{nx}}\right)^{\nu_x\mu_x-1}\exp\left(-\alpha_x\left[\frac{D_x}{D_{nx}}\right]^{\mu_x}\right)\mathrm{d}\left(\frac{D_x}{D_{nx}}\right)$$
$$-\frac{c_{\mathrm{l}}(T-T_0)}{L_{\mathrm{f}}}\frac{1}{\rho_0}\int_0^\infty\left(\frac{\mathrm{d}M(D_x)n(D_x)\mathrm{d}D_x}{\mathrm{d}t}\right)\Bigg|_{xAC_{\mathrm{L}}};$$

for the modified gamma distribution,

$$\frac{1}{\rho_0}\int_0^\infty \frac{\mathrm{d}M(D_x)n(D_x)\mathrm{d}D_x}{\mathrm{d}t}$$
$$= -\frac{1}{\rho_0}\int_0^\infty \frac{2\pi D_x N_{\mathrm{T}x}\mu_x}{\Gamma(\nu_x)}\frac{[K(T-T_0)+\rho_0\psi L_{\mathrm{v}}(Q_{\mathrm{v}}-Q_{\mathrm{s}0})]\left[0.78+0.308N_{\mathrm{sc}}^{1/3}\left(\frac{D_x V_{\mathrm{T}x}}{\nu}\right)^{1/2}\right]}{L_{\mathrm{f}}} \tag{11.33}$$
$$\times\left(\frac{D_x}{D_{nx}}\right)^{\nu_x\mu_x-1}\exp\left(-\left[\frac{D_x}{D_{nx}}\right]^{\mu_x}\right)\mathrm{d}\left(\frac{D_x}{D_{nx}}\right)$$
$$-\frac{c_{\mathrm{L}}(T-T_0)}{L_{\mathrm{f}}}\frac{1}{\rho_0}\int_0^\infty\left(\frac{\mathrm{d}M(D_x)n(D_x)\mathrm{d}D_x}{\mathrm{d}t}\right)\Bigg|_{xAC_{\mathrm{L}}}.$$

The complete integrated gamma distribution parameterization form then is generally written for snow, graupel, frozen drops, and hail as

$$Q_xML_{\mathrm{rw}} = \left\{ \begin{array}{l} \dfrac{-2\pi N_{\mathrm{T}x}\alpha_x^{\nu_x}[K(T-T_0)+\rho\psi L_{\mathrm{v}}(Q_{\mathrm{v}}-Q_{\mathrm{s0}})]}{\rho\Gamma(\nu_x)L_{\mathrm{f}}} \\ \times\left[0.78\Gamma\dfrac{\left(\frac{1+\nu_x\mu_x}{\mu_x}\right)}{\alpha_x^{\left(\frac{1+\nu_x\mu_x}{\mu_x}\right)}}D_{nx}+0.308c_x^{1/2}\nu^{-1/2}N_{\mathrm{sc}}^{1/3}D_{nx}^{(3+d_x)/2}\dfrac{\Gamma\left(\frac{3+d_x}{2\mu_x}+\frac{\nu_x\mu_x}{\mu_x}\right)}{\alpha_x^{\left(\frac{3+d_x}{2\mu_x}+\frac{\nu_x\mu_x}{\mu_x}\right)}}\left(\dfrac{\rho_0}{\rho}\right)^{1/4}\right] \\ -\dfrac{c_{\mathrm{L}}(T-T_0)}{\rho L_{\mathrm{f}}}\left(\sum_m Q_xAC_{\mathrm{L}}\right) \end{array} \right\}, \tag{11.34}$$

where c_x is the coefficient for terminal velocity and d_x is the exponent for terminal velocity, and the summation term is the total number of particle types (rain, drizzle, cloud); ρ_0 is the reference density.

The modified gamma distribution form is

$$Q_xML_{\mathrm{rw}} = \left\{ \begin{array}{l} \dfrac{-2\pi N_{\mathrm{T}x}[K(T-T_0)+\rho\psi L_{\mathrm{v}}(Q_{\mathrm{v}}-Q_{\mathrm{s0}})]}{\rho\Gamma(\nu_x)L_{\mathrm{f}}} \\ \times\left[0.78\Gamma\left(\dfrac{1+\nu_x\mu_x}{\mu_x}\right)D_{nx}+0.308c_x^{1/2}\nu^{-1/2}N_{\mathrm{sc}}^{1/3}D_{nx}^{(3+d_x)/2}\Gamma\left(\dfrac{3+d_x}{2\mu_x}+\dfrac{\nu_x\mu_x}{\mu_x}\right)\left(\dfrac{\rho_0}{\rho}\right)^{1/4}\right] \\ -\dfrac{c_{\mathrm{L}}(T-T_0)}{\rho L_{\mathrm{f}}}\left(\sum_m Q_xAC_{\mathrm{L}}\right) \end{array} \right\} \tag{11.35}$$

and the gamma distribution is given as

$$Q_xML_{\mathrm{rw}} = \left\{ \begin{array}{l} \dfrac{-2\pi N_{\mathrm{T}x}[K(T-T_0)+\rho\psi L_{\mathrm{v}}(Q_{\mathrm{v}}-Q_{\mathrm{s0}})]}{\rho\Gamma(\nu_x)L_{\mathrm{f}}} \\ \times\left[0.78\Gamma(1+\nu_x)D_{nx}+0.308c_x^{1/2}\nu^{-1/2}N_{\mathrm{sc}}^{1/3}D_{nx}^{(3+d_x)/2}\Gamma\left(\dfrac{3+d_x}{2_x}+\nu_x\right)\left(\dfrac{\rho_0}{\rho}\right)^{1/4}\right] \\ -\dfrac{c_{\mathrm{L}}(T-T_0)}{\rho L_{\mathrm{f}}}\left(\sum_M Q_xAC_{\mathrm{L}}\right) \end{array} \right\}, \tag{11.36}$$

where

$$Q_xML_{\mathrm{rw}} = \min(Q_xML_{\mathrm{rw}}, 0.0). \tag{11.37}$$

The number concentration change owing to melting can be approximated by

$$N_{\mathrm{T}x}ML_{\mathrm{rw}} = Q_xML_{\mathrm{rw}}\frac{N_{\mathrm{T}x}}{Q_x}. \tag{11.38}$$

11.4.3.2 Log-normal parameterization

Let's start with the melting equation in mass change form for diameter D_x,

$$\begin{aligned} \frac{\partial M(D_x)}{\partial t} = &-\frac{2\pi D_x[K(T-T_0)+\rho\psi L_{\mathrm{v}}(Q_{\mathrm{v}}-Q_{\mathrm{s0}})]f_{\mathrm{v}}}{L_{\mathrm{f}}} \\ &-\frac{c_{\mathrm{L}}(T-T_0)}{L_{\mathrm{f}}}\left(\frac{\mathrm{d}M(D_x)}{\mathrm{d}t}\right)\bigg|_{xAC_{\mathrm{L}}}. \end{aligned} \tag{11.39}$$

Now the spectral density function for a log-normal distribution is also incorporated,

$$n(D_x) = \frac{N_{\mathrm{T}x}}{\sqrt{2\pi}\sigma_x D_x}\exp\left(-\frac{[\ln(D_x/D_{nx})]^2}{2\sigma_x^2}\right). \tag{11.40}$$

The melting equation for mass (11.39) is expanded in terms of mixing ratio (see Chapter 7 on diffusional growth for methodology) and (11.40) is used to obtain

$$\begin{aligned} Q_x ML_{\mathrm{rw}} = &-\frac{1}{\rho_0}\int_0^\infty \frac{2\pi D_x[K(T-T_0)+\rho\psi L_{\mathrm{v}}(Q_{\mathrm{v}}-Q_{\mathrm{s0}})]f_{\mathrm{v}}}{L_{\mathrm{f}}}\left[0.78+0.308N_{\mathrm{sc}}^{1/3}\left(\frac{DV_{\mathrm{T}}}{\nu}\right)^{1/2}\right] \\ &\times\left[\frac{1}{\sqrt{2\pi}\sigma_x D_x}\exp\left(-\frac{[\ln(D_x/D_{nx})]^2}{2\sigma_x^2}\right)N_{\mathrm{T}x}\right]\mathrm{d}\left(\frac{D_x}{D_{nx}}\right) \\ &-\frac{c_{\mathrm{L}}(T-T_0)}{L_{\mathrm{f}}}\frac{1}{\rho}\int_0^\infty \frac{\mathrm{d}M(D)n(D)}{\mathrm{d}t}\mathrm{d}D\Bigg|_{xAC_{\mathrm{L}}}. \end{aligned} \tag{11.41}$$

The integral is split to obtain,

$$\begin{aligned} Q_x ML_{\mathrm{rw}} = &-\frac{2\pi D_x[K(T-T_0)+\rho\psi L_{\mathrm{v}}(Q_{\mathrm{v}}-Q_{\mathrm{s0}})]f_{\mathrm{v}}}{\rho L_{\mathrm{f}}} \\ &\times\left[\begin{array}{l}\displaystyle\int_0^\infty \frac{0.78}{\sqrt{2\pi}\sigma}\exp\left(-\frac{\ln(D_x/D_{nx})^2}{2\sigma_x^2}\right)N_{\mathrm{T}x}\,\mathrm{d}\left(\frac{D_x}{D_{nx}}\right)+ \\ \displaystyle\int_0^\infty\left(\frac{\rho_0}{\rho}\right)^{1/4}\frac{0.308N_{\mathrm{sc}}^{1/3}\nu^{-1/2}c^{1/2}D^{\frac{1+d_x}{2}}}{\sqrt{2\pi}\sigma_x}\exp\left(-\frac{[\ln(D_x/D_{nx})]^2}{2\sigma_x^2}\right)N_{\mathrm{T}x}\,\mathrm{d}\left(\frac{D_x}{D_{nx}}\right)\end{array}\right] \\ &-\frac{c_{\mathrm{L}}(T-T_0)}{L_{\mathrm{f}}}\frac{1}{\rho_0}\int_0^\infty \frac{\mathrm{d}M(D)n(D)}{\mathrm{d}t}\mathrm{d}D\Bigg|_{xAC_{\mathrm{L}}}. \end{aligned} \tag{11.42}$$

Now the first integral is

$$\begin{aligned} &\int_0^\infty \frac{0.78}{\sqrt{2\pi}\sigma_x}\exp\left(-\frac{\ln(D_x/D_{nx})^2}{2\sigma_x^2}\right)N_{\mathrm{T}x}\,\mathrm{d}\left(\frac{D_x}{D_{nx}}\right) = \frac{0.78N_{\mathrm{T}x}}{\sqrt{2\pi}\sigma_x}\int_0^\infty \exp\left(-\frac{[\ln(D_x/D_{nx})]^2}{2\sigma_x^2}\right)\mathrm{d}\left(\frac{D_x}{D_{nx}}\right) \\ &= \frac{0.78N_{\mathrm{T}x}D_{nx}\sqrt{2\pi}\sigma_x\exp\left(\frac{\sigma_x^2}{2}\right)}{\sqrt{2\pi}\sigma_x} = 0.78N_{\mathrm{T}x}D_{nx}\exp\left(\frac{\sigma_x^2}{2}\right); \end{aligned} \tag{11.43}$$

and the second integral is

$$\frac{0.308N_{\mathrm{T}x}N_{\mathrm{sc}}^{1/3}\nu^{-1/2}c_x^{1/2}}{\sqrt{2\pi}\sigma_x}\left(\frac{\rho_0}{\rho}\right)^{1/4}\int_0^{\infty}D_x^{\frac{1+d_x}{2}}\exp\left(-\frac{[\ln(D_x/D_{nx})]^2}{2\sigma_x^2}\right)\mathrm{d}\left(\frac{D_x}{D_{nx}}\right)$$

$$= 0.308N_{\mathrm{T}x}N_{\mathrm{sc}}^{1/3}\nu^{-1/2}c_x^{1/2}\sigma_x^2 D_{nx}^{\frac{3+d_x}{2}}\left(\frac{\rho_0}{\rho}\right)^{1/4}\exp\left(\frac{[3+d_x]^2\sigma_x^2}{8}\right). \tag{11.44}$$

So the final parameterization is

$$QxML_{\mathrm{rw}} = -\frac{2\pi D_{nx}[K(T-T_0)+\rho\psi L_{\mathrm{v}}(Q_{\mathrm{v}}-Q_{\mathrm{s}0})]f_{\mathrm{v}}}{\rho_0 L_{\mathrm{f}}}$$

$$\times\left[\left(0.78N_{\mathrm{T}x}\exp\left(\frac{\sigma_x^2}{2}\right)\right)+\left(0.308N_{\mathrm{T}x}N_{\mathrm{sc}}^{1/3}\nu^{-1/2}c_x^{1/2}\sigma_x^2 D_{nx}^{\frac{3+d_x}{2}}\left(\frac{\rho_0}{\rho}\right)^{1/4}\right.\right.$$

$$\left.\left.\times\exp\left(\frac{[3+d_x]^2\sigma_x^2}{8}\right)\right)\right]-\frac{c_1(T-T_0)}{L_{\mathrm{f}}}\frac{1}{\rho_0}\int_0^{\infty}\frac{\mathrm{d}M(D)n(D)}{\mathrm{d}t}\mathrm{d}D\Bigg|_{xAC_{\mathrm{L}}}. \tag{11.45}$$

11.5 Soaking and liquid water on ice surfaces

Here a parameterization is discussed to permit liquid water, from hailstone accretion of liquid water during wet growth or from hailstone melting, to soak the porous ice, and thus cause the hailstone to become a mixed-phase particle. Upon further accretion of liquid water or melting, the hailstone may acquire liquid on its surface that may eventually be shed when it reaches a critical mass for the ice-water particle diameter. Reviews of soaking of liquid water as well as liquid water on ice-water surfaces are given by Rasmussen *et al.* (1984) and Rasmussen and Heymsfield (1987a).

Water on surfaces, and water soaked into graupel and hail, in bulk parameterization cloud models are discussed by Meyers *et al.* (1997). The amount of soaking allowed is such that the density of the ice particle in mixed phase reaches 910 kg m^{-3} for a hailstone. Liquid water accreted by graupel also should be allowed to soak, as graupel is often of lower density before melting.

Once a hailstone soaks liquid water to the maximum extent possible so that the density of the particle is at least 910 kg m^{-3}, liquid water then can exist on the hailstone surface during melting (or wet growth). Liquid water can exist on the surface of graupel and frozen drops, but this is generally from melting and not wet growth as they are too small for wet growth. Rasmussen *et al.* (1984) present two possibilities of the amount of water that can exist on

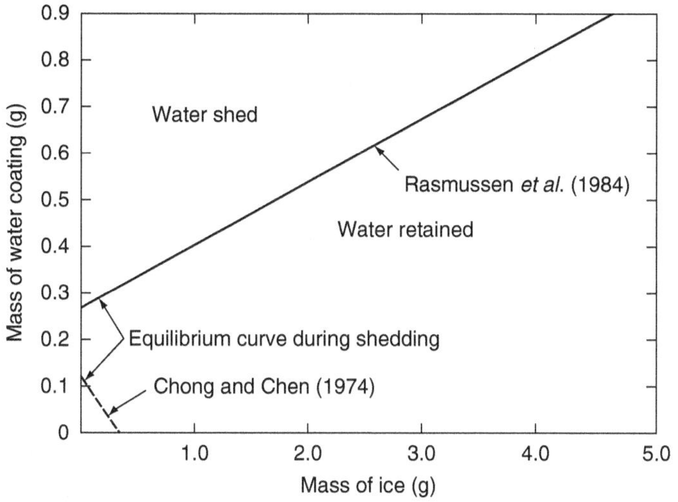

Fig. 11.5. Experimental (Rasmussen *et al.* 1984) and theoretical (Chong and Chen 1974) predictions of equilibrium mass of liquid-water coating for given mass of a spherical, 910 kg m^{-3} density ice core during melting or wet growth. By equilibrium we are referring to the maximum amount of liquid-water mass that can coat the ice core before shedding occurs. Above the respective lines water is shed, while below the respective lines liquid water is retained. (From Rasmussen and Heymsfield 1987a; courtesy of the American Meteorological Society.)

a hailstone surface before being shed from the surface (Fig. 11.5). One of the methods is by Chong and Chen (1974), shows that liquid water on the surface of a hailstone decreases as the hail increases in size, and is based on theoretical consideration. The other method by Rasmussen and Heymsfield (1987a) uses data from Rasmussen *et al.* (1984) to show that the mass of liquid water that can exist on the surface of a hailstone increases with increasing hailstone size. From the data of Rasmussen *et al.* (1984) a critical amount of water can exist on the surface of an ice-water particle before shedding occurs; the amount of this liquid water increases with hail size. The equation for the amount of liquid-water mass that can exist on an ice-water particle of diameter D_{ice} is given by Rasmussen and Heymsfield (1987a). The mass of ice water can be written in terms of density and diameter of ice,

$$M_{\text{liq,crit}}(D) = a + b\frac{\pi}{6}\rho_{\text{ice}}D_{\text{ice}}^3, \tag{11.46}$$

where the excess of the critical water soaked into low-density graupel and low-density hail is then shed from hail with $D > 9$ mm, when particle densities reach or initially are at values of 900 kg m^{-3}. In the derivation of (11.46), it is assumed that particles are spherical. Also, $a = 0.000268$ and $b = 0.1389$.

11.6 Shedding drops from melting hail or hail in wet growth

Rasmussen *et al.* (1984) studied the melting of ice particles in the range 3 to 20 mm in diameter at a constant ambient wind tunnel of temperature $T = 20\,°C$ and a relative humidity of 40%. They showed that shedding or lack of shedding occurred for different sized particles (see Fig. 11.6). Note also the

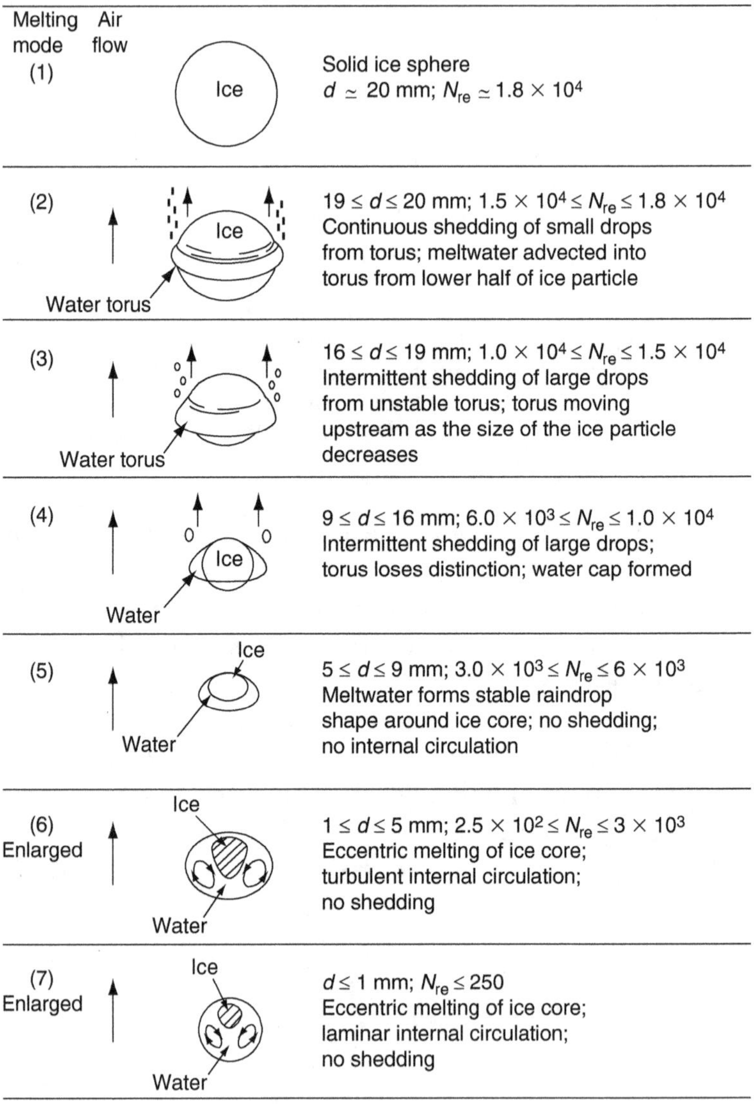

Fig. 11.6. Schematic of melting modes of an initial 2-cm diameter spherical hailstone. Discussion of the modes is given in the text. (From Rasmussen *et al.* 1984; courtesy of the American Meteorological Society.)

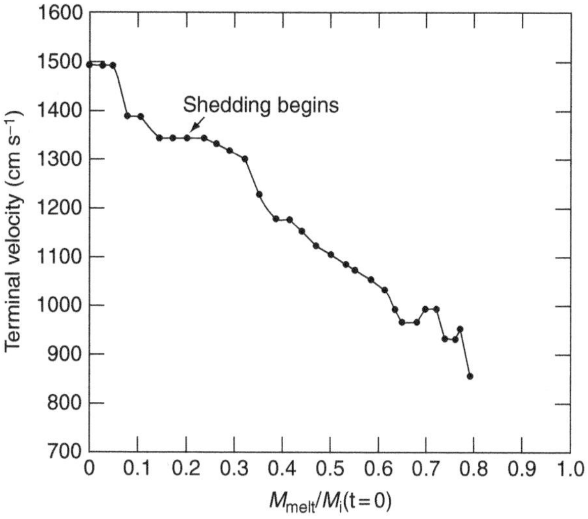

Fig. 11.7. Change in terminal velocity (cm s^{-1}) of a melting ice sphere with an initial diameter of 2 cm, as a function of non-dimensional mass melted; M_{melt} = mass of meltwater. (From Rasmussen *et al.* 1984; courtesy of the American Meteorological Society.)

slope in the terminal velocity (Fig. 11.7) of an initial 2-cm melting hailstone, and a flat zone in the terminal velocity as shedding begins.

For ice particles of $D \simeq 20$ mm ($N_{\text{re}} > 18\,000$), Rasmussen *et al.* describe what they call mode-1 melting. When melting in this mode begins, a small torus or ring of liquid begins to form about the equatorial region of the ice particle and small drops are shed from the torus.

At diameters of $19 < D < 20$ mm ($15\,000 > N_{\text{re}} > 18\,000$), a torus forms owing to melting primarily on the bottom of the ice particle (mode-2 melting). A subsequent tangential stress on the ice bottom allows advection of liquid upward with the flow of meltwater on the ice surface, and then flow separation near the equator, which permits gravity to act downward, keeping the torus at the equatorial region. The torus formation has a profound effect on reducing the terminal velocity of the ice particle. Intermittent shedding of small drops continues.

When 20% of the ice particle's mass has melted larger drops of about 1.5 mm in diameter are shed continuously from the torus, which eventually becomes unstable as it moves downstream (upward) on the ice-particle surface. The instability in the torus results in part from velocity shear and high turbulence very near the surface of the ice particle. Melting continues to occur primarily on the bottom half of the ice particle and results in the ice particle becoming somewhat oblate in shape. As noted by Rasmussen *et al.* this is similar to what Macklin (1963) and Mossop and Kidder (1962) observed on larger ice particles that were melting. Once shedding begins the terminal velocity of the ice particle stabilizes to a near-constant value for liquid fractions of 0.15 and 0.30.

For ice particles of diameter $16 < D < 19$ mm ($10\,000 < N_{re} < 15\,000$) mode-3 melting begins. With the ice particle falling at a slower speed, the stress on the ice particle is reduced and the torus slides down the surface (upstream) of the more oblate ice particle. Drops that are as large as 3 mm are sheared off from the torus in mode-3 melting, and the shedding becomes intermittent. This is because it takes time for meltwater to build up on the ice-particle surface.

As particles continue to melt and become smaller with diameters between $9 < D < 16$ mm ($6000 < N_{re} < 10\,000$), mode-4 melting ensues. In this case, the torus loses its distinction and a water cap forms near the top of the ice particle (pill shaped), from which drops as large as 3 mm are shed. An ice–liquid mixed-phase drop of 10 mm in equivalent diameter may have an equatorial diameter of 14 mm owing to the loss of instability and the initiation of a stabilizing effect on the mixed-phase particle by the meltwater at about $N_{re} = 8500$. Shedding in this smaller size range occurs primarily by bag breakup (Rasmussen *et al.* 1984) producing drops of 4.5 mm in diameter. Collisions with the mixed-phase particles may produce a burst of 300- to 400-micron droplets.

Smaller particles of $5 \leq D \leq 9$ mm ($N_{re} = 6500$) no longer shed drops (mode-5 melting). Blanchard (1950) noted from observations that particles smaller than 9 mm do not shed drops. Melting modes 6 and 7, as shown in Fig. 11.6, do not shed water. They can lose liquid water by collisions with other particles.

The axis ratios for initial sphere diameters of 6.4 mm, 7.7 mm, and 9.2 mm show that as the fraction of ice melts, these ratios quickly decrease to values as small as 0.8 for 9.2-mm particles that have 10% meltwater (Rasmussen *et al.* 1984). For 50% meltwater, the particles have axis ratios of about 0.75 for the smaller initial sizes, and 0.65 for the 9.2 mm particle. Notice that as the particles continue to melt, axis ratios approach 0.55 for all mixed-phase particles.

11.7 Parameterization of shedding by hail particles of 9–19 mm

When the amount of water that exists on the ice surface exceeds the 9–19-mm threshold, water drops are shed with sizes of approximately 0.5 to 2 mm with a mode of around 1 mm. Note that shedding does not occur until hailstone sizes exceed 9 mm. So graupel and frozen drops do not shed drops unless they undergo collisions or accrete enough liquid that they become larger than 9 mm. For this type of parameterization this condition is not permitted. When the liquid-water mass of graupel and frozen drops exceeds 50 percent of the total liquid- and ice-water mass, the particles become melted graupel and frozen drops with ice-water cores until the ice water completely melts. At this point, they are transferred to the meltwater species of rain from graupel and frozen drops. This can be described for a gamma distribution by

$$Q_{x9}AC_y = \frac{0.25\pi E_{xy} N_{\mathrm{T}x} Q_y \Delta \bar{V}_{\mathrm{T}xyQ} \left(\frac{\rho_0}{\rho}\right)^{1/2}}{\Gamma(\nu_x)\Gamma(B_y+\nu_y)} \times \begin{bmatrix} \Gamma\left(2+\nu_x, \frac{0.009}{D_{nx}}\right)\Gamma(3+\nu_y)D_{nx}^2 \\ +2\Gamma\left(1+\nu_x, \frac{0.009}{D_{nx}}\right)\Gamma(4+\nu_y)D_{nx}D_{ny} \\ +\Gamma\left(0+\nu_x, \frac{0.009}{D_{nx}}\right)\Gamma(5+\nu_y)D_{ny}^2 \end{bmatrix}. \quad (11.47)$$

Assuming a gamma distribution, (11.47) becomes

$$Q_{x9}ML_{\mathrm{rw}} = \left\{ \begin{array}{l} \underline{\frac{-2\pi N_{\mathrm{T}x}[K(T-T_0)+\rho\psi L_{\mathrm{v}}(Q_{\mathrm{v}}-Q_{\mathrm{s}0})]}{\rho_0\Gamma(\nu_x)L_{\mathrm{f}}}} \\ \times \begin{bmatrix} 0.78\Gamma\left(1+\nu_x, \frac{0.009}{D_{nx}}\right)D_{nx} + 0.308c_x^{1/2}\nu^{-1/2}N_{\mathrm{sc}}^{1/3}D_{nx}^{(3+d_x)/2} \\ \times\,\Gamma\left(\frac{3+d_x}{2}+\nu_x, \frac{0.009}{D_{nx}}\right)\left(\frac{\rho_0}{\rho}\right)^{1/4} \end{bmatrix} \\ -\frac{C_{\mathrm{L}}(T-T_0)}{\rho_0 L_{\mathrm{f}}}\left(\sum_m Q_x AC_{\mathrm{L}}\right) \end{array} \right\}. \quad (11.48)$$

The amount of liquid water on the surface is given by the sum of the water species collected by hail $D > 0.009$ m plus the amount melted on particles larger than 0.009 m plus what was present on the surface from the previous step. This is written in equation form as

$$Q_{\mathrm{h}1,9}^{n+1} = Q_{\mathrm{h}1,9}^{n} + \Delta t\left[Q_{\mathrm{h}9}ML_m + \sum_m Q_{\mathrm{h}9}AC_{\mathrm{L}}\right], \quad (11.49)$$

where the subscripts hl_9 and h9 represent large hail (hl) greater than 9 mm and large hail or hail water (hw) greater than 9 mm, respectively; the subscript h is generic for hl or hw.

Next, the hail spectrum is broken up into 50 bins that are $\Delta D_h = 0.0005$ m wide. Computing the number concentrations of the bins at each point that has hail using the following (11.50) generates the spectrum of bins,

$$N_{h,i} = \frac{N_{Th}x}{D_{n_h}\Gamma(v)}\left(\frac{D_h}{D_{nh}}\right)^{v_x-1}\exp\left(-\left[\frac{D_h}{D_{nh}}\right]\right)\Delta D_h. \tag{11.50}$$

The critical mass $M_{crit,liq}$ (D) for each bin is given as

$$M_{crit,liq}(D) = 0.000\,268 + 0.1389 a_h D_{h,i}^{b_h}, \tag{11.51}$$

following Rasmussen and Heymsfield (1987a). In (11.50)–(11.51), the index i on D is the bin size index.

For bins with $i < 18$ ($D < 0.009$ m), the liquid remains on the surface until ice reaches 9 mm in diameter and until the maximum allowable liquid has been soaked into the porous ice, if it in fact does. The equation to determine how much new mass is taken on by these smaller hailstones and graupel is

$$Q_{h,kept} = \sum_{i}^{18}\frac{N_{h,i}M_{crit,i}}{\rho}, \tag{11.52}$$

where $Q_{h,kept}$ is the mixing ratio for hailstones retained, where the subscript h is generic for hw or hl. If the particle size becomes larger than $D = 0.009$ m, it loses drops that are 0.001 mm in diameter until it returns to $D = 0.009$ m. For bins with $i > 18$ ($D > 0.009$ m), the mixing ratio becomes

$$Q_{h,kept} = \sum_{i=18}^{51}\frac{N_{h,i}M_{crit,i}}{\rho}. \tag{11.53}$$

The mixing ratio of water collected and previously stored on the ice particles with $D > 0.009$ m is first examined to see if the liquid water can be stored in the ice lattice. If the density is 900 kg m^{-3}, then the amount of mass collected and previously stored is checked to see how much can be stored on the ice particle's surface (11.53). The remainder is shed as 0.001 m diameter drops,

$$Q_{h,shed} = \max\left(Q_{h19}^{n+1} - Q_{h,kept}, 0.0\right). \tag{11.54}$$

At this point any mass that can be shed is done so using the following steps. First Q_{hl9}, the amount of liquid on the surface, is updated; to its maximum if possible. Second, the remainder, if there is mixing ratio, is computed using the mixing ratio and the concentration numbers are given by

$$N_{h,shed} = \left(\frac{6 Q_{h,shed} \rho_0}{\pi \rho_L (0.001)^3} \right). \tag{11.55}$$

The rates for $Q_{h,shed}$ and $N_{h,shed}$ are the values given by (11.54) and (11.55) divided by dt.

11.8 Sensitivity tests with a hail melting model

Rasmussen and Heymsfield (1987b) did sensitivity tests on the melting of hail. The environment is given in Fig. 11.8 with 100% relative humidity at the 0 °C level just above 5 km. Hailstones were "dropped" with initial densities of 450 and 910 kg m^{-3}. Diameters of different hailstones ranged from 0.5 to 3.0 cm (Fig. 11.9a). The rest of Fig. 11.9 pertains to a stone initially 2 cm in diameter. In Fig. 11.9b, from just below the melting level to the ground the axis ratio decreases steadily to nearly 0.6 for the higher-density particle. The lower-density particle does not experience a change in axis ratio until almost 2 km above ground. Presumably the particle is soaking melted liquid-water mass. Figure 11.9c shows the terminal velocities, the lower-density particle

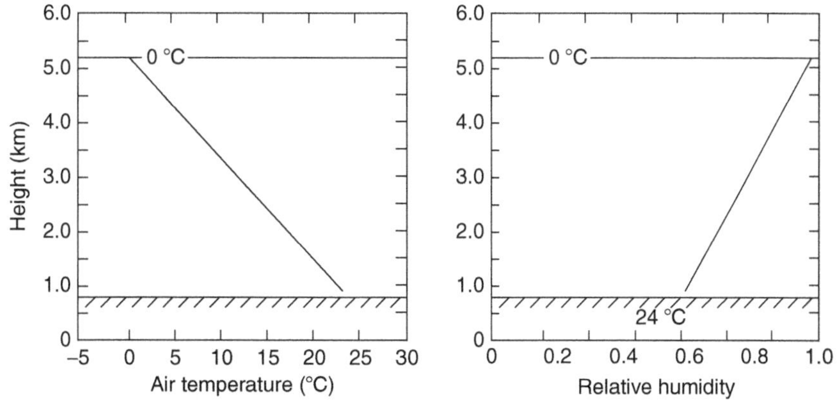

Fig. 11.8. Constant air-temperature (°C) and relative-humidity profiles used in one-dimensional sensitivity runs. The 0 °C level is at 5.2 km above mean sea level at 525 mb, while the ground is at 0.8 km above mean sea level at 920 mb. (From Rasmussen and Heymsfield 1987b; courtesy of the American Meteorological Society.)

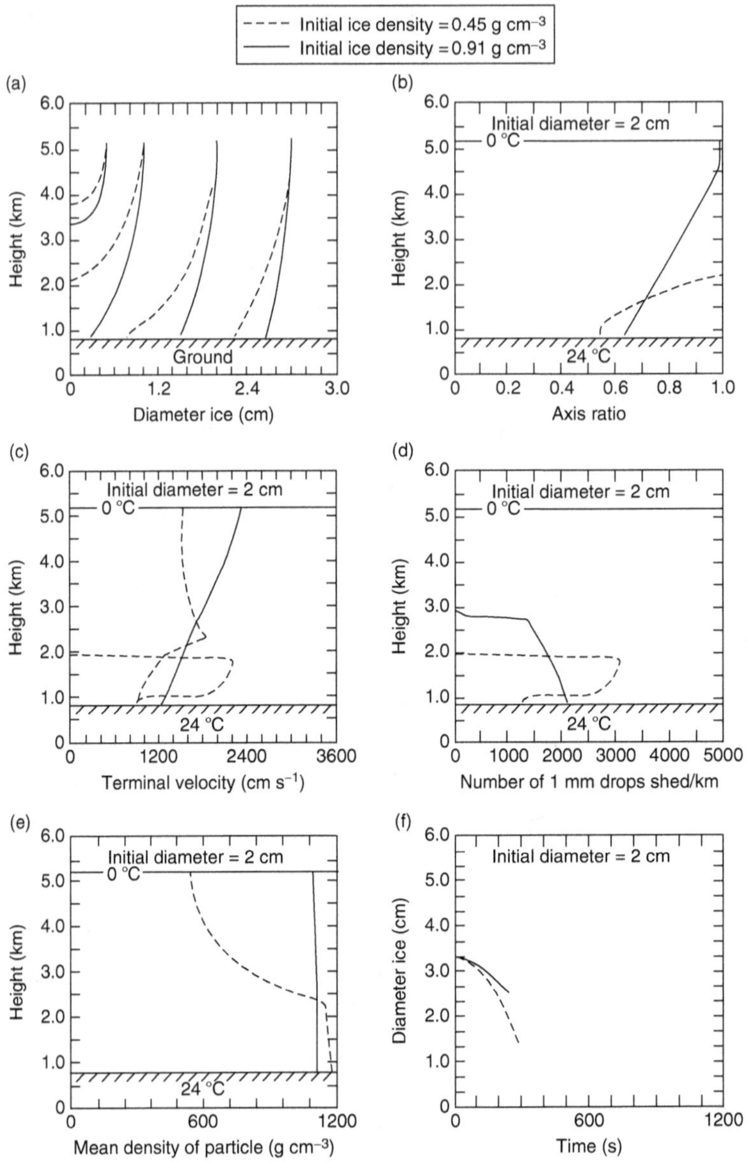

Fig. 11.9. Melting and shedding behavior of spherical ice particles with initial ice densities of 0.450 g cm^{-3} (dashed line) and 0.910 g cm^{-3} (solid line) falling through temperature and relative humidity profiles shown in Fig. 11.8, with updraft/downdraft liquid-water contents set to zero. Plot (a) represents particles with initial diameters of 0.5, 1.0, 2.0, and 3.0 cm, while plots (b)–(f) depict results for particles with 2-cm initial diameter. Plots (a)–(e) show the variation with height of (a) ice diameter cm, (b) axis ratio, minor axis/major axis, (c) terminal velocity (cm s^{-1}), (d) number of 1-mm diameter drops shed per km during melting, and (e) mean density of the particle (g cm^{-3}). Plot (f) shows the diameter of ice (cm) versus time (s) (from the start of melting). (From Rasmussen and Heymsfield 1987b; courtesy of the American Meteorological Society.)

experiencing a transition in the terminal-velocity behavior when its axis ratio begins to change. In Fig. 11.9d the depth of shedding drops starts at 3 km for the higher-density particle, whereas it starts at just below 2 km for the lower-density one. Densification of the low-density hailstone is relatively rapid and begins at the 5.2 km level (Fig. 11.9e). Finally, when the two hailstones reach the ground, their sizes are 1.7 cm for the solid ice particle, and 0.8 cm for the low-density ice (Fig. 11.9f).

12

Microphysical parameterization problems and solutions

12.1 Autoconversion of cloud to drizzle or rain development

Over the past four decades, autoconversion schemes have changed very little. Starting back in the late 1960s two forms emerged, from Berry (1968b) and Kessler (1969). The Berry scheme converted cloud water to rain water, but it did this far too quickly. In fact as soon as cloud was formed the initial Berry (1968b) scheme produced raindrops, which then grew by accretion of cloud drops. Kessler (1969) had a scheme that was designed primarily for warm-rain processes, and was different from the Berry (1968b) scheme in design. The Kessler scheme required that cloud-content amounts reached a certain value before rain was produced. The Kessler was and is still used to an amazingly large extent in cloud models.

Later, some started using a modified form of the Berry 1968 scheme again, but this time with a 2 kg kg^{-1} cut-off value for the cloud-water mixing ratio after which rain would form. However, it had no equation for number concentration and was used in models that only permitted prediction of the third moment, i.e. the mixing ratio (e.g. Lin *et al.* 1983; Ferrier 1994; Gilmore *et al.* 2004a; and Straka and Mansell 2005). After correcting for errors the Berry and Reinhardt (1974b) method became popular in some groups to predict mixing ratio and number-concentration conversion. Cohard and Pinty (2000) showed that, when programmed correctly, this scheme produced reasonable results against quasi-stochastic-growth-equation results. However, most schemes lacked an ability to distinguish between low cloud condensation numbers associated with maritime clouds and high cloud condensation numbers associated with continental clouds without modification. Many have tried to overcome this shortcoming without predicting the number of aerosols. By the 2000s bulk microphysical models had begun to predict aerosols (e.g. Cohard and Pinty 2000; Saleeby and Cotton 2004; Straka *et al.* 2009a).

Nevertheless, with all the efforts put forth none could accomplish what Cotton (1972a) found to be essential; he suggested that there was an "aging" time associated with autoconversion and the clouds with different numbers of cloud condensation nuclei would have different "aging" times for cloud drops to reach what was called raindrop size. Not until Straka and Rasmussen (1997) developed their methodology, was the aging of a parcel of cloudy air known from an Eulerian prognostic equation rather than Lagrangian parcels such as those used by Cotton (1972a). This aging time allows cloud droplets to grow to near to, or larger than, drizzle drop sizes, which would then be the appropriate state for cloud drops to become drizzle drops, or in many models, raindrops. An example of Lagrangian aging time versus actual time is shown in Fig. 12.1a. In addition the Lagrangian distance from an initial location can be predicted as shown in Fig. 12.1b,c,d. The distances are predicted with acceptable accuracy. When the parcel leaves the cloud there is a significant divergence from numerical solutions, as there should be.

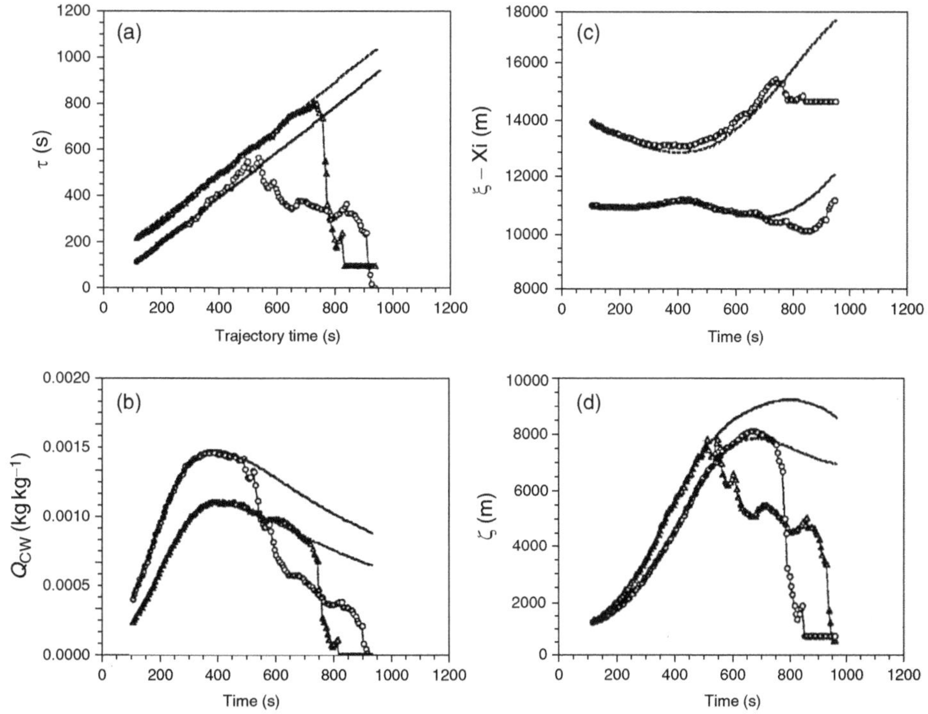

Fig. 12.1. Values of τ, $\bar{Q}_{\text{cw}}$ ξ, and ζ, computed along trajectories (solid lines) and values of these variables integrated by equations for the above variables. The lines with circles correspond to one of the test trajectories, and those through the triangles correspond to the other test trajectory. Where these lines separate from the bold lines the trajectory has left the cloud. (Straka and Rasmussen 1997; courtesy of the American Meteorological Society.)

Another problem that has hindered the development of the cloud autoconversion scheme is that all of the associated physical processes needed to be accounted for, such as self-collection and accretion. Manton and Cotton (1977), Beheng and Doms (1986), Doms and Beheng (1986), Cohard and Pinty (2000), and Seifert and Beheng (2001), to name a few, have tried to incorporate these influences in some fashion or other, but still there has been no explicit aging time associated with the growth of various types of cloud condensation number spectra. The challenge here is the following. Researchers might be advised to try out other methods, rather than using the long-standing parameterizations of Berry (1968b) and Kessler (1969), among numerous others. For example, Straka and Rasmussen (1997) showed what could be done with autoconversion and self-collection processes, along with the mean cloud-water exposure a parcel experienced. In addition the challenge is to be able to adjust the cloud condensation nuclei related to different atmospheric conditions by prognosing aerosols, perhaps by using aerosol bins such as Aitken, large, giant, and ultra-giant nuclei (Straka *et al.* 2009a).

Another challenge is that often autoconversion never does lead to rain, rather it produces what just should be drizzle, but this is treated as raindrops. This is a crucial problem for stratocumulus over oceanic areas in the eastern Pacific Ocean, for example. Thus the author suggests that the models should be able to convert cloud droplets into drizzle drops. Then, when the drizzle drops grow large enough they can be put into a rain category in a bulk microphysical model that uses cloud, drizzle, and rain categories (Straka *et al.* unpublished work; and to some extent Saleeby and Cotton 2004). Bulk microphysical models still have a long way to go in adapting the aging problem solution, and to having a model that uses the appropriate spectrum of cloud condensation nuclei.

12.2 Gravitational sedimentation

The accurate gravitational sedimentation of hydrometeors is a difficult task to undertake. With bulk microphysics parameterizations many different variables might need to have their vertical flux computed, including the mixing ratio, number concentration, and reflectivity (Fig. 12.2). Other variables include some of the many prognostic variables mentioned earlier in the book such as particle density, 'aging' time, etc. There are a couple of approaches that can be taken. First, the variable weighted mean velocity is computed together with the vertical flux with any number of finite difference schemes. The more moments that are predicted usually the more accurate the gravitational

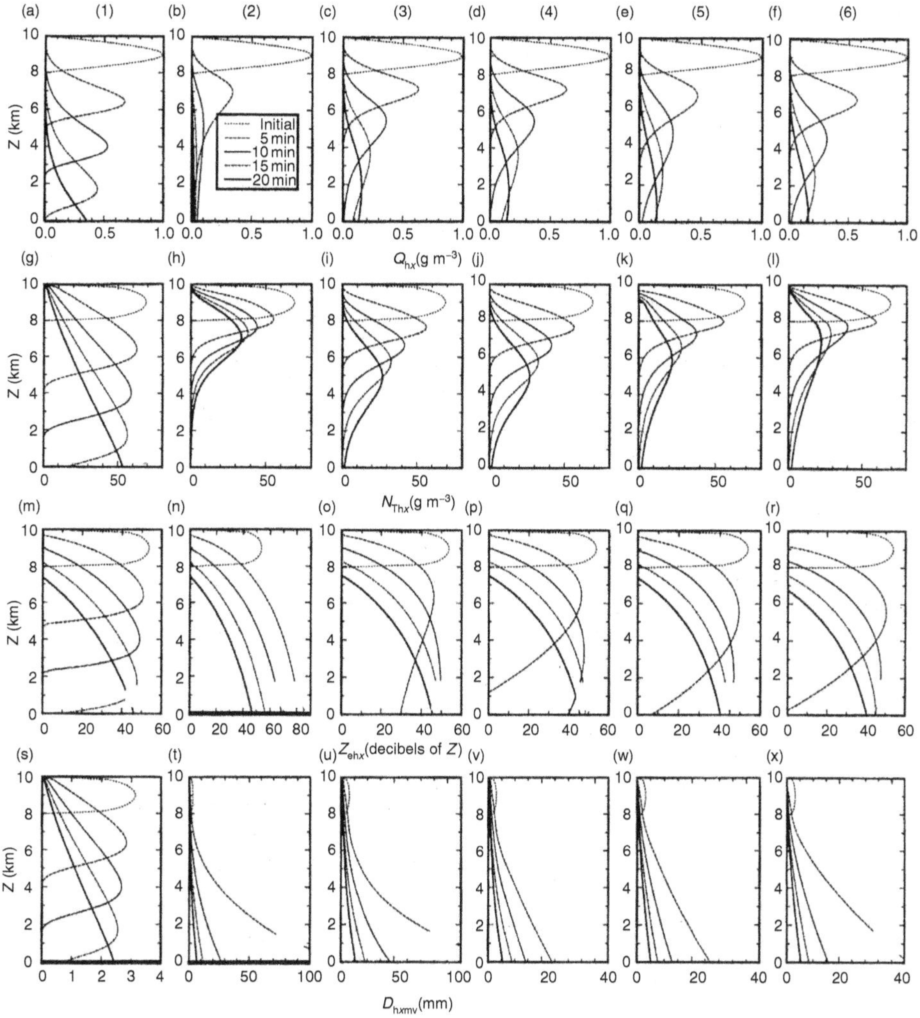

Fig. 12.2. Vertical profiles of (first row) Q_{hx}, (second row) N_{Thx}, (third row) Z_{ehx}, and (fourth row) D_{hxmv} resulting from the sedimentation of hail. The columns are as follows: (1) a one-moment scheme, (2) a two-moment scheme with fixed shape parameter equal to zero (inverse exponential with their equations), (3) a two-moment scheme with fixed shape parameter equal to three, (4) a two-moment scheme with functional shape parameter, (5) a three-moment scheme, and (6) a bin model solution. Profiles in each pane are every 5 minutes between 0 and 20 minutes. The ordinate of each is height above ground level. The abscissa for each panel is indicated under each row. Note the scale changes in n, s, and t. (From Milbrandt and Yau 2005a; courtesy of the American Meteorological Society.)

sedimentation will be (Fig. 12.2; Milbrandt and Yau 2005a). With a three-moment scheme where the shape parameter is computed the most accurate solutions are obtained compared to a bin model. This is well demonstrated by Milbrandt and Yau (2005a). However with this number of moments, at the edges of the precipitation there are often mismatches between the quantities of the moments. These mismatches, or more precisely, errors, occur in the hydrometeor distribution characteristics when computing the vertical flux and advection. Some researchers prefer to just let the mismatches occur, whereas others "clip" or set the variable to zero at locations that out ran the other moment.

Alternatively, one could make bins out of the spectrum of a category of some hydrometeor and compute the vertical flux as described below as would be done for a bin model. Then add up the fluxes of all the bins in a grid zone and add the vertical flux tendency back to the bulk quantity of the category computed in the hybrid bin microphysical model approach. The shortcoming with this method is that each moment may not have any bin model information in it. For example, while the vertical flux is computed with a bin microphysical parameterization, the memory of that bin microphysical parameterization calculation perhaps is lost after each step. The author has found though that this not the case for fallout.

In a bin microphysical model, the gravitational sedimentation is straightforward. First the terminal velocity of each mass/size bin is computed; then the vertical flux, where V_{T} is positive downward. Next the vertical flux is finite-differenced with one of any number of finite difference schemes. With more than one moment it is difficult to guarantee that some grid points or grid zones will always match at the boundaries and have values for vertical-flux computation. In the case where, for example, there is mixing ratio of some category x and no number concentration for the bin for category x then it is suggested that the computation not be made.

12.3 Collection and conversions

Collection has been covered substantially in Chapter 7. However, a few researchers have addressed some ubiquitous and perplexing problems over the past forty years. The problems with bulk hydrometeor microphysics include (a) what to do when number-concentration parameterizations grossly overestimate the number of hydrometeors exchanged; (b) how to include collection efficiencies for known quantities; (c) how to inexpensively solve the Wisner *et al.* (1972) terminal-velocity difference or the $|\Delta V_{\mathrm{T}}|$ problem; and (d) how to better solve the collection equation when it has a lower size end cut-off or an upper size cut-off.

With (a) the state-of-the-art procedure is really quite a simple one, which preserves the slope of the distribution,

$$\frac{\mathrm{d}N_{\mathrm{T}x}AC_y}{\mathrm{d}t} = \frac{\mathrm{d}Q_xAC_y}{\mathrm{d}t}\frac{N_x}{Q_x}. \tag{12.1}$$

A significant problem still exists, and that is to decide when this equation should be used or some other parameterization previously presented.

For collection efficiencies, raindrops collecting cloud droplets is pretty well handled with a polynomial proposed by Proctor (1987) as described in Chapter 7. Alternatively the collision efficiencies proposed by Rogers and Yau (1989) give similar numbers as the polynomial when mean-volume diameters of raindrops and cloud droplets are considered. Alternatively the collection efficiencies developed by Cooper *et al.* (1997) also are very accurate. Cooper *et al.* (1997) multiplied known collision efficiencies and coalescence values. These same values for collection efficiencies can be used for rain collecting snow, and rain collecting graupel. For hail collecting cloud water values from Macklin and Bailey (1966) can be used. There are no known values for hail collecting rain so efficiencies of 1.0 are usually used, but this does not account for splashing, rebounding, or drop breakup. Moreover, hail, graupel, and snow aggregates collecting other ice particles are very poorly known as are ice aggregation efficiencies. Interestingly the values of ice crystals riming cloud droplets are known fairly well owing to the numerical simulation work by Wang and Ji (1992). An excellent discussion of these can be found in Pao K. Wang's (2002) book on *Ice Microdynamics*. One thing is certain, that much wind-tunnel work needs to be done to find the poorly known collection efficiencies.

Next the Wisner *et al.* (1972) velocity-difference problem in the collection equation is questioned. In the early 1990s Verlinde *et al.* (1990) derived analytical equations for continuous collection, general collection, and self-collection. These are expensive to use and it is usually necessary to design lookup tables. Other ways of getting around the zone where terminal velocities of different particles have nearly the same terminal velocity have been devised that reduce errors to some acceptable amount. Collection still occurs in nature, but the equation for $|\Delta V_\mathrm{T}|$ still is in some error (Mizuno 1990; Murakami 1990). These methods are formulated in Chapter 7 and prevent zeros from occurring when they shouldn't. Unfortunately these are only given for a couple of particle-type interactions. Moreover, substantial errors can occur in zones on either side of zero where terminal velocities match. Milbrandt and Yau (2005b) used the formula of Murakami (1990) with all collection interactions. Gaudet and Schmidt (2005) and Seifert and Beheng

(2005) have developed collection schemes that approximately take care of the zero-velocity difference-zone problems. Ultimately hybrid-bin microphysical models may be the best way to handle the collection equations, along with autoconversion and some conversions. But whether this is warranted before more advanced collection efficiencies are developed is of some concern.

Then there is Gaudet and Schmidt's (2007) concern about violating the criteria for numerical stability of various microphysical processes, especially during collection. They advanced the use of numerical bounding of the collection equation solutions. There are many problems with criteria for stability for various microphysical processes. One is checking for over-depletion of a hydrometeor sink; if there is over-depletion, to renormalize the sinks so that they do not over-deplete the collected hydrometeor, for example. A problem that still stands is that if one process overwhelmingly depletes a hydrometeor species, the renormalizing can make other processes essentially zero.

A serious problem is making a decision on how many particles are transferred during a collection process (or any process for that matter). Number concentrations seem horribly difficult to get right without resorting to just preserving the slope of hydrometeors. Perhaps two-moment, hybrid-bin models that incorporate a scheme such as that of Tzivion *et al.* (1987) should be incorporated with solutions made into lookup tables.

Finally, some modelers have made attempts to integrate hail from 5 mm to ∞ rather than from 0 mm to ∞. The purpose is to improve accuracy in collection rates as hail really starts at 5 mm. Some substantial differences, and perhaps improvements, have been claimed by Curic and Janc (1997) concerning collection. Differences in amounts of hail produced using their proposed scheme may be as large as factor of 3 from traditional methods. It is claimed that using the mass-weighted mean terminal velocity proportional to $D^{1/2}$ underpredicts hail terminal velocities by no more than 12% using a density of 900 kg m^{-3} and a drag coefficient of 0.6, which is valid for oblate hail. First the terminal velocity is redefined using partial gamma functions, as given by

$$\overline{V_{\mathrm{Thw}}} = \frac{\int_{D_{\mathrm{hw}^*}}^{\infty} V_{\mathrm{Thw}} Q_{\mathrm{hw}} \mathrm{d}D_{\mathrm{hw}}}{\int_{D_{\mathrm{hw}^*}}^{\infty} Q_{\mathrm{hw}} \mathrm{d}D_{\mathrm{hw}}} \tag{12.2}$$

where D_{hw^*} *is* 0.005 m and $\overline{V_{\mathrm{Thw}}}$ is given by

$$\overline{V_{\mathrm{Thw}}} = \left(\frac{4 g \rho_{\mathrm{hw}} D_{\mathrm{hw}}}{3 C_{\mathrm{d}} \rho} \right)^{0.5}, \tag{12.3}$$

where C_d is the drag coefficient. Now with the incomplete gamma functions for an inverse exponential the equation $\overline{V_{\mathrm{Thw}}}$ is given by

$$\overline{V_{\mathrm{Thw}}} = \frac{\Gamma(4.5)}{\Gamma(4.0)} D_{\mathrm{hw}}^{0.5} \frac{1-\alpha_1}{1-\alpha_2}\left(\frac{4g\rho_{\mathrm{hw}}}{3c_d\rho}\right) \tag{12.4}$$

where α_1 and α_2 are given by

$$\alpha_1 = \frac{\Gamma\left(4.5, \frac{D_{\mathrm{hw}*}}{D_{n\mathrm{hw}}}\right)}{\Gamma(4.5)} \tag{12.5}$$

and

$$\alpha_2 = \frac{\Gamma\left(4.0, \frac{D_{\mathrm{hw}*}}{D_{n\mathrm{hw}}}\right)}{\Gamma(4.0)}. \tag{12.6}$$

In addition $D_{n\mathrm{hw}}$ and $D_{n\mathrm{hw}0}$ are given by

$$D_{n\mathrm{hw}} = D_{n\mathrm{hw}0} f^{0.25} \tag{12.7}$$

and

$$D_{n\mathrm{hw}} = D_{n\mathrm{hw}0}\left(\frac{\rho Q_{\mathrm{hw}}}{\pi\rho N_{\mathrm{Thw}}}\right) \tag{12.8}$$

and

$$f = 1 - \frac{\Gamma\left(4.0, \frac{D_{\mathrm{hw}*}}{D_{n\mathrm{hw}}}\right)}{\Gamma(4.0)}. \tag{12.9}$$

Together the equations above are used to redesign the collection equations of hail collecting, for example, cloud water. The effect is more aggressive hail collection of cloud water, and other hydrometeors for that matter. This occurs by the terminal velocity being some 20 to 40% larger with the proposed technique.

12.4 Nucleation

A significant problem is that few models predict aerosols or more specifically cloud condensation nuclei and ice nuclei. At present bin microphysical models seem to be more likely to predict these nuclei. This probably is because bin microphysical models already predict individual spectral hydrometeor characteristics at each grid point, and the spectral nature of the distributions can easily adapt to different-sized nuclei. Of particular interest are the roles of giant and ultra-giant nuclei in different cloud situations. Thus far bulk microphysical parameters have been able to account for one size of aerosols. However, Saleeby and Cotton (2004) used one model that may be thought of as considering two sizes of cloud-drop aerosols. These then interacted accordingly to their likely physics. In addition the model by Straka *et al.* (unpublished work) predicts four aerosol sizes; however, at present this only has one cloud-droplet size and thus this cannot easily interact physically in a correct way with other hydrometeors. In general, any impact aerosols have on cloud model results, whether bin or bulk microphysical in nature, is probably very important. There are dynamical effects, such as mixing different parcels of air; microphysical effects, such as turbulence influence of droplet growth; or autoconversion for the bulk models. For ice nuclei, bin microphysical models will probably lead the way in first predicting the number of ice nuclei, removing used ice nuclei, and how ice nuclei will nucleate. For both bulk and bin models perhaps an unsolvable problem is how to redistribute ice nuclei and cloud condensation sizes upon complete sublimation and evaporation. In many cases much larger nuclei will be generated, but how much bigger is a major problem to be tackled.

12.5 Evaporation

Some problems appear with evaporation of rain. One is that the calculation of the number concentration in multi-moment models typically employs slope-preserving parameterizations, which are an oversimplification of what happens in nature when compared to bin models. A solution was provided in Chapter 5 where only the droplets that can completely evaporate will do so, whilst all droplets and drops evaporate according to the vapor diffusion equation. Recently, Seifert (2008) reinvestigated the evaporation equation, in particular the use of Lin *et al.*'s (1983) method; he approached the problem using the gamma distribution, and designed an approach based upon the mean-volume diameter of raindrops. Seifert's results were founded upon empirical fits to bin model results similar to those of Hu and Srivastava (1995).

12.6 Conversion of graupel and frozen drops to hail

Another problem is attempting to decide algorithmically when graupel and frozen drops should be converted to hail. The author proposes that using the results of Straka and Rasmussen (1997) to get the aging time, and time-weighted mass exposure and collection of graupel to hail, or frozen drops to hail, seems to be a reasonable means to start a new solution to this problem.

This has its roots in Schoenberg-Ferrier (1994) integrating the continuous growth equation for larger particles for say 120 s and seeing how the spectra of graupel and frozen drops are distributed. The particles larger than 5 mm and 20 mm could be converted to hail, and large hail. The mass and numbers were found using Farley *et al.* (1989).

12.7 Shape parameter diagnosis from precipitation equations

Milbrandt and Yau (2005a,b) used a variable shape parameter that is either determined from an empirical fit or from the growth equations including mixing ratio, number concentration, and reflectivity. More recently Seifert (2008, personal communication) used two moments, mixing ratio and number concentration, to determine the shape parameter or prediction of evaporation. As was seen early in this chapter, the use of the variable shape parameter by Milbrandt and Yau (2005a,b) readily produced more accurate solutions for fallout results from a bin model than either the one-moment or two-moment model. In tests with an empirical form of the shape parameter the results were nearly as good as with prediction of three moments. It is uncertain still how the variable shape parameter influences the accuracy of other microphysical parameterizations.

13

Model dynamics and finite differences

13.1 One-and-a-half-dimensional cloud model

The one-and-a-half-dimensional cloud model has been used by many over the years (e.g. Ogura and Takahashi 1973), in particular to study precipitation processes. Today, testing is the main reason for using this model. The one-and-a-half-dimensional cloud is assumed to be cylindrical with a time-independent updraft radius, a. All variables are assumed to be a function of the vertical coordinate, z, only. The environmental values do not change, whereas those in the cloud do change by advection, mixing, entrainment, and microphysical-related processes.

First, the vertical momentum equation is given following Ogura and Takahashi (1973) as the following, with advection, mixing, and entrainment and buoyancy terms,

$$\begin{aligned}\frac{\partial w}{\partial t} = & -w\frac{\partial w}{\partial z} - \frac{2\alpha^2 w}{a}|w| + \frac{2}{a}u_a(w - w_a) \\ & + g\left(\frac{\theta - \bar{\theta}}{\bar{\theta}} + 0.608\left[Q_{\mathrm{v}} - \bar{Q}_{\mathrm{v}}\right] - \sum_{m=1}^{M} Q_m\right).\end{aligned} \tag{13.1}$$

Here t is the time, g is gravity, u_a is the radial component of velocity, w is the vertical velocity, θ is the potential temperature, subscript e means environment, α^2 is the coefficient for lateral eddy mixing at the perimeter of the cloud, Q_{v} is the vapor mixing ratio, and Q_m is the hydrometeor mixing ratio. The overbars denote environment values. In (13.1) u_a is determined from continuity as given by

$$\frac{2}{a}u_a + \frac{1}{\rho}\frac{\partial(\rho w)}{\partial t} = 0, \tag{13.2}$$

where ρ denotes the air density.

Next it is assumed that for any variable A, that

$$\begin{cases} A_a = \bar{A} & \text{for } u_a < 0 \\ A_a = A & \text{for } u_a > 0, \end{cases} \tag{13.3}$$

where A_a is the value of a at the edge of the updraft. The first three terms on the right-hand side of (13.1) are the vertical advection, lateral eddy mixing between the cloud and the environment, the dynamic entrainment required to satisfy continuity (13.2) between the cloud and environment. The last term on the right-hand side of (13.1) is the buoyancy term, that includes within it virtual potential-temperature differences between the cloud and environment, and drag by the weight of hydrometeors.

Next the thermodynamic equation is given as follows where the terms on the right-hand side of (13.4) are advection, mixing, entrainment, and microphysical source terms (enthalpy of vaporization, freezing, and sublimation as required).

$$\frac{\partial \theta}{\partial t} = -w\frac{\partial \theta}{\partial z} - \frac{2\alpha^2}{a}|w|\left(\bar{\theta} - \theta\right) + \frac{2}{a}u_a(\theta - \theta_a) + \frac{L_V}{c_p \pi}\frac{\delta M}{\delta t}, \tag{13.4}$$

where, π is the Exner function, θ_a is the potential temperature at the edge of the updraft, and $\delta M/\delta t$ is the condensation/evaporation rate for liquid or the deposition/sublimation rate for ice. Freezing and melting are also accounted for by this term. The vapor mixing ratio rate equation is given as

$$\frac{\partial Q_V}{\partial t} = -w\frac{\partial Q_V}{\partial z} - \frac{2\alpha^2}{a}|w|\left(\bar{Q}_V - Q_V\right) + \frac{2}{a}u_a\left(Q_V - Q_{V,a}\right) - \frac{\delta M}{\delta t}, \tag{13.5}$$

where $Q_{V,a}$ is the mixing ratio at the edge of the updraft. Similarly, the hydrometeor equations are

$$\begin{aligned} \frac{\partial Q_m}{\partial t} = &- w\frac{\partial Q_m}{\partial z} + \frac{1}{\rho}\frac{\partial(\rho V_{Tm} Q_m)}{\partial z} - \frac{2\alpha^2}{a}|w|\left(\bar{Q}_m - Q_m\right) \\ &+ \frac{2}{a}u_a\left(Q_m - Q_{m,a}\right) + S_{Q_m} \end{aligned} \tag{13.6}$$

where V_{Tm} is the terminal velocity of species or bin m and S_{Q_m} is the source–sink term for the hydrometeor species m. The zeroth- and third-moment equations can be prognosed using the equation for number concentration N_T, or mixing ratio Q, as a template,

$$\begin{aligned} \frac{\partial(N_{Tm})}{\partial t} = &- w\frac{\partial(N_{Tm})}{\partial z} + \frac{\partial(V_{Tm} N_{Tm})}{\partial z} - \frac{2\alpha^2}{a}|w|(\bar{N}_{Tm} - N_{Tm}) \\ &+ \frac{2}{a}u_a\left(N_{Tm} - N_{Tm,a}\right) + S_{N_m}. \end{aligned} \tag{13.7}$$

13.1.1 Finite differences for the one-and-a-half-dimensional model

First a staggered grid is established. The vertical velocity equation is centered by using second-order quadratic conserving inspace leapfrog differences in time. The overbars denote averages for the finite-difference expressions. This is written as the following,

$$\begin{aligned}\overline{\delta_{2t}w}^{t} &= -\overline{w\delta_z w}^{z} - \frac{2\alpha^2 w}{a} + \frac{2}{a}\overline{u}_a^{z}(w - w_a) \\ &+ g\left[\left(\frac{\overline{\theta}^{z} - \theta}{\theta}\right) - 1 + 0.608\left(\overline{Q}_{\mathrm{V}}^{z} - Q_{\mathrm{V}}\right) - \sum_{1}^{M} \overline{Q}_m^{z}\right].\end{aligned} \tag{13.8}$$

The first term on the left-hand side is expressed with centered second-order finite-difference approximations. The first and last term on the right are also expressed assuming centered second-order approximations at time t. The mixing and entrainment terms are represented in time with forward time differences from $t = t - \Delta t$. The following equations for those variables defined at theta points (e.g. θ, Q_{V}, Q_m, $N_{\mathrm{T}m}$, etc.) are approximated with the same difference techniques, except the vertical flux which is differenced with a forward-in-time, upstream scheme, and the source and sink terms with forward-in-time differences from $t = t - \Delta t$.

The other equations all are expressed in finite-difference form as shown below,

$$\begin{aligned}\overline{\delta\phi_{m,2t}}^{t} &= -\overline{w\delta_z\phi_{m,x}}^{z} + \delta_z V_{\mathrm{T}m}\phi_{m,x} - \frac{2\alpha^2}{a}|\overline{w}^{z}|\left(\bar{\phi}_{m,x} - \phi_{m,x}\right) \\ &+ \frac{2}{a}u_a\left(\phi_{m,x} - \bar{\phi}_{m,x}\right) + S_{\phi_{m,x}}.\end{aligned} \tag{13.9}$$

13.2 Two-dimensional dynamical models

13.2.1 Slab-symmetric model

The slab-symmetric model is a two-dimensional x–z plane or y–z plane model that has been widely used in cloud modeling of squall lines, single-cell convection, and orographic precipitation. The biggest problem with the slab-symmetric model is that the energy can cascade upscale because of the lack of the third dimension. Vortex pairing can generate large-scale vortices on the plane of the model that can be strong enough to make the compensating subsidence more warm than the updraft. This produces warming more strongly than that resulting from enthalpy of vaporization, freezing, and deposition with the associated convective updraft.

The u-momentum equation is given by

$$\frac{\partial u}{\partial t} = -u\frac{\partial u}{\partial x} - w\frac{\partial u}{\partial z} - \frac{1}{\bar{\rho}}\frac{\partial p'}{\partial x} + D_u, \tag{13.10}$$

where D_u is the diffusion term. Similarly the vertical momentum equation is given as follows, where D_w is the diffusion term.

$$\begin{aligned}\frac{\partial w}{\partial t} = & -u\frac{\partial w}{\partial x} - w\frac{\partial w}{\partial z} - \frac{1}{\bar{\rho}}\frac{\partial p'}{\partial z} \\ & + g\left(\frac{(\theta - \bar{\theta})}{\bar{\theta}} + 0.61\left(Q_{\mathrm{v}} - \bar{Q}_{\mathrm{v}}\right) + \frac{c_{\mathrm{v}}\,p'}{c_p\,\bar{p}} - \sum_{m=1}^{M} Q_m\right) + D_w.\end{aligned} \tag{13.11}$$

Here the overbars denote the base state and primes denote perturbations from the base state.

Next is the continuity equation in anelastic form,

$$\frac{\partial(\bar{\rho}u)}{\partial x} + \frac{\partial(\bar{\rho}w)}{\partial z} = 0. \tag{13.12}$$

Then there is the thermodynamic equation, where D_θ is the diffusion and S_θ denotes the source and sink terms,

$$\frac{\partial \theta}{\partial t} = -u\frac{\partial \theta}{\partial x} - w\frac{\partial \theta}{\partial z} + D_\theta + S_\theta. \tag{13.13}$$

The moisture equations include the vapor equation, and any number of mixing-ratio and number-concentration equations to represent any number of species or other aspects of microphysics (i.e. source and sink terms).

$$\frac{\partial Q_{\mathrm{v}}}{\partial t} = -u\frac{\partial Q_{\mathrm{v}}}{\partial x} - w\frac{\partial Q_{\mathrm{v}}}{\partial z} + D_{Q_{\mathrm{v}}} + S_{Q_{\mathrm{v}}}, \tag{13.14}$$

$$\frac{\partial Q_x}{\partial t} = -u\frac{\partial Q_x}{\partial x} - w\frac{\partial Q_x}{\partial z} - \frac{1}{\bar{\rho}}\frac{\partial\left(\bar{\rho}V_{\mathrm{T}Q_x}\right)}{\partial z} + D_{Q_x} + S_{Q_x}, \tag{13.15}$$

and

$$\frac{\partial N_{\mathrm{T}x}}{\partial t} = -u\frac{\partial N_{\mathrm{T}x}}{\partial x} - w\frac{\partial N_{\mathrm{T}x}}{\partial z} + \frac{\partial\left(V_{N_{\mathrm{T}}}N_{\mathrm{T}x}\right)}{\partial z} + D_{N_{\mathrm{T}x}} + S_{N_{\mathrm{T}x}}. \tag{13.16}$$

13.2.1.1 Finite differences for slab-symmetric models

The horizontal- and vertical-velocity equations, and other equations are discretized with centered second-order quadratic differences in space for the advection terms, and lagged forward-in-time differences for the diffusion equations. These are done on an Arakawa-C grid,

$$\overline{\delta_{2t}u}^{t} = -\overline{\bar{u}^{x}\delta_{x}u}^{x} - \overline{\bar{w}^{xz}\delta_{z}u}^{z} - \frac{1}{\bar{\rho}}\delta_{x}p' + D_{u}, \tag{13.17}$$

$$\begin{aligned}\overline{\delta_{2t}w}^{t} = &- \overline{\bar{u}^{xz}\delta_{x}w}^{x} - \overline{w\delta w}^{z} - \frac{1}{\bar{\rho}}\delta_{z}p' \\ &+ g\left(\overline{\frac{(\theta-\bar{\theta})}{\bar{\theta}}}^{z} + 0.61\overline{(Q_{\mathrm{v}}-\bar{Q}_{\mathrm{v}})}^{z} + \frac{c_{v}}{c_{p}}\overline{\left(\frac{p'}{\bar{p}}\right)}^{z} - \sum_{m=1}^{M}\overline{\bar{Q}_{m}}^{z}\right) + D_{w}\end{aligned} \tag{13.18}$$

and

$$\overline{\delta_{2t}\theta}^{t} = -\overline{u\delta_{x}\theta}^{x} - \overline{w\delta_{z}\theta}^{z} + D_{\theta} + S_{\theta}. \tag{13.19}$$

The diffusion terms for the velocity and scalar equations are shown next, where K is the eddy mixing coefficient, defined at the scalar points, and A is described in Chapter 13.3.
For unsaturated conditions

$$R_{i} = \frac{g}{\bar{\theta}_{\mathrm{v}}}(\delta_{z}\theta_{\mathrm{v}})S^{-2}, \tag{13.20}$$

and for saturated conditions

$$R_{i} = \left(A\delta_{z}\theta_{e} - g\sum_{m=1}^{M}\delta_{z}Q_{m}\right)S^{-2}, \tag{13.21}$$

$$S^{2} = 2\left\{(\delta_{x}u)^{2} + (\delta_{z}w)^{2}\right\} + \overline{(\delta_{x}u + \delta_{z}w)^{2}}^{xz}, \tag{13.22}$$

$$D_{u} = \delta_{x}(K\delta_{x}u) + \delta_{z}\left(\bar{K}^{xz}[\delta_{z}u + \delta_{x}w]\right), \tag{13.23}$$

$$D_{v} = \delta_{x}(\bar{K}^{x}\delta_{x}v) + \delta_{z}(\bar{K}^{z}[\delta_{z}v]), \tag{13.24}$$

$$D_{w} = \delta_{x}(\bar{K}^{xz}[\delta_{z}u + \delta_{x}w]) + \delta_{z}(K\delta_{z}w), \tag{13.25}$$

$$D_{Q_{\mathrm{v}}} = \delta_{x}(\bar{K}^{x}\delta_{x}Q_{\mathrm{v}}) + \delta_{z}(\bar{K}^{z}\delta_{z}Q_{\mathrm{v}}), \tag{13.26}$$

$$D_{Q_{m}} = \delta_{x}(\bar{K}^{x}\delta_{x}Q_{m}) + \delta_{z}(\bar{K}^{z}\delta_{z}Q_{m}), \tag{13.27}$$

$$D_{N_{m}} = \delta_{x}(\bar{K}^{x}\delta_{x}N_{m}) + \delta_{z}(\bar{K}^{z}\delta_{z}N_{m}). \tag{13.28}$$

where,

$$K = L^{2}(1 - R_{i})^{1/2}S$$

13.2.2 Axisymmetric model

Assuming cylindrical coordinates (r, φ, z) with the Coriolis parameter (on the f-plane) the three momentum equations are given as

$$\frac{\partial u}{\partial t} = -u\frac{\partial u}{\partial r} - w\frac{\partial u}{\partial z} + \left(f + \frac{v}{r}\right)v - \frac{1}{\rho}\frac{\partial p'}{\partial r} + D_u, \tag{13.29}$$

$$\frac{\partial v}{\partial t} = -u\frac{\partial v}{\partial r} - w\frac{\partial v}{\partial z} - \left(f + \frac{v}{r}\right)u + D_v, \tag{13.30}$$

and

$$\frac{\partial w}{\partial t} = -u\frac{\partial w}{\partial r} - w\frac{\partial w}{\partial z} + g\left(\frac{\theta - \bar{\theta}}{\bar{\theta}} + 0.608\left[Q_{\mathrm{V}} - \bar{Q}_{\mathrm{V}}\right] - \sum_{i=1}^{m} Q_m\right) + D_w. \tag{13.31}$$

The mass continuity is given by

$$\frac{\partial p'}{\partial t} = -c_s^2\left(\frac{\partial(u\bar{\rho})}{\partial r} + \frac{\partial(w\bar{\rho})}{\partial z}\right); \tag{13.32}$$

the first law of thermodynamics by

$$\frac{\partial \theta}{\partial t} = -u\frac{\partial \theta}{\partial r} - w\frac{\partial \theta}{\partial z} + D_\theta + S_\theta + R_\theta; \tag{13.33}$$

and the conservation mixing ratio of water vapor is given by

$$\frac{\partial Q_{\mathrm{V}}}{\partial t} = -u\frac{\partial Q_{\mathrm{V}}}{\partial r} - w\frac{\partial Q_{\mathrm{V}}}{\partial z} + D_{Q_{\mathrm{V}}} + M_{Q_{\mathrm{V}}}. \tag{13.34}$$

The conservation mixing ratio of condensate is given by

$$\frac{\partial Q_m}{\partial t} = -u\frac{\partial Q_m}{\partial r} - w\frac{\partial Q_m}{\partial z} + \frac{1}{\bar{\rho}}\frac{\partial(\bar{\rho}V_{\mathrm{T}}Q_m)}{\partial z} + D_{Q_m} + S_{Q_m}; \tag{13.35}$$

and the number concentration of condensed particles by

$$\frac{\partial N_m}{\partial t} = -u\frac{\partial N_m}{\partial r} - w\frac{\partial N_m}{\partial z} + \frac{\partial(V_{\mathrm{T}}N_m)}{\partial z} + D_{N_m} + S_{N_m}. \tag{13.36}$$

The following equations are a closed set of primitive equations to describe fluid flow and its energetics. The diffusion terms, D, are given as

$$D_u = \frac{1}{r}\frac{\partial(r\tau_{rr})}{\partial r} + \frac{\partial(r\tau_{rz})}{\partial z} - \frac{\tau_{\phi\phi}}{r}, \tag{13.37}$$

$$D_v = \frac{1}{r^2}\frac{\partial\left(r^2\tau_{r\phi}\right)}{\partial r} + \frac{\partial\left(r\tau_{z\phi}\right)}{\partial z}, \tag{13.38}$$

$$D_w = \frac{1}{r}\frac{\partial(r\tau_{rz})}{\partial r} + \frac{\partial(r\tau_{zz})}{\partial z}, \tag{13.39}$$

$$D_\theta = \frac{1}{r}\frac{\partial\left(rF_r^\theta\right)}{\partial r} + \frac{\partial F_z^\theta}{\partial z} \tag{13.40}$$

$$D_{Q_\mathrm{v}} = \frac{1}{r}\frac{\partial\left(rF_r^{Q_\mathrm{v}}\right)}{\partial r} + \frac{\partial F_z^{Q_\mathrm{v}}}{\partial z}, \tag{13.41}$$

and

$$D_{Q_m} = \frac{1}{r}\frac{\partial\left(rF_r^{Q_m}\right)}{\partial r} + \frac{\partial F_z^{Q_m}}{\partial z}. \tag{13.42}$$

The stress tensors are given by the following equations:

$$\tau_{rr} = 2K\frac{\partial u}{\partial r}, \tag{13.43}$$

$$\tau_{\phi\phi} = 2K\left(\frac{u}{r}\right), \tag{13.44}$$

$$\tau_{zz} = 2K\frac{\partial w}{\partial r}, \tag{13.45}$$

$$\tau_{r\phi} = Kr\frac{\partial}{\partial r}\left(\frac{v}{r}\right), \tag{13.46}$$

$$\tau_{rz} = K\left(\frac{\partial u}{\partial z} + \frac{\partial w}{\partial r}\right), \tag{13.47}$$

$$\tau_{z\phi} = 2K\frac{\partial v}{\partial r}, \tag{13.48}$$

$$\tau_r^\chi = -K\frac{\partial \chi}{\partial r}, \tag{13.49}$$

$$\tau_z^\chi = -K\frac{\partial \chi}{\partial z}; \tag{13.50}$$

and

$$S^2 = 2\left[\left(\frac{\partial u}{\partial r}\right)^2 + \left(\frac{u}{r}\right)^2 + \left(\frac{\partial w}{\partial z}\right)^2\right] + \left(\frac{\partial u}{\partial z} + \frac{\partial w}{\partial r}\right)^2 + \left(\frac{\partial v}{\partial r} - \frac{v}{r}\right)^2 + \left(\frac{\partial v}{\partial z}\right)^2, \tag{13.51}$$

where S is the deformation (stress/strain) (Smagorinski 1963), which is a kinematic quantity that describes flow features. The buoyancy term for unsaturated conditions is

$$F_z^\theta = -K\frac{g}{\overline{\theta}_\mathrm{v}}\frac{\partial \theta_\mathrm{v}}{\partial z}, \tag{13.52}$$

whilst for saturated conditions it is

$$F_z^\theta = -K\left(A\frac{\partial\theta_e}{\partial z} - g\frac{\partial Q_T}{\partial z}\right), \tag{13.53}$$

where

$$A = \frac{g}{\bar{\theta}}\left(\frac{1+\frac{L_v Q_v}{R_d T}}{1+0.622\frac{L_v^2 Q_v}{c_p R_d T^2}}\right), \tag{13.54}$$

and

$$K = L^2(1-R_i)^{1/2}S, \tag{13.55}$$

and where K is equal to zero when $R_i > 1$. The Richardson number R_i for unsaturated conditions is given by

$$R_i = \frac{g}{\bar{\theta}_v}\frac{\partial\theta_v}{\partial z}S^{-2}, \tag{13.56}$$

and for saturated conditions by

$$R_i = \left(A\frac{\partial\theta_e}{\partial z} - g\frac{\partial Q_T}{\partial z}\right)S^{-2}. \tag{13.57}$$

where θ_e is equivalent potential temperature.

13.2.2.1 Finite differences for axisymmetric models

The continuous form of the equations written in finite-difference form follows

$$f_u = -\frac{1}{r}\overline{\overline{ur}^r\delta_r u}^r - \frac{1}{\rho r}\overline{\overline{\rho rw}^r\delta_z u}^z + \overline{\left(\frac{v^2}{r}+fv\right)}^r, \tag{13.58}$$

$$f_v = -\frac{1}{r}\overline{ur\delta_r v}^r - \frac{1}{\rho}\overline{\rho w\delta_z v}^z - \left(\frac{v}{r}+fv\right)\frac{\overline{ru}^r}{r}, \tag{13.59}$$

$$f_w = -\frac{1}{r}\overline{\overline{ur}^z\delta_r w}^r - \frac{1}{\rho}\overline{\overline{\rho w}^z\delta_z w}^z + g\left(\frac{\bar{\theta}'^z}{\bar{\theta}} - 1 + 0.608(\bar{Q}_v^z - \bar{Q}_v) - \sum_1^M \bar{Q}_m^z\right), \tag{13.60}$$

$$R_i = \frac{g}{\bar{\theta}_v}(\delta_z\theta_v)S^{-2}, \tag{13.61}$$

$$R_i = \left(A\delta_z\bar{\theta}_e - g\sum_{m=1}^M \delta_z Q_m\right)S^{-2}, \tag{13.62}$$

$$S^2 = 2\left\{\overline{(\delta_r u)^2}^z + \overline{\overline{\left(\frac{u}{r}\right)}^{r2}}^z + \overline{(\delta_z w)^2}^z\right\} + \overline{(\delta_z u + \delta_r w)^2}^{rz}$$
$$+ \overline{\left(\delta_r v - \frac{v}{r}\right)^2}^{rz} + (\delta_z v)^2, \tag{13.63}$$

$$D_u = \frac{1}{r}\delta_r(2r\bar{K}^z\delta_r u) + \delta_z(\bar{K}^z[\delta_z u + \delta_r w]) - 2\bar{K}^{rz}\frac{u}{r^2}, \tag{13.64}$$

$$D_v = \frac{1}{r^2}\delta_r\left(r^2\bar{K}^{rz}\delta_r v - \frac{v}{r}\right) + \delta_z(K\delta_z u), \tag{13.65}$$

$$D_w = \frac{1}{r}\delta_r(r\bar{K}^r[\delta_z u + \delta_r w]) + \delta_z(2\bar{K}^z\delta_z u), \tag{13.66}$$

$$D_\theta = \frac{1}{r}\delta_r(r\bar{K}^{rz}\delta_r\theta) + \delta_z(K\delta_z\theta), \tag{13.67}$$

$$D_{Q_v} = \frac{1}{r}\delta_r(r\bar{K}^{rz}\delta_r Q_v) + \delta_z(K\delta_z Q_v), \tag{13.68}$$

$$D_{Q_m} = \delta_r(rK\delta_r Q_m) + \delta_z(K\delta_z Q_m), \tag{13.69}$$

and

$$D_{N_m} = \delta_r(rK\delta_r N_m) + \delta_z(K\delta_z N_m), \tag{13.70}$$

where K is defined by 13.55 at the w points on a staggered c-grid.

13.2.2.2 Lower boundary conditions for axisymmetric models

One of the more simplistic lower boundary conditions is suggested to capture the influence of the Earth's surface on momentum and heat flux. The lower boundary conditions presented are given by Rotunno and Emanuel (1987) for a hurricane environment. Other formulas are more appropriate for land lower boundaries. The Rotunno and Emanuel (1987) conditions are given for the wind at height $z = \Delta z/2$ by the following relationships

$$\tau_{rz} = c_d u\left(u^2 + v^2\right)^{1/2} \tag{13.71}$$

and

$$\tau_{z\phi} = c_d v\left(u^2 + v^2\right)^{1/2}. \tag{13.72}$$

For heat flux (sensible and latent), the following bulk aerodynamical formulas are used,

$$F_z^\theta = c_\mathrm{d}\left(u^2 + v^2\right)^{1/2}\left(\theta_\mathrm{sfc} - \theta_{\mathrm{d}z/2}\right) \tag{13.73}$$

and

$$F_z^{Q_v} = c_d\left(u^2 + v^2\right)^{1/2}\left(Q_{v,sfc} - Q_{v,dz/2}\right), \tag{13.74}$$

where the subscripts sfc and dz/2 refer to the surface value and the ½ grid point above the surface; the value of the drag coefficient $c_d = 1.1\times10^{-3} + 4\times10^{-5}\,(u^2 + v^2)^{1/2}$ (see Moss and Rosenthal 1975 as suggested by Rotunno and Emanuel 1987).

More complicated, and perhaps better, formulas based on observed fluxes are included in many models (see Stull 1988 for listings of these equations).

13.3 Three-dimensional dynamical model

13.3.1 Theoretical formulation of a three-dimensional dynamical model

The example used here is the Straka Atmospheric Model (SAM) (Straka and Mansell 2005), which is a non-hydrostatic model based on the compressible Navier–Stokes equations for fluid flow. A terrain-following coordinate system is employed. In addition, a three-dimension map factor is used for grid stretching. The model can be applied to flows of air on scales of fractions of meters to hundreds of meters or more. Other physical processes are taken into account such as radiation and microphysics. The latter of these are a primary topic of this book. Parameterizations for other processes can be found in Stensrud (2007).

13.3.2 Analytical equations for the orthogonal Cartesian dynamical model

To begin with, the equations for compressible fluid flow can be written as

$$\frac{\partial u_i}{\partial t} = -\frac{1}{\rho}\frac{\partial(\rho u_j u_i)}{\partial x_j} + \frac{u_i}{\rho}\frac{\partial u_j}{\partial x_j} - \frac{1}{\rho}\frac{\partial p}{\partial x_i} + \frac{1}{\rho}\frac{\partial \tau_{ij}}{\partial x_i} + \epsilon_{ijk}u_j f + \delta_{i3}g \tag{13.75}$$

where the product rule has been used on the advective term to write it in flux form. The orthogonal Cartesian velocity components $u_i = (1, 2, 3)$ are the velocity components in the x-, y-, and z-directions, $g = 9.8$ m s^{-2} is gravitational acceleration, ρ is the density, and p is the pressure. In addition, $f = 2\overline{\Omega}\sin(\phi)$, and $f' = 2\overline{\Omega}\cos(\phi)$, where $\overline{\Omega}$ is the angular frequency of the Earth, which is 7.29×10^{-5} s^{-1}, and ϕ is the latitude. The subgrid stress tensor is given as $\tau_{ij} = \rho\, K_m\, D_{ij}$ and the eddy-mixing coefficient for momentum K_m is defined later. The deformation tensor D_{ij} is defined as

$$D_{ij} = \left(\frac{\partial u_i}{\partial x_j} + \frac{\partial u_j}{\partial x_i} - \frac{2}{3}\delta_{ij}\frac{\partial u_k}{\partial x_k}\right). \tag{13.76}$$

The continuity equation can be written as

$$\frac{\partial \rho}{\partial t} = -\frac{\partial(\rho u_j)}{\partial x_j} \tag{13.77}$$

and the ideal gas equation is

$$P = \rho R_d T_v. \tag{13.78}$$

Lastly, the Poisson equation will be used,

$$\theta_v = T_v \left(\frac{p_0}{p}\right)^{R/c_p}, \tag{13.79}$$

where p_0 is a reference pressure that is 100 000 Pa, $c_p = 1004$ J kg^{-1} K^{-1}, and $c_v = 717$ J kg^{-1} K^{-1}. This fully compressible system permits both acoustic and gravity wave solutions.

Since the thermodynamic quantities vary more rapidly in the vertical than the horizontal direction, they may be written as a sum of a base-state variable which is a function of z only and a perturbation from the base state,

$$\begin{cases} p = \bar{p}(z) + p' \\ \rho = \bar{\rho}(z) + \rho' \\ \theta = \bar{\theta}(z) + \theta' \end{cases}. \tag{13.80}$$

Also, the base state is hydrostatic,

$$\frac{\partial \bar{p}}{\partial z} = -\bar{\rho} g. \tag{13.81}$$

Now (13.80) is substituted into (13.75), using a binomial expansion and neglecting higher-order terms and making use of (13.81) results in

$$\frac{\partial u_i}{\partial t} = -\frac{1}{\bar{\rho}}\frac{\partial(\bar{\rho} u_j u_i)}{\partial x_j} + \frac{u_i}{\bar{\rho}}\frac{\partial u_j}{\partial x_j} - \frac{1}{\bar{\rho}}\frac{\partial p'}{\partial x_i} + \frac{1}{\bar{\rho}}\frac{\partial \tau_{ij}}{\partial x_i} + \epsilon_{ijk} u_j f - \delta_{i3} g \frac{\rho'}{\bar{\rho}} \tag{13.82}$$

and the moist Poisson equation holds for the mean state,

$$\bar{\theta}_v = \bar{T}_v \left(\frac{p_0}{\bar{p}}\right)^{R/c_p}. \tag{13.83}$$

Now (13.79) is used, the natural logarithm taken, and using (13.80), the mean terms are subtracted, resulting in

$$-\frac{\rho'}{\bar{\rho}} \simeq \frac{\theta_v'}{\bar{\theta}_v} - \frac{c_v}{c_p}\frac{p'}{\bar{p}}. \tag{13.84}$$

Substituting (13.84) into (13.82),

$$\frac{\partial u_i}{\partial t} = -\frac{1}{\bar{\rho}}\frac{\partial(\bar{\rho}u_j u_i)}{\partial x_j} + \frac{u_i}{\bar{\rho}}\frac{\partial u_j}{\partial x_j} - \frac{1}{\bar{\rho}}\frac{\partial p'}{\partial x_i} + \frac{1}{\bar{\rho}}\frac{\partial \tau_{ij}}{\partial x_i} + \epsilon_{ijk}u_j f + \delta_{i3}g\left[\frac{\theta'_\mathrm{v}}{\bar{\theta}_\mathrm{v}} - \frac{c_\mathrm{v}}{c_p}\frac{p'}{\bar{p}}\right]. \quad (13.85)$$

Now the buoyancy term can be rewritten to account for liquid- and ice-water effects (Bannon 2002),

$$\frac{\partial u_i}{\partial t} = -\frac{1}{\bar{\rho}}\frac{\partial(\bar{\rho}u_j u_i)}{\partial x_j} + \frac{u_i}{\bar{\rho}}\frac{\partial u_j}{\partial x_j} - \frac{1}{\bar{\rho}}\frac{\partial p'}{\partial x_i} + \frac{1}{\bar{\rho}}\frac{\partial \tau_{ij}}{\partial x_i} + \epsilon_{ijk}u_j f + \delta_{i3}g\left[\frac{\theta'}{\bar{\theta}} + \frac{Q'_\mathrm{v}}{0.608 + \bar{Q}_\mathrm{v}} - \frac{Q'_\mathrm{v} + \sum_{m=1}^{M} Q_{\mathrm{liq+ice}}}{1 + \bar{Q}_\mathrm{v}} - \frac{c_\mathrm{v}}{c_p}\frac{p'}{\bar{p}}\right]. \quad (13.86)$$

13.3.2.1 Pressure equation

Next, an equation is derived for the perturbation pressure. The total derivative of (13.84) is taken

$$-\frac{\mathrm{d}}{\mathrm{d}t}\left(\frac{\rho'}{\bar{\rho}}\right) = \frac{\mathrm{d}}{\mathrm{d}t}\left(\frac{\theta'_\mathrm{v}}{\bar{\theta}_\mathrm{v}}\right) - \frac{c_\mathrm{v}}{c_p}\frac{\mathrm{d}}{\mathrm{d}t}\left(\frac{p'}{\bar{p}}\right). \quad (13.87)$$

Expanding,

$$-\rho'\frac{1}{\bar{\rho}^2}\frac{\mathrm{d}\bar{\rho}}{\mathrm{d}t} - \frac{1}{\bar{\rho}}\frac{\mathrm{d}\rho'}{\mathrm{d}t} = \theta'\frac{1}{\bar{\theta}^2}\frac{\mathrm{d}\bar{\theta}}{\mathrm{d}t} - \frac{1}{\bar{\theta}}\frac{\mathrm{d}\theta'}{\mathrm{d}t} - \frac{c_\mathrm{v}}{c_p}p'\frac{1}{\bar{p}^2}\frac{\mathrm{d}\bar{p}}{\mathrm{d}t} - \frac{c_\mathrm{v}}{c_p}\frac{1}{\bar{p}}\frac{\mathrm{d}p'}{\mathrm{d}t}. \quad (13.88)$$

Now the terms where the mean variable squared is much, much larger than the perturbation terms are neglected and the equation is solved for the pressure term,

$$\frac{\mathrm{d}p'}{\mathrm{d}t} = \frac{c_p}{c_\mathrm{v}}\frac{\bar{p}}{\bar{\theta}_\mathrm{v}}\frac{\mathrm{d}\theta'_\mathrm{v}}{\mathrm{d}t} + \frac{c_p}{c_\mathrm{v}}\frac{\bar{p}}{\bar{\rho}}\frac{\mathrm{d}\rho'}{\mathrm{d}t}. \quad (13.89)$$

Now a conservation equation for θ_v is linearized,

$$\frac{\partial \theta_\mathrm{v}}{\partial t} = -u_j\frac{\partial \theta_\mathrm{v}}{\partial x_j} + \frac{\partial}{\partial x_i}\left(K_\mathrm{h}\frac{\partial \theta_\mathrm{v}}{\partial x_i}\right) + S\theta_\mathrm{v} \quad (13.90)$$

to get an expression for the perturbation quantity,

$$\frac{\mathrm{d}\theta'_\mathrm{v}}{\mathrm{d}t} = -u_j\frac{\partial \bar{\theta}_\mathrm{v}}{\partial x_j} + \frac{\partial}{\partial x_i}\left(K_\mathrm{h}\frac{\partial(\bar{\theta}_\mathrm{v} + \theta'_\mathrm{v})}{\partial x_i}\right) + S(\bar{\theta}_\mathrm{v} + \theta'_\mathrm{v}), \quad (13.91)$$

where K_h is the mixing coefficient for temperature and moisture, along with other scalars.

Also the continuity equation (13.77) is linearized to obtain

$$\frac{\mathrm{d}\rho'}{\mathrm{d}t} = -u_j \frac{\partial \bar{\rho}}{\partial x_j} - (\bar{\rho} + \rho') \frac{\partial u_j}{\partial x_j}. \tag{13.92}$$

Substituting (13.91) and (13.92) into (13.88) gives

$$\begin{aligned} \frac{\mathrm{d}p'}{\mathrm{d}t} = &\frac{c_p}{c_v} \frac{\bar{p}}{\bar{\theta}_v} \left[-u_j \frac{\partial \bar{\theta}_v}{\partial x_j} + \frac{\partial}{\partial x_i} \left(K_h \frac{\partial (\bar{\theta}_v + \theta')_v}{\partial x_i} \right) + S(\bar{\theta}_v + \theta'_v) \right] \\ &+ \frac{c_p}{c_v} \frac{\bar{p}}{\bar{\rho}} \left[-\frac{\partial (\bar{\rho} u_j)}{\partial x_j} - \rho' \frac{\partial u_j}{\partial x_j} \right]. \end{aligned} \tag{13.93}$$

The definition of the sound speed is introduced,

$$\bar{c}_s^2 = R_d \bar{T} \frac{c_p}{c_v} = \frac{\bar{p}}{\bar{\rho}} \frac{c_p}{c_v}. \tag{13.94}$$

Substituting this into (13.93) gives

$$\begin{aligned} \frac{\mathrm{d}p'}{\mathrm{d}t} = &\frac{\bar{\rho}}{\bar{\theta}_v} \bar{c}_s^2 \left[-u_j \frac{\partial \bar{\theta}_v}{\partial x_j} + \frac{\partial}{\partial x_i} \left(K_h \frac{\partial (\bar{\theta}_v + \theta')_v}{\partial x_i} \right) + S(\bar{\theta}_v + \theta'_v) \right] \\ &+ \bar{c}_s^2 \left[-\frac{\partial (\bar{\rho} u_j)}{\partial x_j} - \rho' \frac{\partial u_j}{\partial x_j} \right]. \end{aligned} \tag{13.95}$$

Expanding the total derivative of perturbation pressure, moving the advective term to the right-hand side, and rearranging gives

$$\begin{aligned} \frac{\partial p'}{\partial t} + \bar{c}_s^2 \frac{\partial (\bar{\rho} u_j)}{\partial x_j} = &-u_j \frac{\partial p'}{\partial x_j} + \bar{c}_s^2 \frac{\bar{\rho}}{\bar{\theta}_v} \left[-u_j \frac{\partial \bar{\theta}_v}{\partial x_j} + \frac{\partial}{\partial x_i} \left(K_h \frac{\partial (\bar{\theta}_v + \theta'_v)}{\partial x_i} \right) + S(\bar{\theta}_v + \theta'_v) \right] \\ &- \bar{c}_s^2 \left[\rho' \frac{\partial u_j}{\partial x_j} \right]. \end{aligned} \tag{13.96}$$

Now, the terms on the right-hand side of (13.96) have been shown to be small (Klemp and Wilhelmson 1978) and can be neglected for cloud models. Thus, (13.95) becomes the prognostic equation for perturbation pressure,

$$\frac{\partial p'}{\partial t} = -\bar{c}_s^2 \frac{\partial (\bar{\rho} u_j)}{\partial x_j}. \tag{13.97}$$

13.3.2.2 Density equation

The density of moist air is given by Proctor (1987) approximately as

$$\rho = \frac{P}{R_d T} \left(1 - \frac{Q_v}{0.608 + Q_v} \right) (1 + Q_v + Q_{liq} + Q_{ice}), \tag{13.98}$$

and the hydrostatic equation is given by

$$\frac{\partial p}{\partial z} = -\rho g. \tag{13.99}$$

13.3.2.3 Thermodynamic energy (potential-temperature) equation

The thermodynamic energy equation is represented by an equation for potential temperature, and is given by

$$\frac{\partial \theta}{\partial t} = -\frac{1}{\rho}\frac{\partial(\rho u_i \theta)}{\partial x_j} + \frac{\theta}{\rho}\frac{\partial(\rho u_i)}{\partial x_i} + \frac{\partial}{\partial x_i}\left(\frac{1}{\rho}K_{\mathrm{h}}\frac{\partial \theta}{\partial x_i}\right) + S_\theta. \tag{13.100}$$

The eddy-mixing coefficient for scalars is described later. The last term starting with S_θ is a heating/cooling term for potential temperature.

13.3.2.4 Turbulent kinetic energy equation

The one-and-a-half-order turbulence closure is given by the turbulent kinetic energy equation,

$$\begin{aligned}\frac{\partial E^{1/2}}{\partial t} = &-\frac{1}{\rho}\frac{\partial\left(\rho u_i E^{1/2}\right)}{\partial x_j} + \frac{E^{1/2}}{\rho}\frac{\partial(\rho u_i)}{\partial x_j} + \frac{K_{\mathrm{m}}}{2E^{1/2}}\left(\frac{\partial u_i}{\partial x_j} + \frac{\partial u_j}{\partial x_i}\right) \\ &-\frac{1}{3}\delta_{ij}E^{1/2} - \frac{g}{\theta_0}\left(AK_{\mathrm{h}}\frac{\partial \theta}{\partial x_i} + BK_{\mathrm{h}}\frac{\partial q_w}{\partial x_i}\right) + \frac{\partial}{\partial x_i}\left(2K_{\mathrm{m}}\frac{\partial E^{1/2}}{\partial x_i}\right) - \epsilon.\end{aligned} \tag{13.101}$$

The use of this equation is equivalent to predicting $K_{\mathrm{m}} = c_m LE^{1/2}$ (where c_m is a coefficient ranging from 0.1 to 0.2) as in Klemp and Wilhelmson (1978), except that the mixing length is not specified to be constant. Rather it is a function of stability as follows. First, the sub-grid coefficients are given following Deardorff (1980) as

$$K_{\mathrm{m}} = c_m LE^{1/2}, \tag{13.102}$$

and

$$K_{\mathrm{h}} = (1 + 2L/\Delta s)K_{\mathrm{m}}, \tag{13.103}$$

where,

$$\Delta s = (\Delta x \Delta y \Delta z)^{1/3}. \tag{13.104}$$

Then the mixing length is calculated from,

$$L = 0.76E^{1/2}\left(\frac{\partial \theta_{\mathrm{il}}}{\partial z}\right)^{-1/2} \tag{13.105}$$

when

$$\left(\frac{\partial \theta_{\text{il}}}{\partial z}\right) > 0, \tag{13.106}$$

where θ_{il} is the potential temperature of ice–liquid; otherwise,

$$L = \Delta s. \tag{13.107}$$

The dissipation term is given as

$$\epsilon = \frac{c_\epsilon E^{3/2}}{L}, \tag{13.108}$$

where c_ϵ, a coefficient in turbulent dissipation, goes to 0.19 in the stable limit, with c_ϵ otherwise given by,

$$c_\epsilon = 0.19 + 0.51L/\Delta s. \tag{13.109}$$

It has been suggested that $c_\epsilon = 3.9$ to create a "wall effect" to prevent $E^{1/2}$ from becoming too large near the surface (Deardorff 1980). The terms A and B in (13.101) are given, following Klemp and Wilhelmson (1978), as

$$A = \frac{1}{\theta}\left\{\frac{1 + [(1.608\epsilon L_{\text{v}} Q_{\text{v}})/(R_{\text{d}}T)]}{1 + \left[(0.608 L_{\text{v}}^2 Q_{\text{v}})/(c_p R_{\text{d}} T^2)\right]}\right\}, \tag{13.110}$$

and if $Q_{\text{ice}} > 0$

$$\delta_{i3} g\left(\frac{\overline{u_i''\theta_{\text{v}}''}}{\bar{\theta}_{\text{v}}} - \overline{u_i'' Q_{\text{T}}''}\right) = \delta_{i3} g\left(\frac{BK_{\text{h}}}{\bar{\theta}_{\text{v}}}\right)\left(\frac{\partial \theta_{\text{e,ice}}}{\partial x_i} - \frac{L_{\text{s}} P}{c_p T}\frac{\partial Q_{\text{ice}}}{\partial x_i}\right) + \delta_{i3} g K_{\text{h}} \frac{\partial Q_{\text{T}}}{\partial x_i}, \tag{13.111}$$

where $\theta_{\text{e,ice}}$ is the ice equivalent potential temperature and where

$$B = \frac{1 + 0.608\dfrac{0.622 L_{\text{v}} Q_{\text{v}}}{R_{\text{d}} T}}{1 + \dfrac{0.622 L_{\text{v}}^2 Q_{\text{v}}}{c_p R_{\text{d}} T^2}}. \tag{13.112}$$

13.3.2.5 Moisture equations

There are three variables related to three different moments predicted in this model; the concentration N_{T}, mixing ratio Q, and reflectivity Z, which are related to the zeroth, third and sixth moments, respectively. The purpose of using three different moments is to be able to diagnose the shape parameter at each grid point for each hydrometeor species following Milbrandt and Yau (2005a, b), as explained in Chapter 2. This adds greater freedom for the model to represent real cloud processes that might occur in the atmosphere as shown in solutions of tests compared to analytical solutions and other comparisons by Milbrandt and Yau (2005a, b). They show that the best solutions among single-, double-, and triple-moment schemes are obtained by using these three different moments for predictive equations, and diagnosing, though they call it prognosing, the shape parameter. The continuity equations for N_{T}, Q, and Z are given as

$$\frac{\partial N_{\mathrm{T}x}}{\partial t} = -u_i \frac{\partial N_{\mathrm{T}x}}{\partial x_i} + \frac{\partial}{\partial x_i}\left(K_{\mathrm{h}} \frac{\partial N_{\mathrm{T}x}}{\partial x_i}\right) + \frac{\partial(\bar{V}_{\mathrm{T}N_{\mathrm{T}x}} N_{\mathrm{T}x})}{\partial x_3} + S_{N_{\mathrm{T}x}} \tag{13.113}$$

$$\frac{\partial Q_x}{\partial t} = -\frac{1}{\rho}\frac{\partial \rho (u_i Q)_x}{\partial x_i} + \frac{Q_x}{\rho}\frac{\partial(\rho u_i)}{\partial x_i} + \frac{\partial}{\partial x_i}\left(\rho K_{\mathrm{h}} \frac{\partial Q_x}{\partial x_i}\right) + \frac{1}{\rho}\frac{\partial(\rho \bar{V}_{\mathrm{T}Q_x} Q_x)}{\partial x_3} + S_{Q_x} \tag{13.114}$$

$$\frac{\partial Z_x}{\partial t} = -u_i \frac{\partial Z_x}{\partial x_i} + \frac{\partial}{\partial x_i}\left(K_{\mathrm{h}} \frac{\partial Z_x}{\partial x_i}\right) + \frac{\partial(\bar{V}_{\mathrm{T}Z_x} Z_x)}{\partial x_3} + S_{Z_x}, \tag{13.115}$$

where the "S" terms are source and sink terms.

13.3.3 Equations for the non-orthogonal terrain-following system

The orthogonal Cartesian system is transformed into a non-orthogonal terrain-following system for flow over topography. The vertical coordinate is often called the sigma-z coordinate system owing to similarities to the sigma-p coordinate system. The z coordinate is transformed into a new coordinate, h, where,

$$\eta = \frac{z-h}{1-h/H} = \frac{z-h}{\sqrt{G}}. \tag{13.116}$$

In this equation the Jacobian of the transformation is given by $\sqrt{G} = 1 - h/H$, where h is the height of the topography and $h(x, y)$ and H is the height of the top of the domain. At $h = 0$ the transformed vertical velocity normal to the terrain is $W_c = 0$. In other words where $h = 0 =$ constant surface $W_c = 0$.

The terrain-following coordinate system can be derived using the chain rule to come up with the following definitions such as

$$\begin{aligned} \left.\frac{\partial}{\partial x}\right|_z &= \left.\frac{\partial}{\partial x}\right|_\eta + G^{13}\frac{\partial}{\partial \eta} \\ \left.\frac{\partial}{\partial y}\right|_z &= \left.\frac{\partial}{\partial y}\right|_\eta + G^{23}\frac{\partial}{\partial \eta} \end{aligned} \tag{13.117}$$

with the values of G^{13} and G^{23} given by

$$
\begin{aligned}
G^{13} &= \frac{\partial \eta}{\partial x} = \frac{1}{\sqrt{G}} \left[\frac{\eta}{H} - 1 \right] \frac{\partial h}{\partial x} \\
G^{23} &= \frac{\partial \eta}{\partial y} = \frac{1}{\sqrt{G}} \left[\frac{\eta}{H} - 1 \right] \frac{\partial h}{\partial y}.
\end{aligned}
\tag{13.118}
$$

As pointed out by Clark and Hall (1979) all conservation equations of the form

$$
\frac{\partial \phi}{\partial t} + \frac{1}{\rho}\frac{\partial(\rho u \phi)}{\partial x} + \frac{1}{\rho}\frac{\partial(\rho v \phi)}{\partial y} + \frac{1}{\rho}\frac{\partial(\rho w \phi)}{\partial z} = S_\phi \tag{13.119}
$$

in the non-orthogonal terrain-following system transform into the following in the terrain-following system,

$$
\frac{\partial \phi}{\partial t} + \frac{1}{\sqrt{G}\rho}\frac{\partial\left(\sqrt{G}\rho u \phi\right)}{\partial x} + \frac{1}{\sqrt{G}\rho}\frac{\partial\left(\sqrt{G}\rho v \phi\right)}{\partial y} + \frac{1}{\sqrt{G}\rho}\frac{\partial\left(\sqrt{G}\rho W_c \phi\right)}{\partial z} = S_\phi \tag{13.120}
$$

where W_c is given by

$$
W_c = \frac{\left(w + \sqrt{G}G^{13}u + \sqrt{G}G^{23}v\right)}{\sqrt{G}}. \tag{13.121}
$$

13.3.4 Second-order finite differences

For completeness, the finite-difference forms for the system of equations used are given, similar to that in Clark (1977) and Clark and Hall (1979), but using Schumann operators for differences and derivatives, and noting ρ is defined for simplicity as $\rho\sqrt{G}$. First the u-velocity equation with second-order differences (all differencing is second order) is

$$
\begin{aligned}
&\overline{\delta_{2t} u}^{t} + \frac{1}{\bar{\rho}^x}\delta_x\left(\overline{\bar{\rho}^x u}^x \bar{u}^x\right) + \frac{1}{\bar{\rho}^x}\delta_y\left(\overline{\bar{\rho}^x v}^x \bar{u}^y\right) + \frac{1}{\bar{\rho}^x}\delta_y\left(\overline{\bar{\rho}^\eta W_c}^x \bar{u}^\eta\right) \\
&\quad = -\frac{1}{\bar{\rho}^x}\delta_x(p') - \frac{1}{\bar{\rho}^x}\delta_\eta\left(G^{13}\overline{p'}^{x\eta}\right) + \frac{1}{\bar{\rho}^x}\delta_x\left(\sqrt{G}\tau_{11}\right) + \frac{1}{\bar{\rho}^x}\delta_y\left(\overline{\sqrt{G}}^{xy}\tau_{12}\right) \\
&\quad + \frac{1}{\bar{\rho}^x}\delta_\eta\left(\tau_{13} + \sqrt{G}G^{13}\overline{\tau_{11}}^{x\eta} + \overline{\overline{\sqrt{G}G^{13}}^x \bar{\tau}_{12}^\eta}^y\right).
\end{aligned}
\tag{13.122}
$$

The v-velocity equation is differenced as

$$
\begin{aligned}
&\overline{\delta_{2t} v}^{t} + \frac{1}{\bar{\rho}^y}\delta_x\left(\overline{\bar{\rho}^x u}^y \bar{v}^x\right) + \frac{1}{\bar{\rho}^y}\delta_y\left(\overline{\bar{\rho}^y v}^y \bar{v}^y\right) + \frac{1}{\bar{\rho}^y}\delta_y\left(\overline{\bar{\rho}^\eta W_c}^y \bar{v}^\eta\right) \\
&\quad = -\frac{1}{\bar{\rho}^y}\delta_y(p') - \frac{1}{\bar{\rho}^y}\delta_\eta\left(G^{13}\overline{p'}^{y\eta}\right) + \frac{1}{\bar{\rho}^y}\delta_x\left(\overline{\sqrt{G}}^{xy}\tau_{12}\right) + \frac{1}{\bar{\rho}^y}\delta_y\left(\sqrt{G}\tau_{22}\right) \\
&\quad + \frac{1}{\bar{\rho}^y}\delta_\eta\left(\tau_{23} + \overline{\overline{\sqrt{G}G^{13}}^y \bar{\tau}_{12}^\eta}^x + \sqrt{G}G^{23}\overline{\tau_{22}}^{x\eta}\right).
\end{aligned}
\tag{13.123}
$$

The w-velocity equation is differenced as

$$\overline{\delta_{2t}w}^{t} + \frac{1}{\bar{\rho}^{\eta}}\delta_x\left(\overline{\bar{\rho}^{x}u}^{\eta}\overline{w}^{x}\right) + \frac{1}{\bar{\rho}^{\eta}}\delta_y\left(\overline{\bar{\rho}^{y}v}^{\eta}\overline{w}^{y}\right) + \frac{1}{\bar{\rho}^{\eta}}\delta_\eta\left(\overline{\bar{\rho}^{\eta}W_c}^{\eta}\overline{w}^{\eta}\right)$$

$$= -\frac{1}{\bar{\rho}^{\eta}}\delta_\eta(p') - \frac{1}{\bar{\rho}^{\eta}}\delta_{i3}g\left[\frac{\theta'}{\bar{\theta}} + \frac{Q'_{\mathrm{v}}}{0.608 + \bar{Q}_{\mathrm{v}}} - \frac{c_v}{c_p}\frac{p'}{\bar{p}} - \frac{Q'_{\mathrm{v}} + \sum_{m=1}^{M} Q_m}{1 + \bar{Q}_{\mathrm{v}}}\right] \quad (13.124)$$

$$+ \frac{1}{\bar{\rho}^{\eta}}\delta_x\left(\overline{\sqrt{G}}^{x}\tau_{13}\right) + \frac{1}{\bar{\rho}^{\eta}}\delta_y\left(\overline{\sqrt{G}}^{y}\tau_{23}\right) + \frac{1}{\bar{\rho}^{\eta}}\delta_\eta\left(\tau_{33} + \overline{\overline{\sqrt{G}G^{13}}^{\eta}\bar{\tau}_{13}^{\eta}}^{x} + \overline{\overline{\sqrt{G}G^{23}}^{\eta}\bar{\tau}_{23}^{\eta}}^{y}\right).$$

First the divergence, and then deformation tensors are discretized as follows,

$$\delta = \frac{1}{\sqrt{G}}\left[\delta_x\sqrt{G}^{x}u + \delta_y\sqrt{G}^{y}v + \delta_\eta\sqrt{G}^{\eta}W_c\right], \quad (13.125)$$

$$D_{11} = \frac{2}{\sqrt{G}}\left[\delta_x\left(\overline{\sqrt{G}}^{x}u\right) + \delta_\eta\left(\overline{G^{13}\bar{u}^{\eta}}^{x}\right)\right], \quad (13.126)$$

$$D_{22} = \frac{2}{\sqrt{G}}\left[\delta_y\left(\overline{\sqrt{G}}^{y}v\right) + \delta_\eta\left(\overline{G^{23}\bar{v}^{\eta}}^{y}\right)\right], \quad (13.127)$$

$$D_{33} = \frac{2}{\sqrt{G}}\left[\delta_\eta(w)\right], \quad (13.128)$$

$$D_{12} = \frac{1}{\sqrt{\bar{G}}^{xy}}\left[\delta_y\left(\sqrt{\bar{G}}^{x}u\right) + \delta_x\left(\sqrt{\bar{G}}^{y}v\right) + \delta_\eta\left(\overline{G^{23}}^{x}\bar{u}^{\eta y} + \overline{G^{13}}^{y}\bar{v}^{\eta x}\right)\right], \quad (13.129)$$

$$D_{13} = \frac{1}{\sqrt{\bar{G}}^{x\eta}}\left[\delta_x\left(\sqrt{G}w\right) + \delta_\eta\left(u + \overline{G^{13}\bar{w}^{x}}^{\eta}\right)\right], \quad (13.130)$$

$$D_{23} = \frac{1}{\sqrt{\bar{G}}^{y\eta}}\left[\delta_y\left(\sqrt{G}w\right) + \delta_\eta\left(v + \overline{G^{23}\bar{w}^{y}}^{\eta}\right)\right], \quad (13.131)$$

Then the Reynolds stress tensors are given as

$$T_{11} = \rho K_m D_{11} - \frac{2}{3}\frac{\rho}{\sqrt{G}}\delta^2 - \frac{2}{3}\frac{\rho}{\sqrt{G}}\left(\frac{K_m}{c_m L}\right)^2, \quad (13.132)$$

$$T_{12} = \overline{\rho K_m}^{xy} D_{12}, \quad (13.133)$$

$$T_{13} = \overline{\rho K_m}^{x\eta} D_{13}, \quad (13.134)$$

$$T_{23} = \overline{\rho K_m}^{y\eta} D_{23}, \quad (13.135)$$

$$T_{22} = \rho K_m D_{22} - \frac{2}{3}\frac{\rho}{\sqrt{G}}\delta^2 - \frac{2}{3}\frac{\rho}{\sqrt{G}}\left(\frac{K_m}{c_m L}\right)^2, \tag{13.136}$$

$$T_{33} = \rho K_m D_{33} - \frac{2}{3}\frac{\rho}{\sqrt{G}}\delta^2 - \frac{2}{3}\frac{\rho}{\sqrt{G}}\left(\frac{K_m}{c_m L}\right)^2, \tag{13.137}$$

The finite-difference form of the pressure equation is

$$\begin{aligned}
&\overline{\delta_{2t}p'}^{t} - \left[\bar{u}^x\delta_{2x}p' + \bar{v}^y\delta_{2y}p' + \overline{W}_c^{\eta}\delta_{2\eta}p'\right] + \rho g w \\
&- \rho c_s^2\left[\delta_x u + \delta_y v + \delta_\eta W_c\right] \\
&+ \rho c_s^2\left[\frac{1}{\theta}\frac{\mathrm{d}\theta}{\mathrm{d}t} - \frac{1}{E}\frac{\mathrm{d}E}{\mathrm{d}t}\right].
\end{aligned} \tag{13.138}$$

where E is the sum of the mixing ratios or $1 + 0.608\text{qv} + \sum_{m=1}^{M} Q_m$
The discretized form of the scalar ϕ equations is

$$\begin{aligned}
&\overline{\delta_{2t}\phi}^{t} + \frac{1}{\bar{\rho}^x}\delta_x\left(\bar{\rho}^x\bar{\phi}^x u\right) + \frac{1}{\bar{\rho}^y}\delta_y\left(\bar{\rho}^y\bar{\phi}^y v\right) + \frac{1}{\bar{\rho}^\eta}\delta_\eta\left(\bar{\rho}^\eta\bar{\phi}^\eta W_c\right) = \\
&+\frac{1}{\bar{\rho}^x}\delta_x\left(\overline{\sqrt{G}}^x H_1\right) + \frac{1}{\bar{\rho}^y}\delta_y\left(\overline{\sqrt{G}}^y H_2\right) + \frac{1}{\bar{\rho}^\eta}\delta_\eta\left(H_3 + \overline{\sqrt{G}G^{13}\bar{H}_1^\eta}^x + \overline{\sqrt{G}G^{23}\bar{H}_2^\eta}^y\right),
\end{aligned} \tag{13.139}$$

where H_1, H_2 and H_3 are defined as

$$H_1 = \overline{\left(\frac{\rho K_h}{\sqrt{G}}\right)}^x\left[\delta_x\phi + \delta_\eta\left(G^{13}\bar{\phi}^{x\eta}\right)\right], \tag{13.140}$$

$$H_2 = \overline{\left(\frac{\rho K_h}{\sqrt{G}}\right)}^y\left[\delta_y\phi + \delta_\eta\left(G^{23}\bar{\phi}^{y\eta}\right)\right], \tag{13.141}$$

$$H_3 = \overline{\left(\frac{\rho K_h}{\sqrt{G}}\right)}^\eta\left[\delta_\eta\phi\right], \tag{13.142}$$

with $K_h = 3K_m$
Note that the approximation for W_c is given as

$$W_c = \frac{1}{\sqrt{G}\bar{\rho}^\eta}\left\{\bar{\rho}^\eta w + \left[\overline{\sqrt{G}G^{13}\overline{\bar{\rho}^x u}^\eta}^x + \overline{\sqrt{G}G^{23}\overline{\bar{\rho}^y v}^\eta}^y\right]\right\}. \tag{13.143}$$

13.3.5 Boundary conditions

In three dimensions with and without a terrain-following coordinate model we have to apply various boundary conditions at $z = 0$ and H the ground and top of the model, respectively. The conditions apply especially to the stress tensors. Also at $\eta = 0$ and H the following applies,

$$W_c = \delta_\eta(\bar{\rho}^x u) = \delta_\eta(\bar{\rho}^y v) = 0 \tag{13.144}$$

which approximates a free slip condition and simplifies the finite differencing of the pressure equation tremendously.

$$\left(\overline{\bar{\rho}^\eta \bar{W}_c^\eta \bar{w}^\eta}^\eta\right) = 0 \text{ at } \eta = 0, H. \tag{13.145}$$

The boundary conditions for τ_{ij} at $\eta = 0$ and H are

$$\begin{aligned} \bar{\tau}_{11}^\eta = \bar{\tau}_{22}^\eta = \bar{\tau}_{33}^\eta = \bar{\tau}_{12}^\eta = 0 \text{ at } \bar{\eta} = 0 \\ \tau_{13} = \tau_{23} = \delta_\eta \tau_{22} = \bar{\tau}_{33}^\eta = \delta_\eta \tau_{12} \text{ at } \bar{\eta} = H \end{aligned} \tag{13.146}$$

and the simple drag laws can be incorporated for semi-slip boundary conditions (though more complex, and simpler ones are possible),

$$\begin{aligned} \tau_{13} = 0.5(\rho/G^{1/2})[u\cos(\lambda_x) + w\sin(\lambda_x)]_0 \quad \bar{\eta} = 0 \\ \tau_{23} = 0.5(\rho/G^{1/2})[v\cos(\lambda_y) + w\sin(\lambda_y)]_0 \quad \bar{\eta} = 0. \end{aligned} \tag{13.147}$$

In the equations just given λ_x and λ_y are the angles of inclination of the topography in the x and y directions.

For the turbulent heat and moisture flux terms

$$\begin{aligned} \bar{H}_1{}^\eta = \bar{H}_2{}^\eta = 0 \quad \bar{\eta} = 0, H \\ \bar{H}_3 = 0 \quad \bar{\eta} = 0. \end{aligned} \tag{13.148}$$

Though flux terms with bulk aerodynamic parameterizations (see Section 13.2.2.2) can be applied here for H_3 at $\bar{\eta} = 0$.

13.3.6 Lateral boundary conditions for slab-symmetric models and three-dimensional models

For slab-symmetric models and three-dimensional models, an equation nearly identical to that proposed by Klemp and Wilhelmson (1978) is suggested for u_n,

$$\frac{\partial u_n}{\partial t} = -(u_n + c^*)\frac{\partial u_n}{\partial n} \tag{13.149}$$

where ∂n is the distance in the normal direction, again calculations are made using forward-upstream finite-difference schemes. In addition, that is if $u_n + c^* < 0$, the advection term is set to zero. Similarly, for all other variables, if $u_n < 0$ then the advection is again set to zero. In this equation

c^* is the intrinsic phase velocity of the dominant gravity wave mode moving out through the normal boundary. Clark (1979) wrote that the Klemp and Wilhelmson (1978) scheme led to an increase in the mean vertical velocity in the domain, and proposed an alternative scheme first used by Orlanski (1976). The Orlanski (1976) scheme is an extrapolation method and a mean normal outflow velocity is computed for each layer at each normal boundary. Interestingly, Tripoli and Cotton (1981) did not find this mean vertical velocity drift reported by Clark (1979).

13.3.7 Lateral boundary conditions for axisymmetric models

The normal, lateral boundary condition for axisymmetric models is given below. This equation replaces the equation of motion for u_n,

$$\frac{\partial u_n}{\partial t} = -(u_n + c^*)\frac{\partial u_n}{\partial n} - \left(f + \frac{v}{r}\right)v, \qquad (13.150)$$

where u_n is the normal outflow wind component and c^* is the normal intrinsic gravity wave phase speed. Calculations are made using forward-upstream finite-difference techniques. That is, if $u_n + c^* < 0$, then the advection term is set to zero. Similarly, for all other variables if $u_n < 0$ then the advection is again set to zero. This is a variant of the Klemp and Wilhemson (1978) version suggested by Rotunno and Emanuel (1987).

Appendix

A. Identity proof for integrating log-normal distribution

In this first section we want to find a proof of identity that

$$\int_{-\infty}^{\infty} \exp\,(2b'x)\exp\,\left(-a'x^2\right)\,\mathrm{d}x = \sqrt{\frac{\pi}{a'}}\,\exp\,\left(\frac{b'^2}{a'}\right)$$

We begin with the left-hand side and rearrange,

$$\exp\,(2b'x)\exp\,\left(-a'x^2\right) = \exp\,\left(-a'x^2 + 2b'x\right) \tag{A1}$$

$$= \exp\,\left(-a'\left[x^2 - \frac{2b'}{a'}x\right]\right) \tag{A2}$$

$$= \exp\left(-a'\left[x^2 - \frac{2b'}{a'}x\right]\right) \tag{A3}$$

$$= \exp\,\left(\frac{b'^2}{a'}\right)\exp\left(-a'\left[x - \frac{b'}{a'}\right]^2\right). \tag{A4}$$

Now taking the integral from $-\infty$ to $+\infty$,

$$I = \int_{-\infty}^{\infty} \exp(2b'x)\exp\left(-a'x^2\right)\mathrm{d}x = \int_{-\infty}^{\infty} \exp\left(\frac{b'^2}{a'}\right)\exp\left(-a'\left[x - \frac{b'}{a'}\right]^2\right)\mathrm{d}x \tag{A5}$$

$$I = \int_{-\infty}^{\infty} \exp(2b'x)\exp\left(-a'x^2\right)\mathrm{d}x = \exp\left(\frac{b'^2}{a'}\right)\int_{-\infty}^{\infty} \exp\left(-a'\left[x - \frac{b'}{a'}\right]^2\right)\mathrm{d}x \tag{A6}$$

$$\text{let } p = \sqrt{a'}\left(x - \frac{b'}{a'}\right)$$

$$\therefore \ \mathrm{d}p = \sqrt{a'}\mathrm{d}x \tag{A7}$$

$$\text{or } \mathrm{d}x = \frac{\mathrm{d}p}{\sqrt{a'}}$$

$$\therefore \ I = \frac{1}{\sqrt{a'}}\exp\left(\frac{b'^2}{a'}\right)\int_{-\infty}^{\infty}\exp(-p^2)\mathrm{d}p. \tag{A8}$$

Now we use a trick with dummy variables p and q

$$I^2 = II = \frac{1}{a'}\exp\left(\frac{2b'^2}{a'}\right)\int_{-\infty}^{\infty}\exp(-p^2)\mathrm{d}p\int_{-\infty}^{\infty}\exp(-q^2)\mathrm{d}q, \tag{A9}$$

or

$$I^2 = \frac{1}{a'}\exp\left(\frac{2b'^2}{a'}\right)\int_{-\infty}^{\infty}\exp(-(p^2+q^2))\mathrm{d}p\mathrm{d}q. \tag{A10}$$

We then switch to polar coordinates with

$$\begin{aligned} p &= r\cos\theta \\ q &= r\sin\theta \end{aligned}, \tag{A11}$$

so that it can be written that

$$r^2 = p^2 + q^2$$

and (A12)

$$\mathrm{d}p\mathrm{d}q = r\mathrm{d}r\mathrm{d}\theta,$$

where the limits on r and θ are

$$\begin{aligned} r &: 0 \to \infty \\ \theta &: 0 \to 2\pi. \end{aligned} \tag{A13}$$

Now it can be stated that

$$I^2 = \frac{1}{a'}\exp\left(\frac{2b'^2}{a'}\right)\int_0^{\infty}\int_0^{2\pi}\exp(-(r^2))\mathrm{d}\theta\mathrm{d}r \tag{A14}$$

or

$$I^2 = \frac{2\pi}{a'}\exp\left(\frac{2b'^2}{a'}\right)\int_0^{\infty}\exp(-(r^2))\mathrm{d}r, \tag{A15}$$

and with

$$\begin{aligned} u &= r^2 \\ \mathrm{d}u &= 2r\mathrm{d}r \\ \therefore\ r\mathrm{d}r &= \frac{\mathrm{d}u}{2}, \end{aligned} \tag{A16}$$

$$I^2 = \frac{2\pi}{a'}\exp\left(\frac{2b'^2}{a'}\right)\int_0^{\infty}\int_0^{2\pi}\exp(-(r^2))\mathrm{d}r, \tag{A17}$$

$$I^2 = \frac{2\pi}{a'}\exp\left(\frac{2b'^2}{a'}\right)\int_0^{\infty} -\exp(-u)\frac{\mathrm{d}u}{2}; \tag{A18}$$

$$I^2 = \frac{\pi}{a'}\exp\left(\frac{2b'^2}{a'}\right)[-\exp(-u)]_0^{\infty}; \tag{A19}$$

taking the root of (A19) to get I,

$$I = \sqrt{\frac{\pi}{a'}}\exp\left(\frac{b'^2}{a'}\right). \tag{A20}$$

Therefore,

$$\int_{-\infty}^{\infty}\exp(2b'x)\exp(-a'x^2)\mathrm{d}x = \sqrt{\frac{\pi}{a'}}\exp\left(\frac{b'^2}{a'}\right), \tag{A21}$$

which proves the identity.

B. Gamma function

The gamma function is given from Gradshteyn and Ryzhik (1980) and Abramowitz and Stegun (1964) such that

$$\Gamma(x) = \int_0^{\infty} t^{x-1}\exp(-t)\mathrm{d}t \tag{B1}$$

where Re $x > 0$.

C. Incomplete gamma functions

The formulas for defining the incomplete gamma functions $\gamma(\alpha,x)$ and $\Gamma(\alpha,x)$ are the following,

$$\gamma(\alpha, x) = \int_0^x t^{\alpha-1} \exp(-t)\mathrm{d}t \tag{C1}$$

$$\Gamma(\alpha, x) = \int_x^\infty t^{\alpha-1} \exp(-t)\mathrm{d}t \tag{C2}$$

where Re $\alpha > 0$. The series approximations for these are usually stored in lookup tables in models that use them by the following formulas (though care must be taken at large n so $\Gamma(n+1)$ does not get too large),

$$\gamma(\alpha, x) = \sum_{n=0}^{\infty} \frac{(-1)^n x^{\alpha+n}}{\Gamma(n+1)(\alpha+n)}, \tag{C3}$$

$$\Gamma(\alpha, x) = \Gamma(\alpha) - \sum_{n=0}^{\infty} \frac{(-1)^n x^{\alpha+n}}{n!(\alpha+n)} \quad [\alpha \neq 0, -1, -2, \ldots]. \tag{C4}$$

We note also that

$$\Gamma(\alpha, x) + \gamma(\alpha, x) = \Gamma(\alpha). \tag{C5}$$

Finally, the derivatives are given as

$$\frac{\mathrm{d}}{\mathrm{d}x}\gamma(\alpha, x) = -\frac{\mathrm{d}}{\mathrm{d}x}\Gamma(\alpha, x) = x^{\alpha-1} \exp(-x). \tag{C6}$$

Other integrals that occasionally are used in microphysical parameterizations can be found in Gradshteyn and Ryzhik (1980).

References

Abramowitz, M. and Stegun, I. A. (1964). *Handbook of Mathematical Functions with Formulas, Graphs, and Mathematical Tables.* Dover Publications

Gradshteyn, I. S. and Ryshik, L. M. (1980). *Tables of Integrals, Series, and Products.* Academic Press.

References

Andsager, K., Beard, K. V., and Laird, N. F. (1999). Laboratory measurements of axis ratios for large raindrops. *J. Atmos. Sci.*, **56**, 2673–2683.

Asai, T. (1965). A numerical study of the air-mass transformation over the Japan sea in winter. *J. Meteorol. Soc. Jpn.*, **43**, 1–15.

Asai, T. and Kasahara, A. (1967). A theoretical study of the compensating downward motions associated with cumulus clouds. *J. Atmos. Sci.*, **24**, 487–496.

Auer, A. H. (1972). Inferences about ice nucleation from ice crystal observations. *J. Atmos. Sci.*, **29**, 311–317.

Aydin, K. and Seliga, T. (1984). Radar polarimetric backscattering properties of conical graupel. *J. Atmos. Sci.*, **41**, 1887–1892.

Bailey, M. and Hallett, J. (2004). Growth rates and habits of ice crystals between −20° and −70 °C. *J. Atmos. Sci.*, **61**, 514–544.

Bannon, P. R. (2002). Theoretical foundations for models of moist convection. *J. Atmos. Sci.*, **59**, 1967–1982.

Barge, B. L. and Isaac, G. A. (1973). The shape of Alberta hailstones. *J. Rech. Atmos.*, **7**, 11–20.

Bayewitz, M., Yerushalmi, H. J., Katz, S., and Shinnar, R. (1974). The extent of correlations in a stochastic coalescence process. *J. Atmos. Sci.*, **31**, 1604–1614.

Beard, K. (1976). Terminal velocity and shape of cloud and precipitation drops aloft. *J. Atmos. Sci.*, **33**, 851–864.

Beard, K. V. and Chuang, C. (1987). A new mode for the equilibrium shape of rain drops. *J. Atmos. Sci.*, **44**, 1509–1524.

Beard, K. V. and Ochs, H. T. (1984). Measured collection and coalescence efficiencies for accretion. *J. Geophys. Res.*, **89**, 7165–7169.

Beard, K. V., Kubesh, R. J., and Ochs, H. T. (1991). Laboratory measurements of small raindrop distortion. Part I: Axis ratios and fall behavior. *J. Atmos. Sci.*, **48**, 698–710.

Beheng, K. D. (1981). Stochastic riming of plate like and columnar ice crystals. *Pure Appl. Geophys.*, **119**, 820–830.

Beheng, K. D. (1994). A parameterization of warm cloud microphysical conversion processes. *Atmos. Res.*, **33**, 193–206.

Beheng, K. D. and Doms, G. (1986). A general formulation of collection rates of cloud and raindrops using the kinetic equation and comparison with parameterizations. *Contrib. Atmos. Phys.*, **59**, 66–84.

Berry, E. X. (1967). Cloud droplet growth by collection. *J. Atmos. Sci.*, **24**, 688–701.

Berry, E. X. (1968a). Comments on "Cloud droplet coalescence: Statistical foundations and a one-dimensional sedimentation model." *J. Atmos. Sci.*, **25**, 151–152.

Berry, E. X. (1968b). Modification of the warm rain process. Conference Proceedings, 1st National Conference on Weather Modification, Albany, NY, April 28–May 1, pp. 81–88.

Berry, E. X. and Reinhardt, R. L. (1974a). An analysis of cloud drop growth by collection: Part I. Double distributions. *J. Atmos. Sci.*, **31**, 1814–1824.

Berry, E. X. and Reinhardt, R. L. (1974b). An analysis of cloud drop growth by collection: Part II. Single initial distributions. *J. Atmos. Sci.*, **31**, 1825–1831.

Berry, E. X. and Reinhardt, R. L. (1974c). An analysis of cloud drop growth by collection: Part III. Accretion and self-collection. *J. Atmos. Sci.*, **31**, 2118–2126.

Berry, E. X. and Reinhardt, R. L. (1974d). An analysis of cloud drop growth by collection: Part IV. A new parameterization. *J. Atmos. Sci.*, **31**, 2127–2135.

Bigg, E. K. (1953). The supercooling of water. *Proc. Phys. Soc. London*, **B66**, 688–694.

Blanchard, D. C. (1950). The behavior of water drops at terminal velocity in air. *Trans. Am. Geophys. Union*, **31**, 836–842.

Bleck, R. (1970). A fast approximative method for integrating the stochastic coalescence equation. *J. Geophys. Res.*, **75**, 5165–5171.

Boren, C. F. and Albrecht, B. A. (1998). *Atmospheric Thermodynamics*. Oxford: Oxford University Press.

Bott, A. (1998). A flux method for the numerical solution of the stochastic collection equation. *J. Atmos. Sci.*, **55**, 2284–2293.

Bott, A. (2000). A flux method for the numerical solution of the stochastic collection equation: Extension to two-dimensional particle distributions. *J. Atmos. Sci.*, **57**, 284–294.

Braham, R. Jr. and Squires, P. (1974). Cloud physics 1974. *Bull. Am. Meteor. Soc.*, **55**, 543–586.

Bringi, V. N., Seliga, T. A., and Cooper, W. A. (1984). Analysis of aircraft hydrometeor spectra and differential reflectivity (ZDR) radar measurements during the Cooperative Convective Precipitation Experiment. *Radio Sci.*, **19**, 157–167.

Bringi, V. N., Chandrasekar, V., and Rongrui, X. (1998). Raindrop axis ratios and size distributions in Florida rainshafts: An assessment of multiparameter radar algorithms. *IEEE Trans. Geosci. Remote Sens.*, **36**, 703–715.

Brown, P. S. (1983). Some essential details for application of Bleck's method to the collision-break-up equation. *J. Appl. Meteorol.*, **22**, 693–697.

Brown, P. S. (1985). An implicit scheme for the efficient solution of the coalesence/collision-break-up equation. *J. Comput. Phys.*, **58**, 417–431.

Brown, P. S. (1986). Analysis of the Low and List drop-breakup formulation. *J. Appl. Meteorol.*, **25**, 313–321.

Brown, P. S. (1987). Parameterization of drop-spectrum evolution due to coalescence and breakup. *J. Atmos. Sci.*, **44**, 242–249.

Brown, P. S. (1988). The effects of filament, sheet, and disk breakup upon the drop spectrum. *J. Atmos. Sci.*, **45**, 712–718.

Brown, P. S. (1990). Reversals in evolving raindrop size distributions due to the effects of coalescence and breakup. *J. Atmos. Sci.*, **47**, 746–754.

Brown, P. S. (1991). Parameterization of the evolving drop-size distribution based on analytic solution of the linearized coalescence-breakup equation. *J. Atmos. Sci.*, **48**, 200–210.

Brown, P. S. (1993). Analysis and parameterization of the combined coalescence, breakup, and evaporation processes. *J. Atmos. Sci.*, **50**, 2940–2951.
Brown, P. S. (1997). Mass conservation considerations in analytic representation of raindrop fragment distributions. *J. Atmos. Sci.*, **54**, 1675–1687.
Brown, P. S. (1999). Analysis of model-produced raindrop size distributions in the small-drop range. *J. Atmos. Sci.*, **56**, 1382–1390.
Brown, P. S. and Whittlesey, S. N. (1992). Multiple equilibrium solutions in Bleck-type models of drop coalescence and breakup. *J. Atmos. Sci.*, **49**, 2319–2324.
Bryan, G. H. and Fritsch, J. M. (2002). A benchmark simulation for moist nonhydrostatic numerical models. *Mon. Weather Rev.*, **130**, 2917–2928.
Bryan, G. H. and Fritsch, J. M. (2004). A reevaluation of ice–liquid water potential temperature. *Mon. Weather Rev.*, **132**, 2421–2431.
Byers, H. R. (1965). *Elements of Cloud Physics*. Chicago, IL: The University of Chicago Press.
Carrió, G. G. and Nicolini, M. (1999). A double moment warm rain scheme: Description and test within a kinematic framework. *Atmos. Res.*, **52**, 1–16.
Chandrasekar, V., Cooper, W. A., and Bringi, V. N. (1988). Axis ratios and oscillations of raindrops. *J. Atmos. Sci.*, **45**, 1323–1333.
Chaumerliac, N., Richard, E., Rosset, R., and Nickerson, E. C. (1991). Impact of two microphysical schemes upon gas scavenging and deposition in a mesoscale meteorological model. *J. Appl. Meteorol.*, **30**, 88–97.
Chen, J. P. (1994). Predictions of saturation ratio for cloud microphysical models. *J. Atmos. Sci.*, **51**, 1332–1338.
Cheng, L. and English, M. (1983). A relationship between hailstone concentration and size. *J. Atmos. Sci.*, **40**, 204–213.
Cheng, L., English, M., and Wong, R. (1985). Hailstone size distributions and their relationship to storm thermodynamics. *J. Appl. Meteorol.*, **24**, 1059–1067.
Chong, S. L. and Chen, C. (1974). Water shells on ice pellets and hailstones. *J. Atmos. Sci.*, **31**, 1384–1391.
Clark, T. L. (1973). Numerical modeling of the dynamics and microphysics of warm cumulus convection. *J. Atmos. Sci.*, **30**, 857–878.
Clark, T. L. (1976). Use of log-normal distributions for numerical calculations of condensation and collection. *J. Atmos. Sci.*, **33**, 810–821.
Clark, T. L. (1977). A small-scale dynamical model using a terrain-following coordinate transformation. *J. Comput. Phys.*, **24**, 186–215.
Clark, T. L. (1979). Numerical simulations with a three-dimensional cloud model: Lateral boundary condition experiments and multicellular severe storm simulations. *J. Atmos. Sci.*, **36**, 2191–2215.
Clark, T. L. and Hall, W. (1979). A numerical experiment on stochastic condensation theory. *J. Atmos. Sci.*, **36**, 470–483.
Cober, S. G. and List, R. (1993). Measurements of the heat and mass transfer parameters characterizing conical graupel growth. *J. Atmos. Sci.*, **50**, 1591–1609.
Cohard, J.-M. and Pinty, J. P. (2000). A comprehensive two-moment warm microphysical bulk model scheme: I: Description and tests. *Q. J. Roy. Meteor. Soc.*, **126**, 1815–1842.
Cohard, J.-M., Pinty, J. P., and Suhre, K. (1998). On the parameterization of activation spectra from cloud condensation nuclei microphysical properties. *J. Geophys. Res.*, **105**, 11753–11766.

Cohard, J.-M., Pinty, J. P., and Bedos, C. (2000). Extending Twomey's analytical estimate of nucleated cloud droplet concentrations from CCN spectra. *J. Atmos. Sci.*, **55**, 3348–3357.
Cooper, W. A., Bruintjes, R. T., and Mather, G. K. (1997). Calculations pertaining to hygroscopic seeding with flares. *J. Appl. Meteorol.*, **36**, 1449–1469.
Cotton, W. R. (1972a). Numerical simulation of precipitation development in supercooled cumuli – Part I. *Mon. Weather Rev.*, **100**, 757–763.
Cotton, W. R. (1972b). Numerical simulation of precipitation development in supercooled cumuli – Part II. *Mon. Weather Rev.*, **100**, 764–784.
Cotton, W. R. and Anthes, R. A. (1989). *Storm and Cloud Dynamics.* San Diego, CA: Academic Press.
Cotton, W. R. and Tripoli, G. J. (1978). Cumulus convection in shear flow – three-dimensional numerical experiments. *J. Atmos. Sci.*, **35**, 1503–1521.
Cotton, W. R., Stephens, M. A., Nehrkorn, T., and Tripoli, G. J. (1982). The Colorado State University three-dimensional cloud model – 1982. Part II: An ice phase parameterization. *J. Rech. Atmos.*, **16**, 295–320.
Cotton, W. R., Tripoli, G. J., Rauber, R. M., and Mulvihill, E. A. (1986). Numerical simulation of the effects of varying ice crystal nucleation rates and aggregation processes on orographic snowfall. *J. Climate Appl. Meteor.*, **25**, 1658–1680.
Cotton, W. R. and coauthors (2003) RAMS (2001). Current status and future directions. *Meteor. Atmos. Phys.*, **82**, 5–29.
Curic, M. and Janc, D. (1997). On the sensitivity of hail accretion rates in numerical modeling. *Tellus*, **49A**, 100–107.
Danielsen, E., Bleck, R., and Morris, D. (1972). Hail growth by stochastic collection in a cumulus model. *J. Atmos. Sci.*, **29**, 135–155.
Deardorff, J. W. (1980). Stratocumulus-capped mixed layers derived from a three-dimensional model. *Bound.-Layer Meteorol.*, **18**, 495–527.
DeMott, P. J., Meyers, M. P., and Cotton, W. R. (1994). Parameterization and impact of ice initiation processes relevant to numerical model simulations of cirrus clouds. *J. Atmos. Sci.*, **51**, 77–90.
Dennis, A. and Musil, D. (1973). Calculations of hailstone growth and trajectories in a simple cloud model. *J. Atmos. Sci.*, **30**, 278–288.
Doms, G. and Beheng, K. D. (1986). Mathematical formulation of self collection, auto conversion, and accretion rates of cloud and raindrops. *Meteorol. Rundsch.*, **39**, 98–102.
Drake, J. C. and Mason, B. J. (1966). The melting of small ice spheres and cones. *Q. J. Roy. Meteorol. Soc.*, **92**, 500–509.
Dye, J. E., Knight, C. A., Toutenhootd, V., and Cannon, T. W. (1974). The Mechanism of precipitation formation in northeastern Colorado cumulus, III. Coordinated microphysical and radar observations and summary. *J. Atmos. Sci.*, **29**, 278–288.
English, M. (1973). Alberta hailstorms. Part II: Growth of large hail in the storm. *Meteorol. Monogr.*, **36**, 37–98.
Farley, R. D. (1987). Numerical modeling of hailstorms and hailstone growth. Part II: The role of low-density riming growth in hail production. *J. Appl. Meteorol.*, **26**, 234–254.
Farley, R. D. and Orville, H. (1986). Numerical modeling of hailstorms and hailstone growth. Part I: Preliminary model verification and sensitivity tests. *J. Appl. Meteorol.*, **25**, 2014–2035.
Farley, R. D., Price, P. E., Orville, H. D., and Hirsch, J. H. (1989). On the numerical simulation of graupel/hail initiation via the riming of snow in bulk water microphysical cloud models. *J. Appl. Meteorol.*, **28**, 1128–1131.

Feingold, G., Tzivion (Tzitzvashvili), S., and Levin, Z. (1988). Evolution of raindrop spectra. Part I: Solution to the stochastic collection/breakup equation using the method of moments. *J. Atmos. Sci.*, **45**, 3387–3399.

Feingold, G., Walko, R. L., Stevens, B., and Cotton, W. R. (1998). Simulations of marine stratocumulus using a new microphysics parameterization scheme. *Atmos. Res.*, **47–48**, 505–528.

Feng, J. Q. and Beard, K. V. (1991). A perturbation model of raindrop oscillation characteristics with aerodynamic effects. *J. Atmos. Sci.*, **48**, 1856–1868.

Flatau, P. J., Tripoli, G. J., Verlinde, J., and Cotton, W. R. (1989). The CSU-RAMS cloud microphysical module: General theory and documentation. Technical Report 451. (Available from the Department of Atmospheric Sciences, Colorado State University, Ft. Collins, CO 80523.)

Fletcher, N. H. (1962). *The Physics of Rain Clouds*. Cambridge: Cambridge University Press.

Foote, G. B. (1984). A study of hail growth utilizing observed storm conditions. *J. Appl. Meteorol.*, **23**, 84–101.

Gaudet, B. J. and Schmidt, J. M. (2005). Assessment of hydrometeor collection rates from exact and approximate equations. Part I: A new approximate scheme. *J. Atmos. Sci.*, **62**, 143–159.

Gaudet, B. J. and Schmidt, J. M. (2007). Assessment of hydrometeor collection rates from exact and approximate equations. Part II: Numerical bounding. *J. Appl. Meteorol. Climatol.*, **46**, 82–96.

Gillespie, D. T. (1972). The stochastic coalescence model for cloud droplet growth. *J. Atmos. Sci.*, **29**, 1496–1510.

Gillespie, D. T. (1975). Three models for the coalescence growth of cloud drops. *J. Atmos. Sci.*, **32**, 600–607.

Gilmore, M. S. and Straka, J. M. (2008). The Berry and Reinhardt autoconversion parameterization: A digest. *J. Appl. Meteorol. Climatol.*, **47**, 375–396.

Gilmore, M. S., Straka, J. M., and Rasmussen, E. N. (2004). Precipitation uncertainty due to variations in precipitation particle parameters within a simple microphysics scheme. *Mon. Weather Rev.*, **132**, 2610–2627.

Glickman, T. S. (2000). *The Glossary of Meteorology*, 2nd edn. Boston, MA: American Meteorological Society.

Goddard, J. W. F. and Cherry, S. M. (1984). The ability of dual-polarization radar measurements in rain (copolar linear) to predict rainfall and microwave attenuation. *Radio Sci.*, **19**, 201–208.

Goddard, J. W. F., Cherry, S. M., and Bringi, V. N. (1982). Comparison of dual-polarization radar measurements of rain with groundbased disdrometer measurements. *J. Appl. Meteorol.*, **21**, 252–256.

Golovin, A. M. (1963). The solution of the coagulation equation for cloud droplets in a rising air current. *Isv. Ak. Nk. SSSR (Geophys. Ser.)*, **5**, 783–791.

Greenan, B. J. and List, R. (1995). Experimental closure of the heat and mass transfer theory of spheroidal hailstones. *J. Atmos. Sci.*, **52**, 3797–3815.

Gunn, K. L. S. and Marshall, J. S. (1958). The distribution with size of aggregate snowflakes. *J. Atmos. Sci.*, **15**, 452–461.

Gunn, R. and Kinzer, G. D. (1949). The terminal velocity of fall for raindrops in stagnant air. *J. Meteor.*, **6**, 243–248.

Hall, W. D. (1980). A detailed microphysical model within a two-dimensional dynamic framework: Model description and preliminary results. *J. Atmos. Sci.*, **37**, 2486–2507.

Hall, W. D. and Pruppacher, H. (1976). The survival of ice particles falling from cirrus clouds in subsaturated air. *J. Atmos. Sci.*, **33**, 1995–2006.

Hallet, J. and Mossop, S. C. (1974). Production of secondary ice particles during the riming process. *Nature*, **249**, 26–28.

Hallgren, R. E. and Hosler, C. L. (1960). Preliminary results on the aggregation of ice crystals. *Geophys. Monogr., Am. Geophys. Union*, **5**, 257–263.

Heymsfield, A. J. (1972). Ice crystal terminal velocities. *J. Atmos. Sci.*, **29**, 1348–1357.

Heymsfield, A. J. (1978). The characteristics of graupel particles in northeastern Colorado cumulus congestus clouds. *J. Atmos. Sci.*, **35**, 284–295.

Heymsfield, A. J. and Kajikawa, M. (1987). An improved approach to calculating terminal velocities of plate-like crystals and graupel. *J. Atmos. Sci.*, **44**, 1088–1099.

Heymsfield, A. J. and Pflaum, J. C. (1985). A quantitative assessment of the accuracy of techniques for calculating graupel growth. *J. Atmos. Sci.*, **42**, 2264–2274.

Heymsfield, A. J., Jameson, A. R., and Frank, H. W. (1980). Hail growth mechanisms in a Colorado storm: Part II: Hail formation processes. *J. Atmos. Sci.*, **37**, 1779–1807.

Hitchfield, W. and Stauder, M. (1967). The temperature of hailstones. *J. Atmos. Sci.*, **24**, 293–297.

Hobbs, P. V. (1974). *Ice Physics*. London: Oxford University Press.

Hosler, C. L. and Hallgren, R. E. (1961). Ice crystal aggregation. *Nublia*, **4**, No. 1, 13–19.

Hu, Z. and Srivastava, R. (1995). Evolution of raindrop size distribution by coalescence, breakup, and evaporation: Theory and observations. *J. Atmos. Sci.*, **52**, 1761–1783.

Hubbert, J. V., Bringi, V. N., and Carey, L. D. (1998). CSU-CHILL polarimetric radar measurements from a severe hail storm in Eastern Colorado. *J. Appl. Meteorol.*, **37**, 749–775.

Huffman, P. J. and Vali, G. (1973). The effect of vapor depletion on ice nucleus measurements with membrane filters. *J. Appl. Meteor.*, **12**, 1018–1024.

Jameson, A. R. and Beard, K. V. (1982). Raindrop axial ratios. *J. Appl. Meteorol.*, **21**, 257–259.

Johnson, D. B. and Rasmussen, R. M. (1992). Hail growth hysteresis. *J. Atmos. Sci.*, **49**, 2525–2532.

Jones, D. M. (1959). The shape of raindrops. *J. Atmos. Sci.*, **16**, 504–510.

Joss, J. and Zawadzki, I. (1997). Raindrop distributions again? Preprints, 28th Conference On Radar Meteorology, Austin, TX, 7–12 September, pp. 326–327.

Kajikawa, M. and Heymsfield, A. J. (1989). Aggregation of ice crystals in cirrus. *J. Atmos. Sci.*, **46**, 3108–3121.

Kessler, E. (1969). On the distribution and continuity of water substance in atmospheric circulations. *Meteorol. Monogr.*, No. 32.

Khain, A. P., Ovtchinnikov, M., Pinski, M., Pokrovsky, A., and Krugiliak, H. (2000). Notes on the state-of-the-art numerical modeling of cloud microphysics. *Atmos. Res.*, **55**, 159–224.

Khairoutdinov, M. F. and Kogan, Y. L. (1999). A large eddy simulation model with explicit microphysics: Validation against aircraft observations of a stratocumulus-topped boundary layer. *J. Atmos. Sci.*, **56**, 2115–2131.

Khairoutdinov, M. and Kogan, Y. (2000). A new cloud physics parameterization in a large-eddy simulation model of marine stratocumulus. *Mon. Weather Rev.*, **128**, 229–243.

Kinzer, G. D. and Gunn, R. (1951). The evaporation, temperature, and thermal relaxation-time of free falling waterdrops. *J. Meteorol.*, **8**, 71–83.

Klemp, J. B. and Wilhelmson, R. B. (1978). The simulation of three-dimensional convective storm dynamics. *J. Atmos. Sci.*, **35**, 1070–1096.

Knight, C. A. (1979). Observations of the morphology of melting snow. *J. Atmos. Sci.*, **36**, 1123–1130.
Knight, C. A. and Knight, N. C. (1970). Hailstone embryos. *J. Atmos. Sci.*, **27**, 659–666.
Knight, C. A. and Knight, N. C. (1973). Conical graupel. *J. Atmos. Sci.*, **30**, 118–124.
Knight, C. A. and Knight, N. C. (2001). Hailstorms. In *Severe Convective Storms*, ed. C. A. Doswell, AMS Monograph 50, ch. 6, pp. 223–254.
Knight, C. A. and Miller, L. (1993). First radar echoes from cumulus clouds. *Bull. Am. Meteorol. Soc.*, **74**, 179–188.
Knight, N. C. (1981). The climatology of hailstone embryos. *J. Appl. Meteorol.*, **20**, 750–755.
Knight, N. C. (1986). Hailstone shape factor and its relation to Radar interpretation of hail. *J. Appl. Meteorol.*, **25**, 1956–1958.
Koenig, R. and Murray, F. W. (1976). Ice-bearing cumulus evolution: Numerical simulations and general comparison against observations. *J. Appl. Meteorol.*, **15**, 742–762.
Kogan, Y. L. (1991). The simulation of a convective cloud in a 3-D model with explicit microphysics. Part I: Model description and sensitivity experiments. *J. Atmos. Sci.*, **48**, 1160–1189.
Kogan, Y. L. and Martin, W. J. (1994). Parameterization of bulk condensation in numerical cloud models. *J. Atmos. Sci.*, **51**, 1728–1739.
Komabayasi, M., Gonda, T., and Isono, K. (1964). Lifetime of water drops before breaking and size distribution of fragments. *J. Meteorol. Soc. Jpn.*, **42**, 330–340.
Kovetz, A. and Olund, B. (1969). The effect of coalescence and condensation on rain formation in a cloud of finite vertical extent. *J. Atmos. Sci.*, **26**, 1060–1065.
Kry, P. R. and List, R. (1974). Angular motions of freely falling spheroidal hailstone models. *Phys. Fluids*, **17**, 1093–1102.
Kubesh, R. J. and Beard, K. V. (1993). Laboratory measurements of spontaneous oscillations for moderate-sized raindrops. *J. Atmos. Sci.*, **50**, 1089–1098.
Kumjian, M. R. and Ryzhko, A. V. (2008). Polarimetric signatures in supercell thunderstorms. *J. Appl. Meteorol. Climatol.*, **47**, 1940–1961.
Langlois, W. E. (1973). A rapidly convergent procedure for computing large-scale condensation in a dynamical weather model. *Tellus*, **25**, 86–87.
Laws, J. O. and Parsons, D. A. (1943). The relation of raindrop-size to intensity. *Trans Am. Geophys. Union*, **24**, Part II, 452–460.
Lesins, G. and List, R. (1986). Sponginess and drop shedding of gyrating hailstones in a pressure-controlled icing wind tunnel. *J. Atmos. Sci.*, **43**, 2813–2825.
Lin, Y. L., Farley, R. D., and Orville, H. D. (1983). Bulk parameterization of the snow field in a cloud model. *J. Climate Appl. Meteorol.*, **22**, 1065–1092.
List, R. (1986). Properties and growth of hailstones. *Thunderstorm Dynamics and Morphology*, ed. E. Kessler. Norman, OK: University of Oklahoma Press, pp. 259–276.
List, R. and Gillespie, J. (1976). Evolution of raindrop spectra with collision-induced breakup. *J. Atmos. Sci.*, **33**, 2007–2013.
List, R. and Schemenauer, R. S. (1971). Free-fall behavior of planar snow crystals, conical graupel and small hail. *J. Atmos. Sci.*, **28**, 110–115.
List, R., Rentsch, U. W., Byram, A. C., and Lozowski, E. P. (1973). On the aerodynamics of spheroidal hailstone models. *J. Atmos. Sci.*, **30**, 653–661.
Liu, J. and Orville, H. (1969). Numerical modeling of precipitation and cloud shadow effects on mountain-induced cumuli. *J. Atmos. Sci.*, **26**, 1283–1298.

Liu, Y. and Daum, P. H. (2004). Parameterization of the autoconversion process. Part I: Analytical formulation of the Kessler-type parameterizations. *J. Atmos. Sci.*, **61**, 1539–1548.

Locatelli, J. D. and Hobbs, P. V. (1974). Fall speeds and masses of solid precipitation particles. *J. Geophys. Res.*, **79**, 2185–2197.

Loney, M. L., Zrnic, D. S., Straka, J. M., and Ryzhkov, A. V. (2002). Enhanced polarimetric radar signatures above the melting level in a supercell storm. *J. Appl. Meteorol.*, **41**, 1179–1194.

Long, A. B. (1974). Solutions to the droplet collection equation for polynomial kernels. *J. Atmos. Sci.*, **31**, 1040–1052.

Low, R. D. (1969). A generalized equation for the solution effect in droplet growth. *J. Atmos. Sci.*, **26**, 608–611.

Low, T. and List, R. (1982a). Collision, coalescence and breakup of raindrops. Part I: Experimentally established coalescence efficiencies and fragment size distributions in breakup. *J. Atmos. Sci.*, **39**, 1591–1606.

Low, T. and List, R. (1982b). Collision, coalescence and breakup of raindrops. Part II: Parameterization of fragment size distributions in breakup. *J. Atmos. Sci.*, **39**, 1607–1618.

Ludlam, F. H. (1958). The hail problem. *Nubila*, **1**, 12–96.

Macklin, W. C. (1963). Heat transfer from hailstones. *Quart. J. Roy. Meteorol. Soc.*, **89**, 360–369.

Macklin, W. C. and Bailey, I. H. (1962). The density and structure of ice formed by accretion. *Q. J. Roy. Meteorol. Soc.*, **88**, 30–50.

Macklin, W. C. and Bailey, I. H. (1966). On the critical liquid water concentrations of large hailstones. *Quart. J. Roy. Meteorol. Soc.*, **92**, 297–300.

Magono, C. and Lee, C. W. (1966). Meteorological classification of natural snow crystals. *J. Fac. Sci., Hokkaido Univ.*, Ser. VII, **2**, 321–335.

Mansell, E. R., MacGorman, D. R., Ziegler, C. L., and Straka, J. M. (2002). Simulated three-dimensional branched lightning in a numerical thunderstorm model. *J. Geophys. Res.*, **107** (9), doi: 10.1029/2000JD000244.

Mansell, E. R., MacGorman, D. R., Ziegler, C. L., and Straka, J. M. (2005). Charge structure and lightning sensitivity in a simulated multicell thunderstorm. *J. Geophys. Res.*, **110**, D12101, doi: 10.1029/2004JD005287.

Manton, M. J. and Cotton, W. R. (1977). Formulation of approximate equations for modeling moist convection on the mesoscale. *Technical Report, Colorado State University*, Fort Collins, CD.

Marshall, J. S. and Palmer, W. M. K. (1948). The distribution of raindrops with size. *J. Meteorol.*, **5**, 165–166.

Mason, B. J. (1957). *The Physics of Clouds*. Oxford: Clarendon Press.

Mason, B. J. (1971). *The Physics of Clouds*, 2nd edn. Oxford: Clarendon Press.

Matson, R. J. and Huggins, A. W. (1980). The direct measurement of the sizes, shapes, and kinematics of falling hailstones. *J. Atmos. Sci.*, **37**, 1107–1125.

McDonald, J. (1963). The saturation adjustment in numerical modelling of fog. *J. Atmos. Sci.*, **20**, 476–478.

McFarguhar, G. M. (2004). A new representation of breakup of raindrops and its implications for shapes of raindrop size distributions. *J. Atmos. Sci.*, **61**, 777–792.

McTaggart-Cowan, J. and List, R. (1975). Collision and breakup of water drops at terminal velocity. *J. Atmos. Sci.*, **32**, 1401–1411.

Meyers, M. P., DeMott, P. J., and Cotton, W. R. (1992). New primary ice-nucleation parameterizations in an explicit cloud model. *J. Appl. Meteorol.*, **31**, 708–721.

Meyers, M. P., Walko, R. L., Harrington, J. Y., and Cotton, W. R. (1997). New RAMS cloud microphysics parameterization. Part II. The two-moment scheme. *Atmos. Res.*, **45**, 3–39.

Milbrandt, J. A. and Yau, M. K. (2005a). A multimoment bulk microphysics parameterization. Part I: Analysis of the role of the spectral shape parameter. *J. Atmos. Sci.*, **62**, 3051–3064.

Milbrandt, J. A. and Yau, M. K. (2005b). A multimoment bulk microphysics parameterization. Part II: A proposed three-moment closure and scheme description. *J. Atmos. Sci.*, **62**, 3065–3081.

Mitra, S. K., Vohl, O., Ahr, M., and Pruppacher, H. R. (1990). A wind tunnel and theoretical study of the melting behavior of atmospheric ice particles. IV: Experiment and theory for snow flakes. *J. Atmos. Sci.*, **47**, 584–591.

Mizuno, H. (1990). Parameterization of the accretion process between different precipitation elements. *J. Meteorol. Soc. Jpn.*, **68**, 395–398.

Morrison, H. and Grabowski, W. W. (2007). Comparison of bulk and bin warm rain microphysical models using a kinematic framework. *J. Atmos. Sci.*, **64**, 2839–2861.

Morrison, H. and Grabowski, W. W. (2008). A novel approach for representing ice microphysics in models: Description and tests using a kinematic framework. *J. Atmos. Sci.*, **65**, 1528–1548.

Moss, M. S. and Rosenthal, S. L. (1975). On the estimation of planetary boundary layer variables in mature hurricanes. *Mon. Weather Rev.*, **103**, 980–988.

Mossop, S. C. (1976). Production of secondary ice particles during the growth of graupel riming. *Q. J. Roy. Meteorol. Soc.*, **102**, 25–44.

Mossop, S. C. and Kidder, R. E. (1962). Artificial hailstones. *Bull. Obs. Puy. De Dom*, **2**, 65–79.

Murakami, M. (1990). Numerical modeling of dynamical and microphysical evolution of an isolated convective cloud – the 19 July 1981 CCOPE cloud. *J. Meteorol. Soc. Jpn.*, **68**, 107–128.

Nelson, S. P. (1980). *A Study of Hail Production in a Supercell Storm using Doppler Derived Wind Field and a Numerical Hail Growth Model.* NOAA Technical Memorandum ERL NSSL-89. National Severe Storm Laboratory. (NTIS PB81-17822Q.)

Nelson, S. P. (1983). The influence of storm flow structure on hail growth. *J. Atmos. Sci.*, **40**, 1965–1983.

Nelson, S. P. (1987). The hybrid multicellular–supercellular storm – an efficient hail producer. Part II. General characteristics and implications for hail growth. *J. Atmos. Sci.*, **44**, 2060–2073.

Nickerson, E. C., Richard, E., Rosset, R., and Smith, D. R. (1986). The numerical simulation of clouds, rains and airflow over the Vosges and Black Forest mountains: A meso-β model with parameterized microphysics. *Mon. Weather Rev.*, **114**, 398–414.

Ogura, Y. and Takahashi, T. (1973). The development of warm rain in a cumulus model. *J. Atmos. Sci.*, **30**, 262–277.

Ohtake, T. (1970). Factors affecting the size distribution of raindrops and snowflakes. *J. Atmos. Sci.*, **27**, 804–813.

Orlanski, I. (1976). A simple boundary condition for unbounded hyperbolic flows. *J. Comput. Phys.*, **21**, 251–269.

Orville, H. and Kopp, F. J. (1977). Numerical simulation of the history of a hailstorm. *J. Atmos. Sci.*, **34**, 1596–1618.

Passarelli, R. E. (1978). An approximate analytical model of the vapor deposition and aggregation growth of snowflakes. *J. Atmos. Sci.*, **35**, 118–124.

Passarelli, R. E. and Srivastava, R. C. (1979). A new aspect of snowflake aggregation theory. *J. Atmos. Sci.*, **36**, 484–493.

Pellett, J. L. and Dennis, A. S. (1974). Effects of heat storage in hailstones. Conference Proceedings, *Conference on Cloud Physics, Tucson, AZ*, October 21–24, pp. 63–66.

Pflaum, J. C. (1980). Hail formation via microphysical recycling. *J. Atmos. Sci.*, **37**, 160–173.

Pflaum, J. C., and Pruppacher, H. (1979). A wind tunnel investigation of the growth of graupel initiated from frozen drops. *J. Atmos. Sci.*, **36**, 680–689.

Proctor, F. H. (1987). *The Terminal Area Simulation System. Vol. I: Theoretical Formulation.* NASA Contractor Report 4046, NASA, Washington, DC. [Available from the National Technical Information Service, Springfield, VA, 22161.]

Pruppacher, H. R. and Beard, K. V. (1970). A wind tunnel investigation of the internal circulation and shape of water drops falling at terminal velocity in air. *Q. J. Roy. Meteorol. Soc.*, **96**, 247–256.

Pruppacher, H. R. and Klett, J. D. (1981). *Microphysics of Clouds and Precipitation.* Dordrecht: D. Reidel Publishing.

Pruppacher, H. R. and Klett, J. D. (1997). *Microphysics of Clouds and Precipitation*, 2nd edn. Dordrecht: Kluwer Academic Publishers.

Pruppacher, H. R. and Pitter, R. L. (1971). A semi-empirical determination of the shape of cloud and rain drops. *J. Atmos. Sci.*, **28**, 86–94.

Rasmussen, R. M. and Heymsfield, A. J. (1985). A generalized form for impact velocities used to determine graupel accretional densities. *J. Atmos. Sci.*, **42**, 2275–2279.

Rasmussen, R. M. and Heymsfield, A. J. (1987a). Melting and shedding of graupel and hail. Part I: Model physics. *J. Atmos. Sci.*, **44**, 2754–2763.

Rasmussen, R. M. and Heymsfield, A. J. (1987b). Melting and shedding of graupel and hail. Part II: Sensitivity study. *J. Atmos. Sci.*, **44**, 2764–2782.

Rasmussen, R. and Pruppacher, H. R. (1982). A wind tunnel and theoretical study of the melting behavior of atmospheric ice particles. I: A wind tunnel study of frozen drops of radius < 500 μm. *J. Atmos. Sci.*, **39**, 152–158.

Rasmussen, R. M., Levizzani, V., and Pruppacher, H. R. (1984). A wind tunnel and theoretical study on the melting behavior of atmospheric ice particles: III. Experiment and theory for spherical ice particles of radius > 500 μm. *J. Atmos. Sci.*, **41**, 381–388.

Rauber, R. M., Beard, K. V., and Andrews, B. M. (1991). A mechanism for giant raindrop formation in warm, shallow convective clouds. *J. Atmos. Sci.*, **48**, 1791–1797.

Reinhardt, R. L. (1972). An analysis of improved numerical solution to the stochastic collection equation for cloud drops. Ph.D. Dissertation, University of Nevada.

Reisner, J., Rasmussen, R. M., and Bruintjes, R. T. (1998). Explicit forecasting of supercooled liquid water in winter storms using the MM5 mesoscale model. *Q. J. Roy. Meteorol. Soc.*, **124**, 1071–1107.

Rogers, D. C. (1973). The aggragation of natural ice crystals. M. S. Thesis, Department of Atmospheric Resources, University of Wyoming.

Rogers, R. R. and Yau, M. K. (1989). *A Short Course in Cloud Physics*. Pergamon Press.

Rotunno, R. and Emanuel, K. A. (1987). An air–sea interaction theory for tropical cyclones. Part II: Evolutionary study using a nonhydrostatic axisymmetric numerical model. *J. Atmos. Sci.*, **44**, 542–561.

Rutledge, S. A. and Hobbs, P. V. (1983). The mesoscale and microscale structure and organization of clouds and precipitation in midlatitude cyclones. VIII: A model for the "seeder-feeder" process in warm-frontal rainbands. *J. Atmos. Sci.*, **40**, 1185–1206.

Rutledge, S. A. and Hobbs, P. V. (1984). The mesoscale and microscale structure and organization of clouds and precipitation in midlatitude cyclones. XII: A diagnostic modeling study of precipitation development in narrow cold-frontal rainbands. *J. Atmos. Sci.*, **41**, 2949–2972.

Saleeby, S. M. and Cotton, W. R. (2004). A large-droplet mode and prognostic number concentration of cloud droplets in the Colorado State University Regional Atmospheric Modeling System (RAMS). Part I: Module descriptions and supercell test simulations. *J. Appl. Meteorol.*, **43**, 182–195.

Saleeby, S. M. and Cotton, W. R. (2005). A large-droplet mode and prognostic number concentration of cloud droplets in the Colorado State University Regional Atmospheric Modeling System (RAMS). Part II: Sensitivity to a Colorado winter snowfall event. *J. Appl. Meteorol.*, **44**, 1912–1929.

Saleeby, S. M. and Cotton, W. R. (2008). A binned approach to cloud-droplet riming implemented in a bulk microphysics model. *J. Appl. Meteorol. Climatol.*, **47**, 694–703.

Sauvageot, H. and Lacaux, J. P. (1995). The shape of averaged drop size distributions. *J. Atmos. Sci.*, **52**, 1070–1083.

Schlamp, R. J., Pruppacher, H. R., and Hamielec, H. R. (1975). A numerical investigation of the efficiency with which simple columnar ice crystals collide with supercooled water drops. *J. Atmos. Sci.*, **32**, 2330–2337.

Schoenberg-Ferrier, B. (1994). A double-moment multiple-phase four-class bulk ice scheme. Part I: Description. *J. Atmos. Sci.*, **51**, 249–280.

Schumann, T. E. W. (1938). The theory of hailstone formation. *Q. J. Roy. Meteorol. Soc.*, **64**, 3–21.

Scott, W. T. (1968). Analytical studies of cloud droplet coalescence. *J. Atmos. Sci.*, **25**, 54–65.

Scott, W. T. and Levin, Z. (1975). A comparison of formulations of stochastic collection. *J. Atmos. Sci.*, **32**, 843–847.

Seifert, A. (2008). On the parameterization of evaporation of raindrops as simulated by a one dimensional model. *J. Atmos. Sci.*, **28**, 741–751.

Seifert, A. and Beheng, K. D. (2001). A double-moment parameterization for simulating autoconversion, accretion and self-collection. *Atmos. Res.*, **59–60**, 265–281.

Seifert, A. and Beheng, K. D. (2005). A two-moment cloud microphysical parameterization for mixed phase clouds. Part 1: Model description. *Meteorol. Atmos. Phys.*, doi: 10.1007/s00703-005-0112-4.

Shafrir, U. and Gal-Chen, T. (1971). A numerical study of collision efficiencies and coalescence parameters for droplet pairs with radii up to 300 microns. *J. Atmos. Sci.*, **28**, 741–751.

Simpson, J. and Wiggert, V. (1969). Models of precipitating cumulus towers. *Mon. Weather Rev.*, **97**, 471–489.

Smagorinski, J. (1963). General circulation experiments with the primitive equations. I: The basic experiments. *Mon. Weather Rev.*, **91**, 99–164.

Soong, S. T. (1974). Numerical simulation of warm rain development in an axisymmetric cloud model. *J. Atmos. Sci.*, **31**, 1262–1285.
Soong, S. T. and Ogura, Y. (1973). A comparison between axisymmetric and slab-symmetric cumulus cloud models. *J. Atmos. Sci.*, **30**, 879–893.
Srivastava, R. C. (1971). Size distribution of raindrops generated by their breakup and coalescence. *J. Atmos. Sci.*, **28**, 410–415.
Srivastava, R. (1989) Growth of cloud drops by condensation: A criticism of currently accepted theory and a new approach. *J. Atmos. Sci.*, **46**, 869–887.
Srivastava, R. C. and Coen, J. L. (1992). New explicit equations for the accurate calculation of the growth and evaporation of hydrometeors by the diffusion of water vapor. *J. Atmos. Sci.*, **49**, 1643–1651.
Stensrud, D. J. (2007). *Parameterization Schemes: Keys to Understanding Numerical Weather Prediction Models*. Cambridge: Cambridge University Press.
Stevens, B., Walko, R. L., Cotton, W. R., and Feingold, G. (1996). The spurious production of cloud-edge supersaturations by Eulerian models. *Mon. Weather Rev.*, **124**, 1034–1041.
Stewart, R. E., Marwitz, J. D., Pace, J. C., and Carbone, R. E. (1984). Characteristics through the melting layer of stratiform clouds. *J. Atmos. Sci.*, **41**, 3227–3237.
Straka, J. M. and Mansell, E. R. (2005). A bulk microphysics parameterization with multiple ice precipitation categories. *J. Appl. Meteorol.*, **44**, 445–466.
Straka, J. M. and Rasmussen, E. N. (1997). Toward improving microphysical parameterizations of conversion processes. *J. Appl. Meteorol.*, **36**, 896–902.
Straka, J. M., Zrnic, D. S., and Ryzhkov, A. V. (2000). Bulk hydrometeor classification and quantification using polarimetric radar data: Synthesis of relations. *J. Appl. Meteorol.*, **39**, 1341–1372.
Straka, J. M., Kanak, K. M., and Gilmore, M. S. (2007). The behavior of number concentration tendencies for the continuous collection growth equation using one- and two-moment bulk parameterization schemes. *J. Appl. Meteorol. Climatol.*, **46**, 1264–1274.
Stull, R. B. (1988). *An Introduction to Boundary Layer Meteorology*. Dordrecht: Kluwer Academic Publishers.
Takahashi, T. (1976). Hail in an axisymmetric cloud model. *J. Atmos. Sci.*, **33**, 1579–1601.
Tao, W. K., Simpson, J., and McCumber, M. (1989). An ice-water saturation adjustment. *Mon. Weather Rev.*, **117**, 231–235.
Telford, J. W. (1955). A new aspect of coalescence theory. *J. Atmos. Sci.*, **12**, 436–444.
Thompson, G., Rasmussen, R. M., and Manning, K. (2004). Explicit forecasts of winter precipitation using an improved bulk microphysics scheme. Part I: Description and sensitivity analysis. *Mon. Weather Rev.*, **132**, 519–542.
Tokay, A. and Beard, K. V. (1996). A field study of raindrop oscillations. Part I: Observation of size spectra and evaluation of oscillation causes. *J. Appl. Meteorol.*, **35**, 1671–1687.
Tripoli, G. J. and Cotton, W. R. (1980). A numerical investigation of several factors contributing to the observed variable intensity of deep convection over south Florida. *J. Appl. Meteorol.*, **19**, 1037–1063.
Tripoli, G. J. and Cotton, W. R. (1981). The use of ice-liquid water potential temperature as a thermodynamic variable in deep atmospheric models. *Mon. Weather Rev.*, **109**, 1094–1102.

Twomey, S. (1959). The nuclei of natural cloud formation. Part II: The supersaturation in natural clouds and the variation of cloud droplet concentration. *Geophys. Pure Appl.*, **43**, 243–249.

Twomey, S. (1964). Statistical effects in the evolution of a distribution of cloud droplets by coalescence. *J. Atmos. Sci.*, **21**, 553–557.

Twomey, S. (1966). Computation of rain formation by coalescence. *J. Atmos. Sci.*, **23**, 405–411.

Tzivion (Tzitzvashvili), S., Feingold, G., and Levin, Z. (1987). An efficient numerical solution to the stochastic collection equation. *J. Atmos. Sci.*, **44**, 3139–3149.

Tzivion (Tzitzvashvili), S., Feingold, G., and Levin, Z. (1989). The evolution of raindrop spectra. Part II: Collisional collection/breakup and evaporation in a rainshaft. *J. Atmos. Sci.*, **46**, 3312–3328.

Ulbrich, C. W. (1983). Natural variations in the analytical form of the raindrop size distribution. *J. Climate Appl. Meteorol.*, **22**, 1764–1775.

Ulbrich, C. W. and Atlas, D. (1982). Hail parameter relations: A comprehensive digest. *J. Appl. Meteor.*, **21**, 22–43.

Vali, G. (1975). Remarks, on the mechanism of atmospheric ice nucleation. Proceedings of the 8th International Conference, on Nucleation, Leningrad, 23–29 September, ed. I. I. Gaivoronski, pp. 265–299.

Vali, G. (1994). Freezing rate due to heterogeneous nucleation. *J. Atmos. Sci.*, **51**, 1843–1856.

van Den Broeke, M. S., Straka, J. M., and Rasmussen, E. N. (2008). Polarimetric radar observations at low levels during tornado life cycles in a small sample of classic Southern Plains supercells. *J. Appl. Meteorol. Climatol.*, **47**, 1232–1247.

Verlinde, J. and Cotton, W. R. (1993). Fitting microphysical observations of nonsteady convective clouds to a numerical model: An application of the adjoint technique of data assimilation to a kinematic model. *Mon. Weather Rev.*, **121**, 2776–2793.

Verlinde, J., Flatau, P. J., and Cotton, W. R. (1990). Analytical solutions to the collection growth equation: Comparison with approximate methods and application to cloud microphysics parameterization schemes. *J. Atmos. Sci.*, **47**, 2871–2880.

Walko, R. L., Cotton, W. R., Meyers, M. P., and Harrington, J. Y. (1995). New RAMS cloud microphysics parameterization: Part I. The single-moment scheme. *Atmos. Res.*, **38**, 29–62.

Walko, R. L., Cotton, W. R., Feingold, G., and Stevens, B. (2000). Efficient computation of vapor and heat diffusion between hydrometeors in a numerical model. *Atmos. Res.*, **53**, 171–183.

Wang, P. K. (1985). A convection diffusion model for the scavenging of submicron snow crystals of arbitrary shapes. *J. Rech. Atmos.*, **19**, 185–191.

Wang, P. K. (2002). *Ice Microdynamics*. San Diego, CA: Academic Press.

Wang, P. K. and Ji, W. (1992). A numerical study of the diffusional growth and riming rates of ice crystals in clouds. Preprints volume, 11th International Cloud Physics Conference, August 11–17, Montreal, Canada.

Warshaw, M. (1967). Cloud droplet coalescence: Statistical foundations and a one-dimensional sedimentation model. *J. Atmos. Sci.*, **24**, 278–286.

Wilhelmson, R. and Ogura, Y. (1972). The pressure perturbation and the numerical modeling of a cloud. *J. Atmos. Sci.*, **29**, 1295–1307.

Wisner, C. E., Orville, H. D., and Myers, C. G. (1972). A numerical model of a hail bearing cloud. *J. Atmos. Sci.*, **29**, 1160–1181.

Young, K. C. (1974a). A numerical simulation of wintertime, orographic precipitation: Part I. Description of model microphysics and numerical techniques. *J. Atmos. Sci.*, **31**, 1735–1748.

Young, K. C. (1974b). A numerical simulation of wintertime, orographic precipitation: Part II. Comparison of natural and AgI-seeded conditions. *J. Atmos. Sci.*, **31**, 1749–1767.

Young, K. C. (1975). The evolution of drop spectra due to condensation, coalescence and breakup. *J. Atmos. Sci.*, **32**, 965–973.

Young, K. C. (1993). *Microphysical Processes in Clouds*. London: Oxford University Press.

Ziegler, C. L. (1985). Retrieval of thermal and microphysical variables in observed convective storms. Part I: Model development and preliminary testing. *J. Atmos. Sci.*, **42**, 1487–1509.

Ziegler, C. L., Ray, P. S., and Knight, N. C. (1983). Hail growth in an Oklahoma multicell storm. *J. Atmos. Sci.*, **40**, 1768–1791.

Index

www.ingramcontent.com/pod-product-compliance
Lightning Source LLC
LaVergne TN
LVHW081523060526
838200LV00044B/1982
9780521297592